T0132644

Introduction to
CODING THEORY
Second Edition

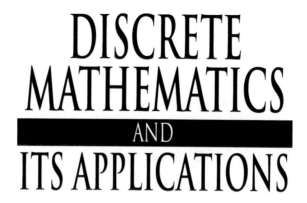

DISCRETE MATHEMATICS AND ITS APPLICATIONS

DISCRETE MATHEMATICS AND ITS APPLICATIONS

Introduction to
CODING THEORY
Second Edition

Jürgen Bierbrauer

Michigan Technological University
Houghton, USA

CRC Press
Taylor & Francis Group
Boca Raton   London   New York

CRC Press is an imprint of the
Taylor & Francis Group, an **informa** business

A CHAPMAN & HALL BOOK

CRC Press
Taylor & Francis Group
6000 Broken Sound Parkway NW, Suite 300
Boca Raton, FL 33487-2742

Printed on acid-free paper
Version Date: 20160414

International Standard Book Number-13: 978-1-4822-9980-9 (Hardback)

Library of Congress Cataloging-in-Publication Data

Names: Bierbrauer, Juergen.
Title: Introduction to coding theory / Jürgen Bierbrauer.
Description: Second edition. | Boca Raton : Taylor & Francis, 2017. | Series: Discrete mathematics and its applications | "A CRC title." | Includes bibliographical references and index.
Identifiers: LCCN 2016017212 | ISBN 9781482299809 (alk. paper)
Subjects: LCSH: Coding theory.
Classification: LCC QA268 .B48 2017 | DDC 003/.54--dc23
LC record available at https://lccn.loc.gov/2016017212

Visit the Taylor & Francis Web site at
http://www.taylorandfrancis.com

and the CRC Press Web site at
http://www.crcpress.com

Printed and bound in the United States of America by Publishers Graphics,
LLC on sustainably sourced paper.

To Stella, Daniel and my mother

Contents

Preface

The theory of error-correcting codes is a branch of discrete mathematics with close ties to other mathematical disciplines, like design theory, combinatorial theory, linear algebra, finite fields, rings, Galois geometry, geometric algebra, algebraic curves over finite fields and group theory. The best known application is in the transmission of messages over noisy communication channels. Other fields of application are to be found in statistics (design of experiments), cryptography (authentication, the design of ciphers) and in many areas of theoretical computer science.

In this textbook we present a self-contained introduction to mathematical coding theory and to its major areas of application. High school algebra and some exposition to basic linear algebra are sufficient as mathematical background. Part I is designed for use in a one semester undergraduate course. A second semester would start with the theory of cyclic codes. In Part II the emphasis is on cyclic codes, applications of codes, liear programming bounds and the geometric description of linear codes. The mathematical tools are developed along the way. Part III offers a brief introduction to some of the basics of the theory of function fields in one variable (algebraic curves) over a finite field of constants, a basic construction of codes (algebraic-geometric codes) and the properties of some interesting families of examples.

A brief overview

The historical origins of coding theory are in the problem of reliable communication over noisy channels. This is a typical problem of the discipline now called **Information Theory.** Both disciplines, **Coding Theory** and **Information Theory,** originated with Claude Shannon's famous 1948 paper [183]. It contains the celebrated **channel coding theorem** (see Chapter 9) which states roughly that good long codes are guaranteed to exist, without giving a clue how to construct them. Closely related is the development of **Cryptography**. Its aim is to ensure reliable communication in the presence of ill-willed opponents. These problems are rather different. In the coding theory scenario we have to overcome a technical problem (the shortcomings of a communication channel), whereas in Cryptography we have to beat an opponent. Nonetheless the mathematical tools used in these two areas have a large intersection. Historically the development of both disciplines was boosted by the efforts of World War II. Another famous paper of Claude Shannon, [184] from 1949, is perceived as the origin of modern cryptography.

The information-theoretic problem prompted the definition of a mathematical structure called **error-correcting code** or simply **code.** Coding theory can be seen as the study of error-correcting codes, their construction, bounds on their parameters, their implementation and so forth. The most important parameter is the **minimum distance.** It measures the code's capability of correcting transmission errors.

Progress in coding theory was slow but steady. One important development was the theory of **cyclic codes,** which is traditionally couched in the language of ring theory. Cyclic codes are particularly useful because they admit a fast decoding algorithm. The theory of cyclic codes is a core topic of Part II. It is developed in Chapter 13, preceded by an introduction to some relevant features of finite fields in Chapter 12. Our approach is different from the traditional approach. It is based on the trace and the action of the Galois group. Ring theory does not come into play at all.

Only the single most famous cyclic code, the binary Golay code, is introduced in Part I, along with a closely related structure, the large Witt design (Chapter 7).

The ties between coding theory and several areas of pure mathematics have grown stronger all the time. The most important insight goes back to the early 1980s. It is the discovery, by Goppa and Manin [97, 138], of a close relationship between codes and **algebraic curves** (in algebraic language **function fields**). Algebraic curves are objects of **number theory** and **algebraic geometry,** mainstream mathematical disciplines with a long and rich history. The observation by Goppa and Manin makes it possible to use these number-theoretic tools for the construction of codes. The theory of those **algebraic-geometric codes** (AG-codes) is the objective of Part III. In fact we develop only some of the basics of the theory of algebraic curves with finite fields of constants, just enough to understand the basic construction of algebraic-geometric codes and to study some interesting families of examples.

Coding theory and **combinatorics** are closely connected. As an example, **block designs** are important objects of modern discrete mathematics. For more information see the **CRC Handbook of Combinatorial Designs** [106]. We will encounter them repeatedly in the text. A formal definition is in Chapter 7, where we also derive the large Witt design from the binary Golay code. Other examples of block designs in the text include projective planes, projective and affine geometry over finite fields (Chapter 17), the small Witt design, which is derived from the ternary Golay code in Section 17.1, and the Denniston arcs in the same section.

Linear codes can be studied from a geometric point of view. From this perspective coding theory can be seen as part of **Galois geometry.** The basic objects of Galois geometry are **affine and projective spaces** defined over finite fields (see Hirschfeld [113] or the beginning of Chapter 17). Linear codes can be equivalently described as sets of points in projective spaces. In many cases the geometric language is more appropriate than the algebraic approach. In Part I we study 3-dimensional codes from this point of view (Chapter 10). This case is particularly easy to understand as the underlying geometrical structures are classical projective planes $PG(2, q)$. The general mechanism is developed and used in Chapter 17. In many cases this leads to a better understanding of the codes. For instance, we use the geometric method to construct the ternary Golay code in Section 17.1. As a natural generalization, the additive codes (network codes) of Chapter 18 are described geometrically by families of subspaces of a fixed projective space. As a special case, the binary quantum codes of Chapter 18 are described by families of lines in binary projective spaces.

Caps are sets of points in projective or affine geometry no three of which are on a line. They are formally equivalent to linear codes of minimum distance $d = 4$. It turns out that caps are best understood from a geometric point of view. This is why we study caps in Chapter 17. The case of caps in projective planes and 3-spaces leads to another link with classical algebra. In fact, parabolic and elliptic quadrics yield canonical examples of caps in those dimensions. We include a self-contained introduction to **geometric algebra** in Section 17.2 which gives a better understanding of those caps.

Duality is emphasized throughout the text. The dual of a linear code with minimum distance d is an **orthogonal array** of **strength** $d - 1$. Originally orthogonal arrays were defined in the framework of design of experiments, in **statistics.** The same is true of block designs. The defining properties of orthogonal arrays and of block designs are both uniformity conditions. They look very similar. Orthogonal arrays can be interpreted as families of **random variables** (functions defined on sample spaces), which satisfy certain statistical independence conditions (see Chapter 6). The strength measures the degree of statistical independence. Such families of random variables are heavily used not only in statistics but also in the theory of **algorithms.** Whenever we construct a good linear code, we also obtain such a statistical object.

Typically in coding theory duality is defined with respect to the usual dot product, the Euclidean bilinear form. However, each non-degenerate **bilinear** (or sesquilinear) **form** defines a notion of duality. An application to the construction of **quantum codes** in Chapter 18 demands the use of a special bilinear form, the symplectic form. This is another motivation for covering the theory of bilinear forms in Chapter 17.

Some of the classical **bounds,** the Singleton bound, the Hamming bound, the Plotkin bound and the Griesmer bound on codes as well as the Bose-Bush bound on orthogonal arrays of strength 2, are derived when they are needed in the text. In fact, there is a multitude of bounds, each of which is better than all the others in certain parameter ranges, both for codes (when the minimum distance is the central parameter) and for orthogonal arrays (when the strength is in the center of attention). A general algebraic mechanism for the derivation of bounds is related to **orthogonal polynomials** (in the case of codes these are the **Kravchouk polynomials**) and **linear programming.** There is a general **linear programming bound.** All the bounds used in the text, with the exception of the Griesmer bound, are special cases of the LP-bound. Chapter 15 is a self-contained introduction to linear programming and contains unified proofs for all those explicit bounds on codes and orthogonal arrays. On the level of LP-bounds there is a relation of duality between bounds on codes and bounds on orthogonal arrays. This is another reason why the notion of an orthogonal array should be seen as the dual of the notion of an error-correcting code, even in the nonlinear case.

This leads us to applications of codes. Traditionally coding theorists are biased towards the information-theoretic application that we encounter so often in this text. It still is one of the major applications. We use it as a motivation in the early chapters, discuss syndrome decoding in Chapter 3 and the decoding algorithm of BCH-codes based on the Euclidean algorithm in Chapter 13. There is, however, a plethora of applications of a completely different nature. Many of them have surfaced in theoretical computer science, in particular in cryptography. **Universal hash families** yield a nice paradigmatic example. One version is presented in Part I, Chapter 6, while a more in depth treatment is in Chapter 16. This chapter is dedicated to applications altogether. They range from statistics and cryptography to numerical integration and the theory of algorithms.

The plan: Part I

Part I forms an elementary introduction to the theory of codes and some typical applications. It assumes only high school mathematics. Some exposition to the basics of linear algebra would be helpful as well.

Chapter 1 introduces some basic concepts like bits and bitstrings and describes the basic problem when messages are sent via a noisy channel. The transmission of pictures from space serves as an illustration. The first steps toward the algebraization of the problem are taken by introducing the field $\mathbb{F}_2$ of two elements and giving the bitstrings of length n the structure of a vector space. We encounter the basic idea of error correction and the basic notion of Hamming distance. This leads to the formal definition of a binary code and of a q-ary code for an arbitrary natural number $q \geq 2$. The binary symmetric channel is the most elementary model which describes how errors are introduced into the message. Basic facts on binomial numbers are reviewed. The sphere-packing bound (or Hamming bound) is our first general bound on codes. The football pool problem is a possible application of ternary ($q = 3$) codes.

Chapter 2 introduces the basics of binary **linear** codes. Key notions are minimum weight, dimension, generator matrix. After a review of basic facts of linear algebra (applied for vector spaces = codes over $\mathbb{F}_2$), we see how block coding works and study the effect on the probability of transmission errors. The dual code is defined with respect to the dot product. The repetition codes and the sum 0 codes are our first pair of dual codes. A check matrix of a code is a generator matrix of its dual code. An important family of codes are the binary Hamming codes. Their duals are the binary Simplex codes. The principle of duality shows how to read off the minimum distance from a check matrix. This leads to the notion of a binary orthogonal array (OA). A linear OA has strength t if and only if its dual has minimum weight $t + 1$.

Chapter 3 generalizes linear codes from the binary ($q = 2$) to the q-ary, where q is an arbitrary prime-power. It is shown how finite fields $\mathbb{F}_q$ of q elements can be constructed. The definition of linear q-ary codes is given and the basic facts generalized from the binary to the q-ary: dimension, generator matrix, dual code, check matrix, principle of duality, orthogonal arrays. Basic methods of linear algebra are reviewed (rank, Gauß elimination, determinants). Mutually orthogonal Latin squares are recognized as special parametric cases of OA. The MacWilliams formula links the weight distribution of a linear code and its dual. Our proof is probabilistic. This motivates the definition of probability spaces. The game of SET leads to natural questions concerning ternary ($q = 3$) linear codes. Syndrome decoding can in principle be used to decode linear codes.

A large and important family of linear codes are the Reed-Solomon codes of Chapter 4. They meet the Singleton bound with equality (they are MDS-codes). Lagrange interpolation shows that they form OA of index $\lambda = 1$. The dual of an RS-code is an RS-code. Mutually orthogonal Latin squares yield non-linear MDS-codes. Covering arrays and their use in software testing are discussed.

Chapter 5 introduces some recursive constructions of codes: shortening and puncturing, the parity check bit for binary codes, the residual code and the Griesmer bound, concatenation and the $(u, u + v)$-construction.

Chapter 6 presents our first application in computer science. The concept of universal hashing is introduced. ϵ-universal hash classes turn out to be formally equivalent with codes.

Chapter 7 is a direct construction of the binary Golay code and the large Witt design. This motivates the definition of t-designs. Classical projective planes $PG(2, q)$ are introduced.

Chapter 8: Meaning and basic properties of the Shannon entropy of a probability space. The binary entropy function. The Jensen inequality as a tool.

In Chapter 9 the concept of asymptotic bounds for infinite families of codes with length $\longrightarrow \infty$ is introduced and the asymptotic version of the Singleton bound given. The Plotkin bound on codes of large distance is proved. It implies an asymptotic bound and the Bose-Bush bound on OA of strength

2. Shannon's channel coding theorem is proved. Basic notions and facts on probability theory (random variable, expectation, variance, Tschebyscheff inequality) are introduced. These are used in the proof. The Gilbert-Varshamov bound is an existence bound on linear codes (certain codes are guaranteed to exist), based on counting arguments. The Justesen codes form an explicit family of codes with asymptotically nontrivial parameters.

Chapter 10 is an introduction to the geometric method. The 3-dimensional codes are described as multisets of points in the projective plane $PG(2, q)$. An application allows the construction of congestion-free networks.

The plan: Part II

Chapter 12 starts with more basic properties of finite fields: Primitive elements, field extensions, the Frobenius automorphism, the Galois group, the trace from a field to a subfield. Trace codes and subfield codes are defined. Delsarte's theorem describes the dual code. We introduce the Galois closure of a linear code with respect to a subfield, prove the second main theorem and sketch the general strategy to construct cyclic codes. Different notions of equivalence of codes are discussed.

Chapter 13 is dedicated to the general machinery of cyclic codes. An example (binary, length 15) is used to illustrate this. This is a subfield code (and a trace code) of a Reed-Solomon code defined over $\mathbb{F}_{16}$. Basic notions of the general construction are cyclotomic cosets, the dimension formula and the BCH-bound on the minimum distance. Parametric examples of cyclic codes are given, as well as an application to fingerprinting. The Roos bound and the van Lint-Wilson method allow improvements on the BCH-bound in special situations. Generator matrices and check matrices of cyclic codes are almost canonically determined. BCH-codes are special cyclic codes. Their decoding algorithm is based on the Euclidean algorithm. Constacyclic codes are generalizations of cyclic codes. They can, however, be described within the theory of cyclic codes. This central chapter ends with two families of particularly good quaternary ($q = 4$) constacyclic codes and a comparison with the traditional ring-theoretic approach to cyclic codes.

Chapter 14 complements Chapter 5 by introducing further recursive constructions of codes. Constructions X and XX can be applied using cyclic codes as ingredients. The covering radius is another basic parameter. It is related to lengthening and has its own applications. We describe an application in steganography and an application in the reduction of switching noise.

Chapter 15 is dedicated to linear programming (LP). The first section is a self-contained introduction to the basics of linear programming. This includes basic notions and results like the simplex algorithm, the Farkas alternative, duality theorems and the principle of complementary slackness. A basic notion, the **Fourier transform,** is introduced in the second section. This section also contains basic properties of the Kravchouk polynomials and the general LP-bound for codes and orthogonal arrays. Several explicit bounds are derived from the LP-bound in the third section. The fourth and last section of Chapter 15 contains a proof of the celebrated bound of four (McEliece, Rodemich, Rumsey, Welch [140]), an asymptotic bound on codes and orthogonal arrays.

Chapter 16 discusses various applications. OA are interpreted as families of random variables with certain independence properties and as perfect local randomizers. We introduce linear shift register sequences and their relation to Simplex codes, describe the role of minimum distance and strength in the construction of block ciphers and the use of OA in two-point based sampling and chips testing. Further topics include the relation between OA and resilient functions, applications of resilient functions for the wire-tap channel and the generation of random bits, applications of OA in the derandomization of Monte Carlo-algorithms, as well as a more detailed study of universal hash classes, their construction from codes and their applications in cryptography (authentication).

Chapter 17 studies the geometric approach to linear codes. Linear codes are described as multisets of points in projective geometry $PG(k-1, q)$. The main theorem determines the minimum distance in terms of hyperplane intersection sizes. The hexacode, ovals and hyperovals, extended Reed-Solomon codes and the Simplex codes are best understood from this point of view. Codes of dimension 2 and 3 are studied. The ternary Golay code is constructed starting from its parameters $[12, 6, 6]_3$. Barlotti arcs and Denniston arcs as well as the corresponding codes are described. Caps are sets of points no three of which are on a line. They are formally equivalent to linear codes of minimum distance $d = 4$. We introduce the theory of bilinear forms and quadratic forms. This yields canonical models of large caps in $PG(2, q)$ (parabolic quadrics) and in $PG(3, q)$ (elliptic quadrics). Direct constructions of caps in $PG(4, q)$ and general bounds on caps in arbitrary dimension are derived. Recursive constructions of caps use as ingredients elliptic quadrics or, in the ternary case, the Hill cap in $PG(5, 3)$.

Chapter 18 introduces codes whose alphabet forms a vector space over a ground field. They generalize the linear codes. Chen projection is a simple recursive construction. Application to caps yields codes which have been used in computer memory systems. Application to Reed-Solomon codes produces codes which have been used in deep space communication. Another recursive construction simplifies the Bose-Bush construction of OA of strength 2. We conclude Section 18.1 with a direct construction of an interesting family of low-dimensional additive codes. A self-contained theory of cyclic additive codes is developed in Section 18.2. It generalizes the approach from Chapter 13. Quaternary additive codes are considered in Section 18.3. Geometrically those are described by multisets of lines in binary projective spaces. We concentrate on codes of short lengths and obtain the best possible parameters for all lengths ≤ 13. Quantum stabilizer codes are by nature additive (q-linear, q^2-ary) codes which are contained in their symplectic dual. In Section 18.3 we develop the basic theory, using the geometric description in terms of systems of lines and applying the cyclic theory. In particular we determine all parameters of binary ($q = 2$) distance $d = 3$ quantum stabilizer codes.

Additive codes have recently reappeared under the name of network codes (see Koetter and Kschischang [126]). The metric used in [126] is very different from what we used so far. It is based on the ranks of pairwise intersections of the subgeometries describing the code. We describe the solution of the smallest non-trivial problem in Section 18.5. It is surprisingly complicated.

The plan: Part III

In the early chapters of Part III we develop some of the basic theory of function fields of transcendence degree 1 over finite fields of constants, using Stichtenoth's by now classical textbook [198]. Highlights are the Riemann-Roch theorem and the Riemann-Hurwitz formula. As motivating examples we use the Klein quartic, hyperelliptic and Artin-Schreier extensions. The basic construction of codes from algebraic curves (AG-codes) is given in Chapter 22, where we describe some important families of AG-codes and an application to universal hashing. Some additional material is collected in the last chapter. A section on list decoding of Reed-Solomon codes has been included as this represents a rather recent development which uses some basic algebra/geometry in a transparent way. The sections on tms-nets and on sphere packings in Euclidean spaces (Chapter 23) are treated somewhat lighter than the applications in Chapter 16. The theory of tms-nets is to be seen in the context of quasi-Monte Carlo algorithms (related to uniformly distributed point sets in Euclidean space). This application of coding theory to numerical integration and the pricing of exotic options is rather surprising. The construction of dense sphere packings is a classical problem in Euclidean space with links not only to discrete mathematics but also to algebra and algebraic geometry. The remaining sections of Chapter 23 have the character of brief survey articles.

How to use this text

Part I arose from several one semester undergraduate courses on coding theory. Chapters 2 to 6 form the core of such an introductory course. The only section which can (and maybe should) be skipped is Section 3.5 on the MacWilliams transform. The remaining time should be dedicated to one or several of the later chapters of Part I. Chapter 7 would be a natural choice as the binary Golay code probably is the most famous of all codes. Usually I cover Chapter 6 as it is the first example of a non-standard application and it can be done in one lecture.

Another choice would be to cover Chapter 8 and some of the material from Chapter 9. These chapters form a unit as the entropy function is used in the asymptotic expressions of Chapter 9.

The canonical starting point for the second semester of a two semester course is the theory of cyclic codes and their implementation, Chapters 12 and 13, as well as Chapter 14. Some of the later parts of Chapter 13 are optional (the application to fingerprinting, the Roos bound, the van Lint-Wilson bound, the comparison with the traditional approach and Section 13.4 on constacyclic codes). Section 13.3 on the decoding algorithm of BCH-codes may be sacrificed as well.

Chapter 17 may be considered another core area of coding theory as it gives a better understanding of many important codes. It would be a pity to sacrifice the main theorem of the first section. Some of the applications in the first section ought to be covered.

From here on there are several choices.

For a thorough introduction to the theory of codes and its links with Galois geometry one might concentrate entirely on Chapter 17.

Another possibility is to cut short on Chapter 17 and to cover several applications from Chapter 16 instead. The sections on tms-nets and on sphere packings in Chapter 23 are then a good choice to round off a graduate course. This also has the advantage that the course ends with an introduction to an exceptional object, the Leech lattice.

A third strategy is to concentrate on the theory of cyclic codes and their applications. Such a course would end with Chapter 18.

The database

Tables on parameters for linear and quantum codes are to be found in M. Grassl's page

http://www.codetables.de

What has changed in the second edition?

The changes are too numerous to be listed exhaustively. A macroscopic change is the addition of a new Part III in the second edition, dedicated to algebraic-geometric codes. A little section in the last chapter of the first edition has turned into a new Chapter 15 of the second edition, containing an introduction to linear programming and the derivation of explicit bounds on codes and orthogonal arrays. Chapter 18 on additive codes and network codes is a completely remodeled version of the corresponding chapter in the first edition. The cyclic case is based on a more general theory, the geometric approach has been greatly expanded and there is an additional section on network codes. The last chapter has undergone a mutation as well. Two of its sections in the first edition have vanished as they turned into chapters of their own. Sections on permutation codes and on highly symmetric codes have been added as well as two sections which should make the text more readable, one on the individual small fields that have been used in the text and another section on the individual short codes constructed in the text.

Textbooks

An early classic among the textbooks is *Algebraic Coding Theory* [11] by E. R. Berlekamp, which presents in particular an excellent introduction to the traditional theory of cyclic codes. The 1977 book *The Theory of Error-Correcting Codes* [142] by MacWilliams and Sloane is considered something like the bible of traditional coding theory. A more recent introduction is J. H. van Lint's *Introduction to Coding Theory* [211], a relatively short and dense text which helped a great deal in attracting pure mathematicians to the area. H. Stichtenoth's book *Algebraic Function Fields and Codes* [198] is a self-contained introduction to the theory of algebraic function fields in one variable (equivalently: algebraic curves) and to the codes derived from them. Among the undergraduate textbooks we mention Vera Pless, *The Theory of Error-Correcting Codes* [165] and R. Hill, *A first course in coding theory* [112]. Hill's book is the first of its kind presenting an introduction to the geometric approach to codes.

Orthogonal Arrays: Theory and Applications [109] by Hedayat, Sloane and Stufken introduces to orthogonal arrays (dual codes) and their applications to the design of experiments.

Acknowledgments

My thanks go to Yves Edel; Wolfgang Ch. Schmid, Peter Hellekalek and the Institut für Mathematik of the University of Salzburg; Fernanda Pambianco, Stefano Marcugini, Massimo Giulietti, Daniele Bartoli, Giorgio Faina and the dipartimento di matematica e informatica of the University of Perugia; Bernd Stellmacher and the Mathematisches Seminar of the University of Kiel; C. L. Chen, Ron Crandall, Ludo Tolhuizen, Albrecht Brandis, Allan Struthers, Michigan Technological University; Gary Ebert; and the National Security Agency.

About the author

Jürgen Bierbrauer received his PhD in mathematics at the University of Mainz (Germany) in 1977, with a dissertation about the theory of finite groups. His mathematical interests are in algebraic methods of discrete mathematics and its applications. He held a position at the University of Heidelberg (Germany) from 1977 to 1994 and the position of full professor at Michigan Technological University in Houghton, Michigan until his retirement in 2015. His non-mathematical interests include Romance languages, literature, and the game of Go.

Part I

An elementary introduction to coding

Chapter 1

The concept of coding

1.1 Bitstrings and binary operations

Basic concepts: bits, bitstrings, transmission of messages, transmitting pictures from space, the Morse code, XORing, the field $\mathbb{F}_2$, the model of message transmission, a first idea of error correction.

The object of coding theory is the transmission of messages over noisy channels. Figure 1.1 shows the standard picture visualizing the situation.

At first we need to understand what the elements of this picture mean: what is a **message,** a **channel,** what is **noise?** Along the way we will encounter more basic notions. In this first chapter some of these will be explained. We start with the message.

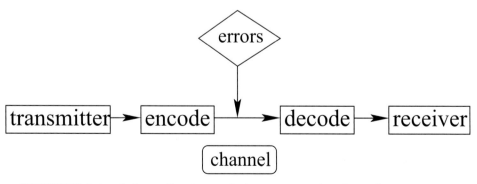

FIGURE 1.1: Information transmission over a noisy channel

If there are 8 possible messages to be sent, say, then we can represent each message as a bitstring of length 3, like 000 or 011 or 110. We will generally assume this has been done and define a **message** to be a bitstring. Here are some examples of bitstrings and their lengths:

bit string	length
000	3
110	3
110011	6
0000011111	10

Transmitting pictures from space

Assume we wish to transmit a photograph from outer space, like one of the pictures of Saturn taken by the Voyager spacecrafts in the early 1980s (*Viger* for Star Trek buffs). The picture is divided into 800×800 pixels; each pixel is assigned one of $256 = 2^8$ degrees of brightness. The brightness of a pixel is thus represented by a bitstring of length 8 and the total black and white picture consists of $800 \times 800 \times 8$ bits. As the picture really is in color, the same photo is transmitted three times, each time through a different color filter. The full color picture will thus be represented by a bitstring of length $3 \times 800 \times 800 \times 8 = 15,360,000$. This is our message. The channel is determined by the properties of space between the spacecraft and the receiver on Earth, above all by the Earth's atmosphere. A certain number of bits will be destroyed. Here we only consider errors of the type that 0 is transmitted and 1 is received or vice versa.

The Morse code

Another illustration for the claim that every message can be represented by bitstrings is the Morse code, which has been in use for telegraphy since the 1840s. It represents each of the 26 letters A, B, ... , Z, each digit $0, 1, \ldots, 9$ as well as the period, the comma and the question mark by a sequence of at most five dots and dashes. The dot stands for a short signal, the dash for a long signal. Dashes are about three times as long as dots. For example, the letter E is represented by a single dot, S is represented by dot-dot-dot and Z is dash-dash-dot-dot.

However, the graphical representation is purely conventional. We can represent a dash by 1, a dot by 0 and obtain a representation of letters and numbers as bitstrings: E=0, S=000, Z=1100, and T=1, Q=1101, V=0001.

Back to the general model

Assume we wish to send one of 8 possible messages (the bitstrings of length 3), for example, message 011. If it should happen along the way (in the channel) that the second bit is flipped (the second coordinate is in error), then 001 will be received.

We want to express this situation in mathematical terminology.

1.1 Definition. $\mathbb{F}_2 = \{0, 1\}$ *with addition*

$$0 + 0 = 0, \ 0 + 1 = 1, \ 1 + 0 = 1, \ 1 + 1 = 0$$

Here the letter $\mathbb{F}$ stands for **field.** It indicates that this tiny structure of only two elements has something in common with the field of real numbers and the complex number field: each of these structures satisfies the same set of axioms. Each of them is a field. We will come back to this topic later.

The idea behind the addition in $\mathbb{F}_2$ is to model the errors in information transmission: a coordinate of a message is in error if and only if 1 is added to the entry in this coordinate. This explains why we must have $1 + 1 = 0$: if a bit is flipped twice, then no error occurs, or: if 1 is flipped, then 0 is received.

Addition in $\mathbb{F}_2$ is also known as XORing, in particular in the computer science literature. Here **XOR** stands for **exclusive or,** a logical operation, where 1 stands for **true** and 0 for **false.** The relation $1 + 1 = 0$ (true or true = false) is what makes it "exclusive": in order for the result to be true, one of the two ingredients has to be true, but not both (the ordinary **or** operation would have $1 + 1 = 1$).

Another motivation for our binary addition comes from arithmetic: if we distinguish even and odd integers, the following familiar rules hold:

$$even + odd = odd + even = odd, \ even + even = odd + odd = even.$$

With $even = 0$ and $odd = 1$ these are exactly the rules of binary addition.

Addition in $\mathbb{F}_2$ describes what happens in each coordinate. Calculation with bitstrings is formalized as follows:

1.2 Definition. $\mathbb{F}_2^n$ *consists of bitstrings* $x = (x_1, x_2, \ldots, x_n)$ *of length n, where $x_i \in \mathbb{F}_2$. Addition in $\mathbb{F}_2^n$ is coordinatewise.*

For example, let $x = 001 \in \mathbb{F}_2^3$ and $y = 101 \in \mathbb{F}_2^3$. Then $x + y = 100$. With this terminology it is easier to express the relation between messages, errors and received messages. Let $x = 001100$ be the message sent. Assume errors occur in coordinates 2 and 4. We can express this by adding the **error vector** $e = 010100$. The received message will then be the sum

$$y = x + e = 011000.$$

Let us do this the other way around: if $x = 010101$ was sent and $y = 010111$ was received, then an error occurred in coordinate 5. This means $e = 000010$.

Return to the situation where we send one of eight messages (the elements of $\mathbb{F}_2^3$), for example, $x = 011$. No matter what the error vector e is, the received message $y = x + e$ is again one of the legitimate messages. There is no way for the receiver to suspect that an error occurred, let alone to correct it. So how can we hope to correct errors?

Error correction: The first idea

Here is the easiest of all error-correcting systems: encode each 0 of the message to be sent as a block 000 and analogously each 1 by 111. If the original message is $x = 011$, the encoded message is now $x' = 000111111$. The receiver knows what encoding scheme has been used. He will therefore divide the received message in blocks of length 3. If such a block is 000 or 111, then decoding is obvious: $000 \mapsto 0$, $111 \mapsto 1$. Assume a block of the received message is 101. The receiver knows that at least one transmission error must have happened. It is a basic assumption that a small number of errors is more probable than many errors. The decoding will therefore be $101 \mapsto 1$ (by majority decision).

The initial picture begins to make sense now. We have seen that messages can be encoded by the transmitter and decoded by the receiver such that the following holds: if not more than one error occurs during transmission (in the channel), then the error will automatically be corrected. What we have used is known as the **repetition code** of length 3.

---

1. We learned how to calculate with bitstrings.

2. $\mathbb{F}_2^n$ consists of the bitstrings of length n.

3. The received message is the binary sum of the sent message and the error vector.

4. The repetition code of length 3 corrects one bit error.

Exercises 1.1

1.1.1. *Compute the sum of* 11001 *and* 01110.

1.1.2. *Assume* 000000 *was sent and two errors occurred. List all possible received messages.*

1.1.3. *Let* $x = 1101$ *be the message to be sent. Encode* x *using the repetition code of length* 3.

1.1.4. *Assume the repetition code of length 3 is used and* 000110111101 *is received. What is the result of decoding?*

1.1.5. *Why does the Morse code represent letters E and T by strings of length* 1, *whereas letters like Q, V, Z are represented by longer bitstrings?*

1.2   The Hamming distance

Basic concepts: The Hamming distance as a metric, the weight of a bitstring.

1.3 Definition. *Let* $x = (x_1, x_2, \ldots, x_n)$ *and* $y = (y_1, y_2, \ldots, y_n)$ *be bitstrings in* $\mathbb{F}_2^n$. *The* **distance** *(or* **Hamming distance***) between* x *and* y *is*

$$d(x, y) = \text{number of coordinates } i \text{ where } x_i \neq y_i.$$

Here are some examples:

$$d(0000, 1111) = 4, \ d(00110, 00101) = 2, \ d(111111, 001100) = 4.$$

Expressed in the context of messages and errors, $d(x, y)$ is the minimum number of errors transforming x into y. In fact, consider the second example above: $x = 00110$ and $y = 00101$ differ in the last two coordinates, $d(x, y) = 2$ and $x + 00011 = y$.

Things get even easier when we use the weight.

1.4 Definition. *The* **weight** $wt(x)$ *of the bitstring* $x \in \mathbb{F}_2^n$ *is the number of nonzero coordinates in* x.

Here are some examples:

$$wt(0000) = 0, \ wt(1111) = 4, \ wt(00110) = 2, \ wt(001101) = 3.$$

The weight of a bitstring is its distance from the all-0 bitstring. If the all-0 bitstring is sent and w errors occur during transmission, then the received message has weight w. If x is sent, e is the error vector and $y = x + e$ is received, then $d(x, y) = wt(e)$.

The Hamming distance is also called the Hamming metric. The general notion of a metric is widely used in mathematics. Here is the definition:

1.5 Definition. *Let* X *be a set. For every pair* $x \in X$, $y \in X$ *let a real number* $d(x, y)$ *be given (the* **distance** *from* x *to* y*). The function* d *is called a* **metric** *if the following are satisfied:*

- $d(x, y) \geq 0$ *for all* x, y.

- $d(y, x) = d(x, y)$ *for all* x, y.

- $d(x, y) = 0$ *if and only if* $x = y$.

- $d(x, z) \leq d(x, y) + d(y, z)$ *for all* x, y, z.

The last requirement is the most important. It is known as the **triangle inequality.** A famous metric is the Euclidean metric in Euclidean space. If, for example, $x = (x_1, x_2)$ and $y = (y_1, y_2)$ are two points in the plane, then their Euclidean distance is $\sqrt{(x_1 - y_1)^2 + (x_2 - y_2)^2}$.

1.6 Theorem. *The Hamming distance is a metric on* $\mathbb{F}_2^n$.

Most of the properties of Definition 1.5 are obvious. Only the triangle inequality is a little interesting. This is left as an exercise.

1. The Hamming distance between two bitstrings of the same length is the number of coordinates where they differ.

2. $d(x, y)$ is the number of bit errors needed to transform x into y.

3. The Hamming distance is a metric.

4. The weight is the distance from the all-0 vector.

5. $d(x, y) = wt(x + y)$.

6. If bitstring x is sent and y is received, then $d(x, y) = wt(e)$ is the weight of the error vector.

Exercises 1.2

1.2.1. *Compute* $d(11001, 01110)$ *and* $d(0000, 0110)$.

1.2.2. *Find* $wt(00110)$ *and* $wt(10111)$.

1.2.3. *List all vectors in* $\mathbb{F}_2^6$ *at distance 3 from* 111000.

1.2.4. *The alphabet has 26 letters. If we want to represent all possible words of length ≤ 3 (all letters, pairs of letters and triples of letters) as bitstrings of the same length n, what is the smallest number n such that this is possible?*

1.2.5. *Prove that the Hamming distance is a metric.*

1.2.6. *Assume x is sent and $y = x + e$ is received. What can we say about $d(x, y)$ and about $wt(e)$ if not more than 3 errors have occurred?*

1.3   Binary codes

Basic concepts: Length, minimum distance. The idea of error correction.

We saw that no errors can be detected or corrected if all elements of $\mathbb{F}_2^n$ are used as messages. The obvious idea is to use only a certain subset. Such subsets will be called **codes.**

Let us send bitstrings of length 6. Instead of using all elements of $\mathbb{F}_2^6$ as (encoded) messages, we use only the following subset:

000000	001011
100110	101101
010101	011110
110011	111000

Such a family of bitstrings of length 6 is also called a **binary code** of length 6. Its elements are **codewords.** In our example we have 8 codewords.

The most important property of this code is the following: any two different codewords are at distance ≥ 3. We say that 3 is the **minimum distance** of the code. Please check for yourself that this is true. The parameters of this binary code are then recorded as $(6, 8, 3)_2$: we have a binary code (indicated by subscript 2), of length 6, consisting of 8 codewords, with minimum distance of 3.

The idea of error correction

Transmitter and receiver agree on the code to be used. Only codewords will be sent. If only one error occurs in the channel, then the received word will be in a ball of radius 1 around a codeword (in the Hamming metric). Assume the code has been chosen such that any two codewords are at distance at least 3. Then the balls of radius 1 do not overlap: if a bitstring has distance 1 from some codeword, then it has a larger distance from any other codeword. In other words, the receiver will decode any vector at distance ≤ 1 from some codeword as that codeword.

In the picture: the whole ball (or call it a disc) of radius 1 is decoded as the center of the ball, or: the received tuple is decoded as the codeword which it resembles most closely. If not more than one error occurred, then this error will be corrected. Observe that Figure 1.2 serves only as an illustration.

The metric in the Euclidean plane is used to illustrate the situation in a rather different metric, the Hamming metric.

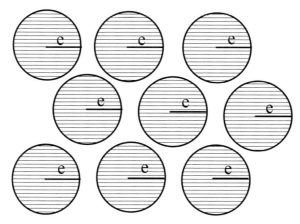

FIGURE 1.2:   Non-overlapping balls centered at codewords

1.7 Definition. *A binary code* $(n, M, d)_2$ *is a set of* M *bitstrings in* $\mathbb{F}_2^n$ *such that any two different elements of the code (codewords) are at distance* $\geq d$. *We call* n *the* **length** *and* d *the* **minimum distance** *of the code.*

A code of minimum distance 5 can correct 2 errors, minimum distance 7 can correct 3 errors and so on. A fundamental problem of coding theory is the following: Given n and d, find the largest M such that there is a code $(n, M, d)_2$.

---

1. A binary code $(n, M, d)_2$ is a collection of M bitstrings of length n such that any two different of these codewords are at Hamming distance at least d.

2. A basic problem of coding theory: determine the maximum M such that an $(n, M, d)_2$-code exists.

3. A code can correct e errors provided its minimum distance is $d \geq 2e + 1$.

4. We saw a $(6, 8, 3)_2$−code.

Exercises 1.3

1.3.1. *If we want to correct 8 bit errors, what would the minimum distance of the code have to be?*

1.3.2. *Using our code* $(6, 8, 3)_2$, *decode the following received vectors:*

$$111100, \ 111011, \ 000001, \ 011110.$$

1.3.3. *Does a code* $(5, 6, 3)_2$ *exist?*

1.4   Error-correcting codes in general

Basic concepts: Basic code parameters. Telegraphy codes as early examples.

The notion of a **binary** code is too narrow, although it is most frequently used in information transmission. Here is the general concept of an (error-correcting) code:

1.8 Definition. *Let $\mathcal{A}$ be a finite set of q elements (the **alphabet**). A q-ary code $\mathcal{C}$ of **length** n is a family of n-tuples with entries in $\mathcal{A}$:*

$$\mathcal{C} \subseteq \mathcal{A}^n.$$

For example, let $q = 3$ and $\mathcal{A} = \{0, 1, 2\}$. Then $\mathcal{A}^4$ consists of the 3^4 tuples of length 4 with entries $0, 1, 2$, like, for example,

$$0000, \ 0102, \ 2221 \text{ or } 2100.$$

A 3-ary code (also called **ternary**) of length 4 consists of a collection of such ternary 4-tuples. The Morse code from Section 1.1 really is a ternary code. The reason is that the individual letters need to be seperated. If we represent a dot by 0, a dash by 1 and a gap between letters as 2, the message SOS will be represented by the ternary word 00021112000.

1.9 Definition. *Let $x = (x_1, x_2, \ldots, x_n)$ and $y = (y_1, y_2, \ldots, y_n)$ be elements (strings, vectors, words) in $\mathcal{A}^n$. The **distance** (or **Hamming distance**) between x and y is defined as*

$$d(x, y) = \text{number of coordinates } i \text{ where } x_i \neq y_i.$$

This is the same as in Definition 1.3. As before, the minimum distance of a code is the minimum of the distances between different codewords, and again the Hamming distance defines a metric.

1.10 Definition. *A q-ary code* $(n, M, d)_q$ *is a set of M strings in* $\mathcal{A}^n$ *(where* $|\mathcal{A}| = q$*) such that any two different elements of the code (codewords) are at distance* $\geq d$*. Call n the* **length** *and d the* **minimum distance** *of the code.*

Here are the words of a ternary code $(4, 27, 2)_3$:

0000	1002	2001	0102	1101	2100	0201	1200	2202
0012	1011	2010	0111	1110	2112	0210	1212	2211
0021	1020	2022	0120	1122	2121	0222	1221	2220

As in the binary case, in order to correct e errors, we need a code of minimum distance at least $2e + 1$. Our code $(4, 27, 2)_3$ will not suffice to correct one error. For example, if 0000 was sent and 0010 received (only one error occurred), the received vector has distance 1 not only from 0000 but also from 0012 and from 0210 and from 2010.

A basic problem of coding theory is the following: given q, n, d, find the maximum number M such that a code $(n, M, d)_q$ exists.

Error detection in telegraphy codes

Error detection is a more modest aim than error correction. It is suitable in situations where the channel is very good. On the rare occasions that an error occurs (and is detected), the receiver can then simply ask for retransmission. The alphabet of the telegraphy codes consists of the 26 letters A, B, ... , Z. The classical commercial codes use five letter groups as codewords. Each trade had its own elaborate codes. The primary aim of these codes was to save transmission time and thus to save money. As an example, take the **Acme Code.** It saves time to send the codeword BUKSI when *Avoid arrest if possible* is intended, and AROJD is shorter than *Please advertise the birth of twins.* It is a little unclear if PYTUO for *Collided with an iceberg* really achieves much in this respect, as such collisions do not happen all the time. In modern terminology, this business of representing messages by short strings is called **Data Compression** or **Source Coding.** It is not our concern in this book.

However, commercial telegraphy codes also took the reliability of message transmission into consideration. A general rule known as the **two-letter differential** stipulated that any two codewords had to differ in at least two letters. This means that each commercial code has minimum Hamming distance ≥ 2, enough to detect single errors. The Acme code also safeguarded against a different type of error: no two codewords (necessarily of Hamming distance 2) may result from each other by transposition of two adjacent letters. For instance, if AHXNO is a codeword (it stands for *Met with a fatal accident in the Acme code*), then HAXNO, AXHNO, AHNXO, AHXON cannot be codewords.

This material is from Chapter 22 of D. Kahn's *The Codebreakers* [122].

---

1. A q-ary code $(n, M, d)_q$ is a collection of M q-ary n tuples (the **codewords**) such that any two different codewords are at Hamming distance at least d.

2. A basic problem of coding theory: given q, n, d, maximize M such that an $(n, M, d)_q$-code exists.

3. A code can correct e errors provided its minimum distance is $d \geq 2e + 1$.

4. We saw a ternary code $(4, 27, 2)_3$.

Exercises 1.4

1.4.1. *Find the smallest length n such that an $(n, 27, 2)_3$ exists.*

1.4.2. *Prove the following: if there is an $(n, M, d)_q$, then there is an $(n + 1, M, d)_q$.*

1.4.3. *Prove the following: If there is an $(n, M, d)_{q-1}$, then there is an $(n, M, d)_q$.*

1.5   The binary symmetric channel

Basic concepts: The BSC, binomial numbers, subsets and paths, the Pascal triangle.

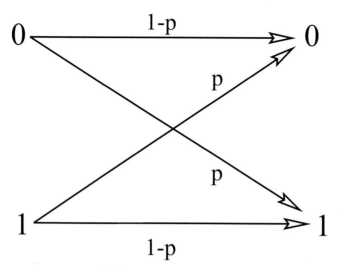

FIGURE 1.3:   The BSC

So how do we model the **noise** mentioned in Section 1.1? In the case of the binary alphabet, the easiest and most widely used model is the binary symmetric channel (BSC). It assumes that there is a certain fixed probability, a number p, where $0 < p < 1$, for a bit transmission error. Clearly a value $p > 1/2$ does not make sense (why?) and $p = 1/2$ would mean that pure noise is received. We can therefore assume $p < 1/2$, and for all practical purposes p will be rather small.

The probability that a sent 0 will be received as 0 is $1 - p$. Likewise the probability that a sent 1 will be received as 1 is $1 - p$. The probability that a sent 0 is received as a 1 is p, just like the probability that a sent 1 is received as 0.

This model of a channel is called **symmetric** because 0 and 1 play symmetric roles. It is also called a **memoryless** channel because the probability of a bit error is independent of the prehistory. Can you think of situations when this model will not be appropriate?

In order to be able to do some combinatorial counting based on the BSC we take a look at the binomial numbers.

Binomials, subsets and paths

The number of bitstrings of length n and weight m is the binomial number $\binom{n}{m}$. This is the number of error patterns that can occur in n bits when the total number of errors is m. For example, the bitstrings of length 4 and weight 2 are

$$0011, \; 0110, \; 0101, \; 1001, \; 1010, \; 1100.$$

Accordingly, $\binom{4}{2} = 6$.

Another interpretation of the binomials uses subsets: identify each coordinate of a bitstring of length n with an element of a set, for example with the numbers $1, 2, \ldots, n$. Each bitstring of length n can then be identified with a subset of $\{1, 2, \ldots, n\}$. If the entry in the corresponding coordinate is 1, we include that element in the subset; if the entry is 0 we will not include it. For example, 0011 corresponds to the subset $\{3, 4\}$, 1001 to the subset $\{1, 4\}$, 1111 to the total set $\{1, 2, 3, 4\}$ and 0000 to the empty set. We see that the bitstrings of length n and weight m correspond precisely to the subsets of m elements of a fixed set with n elements.

A third interpretation of the binomials involves paths in a triangle; see Figure 1.4.

Consider paths starting at the top of the triangle, where in each step the choice is between going southeast or southwest. We may encode this decision by a string of E and W, for example EEWWE for going at first southeast twice, then southwest twice and a final step in southeast direction. Denote the top level by level 0. Then our path will end at level 5, at a node labelled 10 in Figure 1.4. Why that label? Our string with entries E and W is a bitstring in disguise. We can write 1 for E and 0 for W, obtaining bitstring 11001. Each path in the triangle is described by a bitstring. Bitstrings of length n end on level n. Two bitstrings end in the same spot if they have the same length and the same weight. This explains our labels: the label of the node on level n and weight m (start with weight 0 on the western end of the level, end with weight n on the eastern end) is the number of paths ending there. This is the number of bitstrings of length n and weight m, in other words the binomial $\binom{n}{m}$. The endpoint of our path is labeled 10, as $\binom{5}{3} = 10$ is the number of paths ending there. The labels 1 on the western border are the numbers $\binom{n}{0} = 1$ (there is only one bitstring with all entries 0 of any given length); the labels 1 on the eastern border are the numbers $\binom{n}{n} = 1$ (there is only one bitstring with all entries 1).

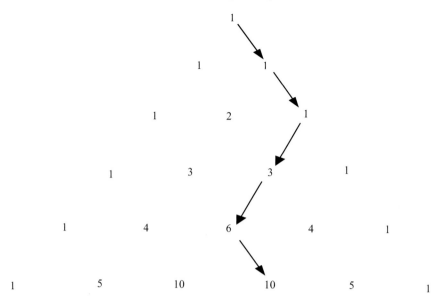

FIGURE 1.4: Pascal's triangle

The triangle is known as the **Pascal triangle.** It can be used to compute the binomials recursively. In fact, each label not on the border is the sum of the labels directly to its northeast and to its northwest, in formulas

$$\binom{n}{m} = \binom{n-1}{m-1} + \binom{n-1}{m} \text{ for } 0 < m < n.$$

A direct formula to compute the binomials is

$$\binom{n}{m} = \frac{n(n-1)\dots(n-m+1)}{m(m-1)\dots 2\cdot 1}.$$

As examples, $\binom{4}{2} = \dfrac{4\cdot 3}{2} = 6$, $\binom{5}{2} = \dfrac{5\cdot 4}{2} = 10$,

$\binom{6}{3} = \dfrac{6\cdot 5\cdot 4}{3\cdot 2} = 20$ and in general $\binom{n}{2} = \dfrac{n(n-1)}{2}$.

Back to the channel

Assume a word (a bitstring) of length n is sent through the channel. The probability that no bit error occurs is $(1-p)^n$. The probability that exactly one error occurs is $np(1-p)^{n-1}$ (there are n choices where the error could happen; the probability of a certain error pattern involving just one error is $p(1-p)^{n-1}$). The probability of a certain error pattern with precisely k errors is $p^k(1-p)^{n-k}$. The number of such error patterns is the number of possibilities of choosing k elements out of n. This is $\binom{n}{k}$. The probability that precisely k errors occur is therefore $\binom{n}{k}p^k(1-p)^{n-k}$. If we want to compute the probability that **at most** k errors happen, we have to sum up these probabilities.

---

1. The binary symmetric channel is the simplest model for noise.

2. If a binary n-tuple is sent via the BSC, the probability that at most k bit errors occur is $\displaystyle\sum_{i=0}^{k} \binom{n}{i}p^i(1-p)^{n-i}$.

While the BSC is conceptually simple, there are other channels which are easier to handle. An example is the binary **erasure channel** where the probability of correct transmission of 0 and 1 is $1-p$ just as in the BSC, but in addition the receiver knows when problems occurred. Formally this can be described as a channel with three possible outputs: $0, 1$ or E, where E stands for **erasure.**

Exercises 1.5

1.5.1. *Compute the probability that no more than one error occurs in the transmission of a bitstring of length 10, when the bit error probability is $p = 10^{-3}$, or $p = 10^{-4}$, or $p = 10^{-5}$.*

1.5.2. *Describe a generalization of the BSC from the binary to the general q-ary case.*

1.5.3. *Sketch a formal picture of the binary erasure channel, analogous to Figure 1.3 for the BSC.*

1.5.4. *Show that, if a code of minimum distance d is used for the erasure channel, then any $d-1$ errors can be corrected.*

1.5.5. *Use the subset interpretation of the binomials to prove the binomial formula*

$$(a + b)^n = \sum_{i=0}^{n} \binom{n}{i} a^i b^{n-i}.$$

1.5.6. *Prove*

$$(1 - b)^n = \sum_{i=0}^{n} (-1)^i \binom{n}{i} b^i.$$

1.6 The sphere-packing bound

and the ternary Hamming code; the football pool problem.

As in all mathematical optimization problems, the problem of finding the maximum number M, such that a code $(n, M, d)_q$ exists, splits into two parts: we have to construct good codes (with high M), and we have to prove upper bounds showing that a higher value of M is impossible. In this section we will prove our first upper bound. It is a direct consequence of the basic idea of error correction.

As a preparation, we need a counting argument (this is a typical problem of elementary combinatorics) of the same type as the argument in Section 1.5. Given a vector $x \in \mathcal{A}^n$ (recall that $\mathcal{A}$ is our alphabet of size q), how many vectors are there at distance $\leq i$ from x? We call this set of vectors the **ball of radius** i with center x.

1.11 Definition. *Consider the space $\mathcal{A}^n$ of vectors of length n, with the Hamming metric, where $|\mathcal{A}| = q$. This is also called a **Hamming space**. The number of vectors at distance $\leq i$ from a given vector is denoted $V_q(i, n)$. We call $V_q(i, n)$ the **volume** of a ball of radius i.*

Fix some distance j and count the vectors at distance precisely j from the given vector. There are $\binom{n}{j}$ choices for the set of coordinates where the entries are different. Once this set is fixed, there are $q - 1$ possibilities for the possible entries in each of these j coordinates. We count $\binom{n}{j}(q-1)^j$. The volume $V_q(i, n)$ is obtained by adding up these numbers, for $j \leq i$. We have seen the following:

1.12 Proposition. *The volume of the ball of radius i is*

$$V_q(i, n) = \sum_{j=0}^{i} \binom{n}{j} (q-1)^j.$$

We are close to our bound. Let $\mathcal{C}$ be a code $(n, M, d)_q$. Let $e = \lfloor (d-1)/2 \rfloor$ (the largest integer less than or equal to $(d-1)/2$). We have chosen e such that $2e + 1 < d$. Our standard argument shows that the balls of radius e centered at the codewords must be disjoint. As each such ball has $V_q(e, n)$ vectors, we must have $MV_q(e, n) \le q^n$ (counting all vectors in these balls, we cannot get more vectors than the whole space contains).

1.13 Theorem (sphere packing bound). *Each q-ary code $\mathcal{C}$ of length n and minimum distance d satisfies*

$$|\mathcal{C}| \le \frac{q^n}{V_q(e, n)}.$$

Here $e = \lfloor (d-1)/2 \rfloor$.

Codes for which equality holds in Theorem 1.13 are known as **perfect codes.** The parameters of perfect codes have all been classified.
Our first code was a $(6, 8, 3)_2$. What is the maximum M for a code $(6, M, 3)_2$? Here $e = 1$. We have $V_2(1, 6) = 7$. The sphere-packing bound says $7M \le 64$, hence $M \le 9$. We claim that a code $(6, 9, 3)_2$ cannot exist. The reader is asked to provide a proof along the lines sketched in Exercises 1.6.4 and 1.6.5.
Here is a nice parameter situation: consider a possible code $(11, 729, 5)_3$. Observe that $729 = 3^6$. We have $e = 2$ and $V_3(2, 11) = 1 + 11 \cdot 2 + \binom{11}{2} \cdot 4 = 1 + 22 + 220 = 243 = 3^5$ (quite a coincidence). It follows that a ternary code $(11, M, 5)_3$ has $M \le 3^{11}/3^5 = 3^6 = 729$. We see that the parameters $(11, 3^6, 5)_3$ are extremal. If such a code exists, then it is perfect. A perfect code $(11, 729, 5)_3$ does indeed exist. It is uniquely determined by its parameters and known as the **ternary Golay code,** named after the Swiss engineer who described it in 1949 (see Golay [94], reprinted in Berlekamp [13]).

A betting system

Here is how the ternary Golay code can be used in a betting system: in most European countries it is popular to bet on the results of football matches (**football** is not to be confused with American football. In the US the game is known by the bizarre name of **soccer**). Each match has one of three possible results: 1= home team wins, 2=guest wins, or 0= a draw. This is why we use ternary codes. There are 11 or more matches in the pool. If there are

11 matches, we can use the 729 codewords of the ternary Golay code as bets. As every vector is at distance ≤ 2 from a codeword, it is guaranteed that no matter what the results of the 11 matches are, one of our bets has at least 9 results correct. Assume there are 13 matches in the pool. Then we may choose two matches whose results seem to be safe and use the Golay code for the remaining 11 matches.

Incidentally, the ternary Golay code is older than Golay. The Finnish journalist Juhani Virtakallio published it in 1947 in the Finnish football journal *Veikaaja* (see Cohen et al. [55]). It is no coincidence that this discovery was made in Finland. This country has a rich tradition in coding theory and related questions.

---

1. The **sphere-packing bound** is our first general bound on codes.

2. Codes meeting it with equality are called **perfect**.

3. The **ternary Golay code** $(11, 729, 5)_3$ is perfect.

Exercises 1.6

1.6.1. *Construct a code* $(4, 2, 3)_2$ *(this is really trivial).*

1.6.2. *Show that there is no* $(4, 3, 3)_2$*-code*

1.6.3. *Construct a code* $(5, 4, 3)_2$ *(hint: use our code* $(6, 8, 3)_2)$*).*

1.6.4. *Show that there is no* $(5, 5, 3)_2$*.*

1.6.5. *Using the preceding exercise, show that there is no* $(6, 9, 3)_2$*.*

1.6.6. *What does the sphere-packing bound tell us about the length n of a binary code* $(n, 2^7, 5)_2$*?*

1.6.7. *Six candidates are examined by 9 referees. Each referee assigns a pass-fail grade to each candidate. Any two referees assign the same grade to not more than 3 of the candidates. Can this really happen?*

1.6.8. *Show that the minimum distance of a perfect code must be odd.*

1.6.9. *Use the sphere-packing bound to show the nonexistence of* $(5, 6, 3)_2$*.*

1.6.10. *Does a* $(7, 9, 3)_2$*-code exist?*

Chapter 2

Binary linear codes

2.1 The concept of binary linear codes

Basic concepts: Dimension, generator matrices, minimum weight. Linear algebra over $\mathbb{F}_2$: basis, rank, linear independence, determinant.

Our binary code $(6, 8, 3)_2$ has an additional structure, which greatly simplifies its description. Consider the following three of its codewords:

$$\boxed{\begin{array}{l} 100110 \\ 010101 \\ 001011 \end{array}}$$

Call them z_1, z_2, z_3. Consider all **linear combinations** of z_1, z_2 and z_3, that is, all vectors of the form $\lambda_1 z_1 + \lambda_2 z_2 + \lambda_3 z_3$, where $\lambda_i \in \mathbb{F}_2$. These linear combinations are different and they are just exactly the words of our code. We call $\{z_1, z_2, z_3\}$ a **basis** of our code and say that the code is **linear.** Another way of seeing this is by the following observation: the sum of any two codewords is a codeword again; the code is closed under sums. We take this as a definition:

2.1 Definition. *A binary code is* **linear** *if it is closed under addition.*

It is a basic fact from linear algebra that each linear code (abstractly: each linear space, each vector space) has a basis. Linear algebra applies to arbitrary fields. Most people are familiar with fields like the rational numbers, the real numbers and the complex numbers, but we can apply the basics of linear algebra to finite fields like $\mathbb{F}_2$ as well. This leads to the following basic fact;

2.2 Theorem. *A binary linear code* $(n, M, d)_2$ *has* $M = 2^k$ *for some* k. *The number* k *is the* **dimension** *of the code. There is a* **basis** $\{z_1, z_2, \ldots, z_k\}$ *of* k *codewords. Each codeword is a linear combination of the* z_i.

There is a general tendency to restrict attention to **linear** codes. One reason is that these are much easier to describe and to work with than codes in general. For example, a binary linear code of dimension k has $M = 2^k$ codewords, but it is uniquely described by a basis, which has only k elements. This is a much more compact representation of the code than a list of all its words.

2.3 Definition. *The parameters of a* k-*dimensional binary linear code of length* n *and minimum distance* d *are written*

$$[n, k, d]_2$$

(the number of codewords is $M = 2^k$*).*

The basic problem of binary linear codes is the following: determine the maximum k such that a code $[n, k, d]_2$ exists. Our code $(6, 8, 3)_2$ is a linear code, a $[6, 3, 3]_2$-code. The compact representation described above leads to the notion of a generator matrix.

2.4 Definition. *Let* C *be a linear code* $[n, k, d]_2$. *A* **generator matrix** G *of* C *is a* (k, n)-*matrix whose rows form a basis of* C.

If we know a generator matrix G, then we know the code. The codewords are just all linear combinations of the rows of G; in other words, the code is the rowspace of G.

Here is another binary linear code. Please check that it is indeed linear. The minimum distance is 3.

0000000	1100110
1101000	0100101
1010100	1000011
0110010	0001110
1110001	1001101
0111100	0101011
1011010	0010111
0011001	1111111

As there are $16 = 2^4$ codewords, its dimension is $k = 4$. The parameters are $[7, 4, 3]_2$. It is known as the **binary Hamming code.** How can we be sure the minimum distance is really 3? We would have to check $\binom{16}{2} = 120$ pairs of codewords. Here is a simplification, valid for linear codes:

2.5 Proposition. *For a binary linear code, the minimum distance equals the minimum of the weights of nonzero codewords (observe that the all-0 word is automatically contained in each linear code).*

PROOF   We know that each weight also is a distance. It suffices to show that the distance between two codewords also is the weight of some codeword. Let x, y be different codewords at distance $d(x, y)$. Addition of the same vector to both words does not change the distance. We add x. This yields $d(x, y) = d(\mathbf{0}, y + x) = wt(y + x)$. Here $\mathbf{0}$ is the all-0 word. As the code is linear, $x + y$ is a codeword again. As $x \neq y$, we have $x + y \neq 0$.   ∎

Because of Proposition 2.5, it suffices to check that each of the 15 nonzero codewords has weight ≥ 3 to check that indeed $d = 3$. Here is a generator matrix:

$$\begin{pmatrix} 1\,0\,0\,1\,1\,0\,1 \\ 0\,1\,0\,1\,0\,1\,1 \\ 0\,0\,1\,0\,1\,1\,1 \\ 0\,0\,0\,1\,1\,1\,0 \end{pmatrix}$$

How can we be sure? At first we check that each row of the matrix is a nonzero codeword. One method of controlling that we have a generator matrix is to make sure that each codeword is indeed a linear combination of rows, equivalently that different linear combinations of rows yield different codewords. We can speed up by using basic facts of linear algebra. One such fact is the following: we have a generator matrix if and only if the matrix has rank equal to the number of rows ($= k$), if and only if there is a (k, k)-submatrix with nonzero determinant. In our case the last check is fastest: the first four columns form a triangular matrix with ones on the diagonal, hence of determinant $= 1$. We repeat:

2.6 Proposition. *Let C be a binary $[n, k]_2$-code (the minimum distance is irrelevant here). Let G be a matrix whose rows are codewords of C. The following are equivalent:*

1. *G is a generator matrix.*

2. *G has rank k (remember that the **rank** of a matrix is a basic term from linear algebra).*

3. *There is a (k, k)-submatrix of nonzero determinant (recall that the **determinant** of a matrix is another basic notion from linear algebra).*

4. *The rows of G are **linearly independent** (yet another such basic notion.)*

1. A binary code is **linear** (or a **subspace** of $\mathbb{F}_2^n$) if it is closed under addition.

2. A binary linear code (subspace) has 2^k codewords; k is the **dimension**.

3. The parameters of binary linear codes: $[n, k, d]_2$.

4. Each binary linear code has a **basis,** consisting of k codewords.

5. A (k, n)-matrix whose rows form a basis is a **generator matrix.**

Exercises 2.1

2.1.1. *Show that our code $[7, 4, 3]_2$ is perfect.*

2.1.2. *Try to decide if an $[8, 4, 4]_2$ exists.*

2.1.3. *Give an example showing that the basis of a code is not uniquely determined.*

2.1.4. *Determine the parameters of the binary linear code generated by the rows of the matrix*

$$\begin{pmatrix} 1\,0\,0\,1\,1\,0\,1 \\ 0\,1\,0\,1\,0\,1\,1 \\ 0\,0\,1\,0\,1\,1\,1 \end{pmatrix}$$

2.1.5. *Compute the parameters $[n, k, d]_2$ of the binary linear code generated by*

$$G = \begin{pmatrix} 1\,1\,0\,0\,0\,0\,1\,1\,0\,1 \\ 0\,0\,1\,1\,0\,0\,1\,0\,1\,1 \\ 0\,0\,0\,0\,1\,1\,0\,1\,1\,1 \end{pmatrix}$$

Find a nonzero codeword of minimum weight.

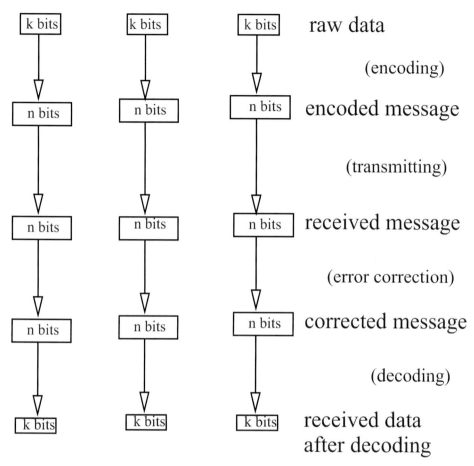

FIGURE 2.1: Block coding

2.2 Block coding

using binary linear codes. Information rate, relative distance.

Binary linear codes are particularly handy for block coding. We use a linear code $\mathcal{C}$ with parameters $[n, k, d]_2$. As there are 2^k codewords, we can find a bijective function (the encoding function) $\alpha : \mathbb{F}_2^k \longrightarrow \mathcal{C}$. There is no reason not to choose this function as a linear function ($\alpha(x + y) = \alpha(x) + \alpha(y)$). Here is a possible encoding function α for our code $[7, 4, 3]_2$. Observe that the inverse α^{-1} is the projection on the last four coordinates.

$0000 \longrightarrow 0000000$	$0110 \longrightarrow 1100110$
$1000 \longrightarrow 1101000$	$0101 \longrightarrow 0100101$
$0100 \longrightarrow 1010100$	$0011 \longrightarrow 1000011$
$0010 \longrightarrow 0110010$	$1110 \longrightarrow 0001110$
$0001 \longrightarrow 1110001$	$1101 \longrightarrow 1001101$
$1100 \longrightarrow 0111100$	$1011 \longrightarrow 0101011$
$1010 \longrightarrow 1011010$	$0111 \longrightarrow 0010111$
$1001 \longrightarrow 0011001$	$1111 \longrightarrow 1111111$

We can describe how block coding works: the incoming data is divided into consecutive blocks of length k. Each such block is therefore a bitstring of length k, an element of $\mathbb{F}_2^k$. Apply the encoding function α to each block. The result is an encoded block of length n. These are transmitted via the channel. Error correction (recall our basic coding idea: $e = \lfloor (d - 1)/2 \rfloor$ bit errors per encoded block can be corrected) produces blocks of length n, which hopefully are the same as the blocks fed into the channel. Finally, the inverse α^{-1} of the encoding function is applied to each block of length n. The result is a block of length k, which hopefully is identical to the corresponding block of raw data. The last two stages of the procedure (error correction and application of α^{-1}) are also collectively known as decoding.

The repetition code of length 3 from Section 1.1 is a part of the general picture: the code has parameters $[3, 1, 3]_2$ and consists of the codewords 000 and 111. The encoding function maps $0 \mapsto 000, 1 \mapsto 111$. This is about as elementary as it gets.

In block coding the length of bitstrings is increased from k to n. That means that k/n of the bits sent through the channel represent pure information, whereas the rest is redundancy that we have cleverly introduced for the purposes of error correction. We want to keep that ratio k/n as large as possible in order not to waste time and money.

2.7 Definition. *The ratio $R = k/n$ of a linear $[n, k, d]_2$-code is its* **information rate**; *the ratio d/n is the* **relative distance**.

Exercises 2.2

2.2.1. *Follow through all stages of block coding for the input string* 00101110 *when transmission errors occur in coordinates* $3, 7, 14$.

2.3 The effect of coding

on the error probability of decoding.

Let us compare the error probabilities with and without coding in a special case, based on the BSC (Section 1.5). The bit error probability is a (small) number p. Use block coding with blocks of length $k = 4$. We want to compute (or bound) the probability P that a given 4-tuple is **not** received correctly after coding and decoding.

Without coding ($n = k = 4$) the probability of correct transmission is $(1 - p)^4 = 1 - 4p + 6p^2 - 4p^3 + p^4 \approx 1 - 4p$, so $P \approx 4p$.

Using our code $[7, 4, 3]_2$ the message will be correctly decoded if not more than 1 bit error occurs. The probability of correct transmission is therefore

$$\geq (1 - p)^7 + 7p(1 - p)^6 \approx (1 - 7p + 21p^2) + (7p - 42p^2) = 1 - 21p^2.$$

In this case we have $P \leq 21p^2$. We pay for this smaller error probability with an information rate $R = 4/7$ (meaning that $3/7$ of transmission time is redundant).

Compare this to our very first code, the repetition code of length 3. We leave it as an exercise to check that, seen as a block code with $k = 4, n = 12$, its error probability is $P \leq 12p^2$. This is a decrease by about a factor 2 with respect to the $[7, 4, 3]_2$, but the information rate has gone down to $1/3$.

There is a code $[11, 4, 5]_2$. If we use it, the probability of correct transmission is

$$\geq (1 - p)^{11} + 11p(1 - p)^{10} + \binom{11}{2} p^2 (1 - p)^9$$

$$\approx \left(1 - 11p + 55p^2 - \binom{11}{3} p^3\right) + (11p - 110p^2 + 495p^3) + (55p^2 - 495p^3)$$

$$\approx 1 - \binom{11}{3} p^3.$$

The error probability is $P \leq 165p^3$.

Let $p = 10^{-4}$. Without coding, the block error probability is $P \approx 4 \cdot 10^{-4}$. With an information rate of $R = 4/7$, we reach $P \leq 2.1 \cdot 10^{-7}$, and, with an information rate of $R = 4/11$, this can be pressed down to $P \leq 1.65 \cdot 10^{-10}$. The repetition code is suddenly looking rather stupid in comparison.

Exercises 2.3

2.3.1. *Compute the block error probability of the repetition code of length 3, seen as a code encoding blocks of length 4 into messages of length 12.*

2.3.2. *Compute information rate and block error probability for a code $[9, 5, 3]_2$ (it exists).*

2.3.3. *Compute information rate and block error probability for a code $[23, 12, 7]_2$ (the **binary Golay code**).*

2.3.4. *Show that the block error probability of an $[n, k, 2e+1]_2$-code is bounded (approximately) by $\binom{n}{e+1}p^{e+1}$.*

2.4   Duality

Basic concepts: Dot product on $\mathbb{F}_2^n$, orthogonality, dual space, binary repetition code, sum zero code, check matrix, P-transform.

As is often the case in mathematics, duality is of great benefit also for coding theory. Everybody knows the dot product from physics. We can use it in our binary context as well.

2.8 Definition. *The **dot product** is defined on $\mathbb{F}_2^n$ by*

$$(x_1, x_2, \ldots, x_n) \cdot (y_1, y_2, \ldots, y_n) = \sum_{i=1}^{n} x_i y_i.$$

Let $x, y \in \mathbb{F}_2^n$. We say that x and y are **orthogonal** ($x \perp y$) if $x \cdot y = 0$. It is obvious that, for every subset $U \subseteq \mathbb{F}_q^n$, the set

$$U^\perp = \{x \mid x \in \mathbb{F}_2^n, x \cdot u = 0 \text{ for all } u \in U\}$$

is a subspace of $\mathbb{F}_2^n$. The following rules hold for all $x, y, z \in \mathbb{F}_2^n$ and $\lambda \in \mathbb{F}_2$

$$\begin{aligned}
x \cdot y &= y \cdot x \\
(x + y) \cdot z &= x \cdot z + y \cdot z \\
(\lambda x) \cdot y &= x \cdot (\lambda y) = \lambda(x \cdot y).
\end{aligned}$$

The dot product is an example of what are called **symmetric bilinear forms** in general (another topic of linear algebra). Here is a basic fact, which will be proved in the more general context of symmetric bilinear forms in Proposition 17.33.

2.9 Lemma. *Let $U \subseteq \mathbb{F}_2^n$ be a linear subspace. Then*

$$dim(U) + dim(U^\perp) = n.$$

In particular we see that, when $\mathcal{C}$ is a linear k-dimensional binary code of length n, then the **dual code** $\mathcal{C}^\perp$ has dimension $n - k$.

2.10 Lemma. *Let $U \subseteq \mathbb{F}_q^n$ be a linear subspace. Then*

$$U^{\perp\perp} = U.$$

PROOF   By definition $U^{\perp\perp} \supseteq U$. It follows from Lemma 2.9 that these spaces have the same dimension. □

Consider $\mathbf{0} = 000\ldots$ (the 0 vector) and $\mathbf{1} = 111\ldots$ the all-1 vector. We have that $U = \{\mathbf{0}, \mathbf{1}\}$ is a one-dimensional subspace of $\mathbb{F}_2^n$. This is the **repetition code**. Its dual $U^\perp$ will therefore have dimension $n-1$. What is this dual? The whole space is orthogonal to the 0 word. The vector $(x_1, x_2, \ldots, x_n)$ is orthogonal to the all-1 vector if

$$x_1 + x_2 + \cdots + x_n = 0.$$

2.11 Definition. *The binary **sum zero code** of length n consists of all binary n-tuples whose entries sum to 0. It is a linear code $[n, n-1, 2]_2$. Its dual is the repetition code.*

Please check that the minimum distance is really 2 (equivalently: it is impossible that two binary vectors at distance 1 both have sum 0).

Is it possible that a binary space is orthogonal to itself ($U \subseteq U^\perp$)? Yes, of course. For example, half of all bitstrings are orthogonal to themselves (see Exercise 2.4.1). This situation is very different from real spaces, where a nonzero vector can never be orthogonal to itself.

Whenever we have a code $[n, k, d]_2$, we can consider its dual. The dimension will be $n - k$, but it is unclear what the minimum distance is going to be. As an example, consider our code $[7, 4, 3]_2$ (the binary Hamming code $\mathcal{H}$ from Section 2.1). Consider the first three rows of the generator matrix G given in that section. They are orthogonal to all rows of G (check that). It follows that these three words generate a three-dimensional code, which is orthogonal to $\mathcal{H}$. As $\mathcal{H}$ has length 7 and dimension 4, $\mathcal{H}^\perp$ has dimension 3. It follows that the first 3 rows of $\mathcal{H}$ form a generator matrix of $\mathcal{H}^\perp$. In particular $\mathcal{H}^\perp \subset \mathcal{H}$. In Exercise 2.1.4 it was shown that $\mathcal{H}^\perp$ is a code $[7, 3, 4]_2$.

2.12 Proposition. *Let $\mathcal{H}$ be the binary Hamming code $[7, 4, 3]_2$.*
Then $\mathcal{H}^\perp \subset \mathcal{H}$ is a code $[7, 3, 4]_2$.

Duality gives us another method to describe a code. Assume a generator matrix H of $\mathcal{C}^\perp$ is given. Then I have a fast method to decide if a bitstring $x \in \mathbb{F}_2^n$ belongs to $\mathcal{C}$. This will be the case if and only if x is orthogonal to all the rows of H. In many instances this description will be handier than the description via a generator matrix.

2.13 Definition. *A generator matrix of the dual code $\mathcal{C}^\perp$ is called a* **check matrix** *for $\mathcal{C}$.*

Here is a little trick (motivated by Gauß elimination from linear algebra) which often simplifies the determination of the check matrix.

2.14 Lemma. *Let a generator matrix for a code $[n, k]_2$ be written in the form* $G = (I|P)$. *Then a check matrix is $H = (P^t|I)$. Here I in G is the (k, k)-unit matrix, the I in H is the $(n - k, n - k)$-unit matrix and P^t is the transpose of P.*

PROOF   We have to show that each row of G is orthogonal to each row of H. How can we express this in matrix notation? Recall the **matrix product:** the entry in row i and column j of the product matrix AB is the dot product of row i of A and column j of B. This means we have to show that $GH^t = 0$ (the 0 matrix). In fact,

$$GH^t = (I|P)(P^t|I)^t = P + P = 0.$$

⬚

Consider our code $[6, 3, 3]_2$ from Section 1.3. It has a generator matrix
$$G = \begin{pmatrix} 100 & 110 \\ 010 & 101 \\ 001 & 011 \end{pmatrix}.$$ This has the form of Lemma 2.14. We obtain as check

matrix $H = \begin{pmatrix} 110 & 100 \\ 101 & 010 \\ 011 & 001 \end{pmatrix}$. Observe that the submatrix P of G has to be transposed to obtain H. In our example P is symmetric $(P = P^t)$.

---

1. Duality is based on the **dot product.**

2. The dual of an $[n, k]_2$-code has dimension $n - k$.

3. A **check matrix** H of C is a generator matrix of the dual $C^\perp$.

4. The sum-zero code $[n, n - 1, 2]_2$ is the dual of the repetition code $[n, 1, n]_2$.

Exercises 2.4

2.4.1. *When is the all-1 word orthogonal to itself?*

2.4.2. *A code is* **self-dual** *if it equals its dual.*
Is there a self-dual $[6, 3, 3]_2$?

2.4.3. *Find a* $(4, 8)$ *matrix in standard form (starting with the unit matrix I) which generates a self-dual code* $(C^\perp = C)$ *with parameters* $[8, 4, 4]_2$.

2.4.4. *Find a generator matrix of the Hamming code* $[7, 4, 3]_2$ *in standard form. Use the P-transform to find a check matrix.*

2.4.5. *Is there a self-dual* $[12, 6, 6]_2$-*code?*

2.4.6. *Is there a* $[12, 6, 6]_2$-*code?*

2.5 Binary Hamming and Simplex codes

and decoding of the Hamming codes.

We learned how to describe binary linear codes using either a generator matrix or a check matrix. It is natural to concentrate on the **rows** of these matrices (they are codewords of the code to be described or of its dual). However, it will turn out to be very profitable to change the point of view and read these matrices **columnwise.**

For every r, let M_r be a matrix whose columns run through all nonzero bitstrings of length r. Clearly M_r is an $(r, 2^r - 1)$ matrix. The order in which the columns are written does not really matter. We will write them in order of increasing weight, and for each weight we order the columns lexicographically:

$$M_3 = \begin{pmatrix} 1001101 \\ 0101011 \\ 0010111 \end{pmatrix}.$$

$$M_4 = \begin{pmatrix} 100011100011101 \\ 010010011011011 \\ 001001010110111 \\ 000100101101111 \end{pmatrix}.$$

Call the code generated by M_r the **Simplex code** $\mathcal{S}_r(2)$. As M_r starts with the unit matrix, it has rank r. The Simplex code therefore has dimension r. How about the minimum distance? Each row of M_r has weight 2^{r-1}. Imagine we write the all-0 vector as the last column. It is then clear that each combination of rows of M_r (hence each nonzero codeword of the Simplex code) has equally many zeroes and ones (2^{r-1} each). It follows that each nonzero codeword of the Simplex code has weight 2^{r-1}. The Simplex codes are **constant-weight codes.** If readers should find the argument not completely convincing, a crystal clear proof is in the first section of Chapter 17. If you do not want to wait that long, simply prove this fact by induction; see Exercise 2.5.4.

2.15 Theorem. *Let $\mathcal{S}_r(2)$ (the binary r-dimensional* **Simplex code***) be the code whose generator matrix has as columns all nonzero bitstrings of length r. Then $\mathcal{S}_r(2)$ is a code $[2^r - 1, r, 2^{r-1}]_2$. Moreover, each nonzero codeword has weight 2^{r-1}.*

We have codes $[7, 3, 4]_2$, $[15, 4, 8]_2$, $[31, 5, 16]_2, \ldots$ The smallest of these is a good old friend already.

Consider the duals of the Simplex codes. The dimension is $2^r - 1 - r = 2^r - (r + 1)$ by Lemma 2.9. How about the minimum distance? This is surprisingly easy to decide. Let $x \in \mathbb{F}_2^{2^r - 1}$. This time M_r is a check matrix. This means that x is a codeword if the sum of the columns of M_r corresponding to entries 1 in x vanishes. If $wt(x) = 1$, then M_r must have a 0 column, a contradiction. If $wt(x) = 2$, then two of the columns of M_r are identical, which is not the case. Can we have $wt(x) = 3$? This means that the sum of some three columns vanishes, equivalently that the sum of two nonzero tuples is a third nonzero tuple. This happens all the time. For example, columns number $1, 2$ and 4 in M_3 sum to 0, implying that $(1, 1, 0, 1, 0, 0, 0)$ is a codeword. We have seen that our code has minimum distance $d = 3$.

2.16 Theorem. *Let* $\mathcal{H}_r(2) = \mathcal{S}_r(2)^\perp$, *known as a binary* **Hamming code.** *Then* $\mathcal{H}_r(2)$ *is a code* $[2^r - 1, 2^r - (r + 1), 3]_2$.

This yields codes $[7, 4, 3]_2$, $[15, 11, 3]_2$, $[31, 26, 3]_2, \ldots$ Again we are already familiar with the smallest member of the family.

2.17 Theorem. *The binary Hamming codes are perfect codes.*

PROOF   Recall from Section 1.6 that a code is perfect if it meets the sphere-packing bound with equality. We have $d = 3$. The volume of the ball of radius 1 is $V_2(1, n) = n + 1 = 2^r$. As $2^n/|\mathcal{H}_r(2)| = 2^r$, we have equality. $\square$

Hamming decoding

An important question which we have not touched on yet is the **decoding algorithm.** How does the receiver compute the codeword that was sent, assuming not more than e errors occurred (here $2e < d$)? The general idea (see Section 1.3) states that we should search in the vicinity of the received word y (at distance $\leq e$). If we find a codeword x in this ball (we are guaranteed not to find more than one), then we decode $y \mapsto x$. How is this search process done in practice? As data comes streaming in, we probably want to decode in real time, so we must use a fast algorithm.

In the case of the Hamming codes this is particularly easy. We illustrate with the familiar Hamming code $[7, 4, 3]_2$. Here is the check matrix again:

$$H = \begin{pmatrix} 1\,0\,0\,1\,1\,0\,1 \\ 0\,1\,0\,1\,0\,1\,1 \\ 0\,0\,1\,0\,1\,1\,1 \end{pmatrix}.$$

Denote the columns of H by $s_1, s_2, \ldots, s_7$. Assume the codeword x was sent. That x is a codeword is equivalent to $xH^t = 0$ (in words: x is orthogonal to the rows of H). For example, take $x = (1, 0, 1, 1, 0, 1, 0)$. As $d = 3$, we assume at most one error occurred. If this error occurred in coordinate number 3, say, then

$$y = (1, 0, 1, 1, 0, 1, 0) + (0, 0, 1, 0, 0, 0, 0) = (1, 0, 0, 1, 0, 1, 0)$$

is the received vector. Here is how the receiver proceeds:
he or she computes $yH^t = (0,0,1)^t$. As $y = x + z$, where z is the error vector,
the receiver knows that $yH^t = xH^t + zH^t = zH^t$. If zH^t is column number
j of H, then the error occurred in that column. The correct codeword is
obtained by adding 1 in coordinate j to x. In our example $(0,0,1) = s_3$.
It follows that $x = y + (0,0,1,0,0,0,0)$.

---

1. Write a matrix M_r with r rows and $2^r - 1$ columns, whose columns are all the nonzero binary r-tuples.

2. Then G is a generator matrix of the Simplex code $\mathcal{S}_r(2)$ (of dimension r) and a check matrix of the Hamming code $\mathcal{H}_r(2)$.

3. $\mathcal{H}_r(2)$ is a perfect code with minimum distance 3.

4. Every nonzero codeword of $\mathcal{S}_r(2)$ has weight 2^{r-1}.

5. There is a particularly easy decoding algorithm for the Hamming codes.

Exercises 2.5

2.5.1. *Using matrix M_3, find at least five different codewords in the Hamming code $[7, 4, 3]_2$.*

2.5.2. *Use M_4 and find at least five different codewords in $\mathcal{H}_4(2)$.*

2.5.3. *Use the binary Hamming code $[7, 4, 3]_2$. Decode the received vectors $y_1 = (1, 1, 0, 1, 1, 0, 0)$, $y_2 = (1, 1, 1, 1, 1, 1, 1)$, and $y_3 = (1, 1, 1, 0, 0, 0, 0)$.*

2.5.4. *Prove by induction on r that each nontrivial linear combination of the rows of M_r (each nonzero word of the Simplex code $\mathcal{S}_r(2)$) has weight 2^{r-1}.*

2.5.5. *Show that the binary Simplex codes $\mathcal{S}_r(2), r \geq 3$ are self-orthogonal (contained in their orthogonals).*

2.6 Principle of duality

Basic concepts: binary orthogonal arrays, the duality of minimum weight and strength.

We have seen the promised principle of duality in action already, when we considered the Simplex and Hamming codes in the preceding section. Let us go over the argument again which led us to conclude that the Hamming codes have minimum distance $d = 3$.

Let H be a check matrix of a binary linear code C with parameters $[n, k, d]_2$ (then H is an $(n - k, n)$ matrix of rank $n - k$). We want to read off d from H. Consider a nonzero bitstring $x = (x_1, x_2, \ldots, x_n)$. Then $x \in C$ if and only if x is orthogonal to all rows of H, equivalently if the linear combination of the columns of H with the x_i as coefficients vanishes. If x has weight w, then the sum of some w columns of H vanishes. In particular d (the minimum weight of C) is the smallest number such that some d columns of H sum to the 0 vector.

2.18 Definition. *A binary matrix H is t–**independent** if no t or less columns of H sum to 0.*

If H is t-independent, then C does not contain a nonzero vector of weight $\leq t$. If H is not t-independent, then C does contain a nonzero vector of weight $\leq t$. This gives us one form of the principle of duality.

2.19 Theorem (principle of duality). *Let $C \subseteq \mathbb{F}_2^n$ be a linear code $[n, k]_2$. Then the following are equivalent:*

- *C has minimum distance $\geq d$.*

- *A check matrix of C is $(d - 1)$-independent.*

Consider

$$H = \begin{pmatrix} 0100011100011101 \\ 0010010011011011 \\ 0001001010110111 \\ 0000100101101111 \\ \overline{1111111111111111} \end{pmatrix}.$$

The 16 columns of H run through all binary 4-tuples (including the 0-tuple), followed by a last entry of 1. The first five columns show that H has rank 5. It is therefore the check matrix of a code $[16, 11, d]_2$. What is d? As there is no 0 column in H, we have $d > 1$. As no two columns in H are equal, we have $d > 2$. The sum of three different columns of H cannot vanish, because of the last entry $(1 + 1 + 1 = 1)$. We have $d \geq 4$. Finally, we find plenty of cases where four columns of H sum to 0. For example, columns number $1, 2, 3$ and 12 sum to the 0-column. This shows

$$x = (1, 1, 1, 0, 0, 0, 0, 0, 0, 0, 0, 1, 0, 0, 0, 0) \in \mathcal{C}.$$

We have seen that the code $\mathcal{C}$ with H as check matrix has parameters $[16, 11, 4]_2$.

Instead of using the check matrix H, we can just as well express the situation basing ourselves on the dual code $\mathcal{C}^\perp$ itself (the code generated by the rows of H). Imagine an array C whose rows are all the codewords of $\mathcal{C}^\perp$. Assume some four columns of H sum to 0. As you run your fingers down these four columns you will not encounter all possible bitstrings of length 4 but only at most half of them: if I know the first three bits, then the last one is uniquely determined. Assume, on the other hand, that H is 3-independent. If I consider the projection onto any set of three columns, then every 3-tuple will be encountered, even the same number of times.

Our matrix M_3 is a generator matrix of the Simplex code $[7, 3, 4]_2$ and a check matrix of the Hamming code $[7, 4, 3]_2$. Let us write down all linear combinations of the rows of M_3, hence all codewords of the Simplex code:

$$C = \begin{array}{|c|c|c|c|c|c|c|}
\hline
0 & 0 & 0 & 0 & 0 & 0 & 0 \\
\hline
1 & 0 & 0 & 1 & 1 & 0 & 1 \\
\hline
0 & 1 & 0 & 1 & 0 & 1 & 1 \\
\hline
0 & 0 & 1 & 0 & 1 & 1 & 1 \\
\hline
1 & 1 & 0 & 0 & 1 & 1 & 0 \\
\hline
1 & 0 & 1 & 1 & 0 & 1 & 0 \\
\hline
0 & 1 & 1 & 1 & 1 & 0 & 0 \\
\hline
1 & 1 & 1 & 0 & 0 & 0 & 1 \\
\hline
\end{array}$$

Observe that, as promised, all the nonzero rows have weight 4. The fact that the Hamming code has $d = 3$, equivalently that its check matrix M_3 is 2-independent, now translates into the following fact: for any $2 = 3 - 1$ columns, if we run our fingers down this pair of columns in C, then every pair of entries (every binary pair) appears the same number of times, twice. This leads to an important combinatorial notion:

2.20 Definition. *A binary array with n columns is a* **binary orthogonal array** *of* **strength** t *if, in the projection onto any set of t columns, each t-tuple of entries occurs the same number λ of times. We write the parameters of such an orthogonal array as*

$$OA_\lambda(t, n, 2).$$

In particular the number of rows is then $\lambda 2^t$. The array C above is an $OA_2(2, 7, 2)$. We record the principle of duality in the following form:

2.21 Theorem (binary principle of duality). *Let $C \subseteq \mathbb{F}_2^n$ be a linear subspace. Then the minimum distance of C and the (maximal) strength of $C^\perp$ are related by*

$$d(C) = t(C^\perp) + 1.$$

So the minimum distance of a binary linear code is precisely one larger than the strength of its dual. For example, the dual of a code $[8, 4, 4]_2$ is a code of dimension 4 (hence with $2^4 = 16$ codewords) with strength $d - 1 = 4 - 1 = 3$. This clearly means that, if we fix any three coordinates and if we fix the 3-tuple in the projection, there will be precisely $\lambda = 16/8 = 2$ codewords of our dual code having this 3-tuple in these three coordinates.

The notion of an orthogonal array originated not in information theory but in statistics. It plays a central role in the design of experiments (see the book by Hedayat, Sloane and Stufken [109]) and has many more applications in theoretical computer science and cryptology. The reader is invited to generalize Definition 2.20 from the binary to the general q-ary case.

Observe that any set of n-tuples has a minimum distance and a strength. One speaks of codes if one is primarily interested in the minimum distance, of orthogonal arrays if strength is the center of interest. The principle of duality shows that these two notions are related by duality in the binary linear case. That means we have another large field of applications for our codes.

---

1. The **strength** is the most important parameter of an **orthogonal array**.

2. The **principle of duality for binary linear codes** states that the minimum distance of C is one more than the strength of its dual.

3. An equivalent formulation of the principle of duality is: the minimum distance of C is the smallest number $t \geq 1$ such that some t columns of a check matrix sum to 0.

Exercises 2.6

2.6.1. *Define q-ary orthogonal arrays for arbitrary q.*

2.6.2. *Show that a binary orthogonal array of strength $t > 1$ also has strength $t - 1$.*

2.6.3. *Find a check matrix of $\mathcal{S}_3(2)$ by applying Lemma 2.14 to M_3. Use this check matrix to prove that $\mathcal{S}_3(2)$ has minimum weight 4.*

2.6.4. *Show that each perfect binary linear code of distance $d = 3$ has the parameters of one of the binary Hamming codes.*

2.6.5. *Show that each perfect binary linear code of distance $d = 3$ is equivalent to one of the binary Hamming codes.*

2.6.6. *Describe an $OA_1(n - 1, n, 2)$ for arbitrary length n.*

Chapter 3

General linear codes

3.1 Prime fields

Basic concepts: Fields, congruences, prime fields.

We have defined codes in general, but considered linear codes only in the binary case. This is not general enough. The tactical advantage when we consider **linear** codes is that linear algebra helps us in finding, describing and applying these structures. Linear algebra works over any **field.** In the binary case the underlying field is $\mathbb{F}_2 = \{0, 1\}$, the smallest of all. Many fruitful code constructions rely on codes defined over a large field and involve "going down" to codes over a small field. Also, there are constructions that work best over larger fields. There is simply no reasonable way to get around linear codes over general finite fields.

The basic properties of a field K are the following: addition and multiplication both define **commutative** groups (in particular $a + b = b + a$ and $a \cdot b = b \cdot a$ for all $a, b \in K$), the collaboration of addition and multiplication is defined by the **distributive law** $(a(b + c) = ab + ac)$. Most important is the following requirement: every nonzero element of K possesses a multiplicative inverse, in formulas: for every $0 \neq a \in K$ there is some $a^{-1} \in K$ such that $a \cdot a^{-1} = 1$. The existence of the inverse has the following important consequence: if $a \neq 0$ and $b \neq 0$, then $a \cdot b \neq 0$ (nonexistence of 0-divisors). For finite fields these conditions (existence of the inverse and nonexistence of 0-divisors) are equivalent. Let us prove one direction:

3.1 Lemma. *If multiplicative inverses exist, then there are no 0-divisors.*

PROOF   Assume $ab = 0$ and $b \neq 0$. We want to show that necessarily

$a = 0$. Multiply both sides by b^{-1}. This yields

$$abb^{-1} = a \cdot 1 = a = 0 \cdot b^{-1} = 0.$$

☐

Some finite fields are easy to construct and to work with:

3.2 Theorem. *Let p be a prime number. Then $\mathbb{F}_p = \mathbb{Z}/p\mathbb{Z}$ is a field (a* **prime field***).*

Let us make sure we know what calculation mod an integer n means. Loosely speaking, we calculate just as in the integers, but we identify n with 0.

3.3 Example. *Let us calculate mod 3. There are only three different numbers mod 3. These are $0, 1, 2$. This is so because $3 = 0, 4 = 3+1 = 1$ and so forth. As $1 + 2 = 3 = 0$, we have that 1 and 2 are additive inverses. This can also be written as $2 = -1$ and $1 = -2$. Also $2^2 = 4 = 1$.*

3.4 Example. *Now calculate mod 4. There are four different numbers mod 4. These are $0, 1, 2, 3$. We have, for example, $2+3 = 5 = 1$ and $2+2 = 4 = 0$, hence $2 = -2$. Where the multiplication is concerned: $2 \cdot 3 = 6 = 2$, $3^2 = 9 = 1$ and $2 \cdot 2 = 4 = 0$. The last formula shows that 2 is a divisor of 0. In particular the numbers mod 4 do not form a field.*

Why is Theorem 3.2 true? As addition and multiplication in $\mathbb{F}_p$ are inherited from the integers, it is clear that distributivity holds and that addition and multiplication are commutative. The only problem is the existence of the inverse (equivalently the absence of 0-divisors). A number is $\neq 0$ mod p if it is not divisible by p. The nonexistence of 0-divisors is therefore equivalent to the following statement: if integers a and b both are not divisible by p, then $a \cdot b$ is not divisible by p. This is clearly true. It is one way of expressing the defining property of a **prime,** and it certainly is not true for composite numbers (see the example for $n = 4$ above).

3.5 Example. *As $p = 101$ is a prime, there is a field $\mathbb{F}_{101} = \mathbb{Z}/101\mathbb{Z}$ of 101 elements. These are the numbers $0, 1, 2, \ldots, 100$. As $2 \cdot 50 = 100 = -1$, we have $2^{-1} = -50 = 51$. In fact, $2 \cdot 51 = 102 = 1$. As $10^2 = 100 = -1$, we have $10^{-1} = -10 = 91$.*

If it is not clear what the underlying prime or more generally integer n is, the notation of **congruences** is used: $a \equiv b \pmod{n}$ means that integers a and b are the same when calculating mod n, equivalently that n divides $b - a$. For example, we have $33 \equiv 17 \pmod 4$ and of course these are $\equiv 1 \pmod 4$. The fact that $30 \equiv 12 \equiv 0 \pmod 2$ simply expresses the fact that both 30 and 12 are even numbers (calculation mod 2 is the distinction between even and odd numbers, as we saw in Section 1.1).

1. For every prime p there is a field $\mathbb{F}_p = \mathbb{Z}/p\mathbb{Z}$ of p elements.

2. We calculate in $\mathbb{F}_p$ just as we calculate with integers, with the simplification that $p = 0$.

3. In the case $p = 2$ this yields precisely the field $\mathbb{F}_2$ which we have used all the time.

Exercises 3.1

3.1.1. *Show that $\mathbb{Z}/6\mathbb{Z}$ is **not** a field (of six elements).*

3.1.2. *Find the multiplicative inverse of 16 mod 101.*

3.1.3. *Find 3^{100} (mod 17).*

3.1.4. *Determine all elements $a \in \mathbb{F}_p$ satisfying $a^2 = 1$.*

3.2 Finite fields

Basic concepts: Field extensions and irreducible polynomials, construction of finite fields, primitive elements.

We have seen in the preceding section that $\mathbb{Z}/p\mathbb{Z} = \mathbb{F}_p$ is a finite field when p is prime. Are there any other finite fields? It is easy to see the following: if n is the number of elements (the **order**) of a field K, then $n = p^r$ must be a power of a prime. Also, each field of order p^r must contain the prime field $\mathbb{F}_p$. The prime p is then called the **characteristic** of K. We do not want to go too deeply into field theory. The following theorem will be accepted without proof:

3.6 Theorem. *For every prime-power p^r there is exactly one field $\mathbb{F}_{p^r}$ of that order.*

We need to know how to construct these fields. As $\mathbb{F}_{p^r}$ contains the prime field $\mathbb{F}_p$, it is an r-dimensional vector space over $\mathbb{F}_p$. We can think of $\mathbb{F}_{p^r}$ as the space $\mathbb{F}_p^r$ of r-tuples with entries in the prime field, with an additional multiplicative structure.

In this section we want to describe how finite fields are constructed and use as an example the smallest case $\mathbb{F}_4$. The first (and in general most difficult) step in the construction of $\mathbb{F}_{p^r}$ is to find a polynomial $X^r + \ldots$ of degree r with coefficients in $\mathbb{F}_p$, which is **irreducible.**

In our example we simply write down all binary polynomials of degree $r = 2$. Fortunately, there are only four such polynomials:

$$X^2, \; X^2 + 1, \; X^2 + X \text{ and } X^2 + X + 1.$$

A polynomial of degree r is irreducible if it cannot be written as the product of two polynomials of smaller degree. Irreducible polynomials play a role, which is analogous to the primes in the context of integers. In our example we see that $X^2 = X \cdot X$, $X^2 + 1 = (X + 1)^2$ and $X^2 + X = X(X + 1)$ are certainly not irreducible. The only remaining candidate is $X^2 + X + 1$. This is an irreducible polynomial.

Assume now we have found an irreducible polynomial

$$f(X) = X^r + a_{r-1}X^{r-1} + \cdots + a_1 X + a_0.$$

The elements of $\mathbb{F}_{p^r}$ are the polynomials of degree $< r$. Addition and multiplication are as usual with polynomials. The important simplification is that $f(X) = 0$. We say we calculate **mod** the polynomial $f(X)$. When doing this it is customary to assign a neutral Greek name to the (former) variable X, say ϵ. The defining equation is therefore $f(\epsilon) = 0$. This means that, whenever a power of ϵ appears with an exponent $\geq r$, we reduce the exponent by using the defining equation

$$\epsilon^r = -a_{r-1}\epsilon^{r-1} - \cdots - a_1\epsilon - a_0.$$

Back to the example $\mathbb{F}_4$. We know that $f(X) = X^2 + X + 1$ is our irreducible polynomial (with coefficients in $\mathbb{F}_2$). As a new name for X when calculating mod $f(X)$ we choose ω. The defining equation is therefore

$$\omega^2 = \omega + 1$$

(observe that in characteristic 2 we have $+ = -$). This equation describes the field structure completely: $\mathbb{F}_4 = \{0, 1, \omega, \omega + 1\}$ and addition is addition of polynomials. The defining equation tells as what ω^2 is. We can go on: $\omega^3 = \omega(\omega + 1) = \omega^2 + \omega = (\omega + 1) + \omega = 1$. As $\omega^3 = 1$, we say that ω has **order** 3. We have written all nonzero field elements as powers of ω. This determines the multiplicative structure.

In general it is not always true that all field elements are powers of X. If it is the case, one speaks of a **primitive** polynomial and calls the corresponding field elements **primitive** elements. Every finite field can be described by a primitive polynomial.

We want to accept that fields $\mathbb{F}_q$ exist (and are uniquely determined) for every prime-power q. It will be assumed that the reader is completely familiar with prime fields and with $\mathbb{F}_4$. Should we encounter any other finite field, it will be constructed following the recipe given in this section.

Although we need it here only over finite fields, the same procedure shows how to construct field extensions in general:

3.7 Theorem. *Let K be a field and $p(X)$ an irreducible polynomial of degree r with coefficients in K. Then $K[X]/(p(X))$ is a field extension of degree r of K.*

Here $((p(X))$ denotes the set of all polynomials that are multiples of $p(X)$ and the factor notation $K[X]/(p(X))$ is a shorthand for the construction we saw: the powers $1, X, \ldots, X^{r-1}$ are a basis of this extension field over K. In Exercise 3.2.5 the reader is asked to construct the complex number field $\mathbb{C}$ as an extension of the reals $\mathbb{R}$ in this way.

---

1. An irreducible polynomial of degree r with coefficients in field K yields a degree r extension field of K.

2. For every prime-power $q = p^r$ there is a finite field $\mathbb{F}_q$ with q elements.

3. It is uniquely determined and contains the prime field $\mathbb{F}_p$.

4. $\mathbb{F}_{p^r}$ is constructed with the help of an irreducible polynomial of degree r with coefficients in $\mathbb{F}_p$.

5. The field $\mathbb{F}_4$ has been described in detail.

6. $\mathbb{F}_4 = \{0, 1, \omega, \omega^2\}$, where $1 + \omega + \omega^2 = 0$ and $\omega^3 = 1$.

Exercises 3.2

3.2.1. *Find all irreducible polynomials of degree 3 with coefficients in $\mathbb{F}_2$.*

3.2.2. *Construct $\mathbb{F}_8$.*

3.2.3. *Is it true that $\mathbb{F}_4 \subset \mathbb{F}_8$?*

3.2.4. *Determine the sum of all elements in $\mathbb{F}_q$.*

3.2.5. *Construct the field of complex numbers as a quadratic extension of the field $\mathbb{R}$ of real numbers. What is the natural choice of a quadratic irreducible polynomial?*

3.2.6. *Construct $\mathbb{F}_9$, using the irreducible polynomial $X^2 - X - 1$.*

3.2.7. *Let $\alpha \in F = \mathbb{F}_{q^r}$. Show that there is a polynomial $f(X)$ of degree $\leq r$ with coefficients in $\mathbb{F}_q$ such that $f(\alpha) = 0$.*

3.2.8. *Let $\alpha \in F = \mathbb{F}_{q^r}$. Show that there is a **unique** monic (highest coefficient $= 1$) polynomial $f(X)$ of smallest degree $s > 0$ such that $f(\alpha) = 0$. The polynomial $f(X)$ is the **minimal polynomial** of α. It is irreducible.*

3.2.9. *Let $\alpha \in F = \mathbb{F}_{q^r}$. Let s be the degree of the minimal polynomial of α, equivalently the smallest number such that the powers $1, \alpha, \ldots, \alpha^s$ are linearly dependent over $\mathbb{F}_q$. Show that the smallest subfield $\mathbb{F}_q(\alpha)$ of F containing $\mathbb{F}_q$ and α has $1, \alpha, \ldots, \alpha^{s-1}$ as basis. Conclude that s must divide r.*

3.2.10. *In the following exercises readers are encouraged to prove some of the basic facts concerning finite fields. Start with the following: let F be a finite field. For every natural number n define $n \cdot 1 = 1 + 1 + \cdots + 1$ as the sum of n copies of 1 in F.*
Prove that there must exist some $n > 0$ such that $n \cdot 1 = 0$.

3.2.11. *Let F be a finite field. Define the **characteristic** of F as the smallest natural number p such that $p \cdot 1 = 0$.*
Prove that the characteristic is a prime number.

3.2.12. *Let p be the characteristic of the finite field F. Prove that $\{0, 1, 2 \cdot 1, \ldots, (p-1) \cdot 1\}$ is a subfield of F and that this subfield is the prime field $\mathbb{F}_p$.*

3.2.13. *Let F be a finite field of characteristic p. Prove that F is a vector space over $\mathbb{F}_p$. Conclude: $|F|$ must be a power of p.*

3.3 Linear codes over finite fields

Basic concepts: Minimum distance, generator matrix. Linear combinations, linear independence, dimension, rank, determinant, Gauß elimination.

3.8 Definition. *Let* $\mathbb{F}_q$ *be the field with* q *elements. A* q-*ary linear code of length* n *and dimension* k *is a linear subspace* $\mathcal{C} \subseteq \mathbb{F}_q^n$ *of vector space dimension* k. *If its minimum distance is* d, *then we record the parameters of* $\mathcal{C}$ *as* $[n, k, d]_q$.

Observe that a q-ary linear code of dimension k has q^k codewords. It contains the all-0 word $0 = (0, 0, \ldots, 0)$.

For fixed q we wish to construct codes $[n, k, d]_q$ with large d, large k and small n. Recall that a code with minimum distance d allows the correction of e transmission errors, when $2e < d$.

All our basic facts concerning binary linear codes carry over to general linear codes. The first benefit we drew from linearity in the binary case was Proposition 2.5, the reduction from minimum distance to minimum weight.

3.9 Proposition. *The minimum distance* d *of a linear code* $[n, k, d]_q$ *equals its minimum weight. Here the weight of a codeword is the number of coordinates with a nonzero entry and the minimum is taken over the nonzero words of the code.*

PROOF This is true because each distance between two different codewords is also the weight of some nonzero codeword: $d(x, y) = wt(x - y)$, and $x - y$ is a codeword because the code is linear. $\Box$

A **generator matrix** G of an $[n, k, d]_q$-code $\mathcal{C}$ is a (k, n)-matrix whose rows form a basis of $\mathcal{C}$. The code $\mathcal{C}$ is recovered from G by taking all linear combinations of rows. If we compare with the binary case, the sums that we used there have to be replaced by **linear combinations** in the general case. Here is a review of some important terms from linear algebra:

3.10 Definition. *Let* V *be a vector space defined over a field* K *and* $v_1, v_2, \ldots, v_m \in V$. *A* **linear combination** *of* $v_1, \ldots, v_m$ *is an expression of the form*

$$\lambda_1 v_1 + \lambda_2 v_2 + \cdots + \lambda_m v_m,$$

where the $\lambda_i \in K$. If there is a linear combination such that $\sum_{i=1}^{m} \lambda_i v_i = 0$ and not all the $\lambda_i = 0$, then the set $\{v_1, \ldots, v_m\}$ is called **linearly dependent.** It is **linearly independent** if it is not linearly dependent.

Recall also that a **basis** of a vector space is a maximal set of linearly independent vectors, and that the **dimension** of V is the number of elements in a basis. Our linear codes are subspaces of $\mathbb{F}_q^n$, the set of all n-tuples. The n-tuples form a vector space of dimension n. The characterization of generator matrices given in Proposition 2.6 for the binary case looks as follows in the general case:

3.11 Proposition. *Let C be an $[n, k]_q$-code and G a matrix whose rows are codewords of C. The following are equivalent:*

1. *G is a generator matrix.*

2. *G has rank k.*

3. *There is a (k, k)-submatrix of nonzero determinant.*

4. *The rows of G are **linearly independent.***

We illustrate with the following code of length 4, with entries in the field $\mathbb{F}_4 = \{0, 1, \omega, \overline{\omega}\}$, where $\overline{\omega} = \omega^2 = \omega + 1$.

0 0 0 0	0 ω $\overline{\omega}$ 1
1 1 1 1	1 $\overline{\omega}$ ω 0
ω ω ω ω	ω 0 1 $\overline{\omega}$
$\overline{\omega}$ $\overline{\omega}$ $\overline{\omega}$ $\overline{\omega}$	$\overline{\omega}$ 1 0 ω
0 1 ω $\overline{\omega}$	0 $\overline{\omega}$ 1 ω
1 0 $\overline{\omega}$ ω	1 ω 0 $\overline{\omega}$
ω $\overline{\omega}$ 0 1	ω 1 $\overline{\omega}$ 0
$\overline{\omega}$ ω 1 0	$\overline{\omega}$ 0 ω 1

It is a linear code (check that linear combinations of codewords are codewords again) and has $16 = 4^2$ elements. This means that its dimension is 2. We can represent it in more compact form by a generator matrix, and hence by a basis. A basis consists of any two linearly independent codewords. If we choose the fifth row $z_1 = (0, 1, \omega, \overline{\omega})$ as one basis element, all scalar multiples of z_1 will be linearly dependent on it, so these cannot be used as second basis elements. These are $0z_1 = (0, 0, 0, 0)$, z_1 itself, $\omega z_1 = (0, \omega, \overline{\omega}, 1)$ and $\overline{\omega} z_1 = (0, \overline{\omega}, 1, \omega)$. We have a choice between the remaining $16 - 4 = 12$ codewords to obtain the second row of a generator matrix. Here is a generator matrix of our code:

$$G = \begin{pmatrix} 1 & 1 & 1 & 1 \\ 0 & 1 & \omega & \overline{\omega} \end{pmatrix}.$$

The list of all 16 codewords shows that $d = 3$. We have a quaternary linear code $[4, 2, 3]_4$.

In case your linear algebra education is biased towards solving systems of linear equations and you are not too familiar with terms like "basis" and "dimension," here is how to proceed:

Let k elements of $\mathbb{F}_q^n$ be given. You want to find the dimension of the code they generate.

1. Write your tuples as rows of a (k, n)-matrix G in some order.

2. Use row operations to bring the matrix in row-echelon form. Call the resulting matrix G'.

3. Row operations will not change the row space (the code). The dimension of the code is the number of nonzero rows of G'. These nonzero rows form a basis of the code.

---

1. q-ary linear codes $[n, k, d]_q$ have been defined.

2. We want large k, large d and small n.

3. Minimum distance equals minimum nonzero weight.

4. The **generator matrix** is a (k, n)-matrix whose rows form a basis of the code.

Exercises 3.3

3.3.1. *Find a generator matrix of a ternary $(q = 3)$ linear code $[4, 3, 2]_3$.*

3.3.2. *Find a generator matrix of a 5-ary code $[5, 3, 3]_5$.*

3.3.3. *Find a generator matrix of a $[5, 2, 4]_4$ code.*

3.3.4. *Find a generator matrix of an $[8, 2, 7]_7$ code.*

3.3.5. *Describe a generator matrix of a $[q + 1, 2, q]_q$ code.*

3.3.6. *Show that a $[q + 2, 2, q + 1]_q$-code cannot exist.*

3.4   Duality and orthogonal arrays

Basic concepts: The dot product on $\mathbb{F}_q^n$, dual space (code), check matrix. The principle of duality for general linear codes. Orthogonal arrays, orthogonal Latin squares. Repetition code and sum zero code, Hamming codes and Simplex codes, P-transform.

There is no reason not to use the dot product in $\mathbb{F}_q^n$.

3.12 Definition. *The* **dot product** *is defined on $\mathbb{F}_q^n$ by*

$$(x_1, x_2, \ldots, x_n) \cdot (y_1, y_2, \ldots, y_n) = \sum_{i=1}^{n} x_i y_i.$$

If $\mathcal{C} \subset \mathbb{F}_q^n$ is a subspace of dimension k, then its dual $\mathcal{C}^\perp$ has dimension $n - k$. A generator matrix H of $\mathcal{C}^\perp$ is called a **check matrix** of $\mathcal{C}$. Also, $\mathcal{C}^{\perp\perp} = \mathcal{C}$.

3.13 Example. *The* **repetition code** $[n, 1, n]_q$ *is the code with the all-1 vector $(1, 1, \ldots, 1)$ as basis. Its dual has dimension $n - 1$. It is the* **sum-0 code**, *whose codewords are the $(x_1, x_2, \ldots, x_n)$ such that $\sum_{i=1}^{n} x_i = 0$. The sum-0 code is an $[n, n - 1, 2]_q$-code. Consider the case $q = 3, n = 4$. The sum-zero code is an $[4, 3, 2]_3$-code. This is the code we saw in Section 1.4. A generator matrix is*

$$G = \begin{pmatrix} 1 & 2 & 0 & 0 \\ 0 & 1 & 2 & 0 \\ 0 & 0 & 1 & 2 \end{pmatrix}.$$

All this is not any different from the binary case. The same is true of the principle of duality.

3.14 Definition. *A matrix H with entries in the field $\mathbb{F}_q$ is* **t-independent** *if any t columns of H are linearly independent (thus the projection onto any t columns of H has full rank t).*

We have seen the following important relation in the binary case already. It is one way of expressing the principle of duality.

3.15 Theorem. *Let $C \subseteq \mathbb{F}_q^n$ be a linear subspace. Then C has minimum distance $\geq d$ if and only if a (every) check matrix of C is $(d-1)$-independent.*

We can express this in terms of the dual code instead of in terms of its generator matrix H. Clearly H is t-independent if and only if, in the restriction of $C^\perp$ to any t columns, any t-tuple occurs the same number of times. This leads to the combinatorial notion of an **orthogonal array** of strength t (see Section 2.6).

Orthogonal arrays

Recall the defining property of an OA of strength t; in the projection onto any set of t columns each possible t-tuple occurs the same number λ of times. The number λ is called the **index** of the array. The parameters of a q-ary OA of strength t with n columns and index λ are recorded as

$$OA_\lambda(t, n, q).$$

The number of rows is λq^t.

An OA of strength 1 is a very weak structure. All we demand is that in each column each entry shows up the same number of times. Most people have encountered OA of strength 2 and index 1, albeit in disguise. A popular structure, in discrete mathematics and statistics as well as in recreational math, are **Latin squares.** A Latin square of order n is by definition an (n, n)-array with entries from an alphabet of size n. Each of the n^2 cells of the array is filled with an entry from the alphabet. The central property (making it Latin) is

- In each row and in each column each entry occurs precisely once, in other words: each row and each column is a permutation of the symbols.

As an example, $L_1 = \begin{pmatrix} 1\,2\,3 \\ 2\,3\,1 \\ 3\,1\,2 \end{pmatrix}$ is a Latin square of order 3. Another such Latin square is $L_2 = \begin{pmatrix} 1\,2\,3 \\ 3\,1\,2 \\ 2\,3\,1 \end{pmatrix}$.

Moreover, the pair L_1, L_2 of Latin squares enjoys a special property: if they are superimposed and in each cell we read the pair of entries (i, j), where i is the entry of L_1 and j is the entry of L_2 in that cell, the nine resulting pairs (i, j) are precisely all possible pairs of entries. For example, the northwest cell yields the pair $(1, 1)$ of entries, the middle cell gives the pair $(3, 1)$ and the southeast cell produces $(2,1)$. The pair $(3,3)$ of entries shows up in the northeast.

In general, two Latin squares of order n are called **orthogonal** if the pairs of entries in cells are precisely all possible pairs of entries. Our squares L_1, L_2 are orthogonal. A set $L_1, L_2, \dots, L_k$ of Latin squares of order n forms a set of MOLS ("mutually orthogonal Latin squares") if any pair is orthogonal. In other words, L_1, L_2 form two MOLS of order 3.

What does this have to do with orthogonal arrays? Consider the following ternary array:

$$
\begin{array}{cccc}
1 & 1 & 1 & 1 \\
1 & 2 & 2 & 2 \\
1 & 3 & 3 & 3 \\
2 & 1 & 2 & 3 \\
2 & 2 & 3 & 1 \\
2 & 3 & 1 & 2 \\
3 & 1 & 3 & 2 \\
3 & 2 & 1 & 3 \\
3 & 3 & 2 & 1 \\
\end{array}
$$

Here the first column individuates the row, the second individuates the column in our Latin squares, in the third column we wrote the entry of L_1, in the fourth column the entry of L_2 in the corresponding cell. For example, the first row of the array says that, in the northwest corner (cell $(1,1)$,) both of our Latin squares have entry 1. The last row of the array gives the information that in the southeast corner L_1 has entry 2, whereas L_2 has entry 1.

This array is an $OA_1(2,4,3)$. We have to check the basic OA property for every pair of columns. For the first two columns this is obvious. For columns $1,3$ this is satisfied as each row of L_1 is a permutation of the symbols. In the same way it follows from the definition of a Latin square, applied to L_1 and L_2, that all pairs of columns are OK, except possibly pair $3,4$. The column pair $3,4$ is OK because L_1, L_2 are orthogonal. For example, why does the pair of entries $(2,3)$ occur in the last two columns? Because we know that there is a cell, exactly one, where L_1 has entry 2 and L_2 has entry 3. This is cell $(2,1)$.

It is clear how this generalizes. A set of k MOLS of order n produces an $OA_1(2, k+2, n)$ (again, the first two columns index the cells, and each of the remaining k columns is determined by one of the Latin squares). This process can be reversed. Given an $OA_1(2, k+2, n)$, interpreting the first two columns as indexing the cells, each further column will describe an (n, n)-array. By the definition of an OA, all these arrays are Latin squares, and they are mutually orthogonal. We have seen that k MOLS of order n are equivalent to an $OA_1(2, k+2, n)$. In other words, the theory of OA of strength 2 and index 1 is identical to the theory of sets of mutually orthogonal Latin squares.

3.16 Theorem. *The following are equivalent:*

- *An $OA_1(2, n, q)$, $n \geq 3$.*

- *A set of $n - 2$ mutually orthogonal Latin squares of order q.*

Back to duality

Here is one of our main theorems:

3.17 Theorem (principle of duality). *Let $C \subseteq \mathbb{F}_q^n$ be a linear subspace. Then the minimum distance of C and the strength of $C^\perp$ are related by*

$$d(C) = t(C^\perp) + 1.$$

As an illustration, consider the code $[4, 2, 3]_4$ given in the preceding section. As it has only 16 codewords, its strength is at most 2. We can check that when we run our fingers down any two columns, each pair of entries occurs precisely once. This means that the strength is indeed 2. As an orthogonal array the parameters are therefore

$$OA_1(2, 4, 4).$$

The principle of duality gives a fancy proof for the fact that the sum-zero code has $d = 2$, as follows: the repetition code has as generator matrix $(1, 1, \ldots, 1)$. This matrix is 1-independent (as it has no 0 entries) but of course not 2-independent. Theorem 3.15 states that the dual of the repetition code has minimum distance $1 + 1 = 2$.

The principle of duality also allows us to construct Hamming codes and Simplex codes in the general q-ary case. The binary case is in Section 2.5.

Hamming codes and Simplex codes

For every prime-power q and dimension r write a matrix $M_r(q)$ with entries in $\mathbb{F}_q$ such that the columns are representatives for the one-dimensional subspaces in $\mathbb{F}_q^r$. This means that

- There is no 0 column.

- No two columns are multiples of each other.

- The matrix is maximal with these properties, meaning that, for every nonzero r-tuple, some nonzero multiple of it is a column of $M_r(q)$.

The number of columns is then $(q^r - 1)/(q - 1)$.

$$M_3(3) = \begin{pmatrix} 1001111001111 \\ 0101200111122 \\ 0010012121212 \end{pmatrix}.$$

This matrix has $13 = (3^3 - 1)/(3 - 1)$ columns of length 3. For each nonzero ternary triple, either this triple or its negative is a column of $M_3(3)$.

Let us use $M_r(q)$ as the check matrix of a code, a q-ary Hamming code. Clearly the length is $(q^r - 1)/(q - 1)$ and the dimension is $(q^r - 1)/(q - 1) - r$. By definition, $M_r(q)$ is 1-independent (no 0 column) and 2-independent (no two columns are multiples of each other). This means that the code generated by $M_r(q)$ has strength 2 (this is the Simplex code) and, by the principle of duality, that the code with $M_r(q)$ as the check matrix has minimum distance 3.

3.18 Theorem. *The **Hamming code** $\mathcal{H}_r(q)$ is the code with check matrix an $(r, (q^r - 1)/(q - 1))$-matrix $M_r(q)$ whose columns are representatives of the one-dimensional subspaces. Then $\mathcal{H}_r(q)$ is an $[\dfrac{q^r - 1}{q - 1}, \dfrac{q^r - 1}{q - 1} - r, 3]_q$ code. The Hamming code is perfect.*

PROOF   It remains to show that the Hamming codes are perfect. We have seen that the minimum distance of $\mathcal{H}_r(q)$ is $d = 3$. In the terminology of the sphere-packing bound Theorem 1.13, we have $e = 1$. The volume of the ball of radius 1 is $V_q(1, n) = 1 + (q - 1)n$. As $n = (q^r - 1)/(q - 1)$, we get $V_q(1, n) = q^r$, hence $|\mathcal{H}_r(q)| \cdot V_q(1, n) = q^n$.   ☐

The Simplex code $\mathcal{S}_r(q) = \mathcal{H}_r(q)^{\perp}$ is a $[\dfrac{q^r - 1}{q - 1}, r, q^{r-1}]_q$ code.

Conclusion

The principle of duality, although rather elementary, is very important. One could almost say it distinguishes coding theorists from the rest of the world. If you really understand it you may (almost) call yourself a coding theorist. We saw that duality motivates the definition of orthogonal arrays. Again, technically this is easy to understand. The theory of linear orthogonal arrays is the theory of dual linear codes, hence of linear codes. However, the strength invites applications which are very different from the application of the minimum distance we considered so far. As it is so important to understand the principle of duality, we sum up and visualize: in order to determine the minimum distance d of a code C, we have to determine the maximum t

such that a check matrix H of C is t-independent. Then $d = t+1$. If we know a generator matrix G of C, we can bring it in standard form at first (using Gauß elimination), then apply P-transformation, meaning the trick described in Lemma 2.14 for the binary case and in Exercise 3.4.2 for the general case. The result is a check matrix.

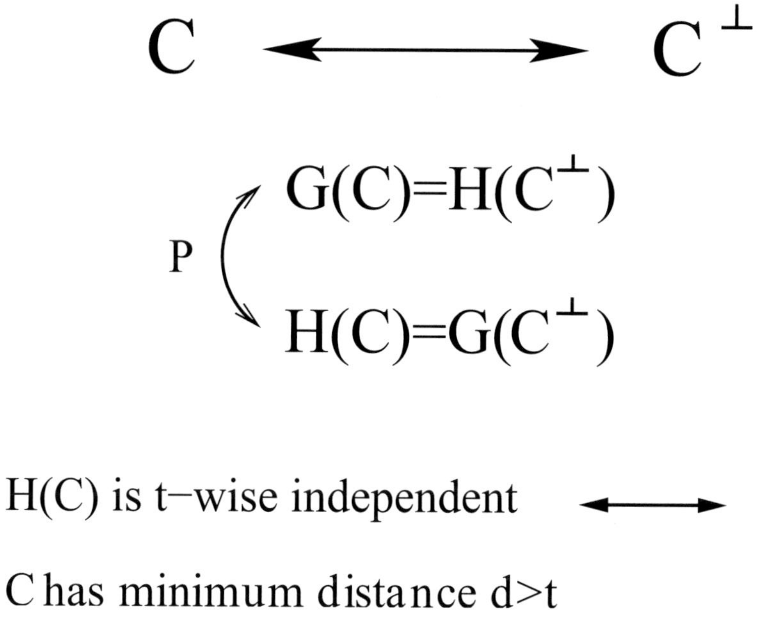

$$C \longleftrightarrow C^{\perp}$$

$$P \left(\begin{array}{c} G(C)=H(C^{\perp}) \\ \\ H(C)=G(C^{\perp}) \end{array} \right.$$

$$H(C) \text{ is } t\text{—wise independent} \longleftrightarrow$$

$$C \text{ has minimum distance } d > t$$

FIGURE 3.1: Principle of duality

code C	$\longrightarrow$	dual code $C^{\perp}$
length n	$\longrightarrow$	length n
dimension k	$\longrightarrow$	complementary dimension $n - k$
strength t	$\longrightarrow$	distance $t + 1$
distance d	$\longrightarrow$	strength $d - 1$

1. The **dual** $\mathcal{C}^{\perp}$ of a linear code $\mathcal{C}$ is defined with respect to the **dot product.** Dual codes have complementary dimensions.

2. The rows of a **generator matrix** of $\mathcal{C}$ form a basis.

3. A **check matrix** of $\mathcal{C}$ is a generator matrix of $\mathcal{C}^{\perp}$.

4. The **strength** is the most important parameter of an **orthogonal array.**

5. The **principle of duality for linear codes** states that the minimum distance of $\mathcal{C}$ is one more than the strength of its dual.

6. Two trivial families of codes, the **sum-zero codes** and the **repetition codes,** are related by duality.

7. Pick a prime-power q and a natural number r.

8. Write a matrix $M_r(q)$ with r rows and $(q^r - 1)/(q - 1)$ columns, whose columns are representatives for the one-dimensional subspaces of $\mathbb{F}_q^r$.

9. Then $M_r(q)$ is a generator matrix of the **Simplex code** $\mathcal{S}_r(q)$ (of dimension r) and a check matrix of the **Hamming code** $\mathcal{H}_r(q)$.

10. $\mathcal{H}_r(q)$ is a perfect code with minimum distance 3.

11. Every nonzero codeword of $\mathcal{S}_r(q)$ has weight q^{r-1}.

Exercises 3.4

3.4.1. *Prove that the dual of the linear repetition code of length n is the sum-zero code.*

3.4.2. *Prove the following generalization of Lemma 2.14: Let a generator matrix for a code $[n, k]_q$ be written in the form $G = (I|P)$. Then a check matrix is $H = (-P^t|I)$. Here I in G is the (k, k)-unit matrix, the I in H is the $(n - k, n - k)$-unit matrix and P^t is the transpose of P.*

3.4.3. *Consider the following quaternary matrix H:*

$$\begin{pmatrix} 1 & 0 & 0 & 1 & 1 & 1 & 0 & 0 & 1 & 1 & 1 & 1 & 1 & 1 & 1 & 1 & 1 & 1 \\ 0 & 1 & 0 & 1 & \omega & \overline{\omega} & 1 & 1 & 1 & 0 & 0 & 0 & 1 & 1 & \omega & \omega & \omega & \overline{\omega} & \overline{\omega} & \overline{\omega} \\ 0 & 0 & 1 & 0 & 0 & 0 & 1 & \omega & \overline{\omega} & 1 & \omega & \overline{\omega} & 1 & \omega & \overline{\omega} & 1 & \omega & \overline{\omega} & 1 & \omega & \overline{\omega} \end{pmatrix}$$

Determine the strength of the orthogonal array generated by H and the parameters of the quaternary code with H as check matrix. Here $\mathbb{F}_4 = \{0, 1, \omega, \omega^2 = \overline{\omega}\}$. What is the name of this code?

3.4.4. *Determine the parameters of the binary code with the following check matrix (this is easy):*

$$\begin{pmatrix} 1 & 1 & 0 & 0 & 1 & 1 & 1 & 1 & 1 & 1 & 0 \\ 0 & 0 & 1 & 1 & 0 & 0 & 1 & 1 & 0 & 1 & 1 \\ 0 & 0 & 0 & 0 & 1 & 1 & 1 & 1 & 1 & 0 & 1 \end{pmatrix}.$$

More interesting is the code with this matrix as generator matrix. Determine its parameters.

3.4.5. *Determine the parameters of the quaternary code with the following check matrix:*

$$\begin{pmatrix} 0 & 0 & 1 & 1 & \omega & \omega \\ 0 & \omega & 0 & \overline{\omega} & \omega & \overline{\omega} \\ 1 & 1 & 1 & 1 & 1 & 1 \end{pmatrix}.$$

3.4.6. *Same problem as before. The check matrix is*

$$\begin{pmatrix} 1 & 0 & 0 & 1 & \overline{\omega} & \omega \\ 0 & 1 & 0 & 1 & \omega & \overline{\omega} \\ 0 & 0 & 1 & 1 & 1 & 1 \end{pmatrix}.$$

Prove your result.

3.4.7. *What is the smallest length n such that a code $[n, 7, 3]_3$ exists? (Hint: think of the Hamming codes.) Show your argument.*

3.4.8. *Construct three pairwise orthogonal Latin squares of order 4 and the corresponding OA of index 1. Determine its parameters when seen as a code.*

3.5   Weight distribution

Basic concepts: The MacWilliams theorem, probability spaces.

In general, the minimum distance of a linear code is far from determining the minimum distance of its dual. The principle of duality relates the minimum distance to a very different parameter of the dual, the strength. There is, however, a finer parameter set related to the error-correcting and error-detecting capability of a linear code, which does determine the corresponding information of the dual code in a unique way. This finer parameter set is the **distance distribution** of a code. We can define it also for not necessarily linear codes.

3.19 Definition. *Let C be a q-ary code of length n. The* **distance distribution** *of C is given by the rational numbers*

$$A_i = \frac{1}{|C|}(\#(x,y) \mid x, y \in C, d(x,y) = i), \ i = 0, 1 \ldots, n.$$

The **distance polynomial** *of C is $W_C(X, Y) = \sum_{i=0}^{n} A_i X^{n-i} Y^i$.*

The distance distribution of a code tells us how many codewords are at a given distance from a given codeword, on average. Here the average is taken over all codewords. Clearly $A_0 = 1$, as each codeword is at distance 0 only from itself. The minimum distance d of the code is the smallest index $i > 0$ such that $A_i > 0$.

In the case of a linear code, A_i simply is the number of codewords of weight i, and $d = d(C)$ still is the smallest positive number such that $A_d \neq 0$.

The distance polynomial (in the linear case one speaks of the **weight polynomial** as well) represents this information in an algebraic fashion. The reason for using this representation is our main theorem in this section. The MacWilliams theorem expresses the distance polynomial of the dual in terms of the distance polynomial of the original code.

This result is attributed to F. J. MacWilliams [141]. It can be seen as a finite (and particularly simple) version of the old **Poisson summation,** see, for example, the theta transformation formula from the theory of modular forms.

3.20 Theorem (MacWilliams formula). *Let C be a linear q-ary code with distance polynomial $W_C(X, Y)$. Then the distance polynomial of the dual code is*

$$\frac{1}{|C|} W_C(X + (q-1)Y, X - Y).$$

We present a probabilistic proof of Theorem 3.20 by Chang and Wolf [46]. It is motivated by the **symmetric channel without memory**. The binary case, the binary symmetric channel, was discussed in Section 1.5.

Probability spaces

As terms like **probability** and **sample space** come in naturally in this context, we introduce these basic notions of probability theory.

3.21 Definition. *Let Ω be a finite set. For every $x \in \Omega$, let a nonnegative number $p(x)$ be given such that*

$$\sum_{x \in \Omega} p(x) = 1.$$

We see p as a mapping with domain Ω and real values. Then (Ω, p) is a (finite) **probability space** *or* **sample space**, *$p(x)$ is the* **probability** *of x.*

Define the probability of a subset $S \subseteq \Omega$ as $p(S) = \sum_{s \in S} p(s)$.

We have seen special cases of this basic notion in earlier sections. In Section 1.5 we saw the particularly simple case of a probability space consisting only of 0 and 1, with probabilities $p(1) = p$ and $p(0) = 1 - p$ (the probability that the symbol is transmitted correctly). In Section 2.3 we used the probabilities of error patterns (bitstrings) of length n. This turns $\mathbb{F}_2^n$ into a probability space. Each error pattern of weight i has probability $p^i(1-p)^{n-i}$.

The generalization from $q = 2$ to general q is obvious now: let $0 \neq u \in \mathbb{F}_q$ be given. There is a fixed (small) probability $p > 0$ such that, whenever $a \in \mathbb{F}_q$ is sent, the received symbol is $a + u$ with probability $p/(q-1)$. In particular the probability that a symbol is received in error is precisely p. Interpret $x \in \mathbb{F}_q^n$ as an error pattern (if $x_i = 0$, then symbol number i is transmitted correctly, if $x_i \neq 0$ then $a + x_i$ will be received if a was sent). The probability of such a particular error pattern x is $P(x) = (\frac{p}{q-1})^{wt(x)}(1-p)^{n-wt(x)}$. This turns $\mathbb{F}_q^n$ into a probability space. The probability of a subset $S \subseteq \mathbb{F}_q^n$ is defined as $P(S) = \sum_{x \in S} P(x)$.

Back to the weight distribution

In the special case $S = \mathcal{C} \setminus \{0\}$ we have that

$$P(S) = P(\mathcal{C}) - (1-p)^n = \sum_{i=1}^{n} A_i(p/(q-1))^i(1-p)^{n-i}$$

is the probability that a codeword is received when a different codeword was sent, in other words, the probability that an undetectable error will happen.

PROOF   of Theorem 3.20. We calculate $P(\mathcal{C})$ in two ways: For one thing we certainly have

$$P(\mathcal{C}) = \sum_{i=0}^{n} A_i(1-p)^{n-i} \cdot \left(\frac{p}{q-1}\right)^i = W_{\mathcal{C}}(1-p, \frac{p}{q-1}).$$

Eventually the dual code and its weight distribution have to come into play. In order to calculate $P(\mathcal{C})$ in a different way, based on the dual code, compute at first

$$\sum_{(x,y)} P(x), \text{where } y \in \mathcal{C}^{\perp}, x \cdot y \neq 0.$$

Elements $x \in \mathcal{C}$ yield no contribution to the sum; each $x \notin \mathcal{C}$ contributes $P(x)(q^{n-k} - q^{n-k-1})$. The sum is therefore

$$(1 - P(\mathcal{C}))q^{n-k-1}(q-1). \tag{3.1}$$

Now fix a word $y \in \mathcal{C}^{\perp}$ of weight w. Its contribution to the sum will depend only on w. We denote it by c_w. If we denote by j the size of the intersection of the supports of y and x, then $c_w = \sum_{j=1}^{w} \binom{w}{j}(1-p)^{w-j}(p/(q-1))^j \cdot f(j)$, where $f(j)$ is the number of j-tuples with nonzero entries whose sum is nonzero. Clearly $f(1) = q - 1, f(j) = (q-1)^j - f(j-1)$. In particular $f(2) = (q-1)^2 - (q-1) = (q-1)(q-2)$ and $f(3) = (q-1)^3 - f(2) = \frac{q-1}{q}\{(q-1)^3 - 1\}$. By induction, $f(j) = \frac{q-1}{q}((q-1)^j + (-1)^{j-1})$ (see the exercises). Thus

$$c_w = \frac{q-1}{q}(1-p)^w \{ \sum_{j=1}^{w} \binom{w}{j}(\frac{p}{1-p})^j -$$

$$- \sum_{j=1}^{w}(-1)^j \binom{w}{j}(\frac{p}{(1-p)(q-1)})^j \}.$$

The sums can be evaluated using the binomial formula. We can start summation at $j = 0$ in both sums, as this generates a summand $+1$ in each which cancels by subtraction. It follows that $c_w = \frac{q-1}{q}(1-p)^w\{(\frac{1}{1-p})^w - (1 - \frac{p}{(1-p)(q-1)})^w\}$, and finally

$$c_w = \frac{q-1}{q}[1 - (1 - \frac{pq}{q-1})^w].$$

Summing over all $y \in \mathcal{C}^\perp$ and comparing with (3.1), we obtain

$$(1 - P(\mathcal{C}))q^{n-k-1}(q-1) = \sum_i A'_i c_i = \frac{q-1}{q}\sum_i A'_i[1 - (1 - \frac{pq}{q-1})^i].$$

Cancelling common factors and using $\sum_i A'_i = |\mathcal{C}^\perp| = q^{n-k}$, this yields

$$P(\mathcal{C}) = \frac{1}{q^{n-k}}\sum_i A'_i(1 - \frac{pq}{q-1})^i = \frac{1}{|\mathcal{C}^\perp|} \cdot W_{\mathcal{C}^\perp}(1, 1 - \frac{pq}{q-1}).$$

We have seen that

$$W_{\mathcal{C}}(1 - p, \frac{p}{q-1}) = \frac{1}{|\mathcal{C}^\perp|} \cdot W_{\mathcal{C}^\perp}(1, 1 - \frac{pq}{q-1}).$$

Putting $X = 1 - p, Y = \frac{p}{q-1}$, we get the desired formula, where $\mathcal{C}$ and its dual have changed places. $\quad\square$

So the weight numbers A'_i of the dual code arise from a certain substitution applied to the weight numbers A_i of the original code. In fact, expand the right hand side of the MacWilliams identity:

$$(X + (q-1)Y)^{n-i}(X - Y)^i = \sum_{k=0}^{n} K_k(i)X^{n-k}Y^k.$$

Here the coefficients $K_k(i)$ can be written as follows:

$$K_k(i) = \sum_{j=0}^{Min(k,i)} (-1)^j \binom{i}{j}\binom{n-i}{k-j}(q-1)^{k-j}.$$

We have seen that the weight numbers A'_k of the dual code satisfy

$$A'_k = \frac{1}{|\mathcal{C}|}\sum_{i=0}^{n} A_i K_k(i).$$

We interpret the numbers $K_k(i)$ as values of a polynomial of degree k. These are the **Kravchouk polynomials:**

3.22 Definition (Kravchouk polynomials). *Given integers n and q, define the Kravchouk polynomial of degree $k \leq n$ as*

$$K_k(X) = \sum_{j=0}^{k}(-1)^j\binom{X}{j}\binom{n-X}{k-j}(q-1)^{k-j}.$$

1. The **distance distribution** of a code records which distances occur between codewords and how often.
   It carries more information than the number of codewords and the minimum distance.

2. The **weight** of a vector is its number of nonzero coordinates.

3. The distance distribution of a linear code agrees with its **weight distribution.**

4. The **weight polynomial** of a linear code of length n is a homogeneous polynomial of degree n in two variables, with the weight numbers as coefficients.

5. The **MacWilliams formula** shows how to calculate the weight distribution (weight polynomial) of $C^\perp$ out of the weight distribution of C.

Exercises 3.5

3.5.1. *Determine the weight polynomial of the repetition code* $[n, 1, n]_q$ *and of the sum-0 code* $[n, n-1, 2]_q$.

3.5.2. *Let $\mathcal{D}$ be a binary linear code with weight distribution (B_i) and $\mathcal{C}$ the code obtained by adding a parity check bit.*
Express the weight distribution of $\mathcal{C}$ in terms of (B_i).

3.5.3. *The Simplex code $[7, 3, 4]_2$ is a constant weight code. Its weight distribution is $A_0 = 1$, $A_4 = 7$. Use the MacWilliams formula to compute the weight distribution of its dual, the binary Hamming code $[7, 4, 3]$, and of the extended Hamming code, the $[8, 4, 4]$ code obtained by adding a parity check bit.*

3.5.4. *Show that each self-dual (equal to its dual) $[8, 4, 4]_2$ code has the same weight distribution as the extended Hamming code $A_0 = A_8 = 1$, $A_4 = 14$.*

3.5.5. *Show that the extended Hamming code is the only self-dual $[8, 4, 4]_2$ code.*

3.5.6. *The weight polynomial in one variable is defined as*

$$P_{\mathcal{C}}(Y) = W_{\mathcal{C}}(1, Y) = \sum_i A_i Y^i.$$

Find the MacWilliams formula for the one variable weight polynomial.

3.5.7. *Let* $[n, k, d]_q$ *be the parameters of a perfect linear code. Determine the number* A_d *of minimum weight codewords.*

3.5.8. *Show that the weight distribution of a perfect* $[n, k, d]_q$ *code is uniquely determined by* n, k, d, q.

3.5.9. *Let* $f(j)$ *be the number of* j*-tuples with nonzero entries in* $\mathbb{F}_q$ *with nonzero sum. Prove that* $f(j) = \frac{q-1}{q}((q-1)^j + (-1)^{j-1})$.

3.6   The game of SET

Basic concepts: $\mathbb{F}_3$-linear independence, some ternary codes.

The game of SET consists of 81 cards. Each card assigns a unique value to each of four attributes. These attributes are

$$NUMBER, \ SHAPE, \ COLOR \text{ and } SHADING.$$

More precisely: each card has one, two or three symbols on it (attribute NUMBER has values one, two, three); these symbols all have the same shape, oval, diamond or squiggle (attribute SHAPE has values oval, diamond, squiggle), they have all the same color, purple, red or green (attribute COLOR has values purple, red, green), and the same shading, either outlined, filled or striped (attribute SHADING has values outlined, filled, striped). We can therefore think of a card as a 4-tuple of attributes. For example, there is a card (two, squiggle, green, filled). This is represented graphically as a card with two green filled squiggles on it. The other way around: the purple card with one outlined oval will be represented as the 4-tuple (one, oval, purple, outlined). We see the representation of the cards as 4-tuples as a first step in a process of simplification and algebraization. The next step is prompted by the observation that the concrete values of the attributes are irrelevant for the combinatorial questions pertaining to the game. If we agree that purple is color number 0, red gets represented by 1 and green by 2, then the value of the attribute COLOR is represented by the number in the third coordinate of the 4-tuple. We do the same for each coordinate (attribute), using the same ordering of the values as above. The cards considered above can now be written in a shorter way:

$$(two, squiggle, green, filled) = (1, 2, 2, 1)$$

$$(one, oval, purple, outlined) = (0, 0, 0, 0)$$

We interpret the cards as the quadruples with entries $0, 1$ or 2 in each coordinate, in other words, each card represents an element of $\mathbb{F}_3^4$.

The SETS

The SETS constitute the fundamental idea of the game. Three cards form a SET if the following holds: for each attribute the values of our three cards are either all the same or all different. For example, the following three cards form a SET:

(two,	squiggle,	green,	filled)
(three,	squiggle,	green,	outlined)
(one,	squiggle,	green,	striped)

They all have the same value in coordinates 2 and 3 (SHAPE and COLOR) and all different values in coordinates 1 and 4 (NUMBER and SHADING). Using the addition in $\mathbb{F}_3^4$, the defining property of SETS can be expressed in a natural way. When three cards form a SET, then by definition the values of each particular attribute are either all the same $(0, 0, 0$ or $1, 1, 1$ or $2, 2, 2)$ or all different $(0, 1, 2$ after reordering). What do all these triples have in common? Their sum is 0 (recall that we calculate mod 3). In fact, three numbers from $\mathbb{F}_3$ sum to 0 if and only if they are all the same or all different. This gives us the desired expression for the defining property of sets:

3.23 Lemma. *Three different elements x, y, z of $\mathbb{F}_3^4$ form a SET if and only if $x + y + z = (0, 0, 0, 0)$.*

We call $(0, 0, 0, 0)$ the zero (card or tuple) and write 0 for it. An important property of SETS is now clearly visible: any two cards are contained in precisely one SET. In fact, consider, for example, the cards $(0, 1, 0, 2)$ and $(1, 2, 0, 2)$. Which card will complete this to a set? Each coordinate entry of the third card has to be such that the sum of the entries of all cards vanishes. This determines the third card. It is in our example $(2, 0, 0, 2)$.

We ask the following natural question: what is the maximum number of cards which do not contain a SET? Let us call such a collection of cards **SET-free**. An equivalent property is that no three of our cards should add to $0 = (0, 0, 0, 0)$. Using a little trick we can express this in terms of linear independence: Let $x = (x_1, x_2, x_3, x_4)$ be a card. Consider the corresponding quintuple $x' = (x_1, x_2, x_3, x_4, 1) = (x, 1)$. If $x+y+z = 0$, then also $x'+y'+z' = 0$, as the entries in the last coordinate add up to 0. In particular, x', y', z' are linearly dependent. Assume conversely that x', y', z' are linearly dependent (and different). The last coordinate shows that the coefficients of this linear combination are three nonzero elements which sum to 0. It follows that these elements are the same (either all $+1$ or all -1). We obtain $x + y + z = 0$. This yields the following translation: A collection of cards x is SET-free if and only if the corresponding collection of 5-tuples $x' = (x, 1)$ has the property that any three are linearly independent. We have an equivalent formulation of the original question. It says: What is the maximum size of a collection of vectors in $\mathbb{F}_3^5$ such that all have last coefficient 1 and no three are linearly dependent? Here is an example:

0111111111	0222222222
1011112222	2022221111
0001021221	0002012112
0011220102	0022110201
1111111111	1111111111

In terms of the original question, the column vectors of this matrix without the 1 at the bottom form a SET-free collection of 20 cards. It can be shown that this is maximal (there is no family of 21 such cards). In terms of coding theory we have a ternary matrix which is 3-independent. By the principle of duality it is a check matrix of a linear code $[20, 15, 4]_3$. Later on we will study such questions in geometric terms; see Chapter 17. For more on the game of SETS see [65].

Exercises 3.6

3.6.1. *How many SETS are there? Show your counting method.*

3.6.2. *Consider the ternary $(4, 10)$-matrix A in the upper left corner of the matrix:*

$$A = \begin{pmatrix} 0111111111 \\ 1011112222 \\ 0001021221 \\ 0011220102 \end{pmatrix}$$

Determine the parameters of the code generated by A and the parameters of its dual.

3.6.3. *Consider the following ternary matrix:*

$$\begin{pmatrix} 1\ 0\ 0\ 0\ 0\ 0\ 2\ 1\ 2\ 1\ 2 \\ 0\ 1\ 0\ 0\ 0\ 1\ 0\ 2\ 1\ 1\ 2 \\ 0\ 0\ 1\ 0\ 0\ 0\ 2\ 2\ 0\ 1\ 1\ 1 \\ 0\ 0\ 0\ 1\ 0\ 0\ 1\ 1\ 1\ 0\ 1\ 1 \\ 0\ 0\ 0\ 0\ 1\ 0\ 1\ 2\ 2\ 2\ 0\ 1 \\ 0\ 0\ 0\ 0\ 0\ 1\ 2\ 1\ 2\ 2\ 1\ 0 \end{pmatrix}$$

Determine the parameters of the ternary code with this check matrix. What are the parameters after erasing one column? Compare with the sphere-packing bound.

3.7   Syndrome decoding

Basic concepts: Cosets, coset leaders, syndrome.

A weak point in the mechanism of error correction is the decoding algorithm. Recall the general idea of block coding (Section 2.2). As the received vectors y stream in, they need to be decoded fast, preferably in real time. The naive method of decoding would be to search through all $\sum_{i=0}^{e}(q-1)^i\binom{n}{i}$ vectors at distance $\leq e$ from y until a codeword is found (we use a code correcting e errors, hence of minimum distance $> 2e$). As we saw in Section 2.3, we want to use strong codes with large e. The length n will then inevitably be large as well. However, the search process will then last all too long. In Section 2.5 we saw a faster decoding method in the case of binary Hamming codes. It is left as an exercise to show that this decoding method generalizes to q-ary Hamming codes. In the present section we present a general method of decoding which uses the linearity of the codes and is easy to understand.

Assume the linear $[n, k, d]_q$ code $\mathcal{C}$ is described by check matrix H. Let x be the codeword which was sent, and $y = x + err$ the received word, where err is the error vector. In Section 2.5 we used the vector yH^t. This vector is known as the **syndrome vector**:

3.24 Definition. *Let H be a check matrix of a code $\mathcal{C}$ with parameters $[n, k, d]_q$, and $y \in \mathbb{F}_q^n$. The **syndrome vector** of y is*

$$\sigma(y) = yH^t \in \mathbb{F}_q^{n-k}.$$

Observe that the entries of $\sigma(y)$ are the dot products of y and the rows of H. In particular, $y \in C$ if and only if $\sigma(y) = 0$ is the 0 vector.

When will two vectors have the same syndrome vector, $\sigma(y) = \sigma(y')$? This means $0 = \sigma(y) - \sigma(y') = \sigma(y - y')$, in other words, $y - y' \in C$. The sets of vectors with the same syndrome vector are called **cosets** of C.

3.25 Definition. *A* **coset** *of the linear code C is a set of vectors of the form $C + y$. Two vectors are in the same coset if they have the same syndrome vector. The number of different cosets is q^{n-k} (if C is an $[n,k]_q$ code).*

What does this have to do with decoding? If $y = x + err$ is the received vector and we decode it (hopefully) as $y \mapsto x = y - err$, then $\sigma(y) = \sigma(err)$. We are therefore looking for a vector with small weight in the same coset as y. These vectors are our candidates for err.

Here is how syndrome decoding works: in each of the q^{n-k} cosets of C, find a vector of smallest weight. Call this element the **coset leader.** If y is received, the following steps are executed:

1. Compute the syndrome vector $\sigma(y)$.

2. In the list find the coset leader ϵ of the corresponding coset.

3. Decode $y \mapsto y - \epsilon$.

Use the extended binary Hamming code $[8, 4, 4]_2$ to illustrate. Consider the matrix

$$G = H = \begin{pmatrix} 1000 & 0111 \\ 0100 & 1011 \\ 0010 & 1101 \\ 0001 & 1110 \end{pmatrix}.$$

Check again our basic facts: the first four columns show that G generates a 4-dimensional code C. As any two rows are orthogonal, we have $C \subseteq C^\perp$, the dimension shows $C = C^\perp$, a self-dual code. We can therefore consider G also as a check matrix H for the same code. C is contained in the sum-0 code (all rows of G have weight 4), and it is clear that there is no codeword of weight 2. This shows that C is an $[8, 4, 4]_2$ code. Observe that the columns of H consist of all 4-tuples of weights 1 or 3. Observe also that C contains the all-1 vector (the sum of all rows of G). It follows that there can be no codewords of weight 6 in C.

The number of cosets is 16. One coset is C itself, with coset leader the 0 word. Each vector of weight 1 is the unique coset leader in its coset (as there are no words of weight 2 in C). This gives us eight more cosets and their leaders. The corresponding syndrome vectors are columns of H, hence all vectors of weights 1 or 3 in $\mathbb{F}_2^4$. If a coset contains a vector of weight 2, then this is the minimum weight in this coset. The six vectors of weight 2 with support in the second half of the coordinates are in pairwise different cosets. This gives us six more cosets. The corresponding syndrome vectors are the vectors of weight 2. Only one coset is still missing. It corresponds to the syndrome vector $(1, 1, 1, 1)$. We can choose $(1, 0, 0, 0, 1, 0, 0, 0)$ as representative. This gives us the following list of coset leaders and syndrome vectors:

coset leader	syndrome vector
00000000	0000
10000000	1000
01000000	0100
00100000	0010
00010000	0001
00001000	0111
00000100	1011
00000010	1101
00000001	1110
00001100	1100
00001010	1010
00001001	1001
00000110	0110
00000101	0101
00000011	0011
10001000	1111

Our code corrects one error. This translates to the fact that each vector of weight ≤ 1 is the uniquely determined leader of its coset. As we choose leaders also from cosets which do not contain a unique leader, an attempt is made to decode also in these cases when more than one error occurred. If this idea is followed systematically, we arrive at **list decoding,** a topic which is developed in Section 23.1.

Assume $y = 11000011$ is received. Compute $\sigma(y) = 1111$. The corresponding coset leader is 10001000. We decode $y \mapsto 01001011$.

1. Let H be a check matrix of the $[n, k, d]_q$ code C.

2. The **syndrome vector** of $y \in \mathbb{F}_q^n$ is $\sigma(y) = yH^t$ (the entries of $\sigma(y)$ are the dot products of y and the rows of H).

3. $y, y' \in \mathbb{F}_q^n$ are in the same **coset** of C if $y - y' \in C$, equivalently if $\sigma(y) = \sigma(y')$.

4. A **leader** of a coset is a vector of smallest weight.

5. **syndrome decoding:**

6. Establish a list of leaders of all q^{n-k} cosets.

7. Decode $y \mapsto x = y - z$, where z is the leader of the coset containing y.

Exercises 3.7

3.7.1. *Use the decoding scheme for* $[8, 4, 4]_2$ *given in this section to decode the following received vectors:*
$y = 00110011$, $y = 11111010$, $y = 10001001$.

3.7.2. *Find coset leaders for the sum-0 codes* $[n, n-1, 2]_q$.

3.7.3. *Find coset leaders for the Hamming code* $[7, 4, 3]_2$.

3.7.4. *Find a check matrix of a quaternary code* $[6, 3, 4]_4$. *Find coset leaders for all cosets.*

Chapter 4

Singleton bound and Reed-Solomon codes

Basic concepts: Singleton bound, Lagrange interpolation, Vandermonde matrices. Latin square codes and OA codes. Covering arrays and software testing.

Here comes our second general bound on codes. Consider a code $(n, M, d)_q$ and arrange the codewords in an array (with M rows and n columns). The trick is the following: pick some $d - 1$ coordinates, say the last ones. Forget these coordinates (in other words, project to the first $n - d + 1$ coordinates). These shorter strings are still different (if two of them agreed, then the codewords they were derived from would have distance $\leq d - 1$, which is not possible by the definition of d). It follows that M cannot be larger than the number of all q-ary $(n - d + 1)$-tuples. We have proved the following bound (see Singleton [190]):

4.1 Theorem (Singleton bound). *Let $\mathcal{C}$ be a q-ary code of length n. Then*

$$|\mathcal{C}| \leq q^{n-d+1}.$$

Codes satisfying this bound with equality are traditionally called **MDS codes.** *Here the letters* **MDS** *stand for* **maximum distance separable.**

In the linear case this simplifies:

4.2 Corollary. *Each linear code $[n, k, d]_q$ satisfies $k + d \leq n + 1$.*

Next we construct a very important family of MDS codes.
Before this can be done, recall the concept of a **polynomial.** A polynomial (in one variable) with coefficients in the field K is an expression of the form

$$f(X) = a_n X^n + a_{n-1} X^{n-1} + \cdots + a_1 X + a_0.$$

If $a_n \neq 0$, we call n the **degree** of $f(X)$. Each polynomial defines a polynomial function ($u \in K$ is mapped to the evaluation $f(u)$). Call $u \in K$ a **root** (or a **zero**) of $f(X)$ if $f(u) = 0$. It is a basic fact that a polynomial of degree n cannot have more than n roots. The reason is that, whenever $f(u) = 0$, we can write $f(X)$ as a product $f(X) = (X - u)g(X)$, where $g(X)$ has degree $n-1$; see Exercises 4.7 to 4.11, where the reader is also asked to prove some of the basic facts concerning a very handy and elementary tool, the **geometric series.**

We describe an array whose rows are the codewords. Let the rows be parametrized by the polynomials $p(X)$ with coefficients in $\mathbb{F}_q$, of degree $< k$ (these form a vector space of dimension k with the **monomials** X^i, $i = 0, 1, \ldots, k-1$ as basis). The columns are parametrized by the elements $u \in \mathbb{F}_q$. The entry in row $p(X)$ and column u is the evaluation $p(u)$. Obviously this defines a linear q-ary code of length q (because linear combinations of polynomials of degree $< k$ are again polynomials of degree $< k$).

4.3 Definition. *For $k \leq q$ the words of the* **Reed-Solomon code** $\mathcal{RS}(k, q)$ *of dimension k over $\mathbb{F}_q$ are parametrized by the polynomials $p(X)$ with coefficients in $\mathbb{F}_q$ of degree $< k$, the coordinates by $u \in \mathbb{F}_q$. The corresponding entry is $p(u)$. The code $\mathcal{RS}(k, q)$ has length q and dimension k.*

Let us consider some examples, at first $\mathcal{RS}(2, 3)$. The columns (coordinates) are indexed by the elements $0, 1, 2$ of the field; the rows of the generator matrix are indexed by 1 (a constant polynomial) and X. The corresponding generator matrix of $\mathcal{RS}(2, 3)$ is $\begin{pmatrix} 1 & 1 & 1 \\ 0 & 1 & 2 \end{pmatrix}$. A nonzero polynomial of degree ≤ 1 (a linear combination of 1 and X) has at most one root, so the minimum weight of the corresponding codeword is $\geq 3 - 1 = 2$. We conclude that $\mathcal{RS}(2, 3)$ is a $[3, 2, 2]_3$ code.

Consider $\mathcal{RS}(2, 4)$. A generator matrix is $\begin{pmatrix} 1 & 1 & 1 & 1 \\ 0 & 1 & \omega & \overline{\omega} \end{pmatrix}$. Again, a polynomial of degree ≤ 1 has at most one zero, so $d \geq 4 - 1 = 3$. It is clear that $d = 3$, so $\mathcal{RS}(2, 4)$ is a $[4, 2, 3]_4$ code. The Reed-Solomon code $\mathcal{RS}(2, 4)$ has been displayed in Section 3.3.

A generator matrix of $\mathcal{RS}(3, 5)$ is $\begin{pmatrix} 1 & 1 & 1 & 1 & 1 \\ 0 & 1 & 2 & 3 & 4 \\ 0 & 1 & 4 & 4 & 1 \end{pmatrix}$ (the last row corresponds to X^2, the entry in column u is u^2). A polynomial of degree ≤ 2 has at most 2 roots, so each codeword has weight $\geq 5 - 2 = 3$. It follows that $\mathcal{RS}(3, 5)$ is a $[5, 3, 3]_5$ code.

Finally, $\mathcal{RS}(4, 7)$, has generator matrix $\begin{pmatrix} 1 & 1 & 1 & 1 & 1 & 1 & 1 \\ 0 & 1 & 2 & 3 & 4 & 5 & 6 \\ 0 & 1 & 4 & 2 & 2 & 4 & 1 \\ 0 & 1 & 1 & 6 & 1 & 6 & 6 \end{pmatrix}$.

A polynomial of degree ≤ 3 has at most three roots, so $d = 7 - 3 = 4$ and $\mathcal{RS}(4, 7)$ is a $[7, 4, 4]_7$ code. In fact it is clear that $d = 4$ and not larger. The polynomial $X(X-1)(X-2)$ has degree 3 and three roots. The corresponding codeword therefore has weight $7 - 3 = 4$. It is $(0,0,0,6,3,4,1)$. As $X(X - 1)(X - 2) = X^3 + 4X^2 + 2X$, this should be the last row plus four times the third row plus the double of the second row of our generator matrix. This is indeed the case.

Matrices of the type that we obtain as generator matrices of Reed-Solomon codes (each column indexed by some field element u, the column consisting of the consecutive powers of u) are known as **Vandermonde matrices.**

It is clear now how to determine the minimum distance of Reed-Solomon codes in general. The number of roots of a nonzero polynomial of degree $< k$ is $\leq k - 1$. It follows that the minimum weight is $\geq q - (k - 1) = q - k + 1$. As there are polynomials of degree $k - 1$ with $k - 1$ different roots (products of different linear factors), we have equality. This shows that $\mathcal{RS}(k, q)$ is a $[q, k, q - k + 1]_q$ code. Comparison with the Singleton bound Theorem 4.1 shows that we have equality: the Reed-Solomon codes are MDS codes.

4.4 Theorem. *The Reed-Solomon code $\mathcal{RS}(k, q)$ is a $[q, k, q - k + 1]_q$ code. Reed-Solomon codes are MDS codes.*

The strength of the Reed-Solomon codes is as large as it could possibly be. Readers who know **Lagrange interpolation** will realize this right away. Lagrange interpolation shows that, for every $k \geq 1$, there is precisely one polynomial of degree $< k$ which takes on given values at k field elements. This statement is valid over any field. As it is of independent interest and rather famous, we formulate and prove it here in this form:

4.5 Theorem (Lagrange interpolation). *Let F be a field. A nonzero polynomial $p(X) \in F[X]$ of degree n has at most n roots.*
If n different elements $x_1, x_2, \ldots, x_n$ in F and n arbitrary elements $y_1, \ldots, y_n$ in F are given, then there is exactly one polynomial $p(X)$ of degree $< n$ with coefficients in F which satisfies

$$p(x_i) = y_i, \quad i = 1, 2, \ldots, n.$$

PROOF If $p(x) = 0$, then $p(X)$ is divisible by the linear factor $X - x$. If $p(X)$ has $n + 1$ different roots $x_1, x_2, \ldots, x_{n+1}$, then $p(X)$ is divisible by the product of the corresponding linear factors. If the degree of $p(X)$ is $\leq n$, it follows that $p(X)$ is the 0 polynomial. This proves the first claim. Consider the second claim. We prove at first the uniqueness of the polynomial. In fact, if $p_1(X)$ and $p_2(X)$ are polynomials both satisfying our conditions, then $p_1(X) - p_2(X)$ has n different roots, hence is the 0 polynomial, by the first claim. The existence of an interpolating polynomial is proved by writing it down:

$$p(X) = \sum_{i=1}^{n} y_i \frac{\prod_{j \neq i}(X - x_j)}{\prod_{j \neq i}(x_i - x_j)}.$$

Theorem 4.5 is particularly obvious, at least over the real numbers, for small values of k. There is precisely one constant polynomial having some prescribed value at a certain point (what else, if it is constant?) For $k = 2$ we obtain the statement that two points with different x-coordinates are on a uniquely determined (nonvertical) line. Things get less obvious at $k = 3$: there is precisely one quadratic polynomial interpolating three points with different x-coordinates; in other words, there is precisely one parabola through these three points. Here it is understood that the parabola may degenerate to a line (if the interpolating polynomial has degree < 2).

In our context we do not need this algebraic proof for Lagrange interpolation. We can prove it using duality. What is the dual of $\mathcal{RS}(k, q)$? If Lagrange interpolation is correct, it should be a $(q - k)$-dimensional code of minimum distance $k + 1$, hence an MDS code again. The only possible candidate we know is the $(q - k)$-dimensional Reed-Solomon code. The following theorem confirms this suspicion.

4.6 Theorem.

$$\mathcal{RS}(k, q)^{\perp} = \mathcal{RS}(q - k, q).$$

PROOF   Observe that $\mathcal{RS}(k, q)$ and $\mathcal{RS}(q - k, q)$ have complementary dimensions, so it suffices to show they are orthogonal. $\mathcal{RS}(k, q)$ is generated by the $X^i, i < k$; the code $\mathcal{RS}(q - k, q)$ is generated by the $X^j, j < q - k$. The dot product of the corresponding codewords is $S = \sum_u u^{i+j}$. Let $l = i + j$. Observe $l \leq q - 2$. There is some $0 \neq c \in \mathbb{F}_q$ such that $c^l \neq 1$, as otherwise the polynomial $X^l - 1$ of degree l would have $q - 1 > l$ roots, which is impossible. We have $c^l S = S$, as cu runs through all nonzero elements of $\mathbb{F}_q$ when u does. It follows that $(c^l - 1)S = 0$. As $c^l \neq 1$, it follows that $S = 0$.   ☐

4.7 Theorem. $\mathcal{RS}(k, q)$ *is an* $OA_1(k, q, q)$.

PROOF   The dual of $\mathcal{RS}(k, q)$ is $\mathcal{RS}(q - k, q)$, of minimum distance $k + 1$. By duality this means that $\mathcal{RS}(k, q)$ has strength k. As it has q^k elements, the claim follows.   ☐

Observe that the interpolation property (there is exactly one polynomial such that . . .) of Theorem 4.5 is equivalent to the fact that the Reed-Solomon code is an OA of index 1.

As q-ary Reed-Solomon codes have length q, we need a relatively large field to obtain interesting codes. Examples for parameters of Reed-Solomon codes are

$$[5, 3, 3]_5, \ [7, 3, 5]_7, \ [8, 4, 5]_8, \ [8, 5, 4]_8, \ [16, 8, 9]_{16}, \ldots$$

Consider 8-ary Reed-Solomon codes. We need the field $\mathbb{F}_8$. Let us just follow the general construction procedure: as $8 = 2^3$, we need an irreducible polynomial $f(X)$ of degree 3 with coefficients in $\mathbb{F}_2$. The constant term cannot vanish (the polynomial would be divisble by X otherwise). This leaves us with four candidates:

$$X^3 + 1, \ X^3 + X^2 + 1, \ X^3 + X + 1 \text{ and } X^3 + X^2 + X + 1.$$

The first and the last of these candidates are reducible, as they satisfy $f(1) = 0$. These polynomials are therefore divisible by $X - 1 = X + 1$. The two remaining candidates are irreducible. We choose

$$f(X) = X^3 + X^2 + 1.$$

Denote by ϵ the image of X in $\mathbb{F}_8$. The elements of $\mathbb{F}_8$ are the polynomials of degree ≤ 2 in ϵ

$$\mathbb{F}_8 = \{0, 1, \epsilon, \epsilon + 1, \epsilon^2, \epsilon^2 + 1, \epsilon^2 + \epsilon, \epsilon^2 + \epsilon + 1\}$$

This defines the addition. The multiplication is determined by the basic equation (equivalent to the choice of $f(X)$)

$$\epsilon^3 = \epsilon^2 + 1.$$

We can multiply any two elements of $\mathbb{F}_8$. The basic equation allows us to get rid of all exponents > 2. Consider the powers of ϵ

$$\epsilon^3 = \epsilon^2 + 1, \ \epsilon^4 = \epsilon^3 + \epsilon = \epsilon^2 + \epsilon + 1.$$

In the same manner we can determine all powers of ϵ

$$\epsilon^5 = \epsilon^3 + \epsilon^2 + \epsilon = \epsilon + 1, \ \epsilon^6 = \epsilon^2 + \epsilon \text{ and } \epsilon^7 = 1.$$

We see once again that knowledge of the irreducible polynomial is equivalent to knowledge of the field.

To write down generator matrices of the 8-ary Reed-Solomon codes, let $0, 1, \epsilon, \ldots, \epsilon^6$ parametrize the columns, in this order. Rows are parametrized by the monomials $1, X, X^2, \ldots$.
The following generator matrix for $\mathcal{RS}(4, 8)$ is obtained:

$$G = \begin{pmatrix} 1 & 1 & 1 & 1 & 1 & 1 & 1 & 1 \\ 0 & 1 & \epsilon & \epsilon^2 & \epsilon^3 & \epsilon^4 & \epsilon^5 & \epsilon^6 \\ 0 & 1 & \epsilon^2 & \epsilon^4 & \epsilon^6 & \epsilon & \epsilon^3 & \epsilon^5 \\ 0 & 1 & \epsilon^3 & \epsilon^6 & \epsilon^2 & \epsilon^5 & \epsilon & \epsilon^4 \end{pmatrix}.$$

We know that $\mathcal{RS}(4, 8)$ is a self-dual code $[8, 4, 5]_8$. Also, the Reed-Solomon codes of different dimensions are contained in each other. The first three rows of G generate $\mathcal{RS}(3, 8)$, the first two rows generate $\mathcal{RS}(2, 8)$ and $\mathcal{RS}(1, 8)$ is the repetition code.

Latin square codes and OA codes

In general, we concentrate on linear codes, as they are easier to construct and to work with. However, there are a number of situations when very good nonlinear codes can be constructed.

Consider an $OA_1(t, n, q)$. It has q^t rows. Let us interpret these rows as words of a q-ary code of length n. Assume two of these codewords agree in t coordinates. Then we have the contradiction that, in the projection onto certain t coordinates, a t-tuple occurs more than once. This shows that any two rows of the orthogonal array are at Hamming distance $\geq n - t + 1$. The rows of an $OA_1(t, n, q)$ therefore form an $(n, q^t, n - t + 1)_q$ code. Comparison with Theorem 4.1 shows that we have an MDS code. It is clear that the converse is true as well.

4.8 Theorem. *The following are equivalent:*

- *An $OA_1(t, n, q)$,*

- *An MDS code $(n, q^t, n - t + 1)_q$.*

Recall from Section 3.4 that an $OA_1(2, n, q)$ is equivalent to a set of $n - 2$ mutually orthogonal Latin squares of order q. Mutually orthogonal Latin squares have been studied for a long time and for a variety of reasons. A survey is in [106]. Theorem 4.8 shows that mutually orthogonal Latin squares are a source of optimal codes. For example, there is a pair of orthogonal Latin squares of order 10, so an MDS code $(4, 100, 3)_{10}$ exists. The existence of 3 MOLS of order 10 is in doubt. This is a famous open problem. It is equivalent to the existence of a $(5, 100, 4)_{10}$ code.

Another famous special case is the nonexistence of a pair of orthogonal Latin squares of order 6, equivalently the nonexistence of a $(4, 36, 3)_6$ code. This was conjectured by L. Euler in 1782. In terms of recreational mathematics it is known as the problem of the 36 officers. Nonexistence was shown by Tarry in 1901 [205]. Stinson's short proof in [199] is coding theoretic.

Covering arrays and software testing

Assume we wish to test a certain software for accuracy. If there are 12 input parameters involved, each with 16 possible parameter values, the total number of possible test runs is $16^{12} = 2^{48}$. This clearly is too much to test, in particular if each single test should be expensive and time consuming. How should the set of test runs be chosen? Let us visualize this set as an array with 12 columns and entries from a 16 set. Each row corresponds to a test run.

One frequently used approach is the following: choose the array of test runs such that, for every pair (or for every triple) of parameters, each possible combination of parameter values does actually occur at least once (is covered). Generalizing from pairwise or triplewise covering to t-wise covering for arbitrary t, this leads to the following definition:

4.9 Definition. *A* **covering array** *of strength t (parameters $CA(N; t, n, q)$) is an array with N rows and n columns, with entries from a q-set, such that in the projection onto any t columns each possible t-tuple of entries appears at least once (is covered).*

In order to save time and money in software testing, we wish to construct covering arrays, most often of strength 2 or 3, with a minimum number of rows.

It is clear that $N \geq q^t$, with equality if and only if there is an $OA_1(t, n, q)$; in other words, covering arrays are generalizations of orthogonal arrays of index 1. In the case of the parameters considered above, we are lucky, as $\mathcal{RS}(2, 16)$ and $\mathcal{RS}(3, 16)$ yield (after throwing away four columns) $OA_1(2, 12, 16)$ and $OA_1(3, 12, 16)$, respectively.

Typically, we will not be that lucky. For example, in the binary case $q = 2$, Reed-Solomon codes will not help us at all. Here is an (optimal) $CA(10; 3, 5, 2)$ (see Chateauneuf and Kreher [47]):

$$
\begin{array}{ccccc}
0 & 0 & 0 & 0 & 0 \\
1 & 1 & 1 & 1 & 1 \\
1 & 0 & 0 & 1 & 1 \\
0 & 1 & 1 & 0 & 0 \\
1 & 1 & 0 & 0 & 1 \\
0 & 0 & 1 & 1 & 0 \\
1 & 1 & 0 & 1 & 0 \\
0 & 0 & 1 & 0 & 1 \\
1 & 0 & 1 & 0 & 0 \\
0 & 1 & 0 & 1 & 1
\end{array}
$$

Most diverse methods are being used for the construction of covering arrays, from combinatorial design theory and coding theory to heuristic search methods. Also, various names are employed, from **qualitatively independent partitions** (Rényi [174]) to **t-independent sets** and **universal sets** (see Goldreich [95]).

Remarks

Reed-Solomon codes were introduced in Reed and Solomon [173]. They are among the most widely used codes. Some applications are discussed in Wicker and Bhargava [219]. There is an attractive application in compact disc technology, where shortened Reed-Solomon codes $[32, 28, 5]_{256}$ and $[28, 24, 5]_{256}$ play a key role. This is also described in Hoffman et al. [116].

---

1. Pick a prime-power q and a dimension $k \leq q$. The **Reed-Solomon code** $\mathcal{RS}(k, q)$ is defined as follows:

2. The coordinates are parametrized by the elements $u \in \mathbb{F}_q$.

3. The codewords are parametrized by the polynomials $p(X)$ with coefficients in $\mathbb{F}_q$, of degree $< k$.

4. The codeword parametrized by $p(X)$ has as entries the evaluations $p(u)$.

5. $\mathcal{RS}(k, q)$ is a code $[q, k, q - k + 1]_q$ and an orthogonal array $OA_1(k, q, q)$.

Exercises 4

4.1. *Prove the following slight generalization of Lagrange interpolation:*
Let $a \geq 0$ be a natural number, $x_1, x_2, \ldots, x_n$ different nonzero $\in F$ and
$y_1, \ldots, y_n$ arbitrary in F. Then there is exactly one polynomial $p(X)$ of degree
$< n$ with coefficients in F which satisfies

$$x_i^a p(x_i) = y_i, \ i = 1, 2, \ldots, n.$$

4.2. *Let $a \geq 0$ be a natural number and q a prime-power. Define a q-ary code*
of length $q - 1$ with coordinates indexed by $0 \neq u \in \mathbb{F}_q$ and with codewords
indexed by the polynomials $p(X)$ of degree $< k$, where $k < q$. The entry of the
codeword indexed by $p(X)$ in coordinate u is $u^a p(u)$.
Determine the dimension, minimum distance and strength of this code.

4.3. *Prove: The dual of a linear MDS code is an MDS code.*

4.4. *Let C be an $[n, k, d]_q$-code. Determine the parameters of $C^\perp$ as an or-*
thogonal array. Derive the Singleton bound (in the case of linear codes).

4.5. *Give another simple proof of the Singleton bound for linear $[n, k, d]_q$ codes*
based on Gauß elimination (no calculations needed).

4.6. *In order to understand why covering arrays are also known as qualita-*
tively independent partitions, consider the $CA(10; 3, 5, 2)$ which we gave as an
example. Interpret the rows as the ground set X (of 10 elements). Then every
column describes a partition of X into two parts.
Express the defining property of a strength 2 covering array in terms of this
family of partitions.

4.7. *Prove the following basic fact: if $f(X)$ is a polynomial with coefficients*
in the field F, and $u \in F$ such that $f(u) = 0$, then $f(X)$ can be written in the
form $f(X) = (X - u)g(X)$, where $g(X)$ is a polynomial.

4.8. *The finite **geometric series** is the identity*

$$(X - 1) \sum_{i=0}^{n-1} X^i = X^n - 1. \ \text{Prove this identity.}$$

4.9. *Let q be a complex number of absolute value $|q| < 1$. Prove*

$$\sum_{n=0}^{\infty} q^n = 1 + q + q^2 + \cdots = 1/(1 - q). \ \text{This is the infinite geometric series.}$$

4.10. *Show the identity $(X - u) \sum_{i=0}^{n-1} X^i u^{n-1-i} = X^n - u^n$.*

4.11. *Solve Exercise 4.7 using the geometric series.*

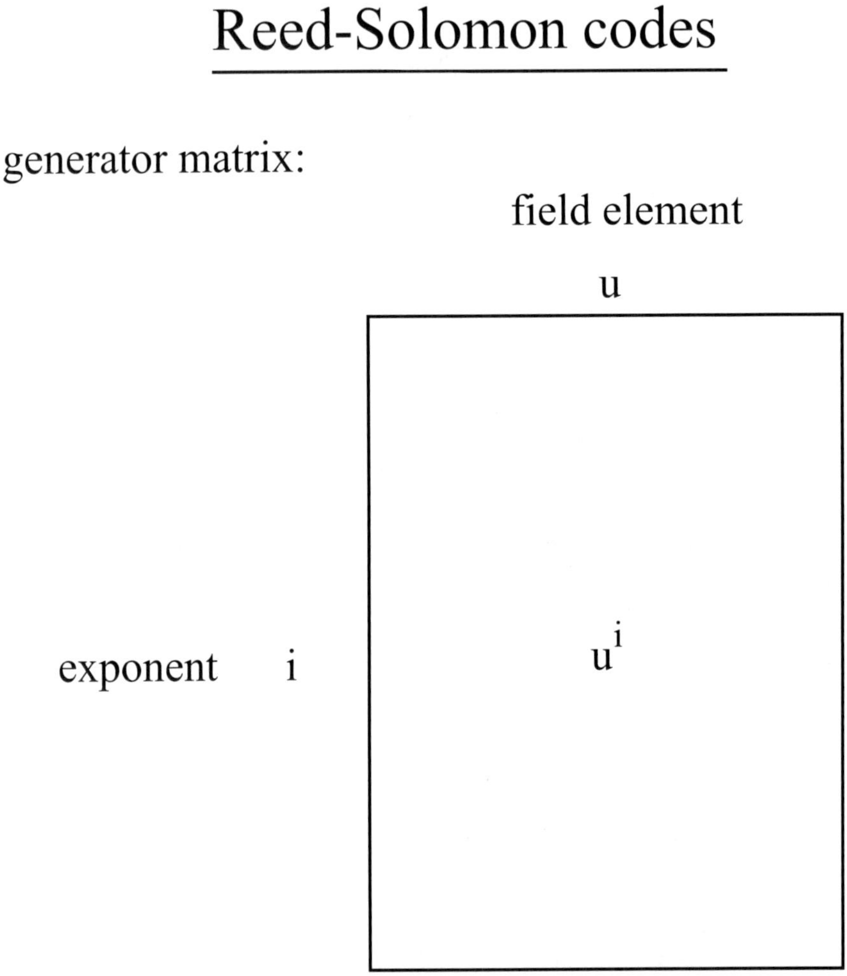

Reed-Solomon codes

generator matrix:

field element

u

exponent   i u^i

FIGURE 4.1:   Reed-Solomon codes, Vandermonde matrices

Chapter 5

Recursive constructions I

5.1 Shortening and puncturing

Basic concepts: Shortening and puncturing, extension and lengthening. The parity check for binary linear codes. Extended binary Hamming codes. Residual codes and Griesmer bound.

So far we have concentrated on constructing codes from scratch, so to speak. One idea led to the Hamming codes, another to the Reed-Solomon codes. Now we want to start studying recursive mechanisms. The most obvious question is: which codes can be found inside a given code?

Let $\mathcal{C}$ be a code $[n, k, d]_q$. Fix a coordinate, say n.
The **shortening** $\mathcal{D}$ of $\mathcal{C}$ is defined by

$$\mathcal{D} = \{(x_1, x_2, \ldots, x_{n-1}) \mid (x_1, x_2, \ldots, x_{n-1}, 0) \in \mathcal{C}\}.$$

That means we consider the subcode of $\mathcal{C}$ consisting of all codewords that end in 0. Then we throw away this superfluous last coordinate. What can we say about the parameters of $\mathcal{D}$? Clearly $\mathcal{D}$ is linear. As it is defined by one linear equation, the dimension typically is $k-1$ (the dimension will be k if all words of $\mathcal{C}$ should happen to be 0 at the last coordinate). Also, $\mathcal{D}$ is a subcode of $\mathcal{C}$. It follows that the minimum weight (=minimum distance) of $\mathcal{D}$ cannot be smaller than d. This describes our first recursive construction:

5.1 Theorem. *If there is a code* $[n, k, d]_q$, *then shortening yields a subcode* $[n-1, k-1, d]_q$.

Here is another elementary idea. This procedure is called **projection** in algebra. In the coding community it is known as **puncturing** or **truncation:** define

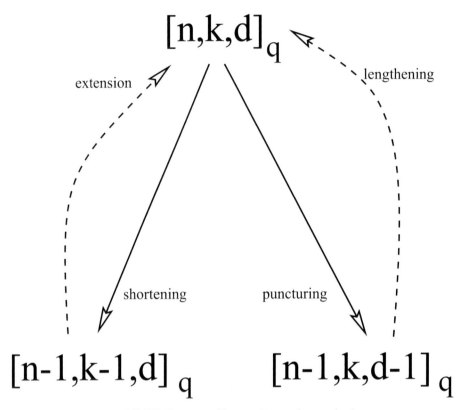

FIGURE 5.1: Shortening and puncturing

$$\mathcal{E} = \{(x_1, x_2, \ldots, x_{n-1}) \mid \text{ there is } x \in \mathbb{F}_q \text{ such that}$$
$$(x_1, x_2, \ldots, x_{n-1}, x) \in \mathcal{C}\}.$$

This means we project the codewords of $\mathcal{C}$ onto the first $n-1$ coordinates. The number of codewords of $\mathcal{E}$ is the same as in $\mathcal{C}$ (if two different words in $\mathcal{C}$ agreed in all but one coordinate, their distance would be 1). In particular, $\mathcal{E}$ has dimension k. This time we cannot expect the minimum distance to remain unchanged. However, as only one coordinate has been removed, the minimum distance will decrease by at most 1.

5.2 Theorem. *If there is a code* $[n, k, d]_q$, *where* $d > 1$, *then puncturing (projection) yields a code* $[n-1, k, d-1]_q$.

For example, given a code $[24, 12, 8]_2$ (it is known as the **binary extended Golay code**), we can construct by repeated shortening and puncturing

$$[23, 12, 7]_2, \ [23, 11, 8]_2, \ [22, 12, 6]_2, \ [22, 11, 7]_2, \ [22, 10, 8]_2, \ldots$$

This raises the interesting question when one of these constructions can be reversed. Given a code $[n-1, k-1, d]_q$, can we find a code $[n, k, d]_q$ such that the original code is obtained by shortening? This process is called **extension.** Clearly it cannot always be possible. The process of shortening shows that parameters $[n, k, d]_q$ are stronger than $[n-1, k-1, d]_q$. It is natural to ask what is the maximum minimum distance of a one-step extension. The covering radius will shed new light on this problem; see Section 14.2.

In the case of puncturing, the situation is analogous. The inverse of puncturing is called **lengthening** and is clearly not always possible. More generally, we will speak of (i-step) **lengthening** of a code $\mathcal{C}$ of length n if the new code of length $n + i$ has the same dimension as $\mathcal{C}$ and projects to $\mathcal{C}$; we speak of (i-step) **extension** if we have a code of length $n + i$ and repeated shortening at some i coordinates produces $\mathcal{C}$. Unfortunately, the terminology in the literature is rather chaotic.

Here is a famous case when lengthening works: let $\mathcal{C}$ be an $[n, k, d]_2$, where d is odd. Append a new coordinate, where the entry is chosen such that the new lengthened codeword has even weight. This means that the entry in the last coordinate is 0 if the weight of the old codeword is even, and the entry in the last coordinate is 1 if the old weight was odd. Clearly the new code is linear and has the same dimension as the old one. Also, in the lengthened code all weights are even. In particular, weight d does not occur. We have a code $[n+1, k, d+1]_2$. The entry in the extra coordinate is known as the **parity check bit.**

5.3 Theorem. *Let $\mathcal{C}$ be a binary code $[n, k, d]_2$, where d is odd. Appending a parity check bit yields a code $[n+1, k, d+1]_2$.*

In particular, a code $[n, k, d]_2$ with odd d exists if and only if a code $[n+1, k, d+1]_2$ exists. In the theory of binary linear codes one can therefore concentrate on even minimum weight. However, this is a specialty of the binary case.

We have seen the binary Hamming code $[7, 4, 3]_2$ in Sections 2.1, 2.2 and 2.5. A generator matrix of $\mathcal{H}_3(2)$ was given in Section 2.1:

$$\begin{pmatrix} 1\,0\,0\,1\,1\,0\,1 \\ 0\,1\,0\,1\,0\,1\,1 \\ 0\,0\,1\,0\,1\,1\,1 \\ 0\,0\,0\,1\,1\,1\,0 \end{pmatrix}$$

After addition of a parity check bit, we obtain the matrix

$$\begin{pmatrix} 1\,0\,0\,1\,1\,0\,1\,0 \\ 0\,1\,0\,1\,0\,1\,1\,0 \\ 0\,0\,1\,0\,1\,1\,1\,0 \\ 0\,0\,0\,1\,1\,1\,0\,1 \end{pmatrix}$$

It is a generator matrix of the so-called **extended Hamming code** $[8, 4, 4]_2$ (alas: we would prefer to call it the lengthened Hamming code). We check that any two rows are orthogonal. This means that the extended Hamming code is orthogonal to itself. As its dimension is just half the length (see Lemma 2.9), we conclude that it equals its dual. Such codes are called **self-dual**.

5.4 Theorem. *The extended binary Hamming code* $[8, 4, 4]_2$ *is self-dual* $(\mathcal{C} = \mathcal{C}^\perp)$.

We can apply the parity check to all the binary Hamming codes $[2^r - 1, 2^r - (r + 1), 3]_2$. This yields the extended (really lengthened) binary Hamming codes:

5.5 Theorem. *The* **extended binary Hamming codes** *have parameters*

$$[2^r, 2^r - (r + 1), 4]_2.$$

Examples are

$$[8, 4, 4]_2, \ [16, 11, 4]_2, \ [32, 26, 4]_2.$$

Residual codes, Griesmer bound

Let $\mathcal{C}$ be an $[n, k, d]_q$ code and $v \in \mathcal{C}$ a codeword of weight d. We can assume the nonzero entries of v are $v_1 = v_2 = \cdots = v_d = 1$. Otherwise permute the coordinates and multiply all codewords by suitable nonzero field elements in the first d coordinates. This produces a code with the same parameters. Choose a generator matrix G whose first row is v. The **residual code** $\mathcal{R}$ is the code generated by the remaining $k - 1$ rows of G, after projection to the last $n - d$ coordinates (where v has zero entries); in other words, $\mathcal{R}$ is generated by the matrix in the southeast corner of G. Clearly $\mathcal{R}$ is a linear code of length $n - d$. It has dimension $k - 1$, as otherwise $\mathcal{C}$ would contain a nonzero codeword of weight $< d$. What can be said about the minimum distance d' of $\mathcal{R}$?

Let $0 \neq x \in \mathcal{C}$ be a linear combination of the last $k - 1$ rows of G whose image in $\mathcal{R}$ (consisting of the last $n - d$ coordinates) has weight w. We want a lower bound on w. Let δ_i be the number of coordinates in the first section where x has entry i, for arbitrary $i \in \mathbb{F}_q$. The codeword $x - iv \in \mathcal{C}$ has weight $w + d - \delta_i$. This weight must be $\geq d$. It follows that $w \geq \delta_i$. As the average value of δ_i is d/q, we have $w \geq \lceil d/q \rceil$, the smallest integer at least as large as d/q. We have seen the following:

5.6 Proposition. *Let $\mathcal{C}$ be an $[n, k, d]_q$ code of precise minimum weight d. The residual code is an $[n - d, k - 1, \lceil d/q \rceil]_q$ code.*

Repeated application of Proposition 5.6 yields $n \geq \sum_{i=0}^{k-1} \lceil d/q^i \rceil$. This is true even if at any point in the process the true minimal distance should be larger than its guaranteed lower bound.

5.7 Theorem (Griesmer bound). *If a linear $[n, k, d]_q$ code exists, then*

$$n \geq \sum_{i=0}^{k-1} \lceil d/q^i \rceil.$$

The Griesmer bound is very useful. It is a bound on linear codes. How does it compare to the Singleton bound? The lower bound on n is a sum of k integers. The largest of these integers is d; all are strictly positive. This implies $n \geq d + (k-1)$. We see that the Griesmer bound implies the Singleton bound Theorem 4.1. The Griesmer bound was proved by Griesmer [99] in the binary case, by Solomon and Stiffler [196] in the general q-ary case.

As first examples, codes $[8, 4, 4]_2$, $[11, 3, 6]_2$ and $[12, 6, 6]_3$ meet the Griesmer bound with equality. In Chapter 17 we will encounter an even more natural proof of the Griesmer bound in terms of the geometric description of linear codes; see Exercise 17.1.8.

---

1. Let $\mathcal{C}$ be a linear code $[n, k, d]_q$.

2. One-step **shortening** produces $[n - 1, k - 1, d]_q$, one-step **projection** yields $[n - 1, k, d - 1]_q$.

3. We speak of a one-step **extension** if $[n + 1, k + 1, d]_q$ can be obtained, of a one-step **lengthening** if $[n + 1, k, d + 1]_q$ can be constructed.

4. The **parity check** works only in the binary case. It constructs $[n + 1, k, d + 1]_2$ from $[n, k, d]_2$ when d is odd.

5. The **extended binary Hamming codes** have parameters $[2^r, 2^r - (r + 1), 4]_2$.

6. The **Griesmer bound:** $n \geq \sum_{i=0}^{k-1} \lceil d/q^i \rceil$ is stronger than the Singleton bound.

Exercises 5.1

5.1.1. *Let C be a binary linear code with odd minimum distance d and $\mathcal{D} \subset C$ the set of codewords of even weight. Prove that $\mathcal{D}$ is a subspace of codimension 1 in C.*

5.1.2. *Given a linear code $[192, 11, 92]_2$ (it exists), make a list of all parameters of codes of lengths 188 and above that can be constructed using iterated shortening and projection.*

5.1.3. *Prove the following: if a code is MDS, then every code derived from it by shortening or projection is MDS as well.*

5.1.4. *Determine the parameters of the three-dimensional binary code generated by*

$$G = \begin{pmatrix} 111000000111100 \\ 000111000110011 \\ 000111111111100 \end{pmatrix}.$$

Add a parity check bit to this generator matrix. What are the parameters of the resulting extended code?

5.1.5. *Let $X = \{1, 2, \ldots, 8\}$. Define a binary matrix M of size $(8, 24)$ as follows: the rows of M are indexed by the elements $x \in X$. The columns of M are indexed by unordered pairs $\{i, j\} \subset X$, where $\{i, j\}$ is none of $\{1, 2\}, \{3, 4\}, \{5, 6\}, \{7, 8\}$ (these forbidden pairs form what is known as a **one-factor**). The entry in row x and column $\{i, j\}$ is 0 if $x = i$ or $x = j$; it is 1 otherwise. Observe that every column of M has weight 6 and every row has weight 18.*
Prove that M has rank 7 and that its row-space is a code $[24, 7, 10]_2$.
This description is from Jaffe [119].

5.1.6. *Given a linear code $[24, 7, 10]_2$, which parameters of 6- or 7-dimensional codes can be constructed using iterated shortening and projection?*

5.1.7. *Consider the matrix*

$$A = \begin{pmatrix} 0111111111 \\ 1011112222 \\ 0001021221 \\ 0011220102 \end{pmatrix}$$

from Section 3.6, which generates a code $[10, 4, 6]_3$. Find generator matrices for the codes derived by shortening and by projection.

5.1.8. *Let C be an $[n, k, d]_q$ code, $x = (x_1, \ldots, x_w, 0, 0, \ldots) \in C^\perp$ a codeword of weight w in the dual code. Apply shortening to C with respect to the first w coordinates. We know that the resulting code has parameters (at least) $[n - w, k - w, d]_q$.*

Show that this shortened code has in fact better parameters.
This is known as **construction Y_1**.

5.1.9. *From Section 3.4 we know the Simplex codes. Their duals are the Hamming codes. Determine the parameters of the codes obtained by applying construction Y_1 to the Simplex codes.*

5.1.10. *Show that a code with minimum distance $d > q$ cannot be MDS.*

5.1.11. *Prove that the Simplex codes meet the Griesmer bound with equality.*

5.1.12. *Prove the following generalization of the residual code construction, which is used in Dodunekov [72] and Groneick and Grosse [100]:*

Let C be an $[n, k, d]_q$ code and $v \in C$ a codeword of weight w. Assume $d - w + \lceil w/q \rceil > 0$.
Then there is an $[n - w, k - 1, d - w + \lceil w/q \rceil]_q$ code.

5.2 Concatenation

and the $(u, u + v)$-construction.

The best general family of codes we have seen thus far is the Reed-Solomon codes. Recall that q-ary Reed-Solomon codes have length q. In particular, these codes are interesting only over relatively large fields. On the other hand, if we are searching for q-ary codes of length $n \leq q$, there is no reason not to use Reed-Solomon codes and what results from shortening or projecting. All these codes $[n, k, n - k + 1]_q$ are MDS codes. How about larger lengths? We promised several times to use codes defined over larger fields and certain methods of "going down" to construct good codes over smaller fields. This was one of our motivations to consider codes over large fields. Concatenation is one such general method.

Let $Q = q^r$. The field $\mathbb{F}_Q$ contains $\mathbb{F}_q$ as a subfield. Start from an arbitrary linear Q-ary code $\mathcal{C}$ with parameters $[N, K, D]_Q$. Each codeword is an N-tuple $(x_1, x_2, \ldots, x_N)$, where $x_i \in \mathbb{F}_Q$. Now $\mathbb{F}_Q$ is a vector space of dimension r over $\mathbb{F}_q$. Every r-dimensional code over $\mathbb{F}_q$ is an r-dimensional $\mathbb{F}_q$-vector space as well. Here is the idea of concatenation: fix a code $\mathcal{D}$ with parameters $[n, r, d]_q$ and a one-to-one mapping, a vector space isomorphism $\alpha : \mathbb{F}_Q \longrightarrow \mathcal{D}$. In order to obtain the new q-ary concatenated code, we replace each field element $u \in \mathbb{F}_Q$ with its image $\alpha(u) \in \mathbb{F}_q^r$. The images of codewords of $\mathcal{C}$ (Q-ary N-tuples) are q-ary nN-tuples.

5.8 Definition. *Let $Q = q^r$ and $\mathcal{C}$ be a linear Q-ary code $[N, K, D]_Q$ (the* **outer code**). *Let $\mathcal{D}$ be a linear $[n, r, d]_q$ (the* **inner code**). *Choose an $\mathbb{F}_q$-isomorphism*

$$\alpha : \mathbb{F}_Q \longrightarrow \mathcal{D} \subseteq \mathbb{F}_q^n.$$

The words of the **concatenated code** *are in bijection with the words of $\mathcal{C}$. If $(x_1, x_2, \ldots, x_n) \in \mathcal{C}$, then the corresponding word of the concatenated code is $(\alpha(x_1), \ldots, \alpha(x_N))$.*

The length of the concatenated code is nN. It has as many codewords as the outer code $\mathcal{C}$. This number is $Q^K = q^{rK}$. The dimension of the concatenated code is therefore rK. Let $\alpha(x)$, $x \in \mathcal{C}$ be a nonzero codeword. As $\mathcal{C}$ has minimum weight D, there are at least D coordinates such that $x_i \neq 0$. For these coordinates, $\alpha(x_i)$ is a nonzero word of the inner code $\mathcal{D}$ and has therefore weight $\geq d$. It follows that $\alpha(x)$ has weight $\geq dD$.

5.9 Theorem. *The concatenated code of an outer code $[N, K, D]_Q$, where $Q = q^r$, and an inner code $[n, r, d]_q$, is a linear q-ary code $[nN, rK, dD]_q$.*

Use concatenation to derive binary codes from quaternary. The inner codes must then be two-dimensional binary codes. Use the Reed-Solomon code $[4, 2, 3]_4$ as the outer code. As the inner code use the sum-0 code $[3, 2, 2]_2$. Its codewords are $000, 110, 101, 011$. We can choose

$$\alpha : 0 \mapsto 000, \ 1 \mapsto 110, \ \omega \mapsto 101, \ \omega^2 \mapsto 011.$$

Consider the codeword of $\mathcal{RS}(2, 4)$ parametrized by the polynomial X. It is $(0, 1, \omega, \overline{\omega})$. Its image under α is 000110101011 of weight 6. The concatenated code is a $[12, 4, 6]_2$ code.

Assume we want to use our $[8, 4, 4]_2$ as the inner code. The outer code will then have to be a 16-ary code. The Reed-Solomon codes yield $[16, k, 17 - k]_{16}$ for all $k \leq 16$. Concatenation yields binary codes $[128, 4k, 68 - 4k]_2$ for all $k \leq 16$. We see that we can construct binary codes of arbitrary lengths by using concatenation with a Reed-Solomon code defined over a large extension field as the outer code and a suitable short inner code.

Let us see how to obtain a generator matrix of the concatenated code. Let G be a generator matrix of the inner code. Denote its rows by $z_1, z_2, \ldots, z_r$. We need another fact from field theory (see Section 3.2): when $\mathbb{F}_{q^r}$ is constructed from the base field $\mathbb{F}_q$, using an irreducible polynomial $f(X)$ of degree r and ϵ is $X \pmod{f(X)}$, then $1, \epsilon, \ldots, \epsilon^{r-1}$ is a basis of $\mathbb{F}_{q^r}$ as a vector space over $\mathbb{F}_q$. Recall that ϵ^r can be written as a linear combination of these smaller powers of ϵ.

We can choose α such that

$$\alpha : 1 \to z_1, \ \epsilon \to z_2, \ldots, \ \epsilon^{r-1} \to z_r.$$

Let now M be a generator matrix of the outer code, with rows $R_1, R_2, \ldots, R_K$. The rows of a generator matrix of the concatenated code are the images under α of

$$R_1, \epsilon R_1, \ldots, \epsilon^{r-1} R_1, \ldots, R_K, \epsilon R_K, \ldots, \epsilon^{r-1} R_K.$$

Consider our first example. Our "large" field $\mathbb{F}_4$ has basis $1, \omega$. The rows of a generator matrix of the inner code $[3, 2, 2]_2$ can be chosen as $z_1 = 110$ and $z_2 = 101$. According to our recipe, this gives us

$$\alpha : 1 \to 110, \ \omega \to 101.$$

The rest is determined by linear combinations: $\alpha(0) = 000$ of course and $\alpha : \overline{\omega} = 1 + \omega \to 110 + 101 = 011$. This is what we had before. We know how a generator matrix of the outer $[4, 2, 3]_4$ can be determined. As $R_1 = (1, 1, 1, 1)$, $R_2 = (0, 1, \omega, \overline{\omega})$, this yields the following generator matrix for the concatenated $[12, 4, 6]_2$:

$$
\begin{array}{r|l}
R_1 \to & 110110110110 \\
\omega R_1 \to & 101101101101 \\
R_2 \to & 000110101011 \\
\omega R_2 \to & 000101011110
\end{array}
$$

The $(u, u + v)$-construction

This is another simple and useful recursive construction. It seems to go back to Sloane and Whitehead [194] and can be generalized to the case of not necessarily linear codes. The basic idea is described by its name.

We start from two linear codes $\mathcal{C}_1, \mathcal{C}_2$, which are defined over the same field and have the same length n. Let u be a typical element of $\mathcal{C}_1$ and v a typical element of $\mathcal{C}_2$. The code we construct has length $2n$. Its typical codeword is $f(u, v) = (u, u + v)$. It is rather clear that the resulting code has as dimension the sum of the dimensions of $\mathcal{C}_1$ and $\mathcal{C}_2$. The point is to control the minimum distance.

5.10 Theorem. *Let $\mathcal{C}_i$, $i = 1, 2$ be codes $[n, k_i, d_i]_q$, where $d_1 \leq d_2$. We can construct a code $[2n, k_1 + k_2, d]_q$, where $d \geq Min\{d_2, 2d_1\}$.*

PROOF   We construct the code $\mathcal{C}$ as the image of $f : \mathcal{C}_1 \oplus \mathcal{C}_2 \longrightarrow \mathbb{F}_q^{2n}$ defined by $f(u,v) = (u, u+v)$. Clearly, f is one-to-one (if $f(u_1, v_1) = f(u_2, v_2)$; then the first coordinate section shows $u_1 = u_2$, the second section shows $u_1 + v_1 = u_2 + v_2$, and, as we know already that $u_1 = u_2$, this implies $v_1 = v_2$). It follows that $dim(\mathcal{C}) = k_1 + k_2$.

In order to bound the minimum weight, it suffices to distinguish two cases. If $v = 0, u \neq 0$, then $wt(u,u) \geq 2d_1$. If $v \neq 0$, then $wt(u, u+v) \geq wt(v) \geq d_2$. This is the decisive observation. In fact, whenever $v_i \neq 0$, it must be that either $u_i \neq 0$ or $(u+v)_i \neq 0$.   ▯

As an example, start from an $[6,3,3]_2$ and apply Theorem 5.10 recursively, with the repetition code in the role of $\mathcal{C}_2$. This yields codes $[12,4,6]_2$, $[24,5,12]_2$, $[48,6,24]_2$, in general $[6 \cdot 2^i, 3+i, 3 \cdot 2^i]_2$.

Start from the trivial code $[2,2,1]_2$. Using the same method as above, we obtain $[4,3,2]_2$, $[8,4,4]_2$, $[16,5,8]_2$ and in general $[2^i, i+1, 2^{i-1}]_2$ for all $i \geq 1$. This yields another construction for the extended binary Hamming code $[8,4,4]_2$. All these codes meet the Griesmer bound Theorem 5.7 with equality.

5.11 Corollary. *The existence of a linear $[n, k, n/2]_q$ code implies $[2^i n, k + i, 2^{i-1} n]_q$ codes for all $i \geq 1$.*

PROOF   This is the special case of Theorem 5.10 when $d_1 \geq n/2$ and the second code is the repetition code.   ▯

---

1. Let $\mathcal{C}$ be a $[N, K, D]_Q$ code, where $Q = q^r$.

2. Let $\mathcal{D}$ be an $[n, r, d]_q$ code.

3. Concatenation produces an $[nN, rK, dD]_q$ code.

4. We can construct good codes of arbitrary length over small fields by using concatenation with Reed-Solomon codes over extension fields as outer codes.

5. Let codes $[n, k_1, d_1]_q$ and $[n, k_2, d_2]_q$ be given.

6. The $(u, u+v)$-construction produces a code $[2n, k_1 + k_2, d]_q$, where $d \geq Min\{d_2, 2d_1\}$.

7. There are families of binary codes $[6 \cdot 2^i, 3 + i, 3 \cdot 2^i]_2$ and $[2^i, i+1, 2^{i-1}]_2$.

Exercises 5.2

5.2.1. *Use* $\mathcal{RS}(4,8)$ *and* $[7,3,4]_2$ *in a concatenation scheme. What are the parameters of the resulting concatenated code?*

5.2.2. *We have already met the following matrix in Exercise 3.4.6. It generates a code* $[6,3,4]_4$, *which is known as the* **hexacode.** *We want to apply concatenation with the hexacode as the outer code. List at least two parameters of binary codes that can be used as inner codes, and the parameters of the resulting concatenated codes.*

$$\begin{pmatrix} 1\ 0\ 0\ 1\ \overline{\omega}\ \omega \\ 0\ 1\ 0\ 1\ \omega\ \overline{\omega} \\ 0\ 0\ 1\ 1\ 1\ 1 \end{pmatrix}$$

5.2.3. *Use concatenation with Reed-Solomon codes over* $\mathbb{F}_{2^r}$ *as outer codes and binary sum-0 codes as inner codes. Determine the parameters of the resulting binary codes.*

5.2.4. *Consider concatenation with Reed-Solomon codes as outer codes and a ternary* $[10,4,6]_3$ *as the inner code. What are the resulting parameters?*

5.2.5. *Construct a 48-dimensional binary linear code by applying concatenation to a Reed-Solomon code as the outer code and* $[23,12,7]_2$ *as the inner code. What are the parameters of this concatenated code?*

5.2.6. *Show how to construct a code* $[128,12,56]_2$ *by concatenation. What do you use as outer and inner codes?*

5.2.7. *Show how to construct a code* $[27,10,10]_3$ *by concatenation. What do you use as outer and inner codes?*

5.2.8. *An* $[11,4,5]_2$ *was promised in Section 2.3. Construct it by concatenation.*

5.2.9. *The* **Reed-Muller codes** $\mathcal{RM}(a,r)$ *are binary codes of length* 2^r, *minimum distance* 2^{r-a} *and dimension* $\sum_{i=0}^{a}\binom{r}{i}$.

Use the $(u,u+v)$*-construction to obtain codes with these parameters. Here* $\mathcal{RM}(0,r)$ *is the repetition code and* $\mathcal{RM}(r,r)$ *is the ambient space.*

5.2.10. *Give several parameters of good short binary linear codes among the family constructed in the third exercise above.*

5.2.11. *Give several examples of Reed-Muller codes with good parameters.*

5.2.12. *Construct* $[2^i \times 14, 4+i, 2^{i-1} \times 14]_2$ *codes for all* $i \geq 1$. *This yields in particular* $[28,5,14]_2, [56,6,28]_2, [112,7,56]_2$.

5.2.13. *Construct a* $[240,8,120]_2$ *code.*

Chapter 6

Universal hashing

Basic concepts: The concept of hashing. Almost universal hash classes and error-correcting codes. Error detection.

So far we have concentrated on one important area of applications for codes, transmission of messages over noisy channels. We mentioned early in this text that the applications of the concept of codes are manifold and certainly not limited to this historically first area. It is time to take a step in this direction.

An important concept in theoretical computer science is **hash functions.** The idea is to find a function $f : X \longrightarrow Y$ from a "large" set X to a "small" set Y with one of the following properties:

- Given $x \in X$, it is computationally infeasible to find $x' \neq x$ such that $f(x') = f(x)$.

- It is computationally infeasible to find $x \neq x'$ such that $f(x') = f(x)$.

Applications in cryptography are described in Stinson [200]. To give just one example, when a long document is to be signed electronically, one applies at first a hash function to produce a much shorter string (a digest of the original document) and then applies the signature scheme to this hash value. The notion of a hash function is very problematic. For one thing, it is clear that numerous collisions (pairs x, x' such that $f(x') = f(x)$) will happen. We want to make it computationally impossible for an opponent to find any. How can one guarantee that this is the case? The answer is: one cannot. In fact, whenever a new hash function is proposed, it takes just a couple of years to break it, that is, to describe a method which results in collisions.

Universal hashing was designed to find a way out of this dilemma, at least for certain applications. Just as for hash functions themselves, there are numerous variants. Let us fix notation:

6.1 Definition. *An $(n, M)_q$ array is an array (a matrix) with n rows and M columns, where each entry is taken from an alphabet $\mathcal{A}$ of size q. We also interpret each row of the array as a mapping $f : X \longrightarrow \mathcal{A}$, where $|X| = M$ (the elements of X parametrize the columns). In this perspective we speak of an $(n, M)_q$ **hash family**.*

The idea behind universal hashing is to use a carefully chosen family of hash functions instead of a single hash function. Whenever a hash function has to be applied, it will be chosen **at random** from the hash family. The expected behavior of this method is determined by structural properties of the hash family. Recall the notion of a probability space as introduced in Section 3.5.

6.2 Definition. *A probability space Ω is **uniform** if each element of Ω has the same probability $1/|\Omega|$.*

As we intend to choose the rows of our arrays (the functions from our hash family) at random, it is natural to consider the rows as elements of a uniform sample space: each of the n rows (functions) has probability $1/n$. The idea that collisions should be unlikely translates into the following requirement: whenever two different columns (elements $x \neq x'$ of the ground set) are fixed and a row f is chosen at random, the probability that $f(x') = f(x)$ should be small. This leads to the following definition:

6.3 Definition. *An $(n, M)_q$ hash family is ϵ-**universal** for some $0 < \epsilon < 1$ if, for any two distinct elements (columns) $x', x \in X$, the number ν of functions (rows) f such that $f(x') = f(x)$ satisfies $\dfrac{\nu}{n} \leq \epsilon$; in other words: the probability of collision is $\leq \epsilon$.*

Here is a toy example:

$$
\begin{array}{|l|}
\hline
01234\ 01234\ 01234\ 01234\ 01234 \\
01234\ 12340\ 23401\ 34012\ 40123 \\
01234\ 23401\ 40123\ 12340\ 34012 \\
01234\ 34012\ 12340\ 40123\ 23401 \\
01234\ 40123\ 34012\ 23401\ 12340 \\
\hline
\end{array}
$$

We see this as a $(5, 25)_5$ hash family, or as a family of five functions from a 25-set to a 5-set. It can be checked that each pair of columns contains at most one collision. For example, a collision of the first column and some other column is equivalent to an entry 0 in that other column. As each column except the first contains at most one entry 0, we have checked pairs involving the first column. Our array is $\frac{1}{5}$-universal.

It is now easy to interpret this notion from the point of view of the Hamming distance. Pick two columns. A collision is a coordinate (row) where they agree. The Hamming distance is the number of coordinates where there is **no collision**. Let d be the Hamming distance between the two columns. The probability of collision is then $\frac{n-d}{n} = 1 - d/n$. This shows how to interpret the notion of collision from our coding point of view: see the columns as words of a q-ary code of length n and minimum distance d. The array is then ϵ-universal, where $\epsilon = 1 - d/n$. We have established the following general equivalence:

6.4 Theorem. *The following are equivalent:*

- *An ϵ-universal $(n, M)_q$ hash family, and*

- *An $(n, M, d)_q$ code, where $\epsilon = 1 - d/n$.*

In fact, in our toy example we simply chose the columns as the codewords of the Reed-Solomon code $\mathcal{RS}(2, 5)$. As $d = 4$, we have $\epsilon = 1 - 4/5 = 1/5$.

The notion of an ϵ-universal hash family really is nothing new for us, but it gives a different interpretation of codes. In this application very small alphabets are not interesting at all. Also, as the ground set X has to be very large, the length n cannot be small. The probability ϵ of collision has to be small. This means that d/n has to be close to 1. It follows that the dimension k of our code (in the linear case) must be relatively small. As $M = q^k$ we see again that q cannot be all that small. Fortunately, we know already a source of good long codes over large alphabets: the Reed-Solomon codes. If they should not suffice, we can use concatenation with Reed-Solomon codes as outer codes.

Assume we want to hash documents of 2^{20} bits (about 1 million bits). As we want to be able to hash any document of this length, we have a ground set of size $2^{2^{20}}$. The hashed value should have a length of only 32 bits. This means that the alphabet has size $q = 2^{32}$. Can we use a q-ary Reed-Solomon code $\mathcal{RS}(k, q)$? We need $q^k \geq 2^{2^{20}}$, or $k \geq 2^{15}$. The probability of collision is then $\epsilon < 2^{15}/2^{32} = 2^{-17}$. This describes a $(2^{32}, 2^{2^{20}})_{2^{32}}$ hash family, which is 2^{-17}-universal. When using Reed-Solomon codes, we will always have $n = q$.

The idea of universal hashing in the computer science literature seems to go back to two influential papers [43, 218] by J. L. Carter and M. N. Wegman. Section 16.7 has more on universal hashing, its applications and links to codes.

An application to error detection

Here is an application of $\epsilon-$universal hash families for the purposes of error detection. Assume we send the long message x to the receiver via a rather good channel. Let y be the received message. In most cases, no errors will occur, hence $y = x$. In order to be able to detect errors (not to correct them), the following method can be used: we choose a function f at random from our $\epsilon-$universal family and send f as well as $f(x)$ to the receiver, using a public channel which will certainly not commit errors. This will not cost all too much, as these messages are much shorter than x (in the example above x has 2^{20} bits, whereas f and $f(x)$ are bitstrings of length 32 each). If $f(y) \neq f(x)$, then the receiver knows that a transmission error has happened. He will ask for retransmission in this case. If $f(y) = f(x)$, the message will be accepted as correct. The probability that this assumption is in error is then limited by ϵ.

Exercises 6

6.1. *Determine the parameters of Reed-Solomon codes when interpreted as $\epsilon-$universal hash families.*

6.2. *Describe an $(8, 16)_2$ hash family, which is $\frac{1}{2}$-universal.*

6.3. *Find a $\frac{2}{5}$-universal $(10, 81)_3$ hash family. Hint: this code has been constructed in Chapter 3. It is closely related to the ternary Golay code.*

6.4. *Does a $\frac{1}{2}$-universal $(6, 9)_2$ hash family exist?*

6.5. *Construct a $(256, 256^{17})_{256}$ hash family, which is $\frac{1}{16}$-universal.*

6.6. *Explain why a code $[4000, 5, 3996]_{4096}$ exists and determine the parameters of the universal hash family derived from it.*

Chapter 7

Designs and the binary Golay code

Basic concepts: Cyclic codes. The binary Golay code. Combinatorial t-designs, the large Witt design, projective planes.

A fruitful source of good codes is the theory of **cyclic codes.** A code is called cyclic if the cyclic shift of any codeword is a codeword again. As an example, consider the binary 23-tuple

$$z_1 = (1, 1, 0, 0, 0, 1, 1, 1, 0, 1, 0, 1, 0, 0, 0, 0, 0, 0, 0, 0, 0, 0, 0)$$

In order to obtain its first cyclic shift, we write first the final entry 0, followed by the first 22 entries:

$$z_2 = (0, 1, 1, 0, 0, 0, 1, 1, 1, 0, 1, 0, 1, 0, 0, 0, 0, 0, 0, 0, 0, 0, 0).$$

The next cyclic shift is

$$z_3 = (0, 0, 1, 1, 0, 0, 0, 1, 1, 1, 0, 1, 0, 1, 0, 0, 0, 0, 0, 0, 0, 0, 0)$$

and so on. Denote by $\mathcal{C}$ the cyclic binary linear code generated by z_1. It is clear that the first 12 cyclic shifts $z_1, z_2, \ldots, z_{12}$ are linearly independent (why?), but z_{13} is a linear combination of $z_1, z_2, \ldots, z_{12}$. It follows that $\mathcal{C}$ has dimension $k = 12$ and that a matrix with $z_1, z_2, \ldots, z_{12}$ as rows is a generator matrix of $\mathcal{C}$.

7.1 Definition. *The* **binary Golay code** $\mathcal{G}_{23}$ *is the binary linear cyclic code generated by z_1 above and its cyclic shifts.*

Here is the generator matrix G :

$$\begin{pmatrix}
1\,1\,0\,0\,0\,1\,1\,1\,0\,1\,0\,1\,0\,0\,0\,0\,0\,0\,0\,0\,0\,0\,0 \\
0\,1\,1\,0\,0\,0\,1\,1\,1\,0\,1\,0\,1\,0\,0\,0\,0\,0\,0\,0\,0\,0\,0 \\
0\,0\,1\,1\,0\,0\,0\,1\,1\,1\,0\,1\,0\,1\,0\,0\,0\,0\,0\,0\,0\,0\,0 \\
0\,0\,0\,1\,1\,0\,0\,0\,1\,1\,1\,0\,1\,0\,1\,0\,0\,0\,0\,0\,0\,0\,0 \\
0\,0\,0\,0\,1\,1\,0\,0\,0\,1\,1\,1\,0\,1\,0\,1\,0\,0\,0\,0\,0\,0\,0 \\
0\,0\,0\,0\,0\,1\,1\,0\,0\,0\,1\,1\,1\,0\,1\,0\,1\,0\,0\,0\,0\,0\,0 \\
0\,0\,0\,0\,0\,0\,1\,1\,0\,0\,0\,1\,1\,1\,0\,1\,0\,1\,0\,0\,0\,0\,0 \\
0\,0\,0\,0\,0\,0\,0\,1\,1\,0\,0\,0\,1\,1\,1\,0\,1\,0\,1\,0\,0\,0\,0 \\
0\,0\,0\,0\,0\,0\,0\,0\,1\,1\,0\,0\,0\,1\,1\,1\,0\,1\,0\,1\,0\,0\,0 \\
0\,0\,0\,0\,0\,0\,0\,0\,0\,1\,1\,0\,0\,0\,1\,1\,1\,0\,1\,0\,1\,0\,0 \\
0\,0\,0\,0\,0\,0\,0\,0\,0\,0\,1\,1\,0\,0\,0\,1\,1\,1\,0\,1\,0\,1\,0 \\
0\,0\,0\,0\,0\,0\,0\,0\,0\,0\,0\,1\,1\,0\,0\,0\,1\,1\,1\,0\,1\,0\,1
\end{pmatrix}$$

The cyclic structure is clearly visible. The addition of a parity check bit to each row of G yields a generator matrix of the so-called extended binary Golay code $[24, 12, 8]_2$. It is easily checked that this code $\mathcal{G}_{24}$ is self-dual: $\mathcal{G}_{24} = \mathcal{G}_{24}^\perp$. In fact, because of the cyclicity, it suffices to check that the first row of the generator matrix is orthogonal to all rows. All rows of the generator matrix of $\mathcal{G}_{24}$ have weight 8. This together with self-duality suffices to prove that all words of $\mathcal{G}_{24}$ have weights divisible by 4 (see Exercise 7.2). We wish to show that weight 4 does not occur.

The reader is urged to find a generator matrix of the form $M = (I|P)$ of $\mathcal{G}_{24}$. Here I is the $(12, 12)$ unit matrix and P is a $(12, 12)$ matrix (see Exercise 7.3). This is not hard, as the left half of G is already in triangular form. It turns out that all rows of M have weight 8. By Lemma 2.14 a check matrix is $(P^t|I)$. As $\mathcal{G}_{24}$ is self-dual, this is also a generator matrix.

Assume now $v \in \mathcal{G}_{24}$, $wt(v) = 4$. Write $v = (v_L, v_R)$, where v_L is the left half of v and v_R is the right half. Our generator matrices show that only the 0-word has $v_L = 0$ or $v_R = 0$. Assume $wt(v_L) = 1$. Then v_L is a row of M. This is impossible, as rows of M have weight 8. If $wt(v_R) = 1$, v_L is a column of P. These have weight 7. This is a contradiction. The only remaining possibility is $wt(v_L) = wt(v_R) = 2$. It follows that v is the sum of two rows of M. It is easy to check that none of these $\binom{12}{2} = 66$ vectors has weight 4.

7.2 Theorem. *The binary Golay code $\mathcal{G}_{23}$ is a $[23, 12, 7]_2$ code. It is a perfect code. The extended binary Golay code is a self-dual $[24, 12, 8]_2$ code.*

The presence of the all-1 word $\mathbf{1}$ (see Exercise 7.1) shows that the only possible weights of codewords of $\mathcal{G}_{24}$ are $0, 8, 12, 16, 24$ and that there are as many words of weight 16 as there are words of weight 8. Define A_i to be the number of codewords of weight i in $\mathcal{G}_{24}$. You are challenged to find the weight distribution of $\mathcal{G}_{24}$ (see Exercise 7.4). In particular, it turns out that $A_8 = 759$.

This can serve as a motivation for a basic structure of modern discrete mathematics, **combinatorial designs.**

7.3 Definition. *Let V be a set of v objects and B a family of subsets of V, where each such subset (a* **block***) has k elements. We say that B is a t-design with* **replication number** λ *if the following holds:*

Each set $T \subset V$ of t elements is in precisely λ blocks. The parameters of such a t-design are written $t - (v, k, \lambda)$. In the case of $\lambda = 1$ one speaks of a **Steiner system** *and writes $S(t, k, v)$ instead of $t - (v, k, 1)$.*

We can derive a 5-design from our code $\mathcal{G}_{24}$, as follows: As underlying set, use the 24 coordinates. Identify each word of weight i with a subset of cardinality i: those coordinates where the entry is 1. Take as blocks the 759 subsets (called **octads**) corresponding to the words of weight 8. This defines a Steiner system $S(5, 8, 24)$ (see Exercise 7.5).

7.4 Theorem. *The 759 octads of the extended binary Golay code $[24, 12, 8]_2$ are the blocks of a design $S(5, 8, 24)$. It is known as the* **large Witt design.**

We have seen that the cyclic code $\mathcal{G}_{23}$ and its lengthening $\mathcal{G}_{24}$ are highly interesting codes. In fact cyclic codes are one of the most widely used families of codes. The traditional theory of cyclic codes uses notation and results from ring theory (rings are an important type of algebraic structure). A different approach uses deeper knowledge of finite fields. The theory of cyclic codes is a core topic of Part II, in particular Chapters 12 and 13.

A particularly important family of designs is the **projective geometries.** In Part II we are going to make heavy use of these structures. The geometric description of linear codes, another core topic of Part II, is based on projective geometries (see in particular Chapter 17).

We start by describing the projective planes $PG(2, q)$: start from an arbitrary finite field $\mathbb{F}_q$. Consider the space $V = \mathbb{F}_q^3$, a three-dimensional space. Define

- **points:** the one-dimensional subspaces of V,

- **lines:** the two-dimensional subspaces of V.

As V has $q^3 - 1$ nonzero vector, and each one-dimensional space has $q - 1$ nonzero vectors, the total number of points is $(q^3 - 1)/(q - 1) = q^2 + q + 1$. The same counting method shows that each line has $(q^2 - 1)/(q - 1) = q + 1$ points.

The most important property of this geometry is the following: any two different points are on precisely one common line. This fact is clear, as any two one-dimensional spaces generate a two-dimensional space. We conclude that $PG(2, q)$, the classical projective plane of order q, is a design $S(2, q + 1, q^2 + q + 1)$.

Design theory is an important branch of modern combinatorial theory. The basic axiom is a uniformity property reminiscent of the definition of orthogonal arrays. In fact, one of the origins of combinatorial *design* theory is in statistics and in particular in the *design of experiments*.

Exercises 7

7.1. *Show that the all-1 word belongs to the extended binary Golay code* $[24, 12, 8]_2$.

7.2. *Let C be a binary linear code, $C \subseteq C^\perp$. Assume there is a basis of C all of whose elements have weights divisible by 4 (in other words, a generator matrix all of whose rows have weight divisible by 4). Show that all weights of codewords are multiples of 4.*

7.3. *Find a generator matrix of the form $M = (I|P)$ for $\mathcal{G}_{24}$.*

7.4. *Determine the weight distribution of $\mathcal{G}_{24}$.*

7.5. *Use the value $A_8 = 759$ to show the following: for each set of 5 of the 24 coordinates of $\mathcal{G}_{24}$ there is exactly one codeword of weight 8 (a so-called* **octad**) *with entries 1 in those 5 coordinates.*

7.6. *Find a picture of the* **Fano plane** $PG(2, 2)$.

7.7. *For every prime-power q, there is a design $S(2, q, q^2)$*
(the **affine plane** *of order q). Determine the number of blocks.*

7.8. *The SETS from Section 3.6 form a 2-design. Find its parameters.*

7.9. *The extended binary Golay code $[24, 12, 8]_2$ has $A_{12} = 2576$ (2576 codewords have weight 12). The corresponding subsets are called* **dodecads.** *The 2576 dodecads form the blocks of a design $2 - (24, 12, \lambda)$. What is λ? In fact, this is a 5-design. What is λ now?*

Chapter 8

Shannon entropy and the basics of information theory

Basic concepts: Idea, definition, basic properties and characterization of Shannon entropy. Link with binomials.

Recall the definition of a sample space from Section 3.5. Such a (finite) sample space is defined by a probability distribution, nonnegative numbers p_i such that $\sum_i p_i = 1$. The entropy is a function which measures the amount of information contained in a sample space.

8.1 Definition. *Let p_i, $i = 1, \ldots, n$ be a probability distribution. The* **Shannon entropy** *H is defined by*

$$H(p_1, p_2, \ldots, p_n) = -\sum_{i=1}^{n} p_i \cdot log(p_i).$$

Here, as always in information theory, the logarithm is taken base 2. We put $0 \cdot log(0) = 0$. Imagine a source which at every unit of time emits one of n symbols, called letters, according to the given probability distribution. That is, the probability for letter i to be chosen is p_i. We claim that the entropy function is a measure for the average amount of information per letter generated by this source. The unit of information is the **bit,** the amount of information in the choice between two equiprobable events: $H(\frac{1}{2}, \frac{1}{2}) = 1$.

8.2 Theorem. *Let $H_n : \mathbb{R}_+^n \longrightarrow \mathbb{R}_+$, $n = 2, 3, \ldots$ be a family of functions satisfying the following properties:*

- *Each H_n is continuous.*

- *$H_2(1/2, 1/2) = 1$ (one bit).*

- *There is a continuous monotonely increasing function*
  $A(x) : \mathbb{R}_+ \longrightarrow \mathbb{R}_+$ *such that* $A(n) = H_n(1/n, \ldots, 1/n)$.

- $H_n(p_1, p_2, \ldots, p_n) = H_2(p_1, 1 - p_1) +$
  $+ (1 - p_1) \cdot H_{n-1}(p_2/(1 - p_1) , \ldots, p_n/(1 - p_1))$.
  Here the numbers p_i *describe a probability distribution.*

Then $H_n(p_1, \ldots, p_n)$ *is the entropy function.*

This theorem justifies why we consider the entropy function from Definition 8.1 as the measure of information we were looking for. The proof is not hard. It constitutes Exercises 8.6 through 8.8.

The whole idea behind the definition of entropy is in the last property of Theorem 8.2. It says that the average amount of information contained in the choice between certain symbols according to a given probability distribution is not changed when the information is revealed in various steps.

For example, consider $H(\frac{1}{2}, \frac{1}{4}, \frac{1}{4})$. Combine the second and third event into one event. The information contained in the choice between the first and the block of the remaining events is $H(\frac{1}{2}, \frac{1}{2}) = 1$. There is a probability of $1/2$ that the result is the second or third event. The choice between events two and three has an entropy of 1 bit again. In order to obtain $H(\frac{1}{2}, \frac{1}{4}, \frac{1}{4})$, we have to add 1 (the contribution of the first stage) and $\frac{1}{2} \cdot 1$, the entropy contained in the choice of the second stage multiplied by the probability $1/2$ that this situation really arises. The result $H(\frac{1}{2}, \frac{1}{4}, \frac{1}{4}) = 1.5$ is in accordance with our formula. As another example consider

$$H(\frac{1}{2}, \frac{1}{4}, \frac{1}{8}, \frac{1}{8}) = \frac{1}{2} + \frac{1}{4}2 + \frac{1}{8}3 + \frac{1}{8}3 = 1.75.$$

It is clear that this entropy should be larger than $H(\frac{1}{2}, \frac{1}{2}) = 1$. Also, it should be expected that the maximum entropy given by a sample space of n points is obtained by the uniform distribution, and hence is $H(\frac{1}{n}, \frac{1}{n}, \ldots) = log(n)$. In the case $n = 4$ this yields a maximum entropy of $log(4) = 2$. Our entropy in the above example is indeed $1.75 < 2$.

Consider the case of four equiprobable events. The corresponding entropy is $H(\frac{1}{4}, \frac{1}{4}, \frac{1}{4}, \frac{1}{4}) = log(4) = 2$. Now divide these four events into two blocks, one single event and a group of the remaining three. By the basic property, the entropy is $H(\frac{1}{4}, \frac{3}{4}) + \frac{1}{4} \cdot 0 + \frac{3}{4}log(3)$. We can solve for $H(\frac{1}{4}, \frac{3}{4})$ and obtain $H(\frac{1}{4}, \frac{3}{4}) = 2 - \frac{3}{4}log(3)$. This is in accordance with the general formula, which reads

$$H(\frac{1}{4}, \frac{3}{4}) = \frac{1}{4}log(4) + \frac{3}{4}log(4/3).$$

The case of a sample space of just 2 points plays a particular role in coding theory.

8.3 Definition. *The* **binary entropy function** *is defined by*

$$h(x) = -xlog(x) - (1 - x)log(1 - x) \ \textit{(for } 0 \le x \le 1).$$

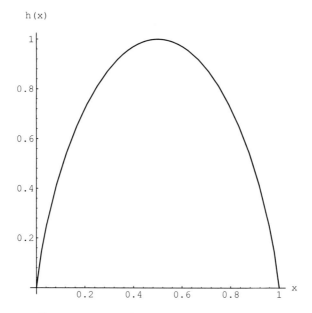

FIGURE 8.1: Binary entropy function

Observe that $h(x) = H(x, 1 - x)$ is indeed a special case of the Shannon entropy. We have $h(0) = 1$, $h(1/2) = 1$, and h is increasing on the interval $[0, 1/2]$.

There is a close relationship between the binary entropy function and binomials. As a tool we use the Stirling formula, which is frequently used in calculus and statistics.

8.4 Lemma. *The **Stirling formula** for factorials states that*

$$n! \sim \sqrt{2\pi n}(n/e)^n,$$

meaning that the quotient of left side and right side approaches 1 for $n \to \infty$.

A proof is in Feller [84]. Now let $n, m \to \infty$ such that $m/n \to p$. We want to check how the binomial coefficient $\binom{n}{m} = \frac{n!}{m!(n-m)!}$ behaves. By the Stirling formula we have

$$lim(\frac{1}{n}log(\binom{n}{m})) = lim(log(n) - \frac{m}{n}log(m) - \frac{n-m}{n}log(n-m)) =$$

$$lim(-\frac{m}{n}log(\frac{m}{n}) - \frac{n-m}{n}log(\frac{n-m}{n})) = h(p).$$

We have seen the following:

8.5 Theorem. *Let $n, m \to \infty$ such that $m/n \to p$. Then*

$$lim(\frac{1}{n}log(\binom{n}{m})) = h(p).$$

For a generalization to multinomials see Exercise 8.9.

Basic properties of Shannon entropy

A basic tool in proving the basic properties of Shannon's entropy function is Jensen's inequality.

8.6 Definition. *A real function $f(x)$ defined on an open interval I is* **concave down** *if, for every $x_0 \in I$, there is a line L through $(x_0, f(x_0))$ which is never under the graph of $f(x)$.*

A function f is concave down if, when biking along its graph from left to right, we lean to the right. This is certainly satisfied if $f''(x) < 0$.

Let $g(x) = f(x_0) + \lambda(x - x_0)$ be the equation of the line in Definition 8.6. By the definition of concavity, $f(x) \le f(x_0) + \lambda(x - x_0)$ for all $x \in I$. Let $x_1 \le x_2 \cdots \le x_n$ be points in I and $p_1, \ldots, p_n$ probabilities of a probability space. Using the inequalities above yields

$$\sum_i p_i f(x_i) \le f(x_0) + \lambda(\sum_i p_i x_i - x_0).$$

Choose in particular $x_0 = \sum_i p_i x_i$; observe that $x_0 \in I$.

8.7 Theorem (Jensen's inequality). *Let $f(x)$ be concave down on the open real interval I, let $x_1 \le x_2 \cdots \le x_n$ be points in I and $p_1, \ldots, p_n$ probabilities of a probability space. Then*

$$\sum_i p_i f(x_i) \le f(\sum_i p_i x_i).$$

Equality holds if and only if $f(x)$ is a linear function (its graph is a straight line) on the interval $[x_1, x_n]$.

The proof for the case of equality is left as Exercise 8.5.

Clearly the *log*-function is concave down on the interval $(0, \infty)$. Choose positive numbers x_i and a probability distribution p_i. An application of Jensen's inequality yields the famous inequality between the arithmetic and the geometric mean:

8.8 Theorem. *Let* $x_i > 0$, $p_i \geq 0$, $i = 1, 2, \ldots, n$, *where* $\sum_i p_i = 1$. *Then*

$$\sum_{i=1}^{n} p_i \cdot x_i \geq \prod_{i=1}^{n} x_i^{p_i}.$$

Equality holds if and only if all the x_i with $p_i \neq 0$ are equal.

Another application of Jensen's inequality to the *log*-function will be used to establish a key property of the entropy:

8.9 Proposition. *Let p_i and q_i be positive numbers, $i = 1, 2, \ldots, n$ such that $\sum_i p_i = 1$. Put $\sum_i q_i = \alpha$. Then the following holds:*

$$\sum_i p_i \cdot log(1/p_i) \leq \sum_i p_i \cdot log(1/q_i) + log(\alpha).$$

Equality holds if and only if $q_i = \alpha p_i$ for all i.

PROOF Apply Jensen's inequality to $x_i = q_i/p_i$. Observe
$$\sum_i p_i x_i = \sum_i q_i = \alpha. \qquad \qquad \square$$

In the case $\alpha = 1$, when the q_i form another probability distribution, this yields an interesting property of the entropy function: consider the expression $\sum_i p_i \cdot log(1/q_i)$, where the p_i form a fixed probability distribution and (q_i) varies through the probability distributions. Proposition 8.9 says that this expression is minimized when $q_i = p_i$, hence by the entropy.

Joint entropy and conditional entropy

Every probability space has its entropy. Another basic notion of probability theory and statistics is random variables.

8.10 Definition. *Let Ω be a probability space. A **random variable** on Ω is a mapping f defined on Ω.*

As every random variable induces a probability distribution, we can also speak of the entropy of a random variable. Let X and Y be random variables with probability distributions $p_i = Pr(X = i)$ and $q_j = Pr(Y = j)$, respectively. Consider further the joint distribution of X and Y, given by the probabilities $p(i, j) = Pr(X = i \text{ and } Y = j)$. Observe that the joint distribution carries more information than the distributions of X and Y taken together. In fact, we may arrange the probabilities $p(i, j)$ in a matrix with rows indexed by i and columns indexed by j. The row sums are the p_i, the column sums are q_j. The other way around it is clear that the matrix is far from being determined by its row- and column sums.

8.11 Definition. *Let X, Y be random variables defined on the same probability space Ω. With the notation just introduced, we call X and Y **statistically independent** if, for every i in the range of X and j in the range of Y, we have*

$$p(i, j) = p_i \times q_j.$$

The meaning behind this notion is that, in the case of independent random variables, information about the value of one of them gives no information whatsoever about the value of the other variable.

We wish to compare the entropies $H(X)$ and $H(Y)$ to the joint entropy $H(X, Y)$. In fact, there are more entropies around.

8.12 Definition. *Let X, Y be random variables defined on the same probability space Ω. The **conditional probability** is*

$$p(i|j) = p(i, j)/q_j.$$

This is the probability that $X = i$ if it is already known that $Y = j$.

Observe that X and Y are independent if and only if $p(i|j) = p_i$ always holds. For every fixed value j of Y, the conditional probabilities $p(i|j)$ form a probability distribution. Denote the corresponding entropy by $H(X|Y = j)$. It is natural to define a conditional entropy $H(X|Y)$ as the weighted sum of these expressions, the weights being the probabilities q_j.

8.13 Definition. *The **conditional entropy** $H(X|Y)$ is defined by*

$$H(X|Y) = \sum_j q_j \cdot H(X|Y = j) = -\sum_{i,j} p(i, j) \cdot log(p(i|j)).$$

Here i, j run through the values of X and Y, respectively.

The conditional entropy is a new notion. It is not the entropy of a probability space. There is no probability space $X|Y$.

The main theorem

The main theorem shows that our notions cooperate in exactly the way our interpretation suggests:

8.14 Theorem. *Let X and Y be random variables defined on the same finite probability space. Then the following relations hold for the entropies:*

- $H(X, Y) \leq H(X) + H(Y)$, *with equality if and only if X and Y are independent.*

- $H(X, Y) = H(X) + H(Y|X)$ *(the chain rule for entropies).*

- $H(X) \geq H(X|Y)$, *with equality if and only if X and Y are independent.*

PROOF Most interesting is the proof of the first statement. If we write out the definitions, we see that we have to prove the following statement: $\sum_{i,j} p(i, j) \cdot log(p(i, j)) \geq \sum_{i,j} p(i, j) \cdot log(p_i q_j)$. This follows from Proposition 8.9 in the case $\alpha = 1$. It also follows from this proposition when equality occurs.

The second statement is a trivial formal manipulation; the third statement follows from the others: $H(X) + H(Y) \geq H(Y, X) = H(Y) + H(X|Y)$. ⬚

Recall that we interpret $H(X)$ as a measure for the average amount of information per letter of a source operating with the corresponding probabilities. The first property of the main theorem says that the information contained in X and Y taken together is at most the sum of the individual informations. The second statement says that the joint information given by X and Y together can be written as the sum of the information given by X and the information given by Y when X is known. The third property says that knowledge of Y cannot increase our uncertainty about X. Exactly then will this knowledge not help when X and Y are independent.

Finally, we come to a natural notion, which is not to be found in Shannon's classical papers, mutual information:

8.15 Definition. *The **mutual information** is*
$I(X, Y) = H(X) - H(X|Y)$.

Observe that this notion is symmetric in X and Y : $I(X, Y) = H(X, Y) - H(Y|X) - H(X|Y) = I(Y, X)$. Does this notion agree with our interpretation? $H(X)$ is the uncertainty about X, and $H(X|Y)$ is the remaining uncertainty about X once Y is known. We should therefore interpret the difference as the amount of information Y has disclosed about X.

The **law of large numbers** provides another interpretation of the entropy function. Let a sample space with probabilities $p_1, p_2, \ldots, p_r$ be given. As before, interpret this space as a source of information, where at every point of time one of r symbols is generated according to the probability distribution. Consider signals of length N (N-tuples of letters) generated by this source. Let h_i be the frequency of letter $i \in \{1, 2, \ldots, r\}$. The law of large numbers asserts the following:

For every pair $\epsilon, \delta > 0$ there is an $N_0 = N_0(\epsilon, \delta)$ such that for every $N \geq N_0$ the set of signals of length N can be partitioned in two parts:

- a set of total probability $< \epsilon$, and

- a set all of whose elements satisfy

$$|h_i/N - p_i| < \delta \text{ for all } i.$$

The law of large numbers follows from the Tschebyscheff inequality, Theorem 9.10 in the next chapter. A proof of the law of large numbers will be sketched in the exercises section of Chapter 9.

The interpretation is as follows: if we pick a long signal at random, then we are almost sure that the relative frequencies of the letters are close to the probabilities. If S is such a signal of length N, then its probability is $p = Pr(S) = \prod_i p_i^{h_i}$. It follows that $log(p) = \sum_i h_i \cdot log(p_i)$, hence $log(p)/N = \sum_i (h_i/N) \cdot log(p_i)$. If S is one of the probable signals, then this expression is close to $\sum_i p_i \cdot log(p_i) = -H(p_1, \ldots, p_n)$. This is another interpretation of the entropy function.

This property is known in information theory as the **asymptotic equipartition property,** AEP; see Cover and Thomas [62].

1. The entropy function (Shannon entropy)
$H(p_1, p_2, \ldots, p_r) = -\sum_{i=1}^{r} p_i \cdot log(p_i)$ (where log is base 2) is a measure for the average amount of information per symbol of a source which generates r symbols with probabilities p_i.

2. Related notions are the **joint entropy** of several sources, the **conditional entropy** $H(X|Y)$ of source X with respect to source Y and the **mutual information**
$I(X, Y) = H(X) - H(X|Y) = I(Y, X)$.

3. The main theorem relates these notions in a natural way.

4. An important elementary tool is **Jensen's inequality.**

5. The typical **signals** (sequences of symbols) of large length N generated by the source have a probability p such that $\frac{1}{N}log(1/p)$ is close to $H(p_1, p_2, \ldots, p_n)$. This is another interpretation of the entropy.

6. The binary entropy function is $h(x) = H(x, 1 - x)$.

7. If $m/n \to p$, then $\frac{1}{n}log(\binom{n}{m}) \to h(p)$.

Exercises 8

8.1. *What is the relationship between* $H(p_1, p_2, p_3, p_4)$ *and*
$H(\frac{p_1}{2}, \frac{p_1}{2}, \frac{p_2}{2}, \frac{p_2}{2}, \frac{p_3}{2}, \frac{p_3}{2}, \frac{p_4}{2}, \frac{p_4}{2})$?

8.2. *What is larger (contains more information),* $H(\frac{1}{2}, \frac{1}{4}, \frac{1}{8}, \frac{1}{8})$ *or*
$H(\frac{1}{3}, \frac{1}{3}, \frac{2}{9}, \frac{1}{9})$? *Compute both entropies and compare.*

8.3. *Compare the following experiments:*

- **A:** *Three possible events with probabilities* 0.5, 0.25, 0.25

- **B:** *Three possible events, each with probability* 1/3.

Which of these experiments yields a larger amount of information? Compute the entropies and compare.

8.4. *Same type of problem as before, with*

- **A:** *Five possible events occur with probabilities* $\frac{1}{4}, \frac{1}{4}, \frac{1}{4}, \frac{1}{8}, \frac{1}{8}$.

- **B:** *Three possible events occur with probabilities* $\frac{1}{2}, \frac{1}{3}, \frac{1}{6}$.

8.5. *Prove the statement concerning the case of equality in Jensen's inequality (Theorem 8.7).*

8.6. *In the language of Theorem 8.2 prove* $A(s^m) = A(s^{m-1}) + A(s)$
for natural numbers s, m *and conclude* $A(s^m) = mA(s)$.

8.7. *Given natural numbers* s, t, n *define* m *such that*

$$s^m \le t^n < s^{m+1}.$$

With this notation prove

$$\frac{m}{n} \le \frac{A(t)}{A(s)} < \frac{m+1}{n}$$

and also

$$\frac{m}{n} \le \frac{log(t)}{log(s)} < \frac{m+1}{n}$$

Conclude $A(t) = log(t)$, *where* log *is the binary logarithm.*

8.8. *Prove the formula for H in the case of rational probabilities. It is clear that, because of continuity, this suffices. So let $p_i = m_i/m$, where m_i and m are natural numbers, $\sum_{i=1}^r m_i = m$.*
Hint: Use the central property of H in an appropriate situation to obtain

$$A(m) = H(p_1, \ldots, p_r) + \sum_{i=1}^r p_i A(m_i).$$

8.9. *The multinomial coefficient $\binom{n}{m_1, m_2, \ldots, m_k}$ is defined as*

$$\frac{n!}{m_1! m_2! \ldots m_k!} \quad \text{(here } m_1 + m_2 + \cdots + m_k = n\text{)}.$$

Prove that

$$lim(log(\binom{n}{m_1, m_2, \ldots, m_k}))/n) = H(p_1, p_2, \ldots, p_k)$$

where $n, m_i \longrightarrow \infty$ such that $m_i/n \longrightarrow p_i$.

Chapter 9

Asymptotic results

Basic concepts: Relative distance, information rate. Asymptotic bounds on codes. Asymptotic Singleton bound. Plotkin bound and Bose-Bush bound. Shannon's channel coding theorem. Gilbert-Varshamov bound. Justesen codes.

Let $\mathcal{C}$ be a code $[n, k, d]_q$. Recall the **relative minimum distance** $\delta = d/n$, and the **information rate** $R = k/n$ from Section 2.2. It was the basic idea of block coding to encode k-tuples as n-tuples in such a way that error correction is possible (because d is large). We are therefore interested in families of codes such that d is relatively large with respect to n (hence δ is bounded away from 0) while keeping R as close to 1 as possible.

9.1 Definition. *A pair (δ, R) is* **asymptotically reachable** *by linear q-ary codes if there is an infinite family of such codes with length $n_i \longrightarrow \infty$, relative minimum distances δ_i and information rates R_i such that $\lim_i \delta_i \geq \delta$ and $\lim_i R_i \geq R$.*

The asymptotic existence problem for linear q-ary codes is then the following: let $\delta \leq 1$ and $R \leq 1$. Decide if the pair (δ, R) is asymptotically reachable or not. When δ is given, we want R as large as possible. When δ increases the maximum R has to decrease. Exercise 9.5 shows that rate $R > 0$ can be reached only if $\delta < (q - 1)/q$. This is an upper bound (a pessimistic statement). It excludes pairs (δ, R) : such a pair cannot be reached if $\delta \geq (q - 1)/q$ and $R > 0$. The Singleton bound from Chapter 4 states $k + d \leq n + 1$. Divide by n. This yields the following:

9.2 Theorem (asymptotic Singleton bound). *If (δ, R) is asymptotically reachable by q-ary linear codes, then $R + \delta \leq 1$.*

Plotkin bound and Bose-Bush bound

The Plotkin bound is a bound on codes of very large minimum distance. The proof is based on an elementary method, **quadratic counting**.

9.3 Theorem (Plotkin bound). *Let C be a q-ary code of length n and minimum distance d. Assume $d > \dfrac{q-1}{q} \cdot n$. Then*

$$|\mathcal{C}| \leq qd/(qd - (q-1)n).$$

PROOF   Count the triples (x, y, j), where x and y are different codewords and j is a coordinate, where x and y differ. Let M be the number of codewords. It follows from the definition of d that the number of such triples is $\geq M(M-1)d$. Let us count each coordinate seperately now. So fix a coordinate and let m_i denote the number of codewords having entry i in the given coordinate. We have $\sum_i m_i = M$, and the contribution of the coordinate to our counting problem is $\sum_i m_i(M - m_i) = M^2 - \sum_i m_i^2$. We need an upper bound on this expression, hence a lower bound on $\sum_i m_i^2$. This bound follows from the classical **Cauchy-Schwartz inequality.** An easy proof is sketched in Exercise 9.9. We need the following inequality, which is rather clear intuitively: if the sum M of q nonnegative numbers m_i is fixed, then the sum $\sum_i m_i^2$ of their squares is minimized by choosing all m_i to be equal, hence $= M/q$. In that case $\sum_i m_i^2 = M^2/q$. It follows that, in our situation, the contribution of each coordinate is $\leq M^2 - M^2/q = M^2(q-1)/q$.
Taking both bounds together, we obtain $M(M-1)d \leq nM^2(q-1)/q$. Solving for M, the theorem is obtained.   ▯

It may seem at first that the Plotkin bound is of limited interest, as it applies only to codes of extremely large minimum distance. This is not really true. Our next result is a general asymptotic bound which follows from the Plotkin bound via a little trick (shortening). We will then derive a classical bound on orthogonal arrays of strength 2 from the Plotkin bound. Finally, we remark that codes with very large distances have their own areas of application. We saw one type of application (universal hashing) in Chapter 6; more is to come in Part II.

9.4 Theorem (asymptotic Plotkin bound). *Points (δ, R), which are asymptotically reachable by linear q-ary codes, are below or on the line joining the points $(0, 1)$ and $(\frac{q-1}{q}, 0)$.*

PROOF   The Plotkin bound shows that, for $\delta > (q-1)/q$, it is impossible to reach (δ, R) by linear q-ary codes for any $R > 0$. Another proof for this elementary fact is in Exercise 9.5. Now let $\delta < (q-1)/q$. We want to show that $R \le 1 - \frac{q}{q-1}\delta$. Assume (δ, R) can be reached, where $R \ge 1 - \frac{q}{q-1}\delta + c$ for some $c > 0$ and let $[n, k, d]_q$ be the parameters of a code in a corresponding family. Apply repeated shortening. This yields codes $[n-i, k-i, d]_q$. Choose i minimal such that $\dfrac{d}{n-i} > \dfrac{q-1}{q}$ (so that the Plotkin bound is applicable). Equivalently, i is the smallest integer such that $i > n - \frac{q}{q-1}d$. The assumption on R guarantees that $i \le k$. The Plotkin bound yields

$$q^{k-i} \le \frac{qd}{qd - (q-1)(n-i)} \le qd.$$

Take log_q and divide by n. The right side is asymptotically 0. The left side is asymptotically $R - (1 - \frac{q}{q-1}\delta)$. The claim $R \le 1 - \frac{q}{q-1}\delta$ follows. □

We can now forget the asymptotic Singleton bound. The asymptotic Plotkin bound is stronger.

The promised application to orthogonal arrays of strength 2 is based on the following construction of a code with large minimum distance from OA of strength 2.

9.5 Proposition. *If an $OA_\lambda(2, n, q)$ exists, we can construct a $(q^2\lambda - 1, n(q-1) + 1, q(q-1)\lambda)_q$ code. More precisely, any two different codewords are at distance $q(q-1)\lambda$.*

PROOF Write the alphabet as $\{\infty\} \cup \mathbb{Z}_{q-1}$. That is, one element (we call it ∞) plays a special role; the remaining $q-1$ symbols are identified with the integers mod $q-1$. Extend the addition on $\mathbb{Z}_{q-1}$ by $\infty + a = \infty + \infty = \infty$. Bring the OA in a standard form such that the first entry in each column is ∞. This can be done either by suitably renaming the entries in each column or by reordering the rows. Each pair (s, a), where s is a column of the OA and $a \in \mathbb{Z}_{q-1}$, gives rise to a codeword $s + a$, where a is added to each entry of s. This gives us $n(q-1)$ codewords. Another codeword is the word with all entries ∞. We have a q-ary code with $M = n(q-1) + 1$ words. As the first entry of each word is ∞, we will omit this coordinate. We have a code of length $q^2\lambda - 1$. Any two different words are at distance $q^2\lambda - q\lambda$. □

Application of Theorem 9.3 to the code derived from an $OA_\lambda(2, n, q)$ yields

$$M = n(q-1) + 1 \le \frac{q^2(q-1)\lambda}{q^2(q-1)\lambda - (q-1)(q^2\lambda - 1)} = q^2\lambda.$$

This is the classical Bose-Bush bound on OA of strength 2.

9.6 Theorem (Bose-Bush). *If an $OA_\lambda(2, n, q)$ exists, then*

$$n(q-1) + 1 \le q^2\lambda$$

(the number of rows is at least $n(q-1)+1$, where n is the number of columns).

An $OA_2(2, 8, 4)$

We illustrate with case $OA_\lambda(2, 8, 4)$. By Theorem 9.6 we have $\lambda \ge 2$. Can $OA_2(2, 8, 4)$ be constructed? Choose $\mathbb{F}_4$ as the alphabet. There can be no linear construction, as $\lambda = 2$ is not a power of 4. Our OA will have $32 = 2^5$ rows. The maximum in the way of linearity is linearity with respect to $\mathbb{F}_2$. This would mean we write a "generator matrix" consisting of five rows. The rows of the OA will all be $\mathbb{F}_2$-linear combinations (equivalently all sums) of these five rows.

It can be expected that linear mappings $\Phi : \mathbb{F}_8 \to \mathbb{F}_4$ will come in. Here we view $\mathbb{F}_4$ and $\mathbb{F}_8$ as vector spaces over $\mathbb{F}_2$, and Φ is a mapping from a three-dimensional space onto a two-dimensional space, described by a $(3, 2)$-matrix with entries in $\mathbb{F}_2$.

9.7 Proposition. *Let $\Phi : \mathbb{F}_8 \to \mathbb{F}_4$ be a surjective (onto) $\mathbb{F}_2$ linear mapping. Consider a $(32, 8)_4$ array $\mathcal{A}$, where the columns are indexed by $x \in \mathbb{F}_8$, the rows are indexed by pairs (a, b) where $a \in \mathbb{F}_8, b \in \mathbb{F}_4$, and with entry $\Phi(ax)+b \in \mathbb{F}_4$ in row (a, b), column x. Then $\mathcal{A}$ is an $OA_2(2, 8, 4)$.*

PROOF   Let two columns x_1, x_2 and a pair (c_1, c_2) of entries be given. We have to show that there are precisely two rows (a, b) with entry c_1 in column x_1 and entry c_2 in column x_2. This means that we have to count the solutions (a, b) of the system of equations $\Phi(ax_1) + b = c_1$, $\Phi(ax_2) + b = c_2$. Form the difference of the two equations and use the additivity of Φ to obtain the equation $\Phi(a(x_1 - x_2)) = c_1 - c_2$. For each solution a of this equation there is precisely one b such that (a, b) is a solution of the system. As Φ is $\mathbb{F}_2$ linear and onto, each element in $\mathbb{F}_4$ has precisely two preimages. It follows that there are precisely two elements $u_1, u_2 \in \mathbb{F}_8$ such that $\Phi(u_1) = \Phi(u_2) = c_1 - c_2$. It follows that $a(x_1 - x_2) \in \{u_1, u_2\}$. As $x_1 \ne x_2$ (we are considering different columns), we obtain two solutions for $a : a_1 = u_1/(x_1 - x_2)$, $a_2 = u_2/(x_1 - x_2)$. ▯

Choose Φ such that $\Phi(1) = 1$, $\Phi(\epsilon) = \omega$, $\Phi(\epsilon^2) = 0$, hence

$$\Phi : 1, \epsilon^3 \mapsto 1 \quad \epsilon, \epsilon^5 \mapsto \omega \quad \epsilon^4, \epsilon^5 \mapsto \overline{\omega} \quad 0, \epsilon^2 \mapsto 0.$$

As a basis of our OA we choose (of course) the rows corresponding to pairs $(1,0), (\epsilon, 0), (\epsilon^2, 0), (0, 1), (0, \omega)$. The following matrix is obtained. Recall that the rows of the OA consist of all $\mathbb{F}_2$ linear combinations of these five rows.

$(a,b)\backslash x$	0	1	ϵ	ϵ^2	ϵ^3	ϵ^4	ϵ^5	ϵ^6
$(1,0):$	0	1	ω	0	$\overline{\omega}$	$\overline{\omega}$	$\overline{\omega}$	ω
$(\epsilon,0):$	0	ω	0	1	$\overline{\omega}$	ω	ω	1
$(\epsilon^2,0):$	0	0	1	$\overline{\omega}$	ω	1	1	ω
$(0,1):$	1	1	1	1	1	1	1	1
$(0,\omega):$	ω	ω	ω	ω	ω	ω	ω	ω

By Proposition 9.5 we can derive a (nonlinear) $(31, 25, 24)_4$-code. This is quite a good code. Observe that a linear $[31, 3, 24]_4$ cannot exist, because of the Griesmer bound. It is then natural to ask if a $(31, 2^5, 24)_4$ exists which forms a vector space over $\mathbb{F}_2$. In fact we will see in Chapter 18 that such an additive $[31, 2.5, 24]_4$ code does exist. It is obtained by shortening a $[32, 3.5, 24]_4$ code constructed in Corollary 18.45.

In Section 16.7 the $OA_2(2, 8, 4)$ will be embedded in an infinite family of strength 2 orthogonal arrays, which can be used in the construction of certain universal hash families.

Shannon's channel coding theorem

Shannon's channel coding theorem from 1948 (see [183]) is generally regarded as the starting point of coding theory. He uses elementary probabilistic arguments to prove a theorem, which says, loosely speaking, that good codes of large length always exist. We will give a light treatment of the binary case, following van Lint [211].

Before we go on let us review some more relevant basic notions and results from probability theory. Probability spaces and random variables are used throughout this text.

9.8 Definition. *Let $f : \Omega \longrightarrow \mathbb{R}$ be a real-valued random variable defined on the finite probability space Ω. Denote by p_i the probability of $i \in \Omega$. The* **expectation** *of f is the real number*

$$E(f) = \sum_{i \in \Omega} p_i f(i).$$

The **variance** *is defined as*

$$Var(f) = \sum_{i \in \Omega} p_i (f(i) - E(f))^2 = E((f - E(f))^2).$$

It follows from the definition that $Var(f) \geq 0$. If we multiply out the square in the definition, we obtain

$$Var(f) = \sum_i p_i f(i)^2 - 2E(f) \sum_i p_i f(i) + E(f)^2 \sum_i p_i$$

$$= E(f^2) - 2E(f)^2 + E(f)^2 = E(f^2) - E(f)^2.$$

9.9 Lemma. $Var(f) = E(f^2) - E(f)^2 \geq 0$.

Going back to the definition of the variance, let us sum not over all $i \in \Omega$, but only over those satisfying $|f(i) - E(f)| \geq \lambda$, for some $\lambda \geq 0$. We obtain $Var(f) \geq \lambda^2 \times Prob(|f(i) - E(f)| \geq \lambda)$. This proves a basic inequality:

9.10 Theorem (Tschebyscheff inequality). *Let* $f : \Omega \longrightarrow \mathbb{R}$ *be a real-valued random variable. Then*

$$Prob(|f(i) - E(f)| \geq \lambda) \leq \frac{Var(f)}{\lambda^2}.$$

This reveals part of the meaning of the variance. If the variance is small, then large deviations from the expectation do not occur with high probability.

Here is the scenario for Shannon's theorem: The model for the information channel is the binary symmetric channel, with symbol error probability p. The strategy is **block coding**. Choose a length n and a code $\mathcal{C} \in \mathbb{F}_2^n$ with M elements. Assume that all codewords occur with the same probability $1/M$. Assume $x \in \mathcal{C}$ is sent and $y \in \mathbb{F}_2^n$ is received. No error will occur provided x is the unique codeword closest to y. For each codeword $x \in \mathcal{C}$ let $P(x)$ be the probability that, when x is sent, the received word is not decoded correctly. The error probability of code $\mathcal{C}$ is defined as the average value $P(\mathcal{C}) = \frac{1}{M} \sum_x P(x)$. Finally,

$$P^*(M, n, p) = min\{P(\mathcal{C})\},$$

where $\mathcal{C}$ varies over all codes $\mathcal{C} \subseteq \mathbb{F}_2^n$, $|\mathcal{C}| = M$.

With this terminology the channel coding theorem states:

9.11 Theorem. *Let* $R < 1 - h(p)$, *where* $h(p) = H(p, 1-p)$ *is the binary entropy function, and* M_n *such that* $\dfrac{log_2(M_n)}{n} \leq R$. *Then*

$$P^*(M_n, n, p) \longrightarrow 0 \text{ for } n \to \infty.$$

This implies the following: as long as the rate of our codes is within the given bound, it is always guaranteed that a code of large length n exists, which makes the probability of a decoding error arbitrarily small.

PROOF   Denote by $P(x|y)$ the probability that x was sent when y is received. Clearly,

$$P(x|y) = p^d q^{n-d} = P(y|x), \text{ where } d = d(x, y), \ q = 1 - p.$$

Consider the random variable $F(y)$, the number of errors. We have
$$F = \sum_{i=1}^{n} F_i, \text{ where } F_i = \begin{cases} 1 & \text{if symbol } y_i \text{ is incorrect,} \\ 0 & \text{if symbol } y_i \text{ is correct.} \end{cases}$$
By definition, $E(F_i) = p$. Because of the additivity of the expectation (see Exercise 9.1) $E(F) = np$. We have $\sigma^2 = Var(F) = npq$ (see Exercise 9.2).

For some small fixed $\epsilon > 0$, choose $b = \sqrt{\dfrac{\sigma^2}{\epsilon/2}}$. Theorem 9.10 proves that $Prob(|F - np| > b) \leq \epsilon/2$. We can choose n large enough such that $\rho = \lfloor np + b \rfloor < n/2$.

Introduce the function
$$f(u, v) = \begin{cases} 0 & \text{if } d(u, v) > \rho \\ 1 & \text{if } d(u, v) \leq \rho \end{cases}$$

(the points at distance $> \rho$ are taken care of by Tschebyscheff). Each $x \in \mathcal{C}$ defines a function g_x, where
$$g_x(y) = 1 - f(y, x) + \sum_{x' \neq x} f(y, x').$$

The main property of g_x is the following:
$$g_x(y) = \begin{cases} 0 & \text{if } x \text{ is the only codeword at distance } \leq \rho \text{ from } y \\ \geq 1 & \text{otherwise} \end{cases}$$

This property shows that the probability $P(x)$ that x is not decoded correctly is bounded by $\sum_y P(y|x) g_x(y)$. It follows that
$$P(x) \leq \sum_y P(y|x)(1 - f(y, x)) + \sum_y \sum_{x' \neq x} P(y|x) f(y, x').$$

The first summand is $Prob(d(y, x) > \rho) \leq \epsilon/2$. We have
$$P(\mathcal{C}) \leq \frac{\epsilon}{2} + \frac{1}{M} \sum_x \sum_y \sum_{x' \neq x} P(y|x) f(y, x').$$

In these sums x, x' vary in the code, y in $\mathbb{F}_2^n$. So far we have worked with a fixed code $\mathcal{C}$. Consider now the average of this expression when $\mathcal{C}$ varies over all such codes. Rewrite the second sum above as
$$\frac{1}{M} \sum_y \left(\sum_{x'} f(y, x') \sum_x P(y \mid x) \right).$$

Then $\sum_{x'} f(y, x')$ has an average value of $V_2(\rho, n)(M - 1)/2^n$ (here $V_2(\rho, n)$ is the volume of the ball of radius ρ; see Definition 1.11), and the average value of the sum is therefore bounded by
$$\frac{1}{M} V_2(\rho, n)(M - 1)/2^n \sum_x \sum_y P(y \mid x) = V_2(\rho, n)(M - 1)/2^n.$$

It follows that

$$P^*(M, n, p) \leq \frac{\epsilon}{2} + (M - 1)V_2(\rho, n)/2^n.$$

In order to bound the second term, take logarithms, divide by n, take the limit for $n \to \infty$. The first term $M - 1$ yields a limit $\leq R$. As $lim(\rho/n) = p$, it follows from Theorem 8.5 that the second term $V_2(\rho, n)$ yields a limit of $\leq h(p)$. It follows that the limit of the entire second term above under this process is $\leq R + h(p) - 1$, which is negative. This in turn implies that the second term can be made arbitrarily small, for example, $< \epsilon/2$. ∎

The Gilbert-Varshamov bound

The Gilbert-Varshamov bound proves the existence of certain good linear codes without actually constructing these codes. It is based on a simple counting argument.

Recall the volume $V_q(i, n) = \sum_{j=0}^{i} \binom{n}{j}(q - 1)^j$ of the ball of radius i as introduced in Section 1.6. Let $\mathcal{C}$ be a code $[n, k-1, d]_q$. As $\mathcal{C}$ has q^{k-1} elements, it follows that if $q^{k-1}V(d-1, n) < q^n$, then there is a vector $v \in \mathbb{F}_q^n$ which has distance $\geq d$ from every codeword $\in \mathcal{C}$. The code generated by $\mathcal{C}$ and v has dimension k and minimum distance d. In order to construct $\mathcal{C}$ we can use the same argument, based on a code with dimension $k - 2$. The condition is then automatically satisfied. We see that we can start from the 0 code and apply the same argument, constructing codes of increasing dimension as long as the condition $q^{k-1}V(d - 1, n) < q^n$ is satisfied. We have proved the following:

9.12 Theorem (Gilbert-Varshamov bound). *If*
$V_q(d - 1, n) < q^{n-k+1}$, *then an* $[n, k, d]_q$ *code exists.*

The proof of a slightly improved version of this bound is left as Exercise 9.11: if $V_q(d - 2, n - 1) < q^{n-k}$, then $[n, k, d]_q$ exists. Theorem 9.12 is also known as the Gilbert bound, whereas the slightly stronger result of Exercise 9.11 is the Varshamov bound.

Consider case $q = 2, d = 5$. The existence of $[n, k, 5]_2$ is guaranteed if $n + \frac{(n-1)(n-2)}{2} + \frac{(n-1)(n-2)(n-3)}{6} < 2^{n-k}$, equivalently if $\frac{n}{6}(n^2 - 3n + 8) < 2^{n-k}$. In the case $n = 15$ this is satisfied for $k = 6$, so $[15, 6, 5]_2$ exists. In the case $n = 30$ the left side is $4090 < 2^{12} = 2^{30-18}$, so $[30, 18, 5]_2$ exists. A quaternary example is $[87, 56, 13]_4$, whose existence is guaranteed by the Gilbert-Varshamov bound. No quaternary code with larger distance, larger dimension or shorter length is known.

The Gilbert-Varshamov bound implies an **asymptotic** existence theorem, as follows. Take log_q on both sides, divide by n and take the limit. It is clear (why?) that the largest term for $i = d - 1$ is dominating on the left side: we can replace the left side by this term and obtain the asymptotic contribution $log_q(2)h(\delta)$ (see Theorem 8.5). The second factor $(q-1)^{d-1}$ yields an asymptotic contribution $\delta log_q(q-1)$; the right side yields $1-R$. We have proved the following **asymptotic Gilbert-Varshamov bound:**

9.13 Theorem. *If $\delta \leq (q-1)/q$ and*

$$R \leq 1 - log_q(2)h(\delta) - \delta log_q(q-1),$$

then (δ, R) is asymptotically reachable.

The Justesen codes

The Gilbert-Varshamov bound has the drawback that it is a purely existential bound. We know these codes exist but we do not have a clue how to construct them. This is why we include a section on the Justesen codes (see [120]), which provide a rather elementary construction of asymptotically nontrivial families of codes. We restrict to the binary case.

Let $q = 2^m$. Justesen's idea is based on concatenation with a $q-$ary Reed-Solomon code as the outer code (see Chapter 4). Let $\mathbb{F}_q = \{u_1, u_2, \ldots, u_q\}$. Recall that the codewords of the Reed-Solomon code $\mathcal{RS}(K, q)$ are indexed by the polynomials $p(X) \in \mathbb{F}_q[X]$ of degree $< K$, the corresponding codeword being

$$(p(u_1), p(u_2), \ldots, p(u_q)).$$

The Justesen code $\mathcal{J}(K, q)$ has the same number of codewords as $\mathcal{RS}(K, q)$. The codeword of $\mathcal{J}(K, q)$ indexed by $p(X)$ is

$$(p(u_1), u_1 p(u_1), p(u_2), u_2 p(u_2), \ldots, p(u_q), u_q p(u_q)).$$

Here we interpret each entry not as an element of $\mathbb{F}_q$, but as a binary m-tuple (this is a special case of the idea of concatenation, see Section 5.2 where the inner code is the full space $[m, m, 1]_2$). It follows that $\mathcal{J}(K, q)$ is a binary linear code of length $2mq$ and dimension mK. Fix a **rate** R, where $0 < R < 1/2$ and choose K minimal such that $\frac{K}{2q} \geq R$. It follows that the minimum distance of $\mathcal{RS}(K, q)$ is $q - K + 1 \approx q(1 - 2R)$.

The whole point of the construction is the bound on the minimum distance d of $\mathcal{J}(K, q)$. It is derived from the obvious fact that the binary $(2m)-$tuples $(p(u_i), u_i p(u_i))$ and $(p(u_j), u_j p(u_j))$ are **different** provided $i \neq j$ and both $p(u_i)$ and $p(u_j)$ are nonzero. This is the reason behind the definition. It follows that the weight of any nonzero codeword of $\mathcal{J}(K, q)$ is at least the sum of the weights of the $q - K + 1$ lowest-weight $2m-$tuples. This gives us the following bound: if $\sum_{j=0}^{r} \binom{2m}{j} \leq q(1 - 2R)$, then $d \geq \sum_{j=0}^{r} j\binom{2m}{j}$. The usual process (take logarithms, divide by $2m$, take the limit) shows (for large m) $h(\frac{r}{2m}) = 1/2$ or $r = 2mh^{-1}(1/2)$, and $d \geq rq(1-2R) \approx 2mq(1-2R)h^{-1}(1/2)$, hence $\delta \geq (1 - 2R)h^{-1}(1/2)$. We conclude:

9.14 Theorem. *Let $0 < R < 1/2$. The Justesen codes $\mathcal{J}(K, 2^m)$, where K is minimal such that $K/2^{m+1} \geq R$, form a family of binary linear codes of length $m2^{m+1}$, rate $\geq R$ and relative minimum distance δ asymptotically not inferior to $(1 - 2R)h^{-1}(1/2)$. Here $h^{-1}(1/2) \approx 0.11$.*

In the asymptotic picture this gives a line from $(0, 1/2)$ on the R-axis to $(0.11, 0)$ on the δ-axis.

1. (δ, R) is **asymptotically reachable** by linear q-ary codes if there is a family of such codes with lengths $n \to \infty$ and rate $\geq R$, relative distance $\geq \delta$ in the limit.

2. The **Gilbert-Varshamov bound** is an elementary nonconstructive existence theorem for linear codes.

3. Most of the bounds on codes derived earlier yield asymptotic upper bounds (nonexistence theorems).

4. The Gilbert-Varshamov bound yields a nonconstructive asymptotic existence theorem.

5. The **Plotkin bound** on codes of large lengths is based on quadratic counting.

6. The **Bose-Bush bound** on q-ary strength 2 orthogonal arrays: the number of rows is $\geq (q - 1)n + 1$, where n is the number of columns.

7. The **Justesen codes** use concatenation and Reed-Solomon codes. They yield a constructive asymptotic lower bound.

8. The **channel coding theorem** guarantees the existence of good long codes: if p is the symbol error probability of the binary symmetric channel, there exist long codes of all rates $R < 1 - h(p)$ with arbitrarily small decoding error probability.

Exercises 9

9.1. *Prove the additivity of the expectation: if f and g are real-valued random variables defined on the same probability space, then $E(f + g) = E(f) + E(g)$.*

9.2. *Let $F = F_1 + \cdots + F_n$ be the real-valued random variable considered in the proof of the channel coding theorem. We have $E(F) = np$.*
Prove that the variance of F is $\sigma^2 = npq$, where $q = 1 - p$.

9.3. *Determine the maximum dimension k such that the GV-bound guarantees the existence of a code $[8, k, 4]_2$.*

9.4. *Determine the maximum dimension k such that the GV-bound guarantees the existence of a code $[6, k, 4]_3$.*

9.5. *Let C be a code $[n, k, d]_q$. Prove that $\frac{d}{n} \leq (q-1)q^{k-1}/(q^k - 1)$. This can be done by a simple counting argument. Conclude that if (δ, R) is asymptotically reachable by linear codes and $R > 0$, then $\delta \leq (q - 1)/q$.*

9.6. *Let $h(x)$ be the binary entropy function. Determine the derivative $h'(x)$.*

9.7. *The q-ary entropy function is defined as*

$$H_q(\delta) = log_q(2)h(\delta) + \delta log_q(q - 1).$$

In particular $H_2 = h$. The right side of the asymptotic GV-bound is $1 - H_q(\delta)$.
Prove that H_q is increasing for $0 \leq \delta \leq (q-1)/q$. Determine $H_q((q-1)/q)$.

9.8. *Determine the maximum dimension k such that the Gilbert-Varshamov bound guarantees the existence of a code $[10, k, 5]_4$. Show your calculation.*

9.9. *Let the natural number n and the positive real number S be given. Consider nonnegative numbers $x_i, i = 1, 2, \ldots, n$ such that $\sum_{i=1}^{n} x_i = S$.*
Prove: the expression $\sum_{i=1}^{n} x_i^2$ is minimized by the choice $x_i = S/n$.
Determine this minimum.
Hint: Let some n-tuple $(x_1, x_2, \ldots, x_n)$ be given such that not all the x_i equal S/n. Pick i such that $x_i > S/n$ and j such that $x_j < S/n$. Show that you can replace x_i by $x_i - \epsilon$ and x_j by $x_j + \epsilon$ for some $\epsilon > 0$ and obtain a tuple with a smaller sum of squares.

9.10. *Prove an asymptotic version of the sphere-packing bound (for q-ary linear codes): if (δ, R) is asymptotically reachable, then $R \leq 1 - H_q(\delta/2)$.*

9.11. *Prove another version of the Gilbert-Varshamov bound (see MacWilliams and Sloane [142]):*
If $V_q(d - 2, n - 1) < q^{n-k}$, then an $[n, k, d]_q$-code exists. Hint: assume at first that $[n - 1, k - 1, d]_q$ exists and use a check matrix.

9.12. *Prove that the version of the GV-bound proved in the preceding exercise is stronger than the original GV-bound of Theorem 9.12.*

9.13. *Use the method of Theorem 9.12 to prove a Gilbert-Varshamov bound for not necessarily linear q-ary codes.*

Chapter 10

Three-dimensional codes and projective planes

Basic concepts: Projective planes and geometric description of three-dimensional codes. Application to congestion-free networks.

In this chapter we want to learn how to interpret linear codes geometrically. The starting point is a generator matrix of the code. We limit ourselves to the case of three-dimensional codes. The underlying geometry is the projective plane $PG(2, q)$, see Chapter 7. Recall the basic data:

We call **points** of $PG(2, q)$ the one-dimensional subspaces of $\mathbb{F}_q^3$, the two-dimensional subspaces are called **lines.** There are $q^2 + q + 1$ points and equally many lines. The lines form the blocks of a design, a Steiner system $S(2, q + 1, q^2 + q + 1)$ (in particular each line has $q + 1$ points). The smallest projective plane is $PG(2, 2)$ (7 points, 7 lines, each line has 3 points, each point is on 3 lines, each pair of points is on precisely 1 line). This binary projective plane $PG(2, 2)$ is also known as the **Fano plane.**

Each nonzero vector (x, y, z) of $PG(2, q)$ generates a one-dimensional subspace, hence a point. We can label the points by these triples. The difference between triples and points is that triples, which are scalar multiples of each other, define the same point. The Fano plane is depicted in Figure 10.

As an example, consider the binary code with generator matrix

$$G = \begin{pmatrix} 1 & 0 & 0 & 1 & 1 & 0 & 1 \\ 0 & 1 & 0 & 1 & 0 & 1 & 1 \\ 0 & 0 & 1 & 0 & 1 & 1 & 1 \end{pmatrix}.$$

ere is the link to geometry: we consider the **columns** of G as points of $PG(2, 2)$. In this case each point of the Fano plane occurs precisely once as a column. A geometer would identify this code $\mathcal{C}$ (or this generator matrix)

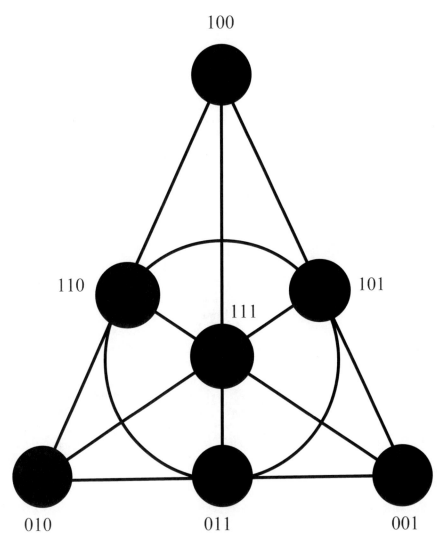

FIGURE 10.1: The Fano plane

with the Fano plane itself. What is the minimum distance d of this code $[7, 3, d]_2$, or, more generally, what are the weights of codewords?

Denote the rows of G as z_1, z_2, z_3. The codewords are the linear combinations $\lambda_1 z_1 + \lambda_2 z_2 + \lambda_3 z_3$. Fix a coordinate.

The corresponding column of G is a point P of $PG(2, 2)$. When will our codeword have entry 0 in column P? Let $P = (x, y, z)$. Entry 0 means $\lambda_1 x + \lambda_2 y + \lambda_3 z = 0$, or $(\lambda_1, \lambda_2, \lambda_3) \cdot P = 0$.

Let us sum up: each coordinate is indexed by a point P of $PG(2, 2)$ (the corresponding column of the generator matrix). Each nonzero codeword is given by a nonzero triple $(\lambda_1, \lambda_2, \lambda_3)$ (the codeword is $\lambda_1 z_1 + \lambda_2 z_2 + \lambda_3 z_3$). This codeword has entry 0 in this coordinate if and only if the dot product vanishes:

$$(\lambda_1, \lambda_2, \lambda_3) \cdot P = 0.$$

The dot product represents one nontrivial linear equation on $P = (x, y, z)$. These points P are therefore precisely the points of a two-dimensional subspace, in other words, of a **line.**

Sum up again: each nonzero codeword defines a line l (scalar multiples define the same line). Each coordinate is indexed by a point P. The corresponding entry of the codeword vanishes ($= 0$) if $P \in l$. The weight of the codeword is therefore the number of points **which are not on** l.

This description is true for any three-dimensional code. In our example the 7 coordinates are described by the 7 points of the Fano plane. As each line has 3 points, there are 4 points outside each line. We have seen that our code is a $[7, 3, 4]_2$ code, more precisely each nonzero word of our code has weight 4.

We know this code. It is the Simplex code $\mathcal{S}_3(2)$ from Section 2.5; for general q see Section 3.4.

In general there is no reason why points of $PG(2, q)$ should not occur several times as columns of a generator matrix. We should therefore describe every three-dimensional code by a weight function w defined on the points of $PG(2, q)$, where $w(P)$ tells us how many columns of a generator matrix give point P. The length n of the code is then $n = \sum_P w(P)$. More importantly, we understand how to express the minimum distance d (and more generally all weights of codewords) geometrically: For each line l define the weight of l by

$$w(l) = \sum_{P \in l} w(P).$$

The weights of codewords are then the numbers $n - w(l)$, where l varies through the lines, and the minimum distance is therefore $d = n - Max\{w(l)\}$.

10.1 Example. *The properties of the Simplex codes are particularly easy to understand in this geometric language. The three-dimensional Simplex code by definition has length $q^2 + q + 1$. Each point of $PG(2, q)$ appears once as a column ($w(P) = 1$). This means that every line has weight $q + 1$ and that*

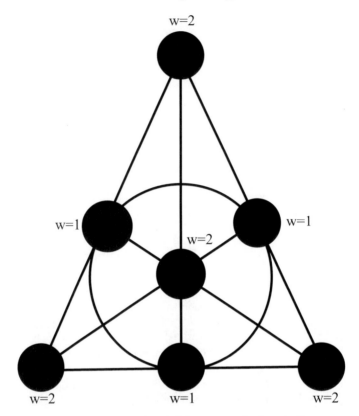

w=2

w=1   w=2   w=1

w=2   w=1   w=2

FIGURE 10.2:   A binary code $[11, 3, 6]_2$

every nonzero codeword has weight $(q^2 + q + 1) - (q + 1) = q^2$. In particular, the code parameters are $[q^2 + q + 1, 3, q^2]_q$.

Here is another binary example. We start from the Fano plane and use the weights given in Figure 10.2. The length of the code is the sum of all weights: $n = 11$. Also, we see that the lines have weights ≤ 5. This shows that we have a code $[11, 3, 6]_2$. A corresponding generator matrix is obtained by writing each point P as often as a column as $w(P)$ indicates ($w(P) = 0$ means that this point does not occur, $w(P) = 1$ means it occurs once, $\ldots$). In our case we obtain the generator matrix

$$\begin{pmatrix} 1 & 1 & 0 & 0 & 1 & 1 & 1 & 1 & 1 & 1 & 0 \\ 0 & 0 & 1 & 1 & 0 & 0 & 1 & 1 & 0 & 1 & 1 \\ 0 & 0 & 0 & 0 & 1 & 1 & 1 & 1 & 1 & 0 & 1 \end{pmatrix}.$$

The geometric description of general linear codes is studied in Chapter 17.

Three-dimensional codes and congestion-free networks

Consider the following scenario: There are n network sites to be connected by broadcast media. A **link** or **bus** is a subset of the sites. In order to avoid congestion, it is required that every two sites appear together on at least one link. Each site is equipped with a limited number of communication ports and hence can appear on at most some fixed number r of the links. Also, each link cannot accomodate more than k sites. This construction problem for network topologies has been studied extensively. For an introduction, literature and combinatorial-geometric links, see Colbourn [57].

Assume there is a code $[n, 3, n - k]_q$. By the geometric description, this is equivalent with a multiset of n points in $PG(2, q)$ such that no more than k are on a line. This satisfies the requirements. It can even be shown that this construction is the best possible when $r - 1$ is a prime-power, at least in an asymptotic sense. As an illustration consider Figure 10.2. This represents a network for 11 sites such that each site is on precisely 3 links, each link has at most 5 sites and any 2 sites occur in a common link.

---

1. The general notion underlying the geometric method for three-dimensional codes is that of **projective planes** $PG(2, q)$ (of **order** q).

2. The **points** of $PG(2, q)$ are the one-dimensional subspaces of $\mathbb{F}_q^3$.

3. The **lines** of $PG(2, q)$ are the two-dimensional subspaces of $\mathbb{F}_q^3$.

---

1. The **geometric method:** view the columns of a generator matrix G as a family of points from $PG(2, q)$.

2. Write $w(P)$ if point P occurs $w(P)$ times as a column of G.

3. w is a function defined on the points of $PG(2, q)$ with non-negative integer values (w describes a multiset of points).

4. The length is $n = \sum_P w(P)$; the dimension is 3.

5. For every line l, let $w(l) = \sum_{P \in l} w(P)$.

6. The minimum distance is $d = n - max_l w(l)$.

Exercises 10

10.1. *Construct a code* $[14, 3, 8]_2$ *by the geometric method.*

10.2. *Describe the construction of codes* $[q^2, 3, q^2 - q]_q$
by the geometric method.

10.3. *Under which condition on the weight function* w *will the code really have dimension 3?*

10.4. *The* **projective line** $PG(1, q)$ *consists of the* $q + 1$ *one-dimensional subspaces* (**points**) *of* $\mathbb{F}_q^2$. *Describe the parameters of codes* $[n, 2, d]_q$ *using geometric language.*

10.5. *Given* d, *determine the smallest* n *such that an* $[n, 2, d]_q$ *code exists.*

10.6. *Write the generator matrix of a code* $[10, 2, 8]_7$ *and show that the parameters are as claimed.*

10.7. *Find a description of four-dimensional codes, which is analogous to our geometric description of three-dimensional codes.*

10.8. *Find a generator matrix of a code* $[17, 3, 9]_2$.

10.9. *Imagine you have 28 points in* $PG(2, 8)$ *such that no line contains more than four of these points.* (*Such a point set exists. It is a maximal arc; see Theorem 17.26.*) *Use these points as columns of a generator matrix.*
    What are the parameters of the code? Give a brief explanation.

10.10. *Show that the maximum minimum distance of a* $[9, 3, d]$ *code is* $d = 4$.

Chapter 11

Summary and outlook

We have reached a point where we understand the basic notions and objectives of coding theory. The most important family introduced thus far is the Reed-Solomon codes of Chapter 4. The recursive techniques of Chapter 5, in particular concatenation, allow the construction of large classes of codes. This allows the construction of universal hash classes as well (see Chapter 6). Where the classical application of codes for the transmission of messages via noisy channels is concerned, we know the principle of syndrome decoding (Section 3.7) but we do not have a really effective algorithm yet.

Part II starts with an introduction to cyclic codes. The theory leading to the construction of these codes and to bounds on their parameters is a core topic of classical coding theory. It is hoped that the approach presented here is easier to understand and more readily generalizable. The appetizer (Chapter 7) given in Part I is not sufficient to convey an impression. The construction of cyclic codes utilizes the Reed-Solomon codes. In fact, cyclic codes are trace codes (equivalently subfield codes) of suitably shortened Reed-Solomon codes. An important feature of cyclic codes is that they possess an effective decoding algorithm. All this constitutes the content of the first two chapters of Part II. This is followed by the introduction, in Chapter 14, to more recursive constructions as well as a new parameter, the covering radius, which has its own area of applications.

Chapter 15 presents a self-contained introduction to the basics of linear programming. Unified proofs for several explicit bounds on codes and OA are obtained as an application. Chapter 16 collects some areas of application which are different from the classical application for information transmission. We start from the statistical interpretation of orthogonal arrays (dual codes) as families of random variables with a statistical independence property (Section 16.1).

Linear shift register sequences (Section 16.2) are closely related to the Simplex codes. Section 16.3 discusses the nature of cryptography and the use of codes in the construction of block ciphers. Two-point based sampling (Section 16.4) is a principle which can be seen as an application of OA of strength 2. An application is the testing of chips. Resilient functions have multiple applications (see Section 16.5), among others in the construction of stream ciphers offering resistance to correlation attacks. There is a strong relationship to OA. Section 16.6 discusses an application of Reed-Solomon codes for

131

the purposes of derandomization of Monte Carlo algorithms. In Part I it was shown how codes can be interpreted as universal hash classes. In the final section of Chapter 16 we return to a more in-depth study of universal hashing and its applications.

Chapter 17 is devoted to the description of linear codes as point sets in projective spaces defined over finite fields. Many highly structured codes are best understood from the geometric point of view. The general mechanism as well as a variety of applications are in Section 17.1. Among the applications we consider codes of low dimension, the ternary Golay codes, the extended Reed-Solomon codes, oval and hyperoval codes, Barlotti codes and Denniston codes. Section 17.2 presents a self-contained introduction to geometric algebra over finite fields, the theory of bilinear, sesquilinear and quadratic forms. A first application concerns caps (Section 17.3). These are sets of points no three of which are on a line. They are equivalent to linear codes of minimum distance 4.

In Chapter 18 we consider a generalization of linear codes. Additive codes are linear over a subfield of their alphabet. Another way of saying this is: the alphabet is considered not as a field but as a vector space over some ground field. In Section 18.1 we discuss some recursive and direct constructions as well as applications to computer memory systems and deep-space communication. In Section 18.2 the mechanism of cyclic codes is generalized to additive codes. The geometric approach to quaternary additive codes is used in Section 18.3. In another specialization step, the theory of quantum stabilizer codes is developed in Section 18.4. Chapter 18 and Part II close with an introduction to network codes, Section 18.5.

Part II

Theory and applications of codes

Chapter 12

Subfield codes and trace codes

12.1 The trace

Basic concepts: Field extensions, primitive polynomials. The Frobenius automorphism, the Galois group, the trace.

A general construction idea for good codes over some field $\mathbb{F}_q$ is to first construct a code over a larger field $\mathbb{F}_{q^r}$ and then to go down in some way from the large field to the small field. The most fruitful methods of "going down" are the trace codes and the subfield codes. Before we can discuss these methods, we need to know more about finite fields, in particula,r about pairs of finite fields contained in one another.

We saw that, for every prime p and every natural number r, there is a field $\mathbb{F}_{p^r}$ of order p^r. It can be constructed as an extension of the prime field $\mathbb{F}_p$, using an irreducible polynomial of degree r with coefficients in $\mathbb{F}_p$. In fact, we chose not to give all the details, as this would have led us too deeply into field theory. We chose to accept without proof that all the required irreducible polynomials exist over finite fields (not over all fields: there is, for example, no irreducible polynomial of odd degree over the reals) and also that, for each order p^r, there is only one field.

Consider fields $\mathbb{F}_{p^s}$ and $\mathbb{F}_{p^r}$ of the same characteristic p, where $s \leq r$. When is $\mathbb{F}_{p^s} \subset \mathbb{F}_{p^r}$? If this is the case, then the larger field $\mathbb{F}_{p^r}$ is in particular a vector space over $\mathbb{F}_{p^s}$ (the field axioms are stronger than the vector space axioms), of some dimension d. Then $p^r = (p^s)^d = p^{sd}$. It follows that s divides r, and the dimension is $d = r/s$. If, on the other hand, $s|r$, we can choose an irreducible polynomial of degree $d = r/s$ with coefficients in $\mathbb{F}_{p^s}$ and construct $\mathbb{F}_{p^r}$ as an extension field of $\mathbb{F}_{p^s}$, by the method from Section 3.2 (we believe that these irreducible polynomials always exist).

This shows the following:

12.1 Proposition. *We have $\mathbb{F}_{p^s} \subset \mathbb{F}_{p^r}$ if and only if s divides r.*

In fact, there are good and not so good irreducible polynomials.

12.2 Definition. *Let $f(X) \in \mathbb{F}_q[X]$ be irreducible of degree r, and $\epsilon \in \mathbb{F}_{q^r}$ the element corresponding to X. We say that $f(X)$ is a **primitive polynomial** if $1 = \epsilon^0, \epsilon, \ldots, \epsilon^{q^r-2}$ are different, equivalently if the powers of ϵ are all nonzero elements of $\mathbb{F}_{q^r}$. An element whose powers run through all nonzero field elements is called a **primitive element.***

It can be shown that primitive polynomials always exist. They are also known as maximum exponent polynomials. In practice one always works with primitive polynomials. Should we come across an irreducible polynomial, which is not primitive, we will throw it away and replace it with a primitive polynomial.

12.3 Example. *$f(X) = X^2 + 1$ certainly is an irreducible polynomial over $\mathbb{F}_3$. In fact, as $f(0) = 1$, $f(1) = f(2) = 2$, it has no roots and therefore no linear factor. However, we will not use the polynomial to describe $\mathbb{F}_9$. Let $\epsilon = X \mod f(X)$. Then $\epsilon^2 = -1 = 2$ and therefore $\epsilon^4 = 1$. This shows that $f(X)$ is not primitive.*

The existence of primitive elements has important consequences. Let $\epsilon \in \mathbb{F}_q$ be primitive. Then $\epsilon^{q-1} = 1$. Each $0 \neq x \in \mathbb{F}_q$ has the form $x = \epsilon^i$ for some i. It follows that $x^{q-1} = \epsilon^{i(q-1)} = 1^i = 1$ and $x^q = x$ for all $x \in \mathbb{F}_q$ (including $x = 0$).

12.4 Theorem. *Each nonzero element $x \in \mathbb{F}_q$ satisfies $x^{q-1} = 1$. Each element $x \in \mathbb{F}_q$ satisfies $x^q = x$. An element $x \in \mathbb{F}_{q^r}$ is in $\mathbb{F}_q$ if and only if $x^q = x$.*

PROOF Only the last statement still needs a proof. We know already that $x^q = x$ for all $x \in \mathbb{F}_q$. Consider the polynomial $X^q - X$ as a polynomial with coefficients in $\mathbb{F}_{q^r}$. It has degree q and we know q of its roots, the elements of $\mathbb{F}_q$. A polynomial of degree q cannot have more than q roots. ∐

Let F be a finite field of characteristic p. The mapping $x \mapsto \phi(x) = x^p$ has surprising properties. It is known as the **Frobenius automorphism.** The multiplicative structure of F is respected: $\phi(xy) = \phi(x)\phi(y)$. What happens if ϕ is applied to a sum? The binomial formula

$$(x+y)^p = \sum_{i=0}^{p} \binom{p}{i} x^i y^{p-i}$$

is of course valid in all fields. The decisive observation is that the integers $\binom{p}{i}$ are multiples of p except for $i = 0$ and $i = p$. In our field of characteristic p, these coefficients are therefore equal to 0. This means that all but two terms of the sum vanish, and we obtain the simple formula

$$(x + y)^p = x^p + y^p \text{ in fields of characteristic } p.$$

Some call this the **freshman's dream.** It means that ϕ respects both the multiplicative and the additive field structure. It is clear that ϕ is a one-to-one mapping.

12.5 Proposition. *For each finite field F of characteristic p, the* **Frobenius automorphism** $\phi(x) = x^p$ *is a field automorphism. In other words, ϕ is a one-to-one mapping satisfying*

$$\phi(x + y) = \phi(x) + \phi(y) \text{ and } \phi(xy) = \phi(x)\phi(y)$$

for all $x, y \in F$.

It follows from Theorem 12.4 that ϕ acts trivially on $\mathbb{F}_p$: it maps each element of the prime field to itself.

We continue to consider a field extension $\mathbb{F}_q \subset \mathbb{F}_{q^r} = F$, where $q = p^f$. As ϕ is a field automorphism of F, the same is true of the f-th power of ϕ. This is the mapping $x \mapsto x^q$. By Theorem 12.4 it acts trivially on $\mathbb{F}_q$. Repeated application yields automorphisms $x \mapsto x^{q^i}$. Theorem 12.4 shows that, for $i = r$, the identity automorphism of F is obtained.

12.6 Definition. *Let g_i be the field automorphism of $F = \mathbb{F}_{q^r}$ defined by*

$$g_i(x) = x^{q^i}, \ i = 0, 1, \ldots r - 1.$$

Call $G = \{g_0, g_1, \ldots, g_{r-1}\}$ the **Galois group** *of $\mathbb{F}_{q^r}$ over $\mathbb{F}_q$.*

In fact, g_0 is the identity, g_i is the i-th power of g_1 (in the sense of applying g_1 i times) and $g_r = g_0$. In particular, G is closed under multiplication. It consists of the powers of g_1. The inverse of g_1 is g_{r-1}, and in general the inverse of g_i is g_{r-i}. Groups which consist of the powers of one element are known as cyclic groups. The Galois group is cyclic of order r. The term that G is the Galois group **over** $\mathbb{F}_q$ reflects the fact that each element of G maps each element of the ground field $\mathbb{F}_q$ to itself.

The Galois group allows us to go down from the extension field F to the ground field $\mathbb{F}_q$, as follows. Let $x \in F$. Consider the images of x under the elements of the Galois group and add up:

$$y = g_0(x) + g_1(x) + \cdots + g_{r-1}(x).$$

Apply g_1 to y :

$$g_1(y) = y^q = g_1(x) + g_2(x) + \cdots + g_r(x).$$

However, $g_r = g_0$, and so we are adding up the same terms as in the sum defining y. This shows $y^q = y$. By Theorem 12.4 we have $y \in \mathbb{F}_q$. The mapping $x \mapsto y$ is known as the **trace** from $F = \mathbb{F}_{q^r}$ to $\mathbb{F}_q$.

12.7 Definition. *Let* $\mathbb{F}_q \subset \mathbb{F}_{q^r} = F$ *be an extension of finite fields. The* trace

$$tr = tr_{q^r \mid q} : F \to \mathbb{F}_q$$

is defined by

$$tr(x) = x + g_1(x) + \cdots + g_{r-1}(x) = x + x^q + x^{q^2} + \cdots + x^{q^{r-1}}.$$

It follows from the definition of the trace that it is additive ($tr(x_1 + x_2) = tr(x_1) + tr(x_2)$) and $\mathbb{F}_q$-linear ($tr(\lambda x) = \lambda tr(x)$ for $\lambda \in \mathbb{F}_q$). In particular, tr is a linear mapping from the r-dimensional vector space F to the one-dimensional vector space $\mathbb{F}_q$ (linear over $\mathbb{F}_q$).

Let us check the trace in our favorite finite extension fields.

12.8 Example. *Let* $tr = tr_{4\mid 2} : \mathbb{F}_4 \to \mathbb{F}_2$. *We have* $tr(0) = tr(1) = 0$ *and* $tr(\omega) = \omega + \omega^2 = 1$, *likewise* $tr(\omega^2) = 1$.

12.9 Example. *Let* $tr = tr_{8\mid 2} : \mathbb{F}_8 \to \mathbb{F}_2$. *As always* $tr(0) = 0$, *but* $tr(1) = 3 = 1$. *Also*

$$tr(\epsilon) = \epsilon + \epsilon^2 + \epsilon^4 = \epsilon(1 + \epsilon + \epsilon^3) = \epsilon(\epsilon^5 + \epsilon^3) = \epsilon^4(1 + \epsilon^2) = 1$$

(miraculously). The elements of trace 0 are $0, \epsilon^3, \epsilon^5, \epsilon^6$; *the remaining four elements have trace 1.*

---

1. $\mathbb{F}_{p^s} \subset \mathbb{F}_{p^r}$ if and only if s divides r.

2. An element $\epsilon \in F$ is **primitive** if every nonzero element of F is a power of ϵ.

3. Every finite field has primitive elements.

4. An element of $\mathbb{F}_{q^r}$ is in $\mathbb{F}_q$ if and only if it is a root of the polynomial $X^q - X$.

5. The Frobenius automorphism $\phi : \mathbb{F}_{p^r} \to \mathbb{F}_{p^r}$ is defined by $\phi(x) = x^p$. It is a field automorphism and fixes precisely the elements of $\mathbb{F}_p$.

6. The **Galois group** G of $\mathbb{F}_{q^r}$ over $\mathbb{F}_q$ is cyclic of order r. It consists of the powers of g_1, where $g_1(x) = x^q$ is a power of the Frobenius automorphism.

7. The **trace** $tr : \mathbb{F}_{q^r} \to \mathbb{F}_q$ is defined by $tr(x) = \sum_{g \in G} g(x)$ (G is the Galois group).

8. The trace is an $\mathbb{F}_q$-linear mapping.

Exercises 12.1

12.1.1. *What is the dimension of $\mathbb{F}_{2^{12}}$ as a vector space over $\mathbb{F}_8$?*

12.1.2. *Show that every irreducible polynomial of degree 5 over $\mathbb{F}_2$ must be primitive.*

12.1.3. *Let ϵ be a primitive element of $\mathbb{F}_q$, where q is odd. How can you read off from the exponent j if ϵ^j is a square in $\mathbb{F}_q$?*
How many nonzero squares are there in $\mathbb{F}_q$?

12.1.4. *The* **norm** *N from $\mathbb{F}_{q^r}$ to $\mathbb{F}_q$ is defined by*

$$N(x) = x \cdot g_1(x) \cdot \cdots \cdot g_{r-1}(x) = \prod_{g \in G} g(x).$$

Prove that indeed $N(x) \in \mathbb{F}_q$ for every $x \in \mathbb{F}_{q^r}$.

12.1.5. *Let $N : \mathbb{F}_{q^2} \to \mathbb{F}_q$ be the norm. How many elements $x \in \mathbb{F}_{q^2}$ have $N(x) = 1$?*

12.1.6. *Show that there are exactly q^{r-1} elements in $\mathbb{F}_{q^r}$ whose trace is 0. Continue to show that this* **kernel** *$ker(tr)$ of the trace is an $(r-1)$-dimensional subspace of F when F is considered as a vector space over $\mathbb{F}_q$.*

12.1.7. *Determine the kernel $ker(tr)$ in the cases $tr : \mathbb{F}_4 \to \mathbb{F}_2$ and $tr : \mathbb{F}_8 \to \mathbb{F}_2$.*

12.1.8. *Let $tr : \mathbb{F}_{q^r} \to \mathbb{F}_q$. Under which conditions on q, r is it true that $\mathbb{F}_q \subseteq ker(tr)$? When do we have $ker(tr) = \mathbb{F}_q$?*

12.1.9. *Determine all values $tr(x)$ for $tr : \mathbb{F}_9 \to \mathbb{F}_3$.*

12.1.10. *Let $\alpha \in \mathbb{F}_{q^r}$. We studied the minimal polynomial in a series of exercises in Section 3.2.*
Prove that α and α^q have the same minimal polynomial.

12.1.11. *Prove $tr(\alpha^q) = tr(\alpha)$, where $tr : \mathbb{F}_{q^r} \to \mathbb{F}_q$.*

12.1.12. *Let $\alpha \in \mathbb{F}_{q^r}$. Show that $\mathbb{F}_q(\alpha) = \mathbb{F}_{q^r}$ if and only if the images $g(\alpha)$ under the elements $g \in G$ are pairwise different.*

12.1.13. *Why is the trace called the trace? The reason is that $tr(\alpha)$ is the trace of the $\mathbb{F}_q$-linear mapping : $\mathbb{F}_{q^r} \to \mathbb{F}_{q^r}$ defined by multiplication with α. We want to prove this in a special case.*
Let $F = \mathbb{F}_{q^r} = \mathbb{F}_q(\alpha)$. Prove that the $\mathbb{F}_q$-linear mapping : $F \to F$ defined by $x \mapsto \alpha x$ has trace $tr(\alpha)$.

12.1.14. *Show that $X^3 - X - 1$ is an irreducible polynomial over $\mathbb{F}_3$. Decide if it is primitive.*

12.1.15. *Show that $X^3 - X^2 + 1$ is an irreducible polynomial over $\mathbb{F}_3$. Use it to construct $\mathbb{F}_{27}$. Determine $ker(tr)$, where tr is the trace down to $\mathbb{F}_3$.*

12.1.16. *Prove the* **transitivity of the trace:** *if $\mathbb{F}_q \subset L = \mathbb{F}_{q^s} \subset F = \mathbb{F}_{q^r}$, then*

$$tr_{F|\mathbb{F}_q}(x) = tr_{L|\mathbb{F}_q}(tr_{F|L}(x)) \text{ for all } x \in F.$$

12.2   Trace codes and subfield codes

and Delsarte's duality theorem.

The trace gives us a method to go down from a code defined over an extension field to a code defined over the ground field. As in Section 12.1, consider the field extension $\mathbb{F}_q \subset \mathbb{F}_{q^r} = F$, with the corresponding trace tr. Let $x = (x_1, x_2, \ldots, x_n)$ be a codeword of a linear code of length n defined over F. Applying the trace in each coordinate yields the codeword $tr(x) = (tr(x_1), tr(x_2), \ldots, tr(x_n))$, still of length n but with entries in $\mathbb{F}_q$. We apply this to every codeword and obtain the trace code.

12.10 Definition. *Let $F = \mathbb{F}_{q^r}$ and C be an F-linear code of length n. Let $tr = tr : F \longrightarrow \mathbb{F}_q$ be the trace. The* **trace code** $tr(C)$ *is defined as the set of all*

$$(tr(x_1), tr(x_2), \ldots, tr(x_n)) \text{ where } (x_1, x_2, \ldots, x_n) \in C.$$

The parameter we can control is the strength:

12.11 Theorem. *Let $C \subset \mathbb{F}_{q^r}^n$ be a linear code of strength t. Then $tr(C)$ has strength t (and it may have even higher strength).*

PROOF   Fix t coordinates (say the first t coordinates) and for each $i = 1, \ldots, t$ an entry $c_i \in \mathbb{F}_q$. We have to find a codeword $x = (x_1, x_2, \ldots, x_n) \in C$ such that $tr(x_i) = c_i$ for $i = 1, \ldots, t$. For each i choose $y_i \in F$ such that $tr(y_i) = c_i$. As C has strength t, we can find $x \in C$ such that $x_i = y_i$ for

$i = 1, \ldots, t$. It follows that $tr(x)$ has the required entries c_i in coordinates $i = 1, \ldots, t$. The fact that each such tupel occurs equally often follows from $\mathbb{F}_q$-linearity. □

We will consider the trace code when we wish to control the strength. In contrast, the minimum distance of the trace code is not that easily controlled in general.

The trace code is in close relationship with another method of "going down", the subfield code.

12.12 Definition. *Let $F = \mathbb{F}_{q^r}$ and $\mathcal{C}$ be an F-linear code of length n. The* **subfield code** $\mathcal{C}_{\mathbb{F}_q}$ *consists of the codewords of $\mathcal{C}$ all of whose entries happen to be in $\mathbb{F}_q$, formally $\mathcal{C}_{\mathbb{F}_q} = \mathcal{C} \cap \mathbb{F}_q^n$.*

It is obvious that trace codes and subfield codes are linear codes over the ground field $\mathbb{F}_q$, and it is equally obvious how to control the minimum distance of the subfield code.

12.13 Theorem. *The minimum distance of the subfield code of $\mathcal{C}$ is at least as large as the minimum distance of $\mathcal{C}$ itself.*

PROOF   This does not really merit a proof. Observe simply that each word of the subfield code $\mathcal{C}_{\mathbb{F}_q}$ is also a word of $\mathcal{C}$. □

The relationship between trace code and subfield code is clarified by a famous theorem due to Delsarte, which is the main result of the present section. There is a relation of duality.

12.14 Theorem (Delsarte). *Let $F = \mathbb{F}_{q^r}$ and $\mathcal{C}$ an F-linear code of length n. Then the following holds:*

$$tr(\mathcal{C})^\perp = (\mathcal{C}^\perp)_{\mathbb{F}_q}.$$

PROOF   Let $c = (c_1, c_2, \ldots, c_n) \in \mathbb{F}_q^n$. Then $c \in tr(\mathcal{C})^\perp$ if and only if

$$\sum_{i=1}^n c_i tr(x_i) = 0 \text{ for all } x = (x_1, x_2, \ldots, x_n) \in \mathcal{C}.$$

Because of the linearity of the trace, this is equivalent to $tr(\sum_{i=1}^n c_i x_i) = 0$. We claim that this is equivalent to $\sum_{i=1}^n c_i x_i = 0$. In fact, if this was not the case, then, because of the F-linearity of $\mathcal{C}$, it would follow that tr is the 0 mapping, a contradiction. We conclude that $c \in tr(\mathcal{C})^\perp$ if and only if $c \in \mathcal{C}^\perp$ and all $c_i \in \mathbb{F}_q$. This is our claim. □

Observe that the duals are taken in different spaces, one in F^n, the other in $\mathbb{F}_q^n$. The Delsarte theorem says in words that the dual of the trace code is the subfield code of the dual. Our main objective is a study of subfield codes and trace codes of Reed-Solomon codes and codes obtained from puncturing Reed-Solomon codes. These are essentially the cyclic codes. Start from a certain (punctured) Reed-Solomon code $\mathcal{B}$ defined over F. We aim at the code $\mathcal{C} = (\mathcal{B}^\perp)_{\mathbb{F}_q}$, the subfield code of the dual. By Delsarte's theorem, $\mathcal{C}$ can also be described as the dual of the trace code. As we know the strength of $\mathcal{B}$, we know the strength of the trace code by Theorem 12.11. By duality we know the minimum distance of $\mathcal{C}$. The remaining problem is to determine the dimensions of these codes.

1. Let $\mathcal{C}$ be a linear code over $F = \mathbb{F}_{q^r}$.

2. The trace code $tr(\mathcal{C})$ is linear over $\mathbb{F}_q$ and has at least the strength of $\mathcal{C}$.

3. The subfield code $\mathcal{C}_{\mathbb{F}_q}$ consists of the codewords of $\mathcal{C}$ all of whose entries are in $\mathbb{F}_q$. It has at least the minimum distance of $\mathcal{C}$.

4. Trace code and subfield code are related by Delsarte's theorem: the dual of the trace code equals the subfield code of the dual.

Exercises 12.2

12.2.1. *Determine the trace code and the subfield code of* $\mathcal{RS}(2,4)$ *(clearly down to* $\mathbb{F}_2$*).*

12.2.2. *Determine the trace code and the subfield code of* $\mathcal{RS}(2,9)$ *(down to* $\mathbb{F}_3$*).*

12.3 Galois closed codes

Basic concepts: Galois closure, its trace code and subfield code.

It is time to bring the Galois group into play. We saw it in Section 12.1, Definition 12.6. The Galois group G consists of all powers of g_1, where g_1 is defined by $g_1(x) = x^q$. Let $F = \mathbb{F}_{q^r}$ and $\mathcal{C}$ be an F-linear code of length n. Consider the code $\mathcal{C}^q$, the image of $\mathcal{C}$ under g_1 (acting coordinatewise). It is clear that $\mathcal{C}^q = g_1(\mathcal{C})$, being an image of $\mathcal{C}$ under a field automorphism, has the same structure as $\mathcal{C}$. In particular, it has the same length, dimension, minimum distance, strength and distance distribution. Also, as $tr(x^q) = tr(x)$, where $tr : F \to \mathbb{F}_q$ is the trace, the trace codes do not only have the same structure, they are identical:
$$tr(\mathcal{C}^q) = tr(\mathcal{C}).$$
Even better, consider the code $\mathcal{C} + \mathcal{C}^q$ generated by $\mathcal{C}$ and $\mathcal{C}^q$. Its trace code is of course still $tr(\mathcal{C})$. We can go on and increase the parent code by including images of codewords under g_1. The resulting code will still have the same trace code as $\mathcal{C}$. This process will end when we have found a code which is closed under g_1 : each image of a codeword under g_1 is a codeword. We arrive at the natural notion of a Galois closed code:

12.15 Definition. *Let $F = \mathbb{F}_{q^r} \supset \mathbb{F}_q$ and $\mathcal{C}$ be an F-linear code. We call $\mathcal{C}$* **Galois closed** *if $\mathcal{C} = \mathcal{C}^q$.*

Let $\mathcal{C}$ be an F-linear code. The **Galois closure** *$\tilde{\mathcal{C}}$ of $\mathcal{C}$ is the code generated by all its images under the elements of the Galois group G, equivalently; $\tilde{\mathcal{C}}$ is the smallest Galois closed code containing $\mathcal{C}$.*

We have seen that the trace code of the Galois closure $\tilde{\mathcal{C}}$ is the same as the trace code of $\mathcal{C}$:

12.16 Theorem.
$$tr(\tilde{\mathcal{C}}) = tr(\mathcal{C}).$$

Because of Theorem 12.16, when considering the trace code we may as well replace the original parent code $\mathcal{C}$ by its Galois closure. This has its advantages, as the following important theorem shows. It is the second basic theorem on trace codes and subfield codes, together with Delsarte's Theorem 12.14.

12.17 Theorem. *Let C be a linear code over $F = \mathbb{F}_{q^r}$ and $tr : F \to \mathbb{F}_q$ the trace. Assume C is Galois closed ($C^q = C$). Then the following hold:*

- *$tr(C) = C_{\mathbb{F}_q}$.*

- *The $\mathbb{F}_q$-dimension of $tr(C)$ equals the F-dimension of C.*

- *The minimum distance of $tr(C)$ equals the minimum distance of C.*

- *The strength of $tr(C)$ equals the strength of C.*

PROOF The definition of the trace shows that $tr(C) \subseteq C_{\mathbb{F}_q}$. Exercise 12.3.6 implies $dim_{\mathbb{F}_q}(C_{\mathbb{F}_q}) \leq dim_F(C)$. The last link in the chain is provided by Theorem 12.14. If the length of C is n, then

$$dim(tr(C)) = dim(((C^{\perp})_{\mathbb{F}_q})^{\perp}) = n - dim((C^{\perp})_{\mathbb{F}_q}) \geq n - dim_F(C^{\perp})$$

$$= dim_F(C) \geq dim(C_{\mathbb{F}_q}).$$

It follows that we have equality all along.

As $tr(C)$ is a subcode of C, its minimum distance is at least as large as that of C. Let $x \in C$ of weight $w \neq 0$. After suitable scalar multiplication, we can assume that x has an entry with a nonzero trace. It follows that $tr(x)$ has nonzero weight $\leq w$. This shows the equality of the minimum distances. The equality of the strengths follows from Delsarte's theorem. ꟷ

The strategy

Let B be an F-linear code ($F = \mathbb{F}_{q^r}$), obtained by puncturing a Reed-Solomon code of dimension t. We have all the ingredients to determine the parameters of $C = tr(B)^{\perp}$, the dual of the trace code of B. By Delsarte's theorem, we can describe C alternatively as $C = (B^{\perp})_{\mathbb{F}_q}$, the subfield code of the dual of B. As B has strength t, it follows from duality that C has minimum distance $d \geq t + 1$.

Theorem 12.17 suggests how the dimension can be determined: find the Galois closure $\tilde{B}$ of the F-linear code B. By Theorem 12.17 the F-dimension of $\tilde{B}$ equals the $\mathbb{F}_q$-dimension of $tr(B)$.

The dimension of C follows by duality. In the beginning of the next chapter we consider an example for this procedure.

1. Let $\mathbb{F}_q \subset \mathbb{F}_{q^r} = F$ and $\mathcal{C}$ be an F-linear code.

2. The **Galois closure** $\tilde{\mathcal{C}}$ is the smallest code containing $\mathcal{C}$ which is mapped to itself by g_1, where $g_1(x) = x^q$.

3. Equivalently, $\tilde{\mathcal{C}}$ is the code generated by all images of $\mathcal{C}$ under the Galois group.

4. We have $tr(\tilde{\mathcal{C}}) = \tilde{\mathcal{C}}_{\mathbb{F}_q} = tr(\mathcal{C})$, and the $\mathbb{F}_q$-dimension of this code equals the F-dimension of $\tilde{\mathcal{C}}$.

5. The minimum distance and strength remain unchanged when passing from a Galois closed code to its trace code.

Exercises 12.3

12.3.1. *Determine the Galois closure of* $\mathcal{RS}(2,4)$ *(over* $\mathbb{F}_2$*).*

12.3.2. *Determine the Galois closure of* $\mathcal{RS}(2,9)$ *(over* $\mathbb{F}_3$*).*

12.3.3. *The situation is* $\mathbb{F}_q \subset F = \mathbb{F}_{q^r}$*. Prove the following general bound: if* $dim_F(\mathcal{C}) = k$*, then the Galois closure* $\tilde{\mathcal{C}}$ *has dimension at most* rk*.*

12.3.4. *Let* $\mathcal{C}$ *be Galois closed over* $F = \mathbb{F}_{q^r}$*, of minimum distance* d*. Show that each codeword of weight* d *in* $\mathcal{C}$ *is a scalar multiple of a codeword of* $tr(\mathcal{C})$*.*

12.3.5. *Let* $\mathcal{C}$ *be linear over* $F = \mathbb{F}_{q^r}$ *and* $\mathcal{U} \subset \mathcal{C}$ *a Galois closed subcode. Prove*

$$dim_{\mathbb{F}_q}(tr(\mathcal{C})) \leq r \cdot dim(\mathcal{C}) - (r-1)dim(\mathcal{U}).$$

12.3.6. *Show that codewords of the subfield code which are linearly independent over the subfield* $\mathbb{F}_q$ *are also linearly independent over the large field* $F = \mathbb{F}_{q^r}$*.*

12.3.7. *(see Giorgetti and Previtali [92]). Let* $\mathbb{F}_q \subset F = \mathbb{F}_{q^r}$*. Let* D *be an* $\mathbb{F}_q$*-linear code. Define the tensor product* $D \otimes_{\mathbb{F}_q} F$ *by extension of constants: a basis of* D *is seen as a basis of* $D \otimes_{\mathbb{F}_q} F$ *over* F*.*
Prove that the mapping $D \mapsto D \otimes_{\mathbb{F}_q} F$ *defines a one-to-one correspondence between the* $\mathbb{F}_q$*-linear codes and the Galois closed* F*-linear codes.*

12.4 Automorphism groups

Basic concepts: Equivalence under permutations, the automorphism group, monomial equivalence.

When will we consider two codes as essentially the same? Such codes will be called equivalent or also **isomorphic** ("of the same shape"). This question is closely related to the automorphism group of a code. Surprisingly, there are several different notions of equivalence. The most popular one uses only permutations of coordinates.

Equivalence under permutations

Recall that a **permutation** on n objects is a bijective (one-to-one and onto) mapping from an n-set to itself. For example, the mapping

$$\sigma : 1 \mapsto 2, \ 2 \mapsto 1, \ 3 \mapsto 4, \ 4 \mapsto 3$$

is a permutation of the set $\{1, 2, 3, 4\}$. A permutation on n objects acts in a natural way on n-tuples, by permuting the coordinates. For example, the permutation σ above maps

$$\sigma : (1, 0, 1, 0) \mapsto (0, 1, 0, 1) \text{ and } (1, 0, 1, 1) \mapsto (0, 1, 1, 1).$$

Should the entries be in $\mathbb{F}_4$, for example, we have

$$\sigma : (1, 0, \omega, 1) \mapsto (0, 1, 1, \omega) \text{ and } (\omega, 1, \overline{\omega}, 1) \mapsto (1, \omega, 1, \overline{\omega}).$$

Everybody agrees that the image of a code under a permutation of coordinates has the same structure as the code itself: $\mathcal{C}$ and $\sigma(\mathcal{C})$ have the same length, number of words, minimum distance, distance distribution, strength, you name it. We do not want to distinguish between these two codes. They are equivalent (with respect to permutations of coordinates). A formal definition would be to call code $\mathcal{C}$ equivalent to code $\mathcal{D}$ if there exists a permutation σ of the coordinates such that $\mathcal{D} = \sigma(\mathcal{C})$. It follows that $\mathcal{D}$ is equivalent to $\mathcal{C}$

as well (use the inverse permutation). Also, if C is equivalent to D, and D is equivalent to $\mathcal{E}$, then C is equivalent to $\mathcal{E}$. All codes equivalent to some code C form an **equivalence class.** The number of all permutations of an n-set is $n!$ The group S_n of all permutations on n objects is the **symmetric group.**

Such a clearly defined notion of equivalence is really helpful. As an example, if I want to decide the existence of a linear code with certain parameters $[n, k, d]_q$, I can assume the generator matrix starts with a (k, k)-unit matrix, that is, $G = (I|A)$ where A is a $(k, n - k)$-matrix. Why is that so? Assume a code with these parameters exists, with generator matrix G. There must be some k columns of G, which are linearly independent. Choose a permutation σ mapping these k coordinates into the first k coordinates. The image of my code under σ is an equivalent code with a generator matrix whose k first columns are linearly independent. After Gauß elimination, we find a generator matrix of the form $G = (I|A)$.

It should now be clear when two codes are equivalent under permutations of coordinates. A highly interesting situation occurs if we find a permutation σ such that $\sigma(C) = C$ (in words: the image of our code under σ is the code itself). In this case we call σ an **automorphism** of C. The automorphisms of C form a subgroup, the **stabilizer** or automorphism group of C. We saw a large example in Chapter 7. The generator matrix for the binary Golay code $[23, 12, 7]_2$ given there has the special feature that all rows are cyclic shifts of the first row. In other words, the cyclic permutation

$$\sigma : 1 \mapsto 2, \ 2 \mapsto 3, \ldots 22 \mapsto 23, \ 23 \mapsto 1$$

is an automorphism of $\mathcal{G}_{23}$.

It is clear that the identity permutation (fixing all coordinates) always is an automorphism. If σ is an automorphism, then the inverse permutation σ^{-1} is an automorphism. Also, if σ_1 and σ_2 both are automorphisms, then the concatenation $\sigma_2 \circ \sigma_1$ is an automorphism as well. In other words, the set of all automorphisms of a code C is a group, the automorphism group.

12.18 Definition. *The* **permutation automorphism group** *of a code* C *of length* n *is the group of all permutations* σ *of the coordinates which satisfy* $\sigma(C) = C$.

As an example, consider the following generator matrix of an octal (8-ary) four-dimensional code of length 16 :

$$\begin{pmatrix} 1\ 0\ 0\ 0 & a_1\ a_2\ a_3\ a_4 & a_1^2\ a_2^2\ a_3^2\ a_4^2 & a_1^4\ a_2^4\ a_3^4\ a_4^4 \\ 0\ 1\ 0\ 0 & a_2\ a_1\ a_4\ a_3 & a_2^2\ a_1^2\ a_4^2\ a_3^2 & a_2^4\ a_1^4\ a_4^4\ a_3^4 \\ 0\ 0\ 1\ 0 & a_3\ a_4\ a_1\ a_2 & a_3^2\ a_4^2\ a_1^2\ a_2^2 & a_3^4\ a_4^4\ a_1^4\ a_2^4 \\ 0\ 0\ 0\ 1 & a_4\ a_3\ a_2\ a_1 & a_4^2\ a_3^2\ a_2^2\ a_1^2 & a_4^4\ a_3^4\ a_2^4\ a_1^4 \end{pmatrix}$$

Here $(a_1, a_2, a_3, a_4) = (\epsilon^6, \epsilon^2, \epsilon, 1)$ and the same representation of $\mathbb{F}_8$ as in Chapter 4 is used. The parameters of this code are $[16, 4, 12]_8$, but we do not want to prove this here.

Some automorphisms are clearly visible. The permutation

$$\sigma_1 : 1 \leftrightarrow 2, \ 3 \leftrightarrow 4, \ldots 15 \leftrightarrow 16$$

maps the first row to the second and back, the third row to the fourth and back. Similarly,

$$\sigma_2 : 1 \leftrightarrow 3, \ 2 \leftrightarrow 4, \ldots 13 \leftrightarrow 15, \ 14 \leftrightarrow 16$$

is an automorphism. It becomes obvious that our notation for permutations is rather clumsy. Here is a better notation:

$$\sigma_1 = (1,2)(3,4)(5,6)(7,8)(9,10)(11,12)(13,14)(15,16)$$

$$\sigma_2 = (1,3)(2,4)(5,7)(6,8)(9,11)(10,12)(13,15)(14,16)$$

The rule is: inside each parenthesis we map to the right neighbor, and the end of the parenthesis is mapped back to the beginning. Each permutation can be written in this **cycle notation.** The concatenation $\sigma_2 \circ \sigma_1$ (apply at first σ_1, then σ_2) is also an automorphism, in cycle notation

$$\sigma_2 \circ \sigma_1 = (1,4)(2,3)(5,8)(6,7)(9,12)(10,11)(13,16)(14,15).$$

The $\circ$ can also be omitted. It is easy to check that $\sigma_2\sigma_1 = \sigma_1\sigma_2$. These four automorphisms, the identity permutation, σ_1, σ_2 and $\sigma_2\sigma_1$ form a group, a subgroup of the automorphism group.

Monomial equivalence in the linear case

There is a more general notion of equivalence for q-ary linear codes. We call it **monomial equivalence** in the linear case. The group of motions is larger: we are allowed to multiply by nonzero elements of the field in each coordinate, and we permute the coordinates. The corresponding group has order (number of elements) $(q-1)^n n!$
Each such mapping respects most of the basic properties of codes: length, number of codewords, distance distribution, strength.

12.19 Definition. *The **monomial automorphism group** (or simply the **automorphism group**) of a linear code C of length n is the group of all monomial operations σ of the coordinates which map C to C.*

So why is this larger group not always used; in other words, why is the restricted notion of equivalence based on permutations of coordinates alone so popular? Probably there are two reasons. One is that permutations of coordinates are easier to work with. General monomial operations are much harder to see. A more serious reason is that there are important invariants of linear codes which are respected by permutations of coordinates but not by monomial operations. The most important are the dimensions of trace codes and subfield codes with respect to some subfield. It is clear that permutations of coordinates do respect those, and it is easy to find examples which show that general monomial operations do not. In other words, it can happen that the subfield code of a code and the subfield code of a monomial image of the code have different dimensions. Observe that this is a problem only if q is not a prime. Also, in the binary case (linear) monomial equivalence and permutation equivalence coincide. We consider monomial equivalence and the automorphism group of Definition 12.19 the most natural notion in the case of linear codes.

General monomial equivalence

An even more general notion of monomial equivalence can be used when linearity is not an issue.

We allow an arbitrary permutation of the alphabet in each coordinate as well as an arbitrary permutation of the coordinates. The corresponding group has order $(q!)^n n!$. Clearly, it can happen that these general monomial operations map linear codes (if q is a prime-power) to nonlinear codes.

Cyclic codes

A code of length n is **cyclic** if it admits an automorphism in the group of permutations of coordinates which consists of one single n-cycle. By passing to an equivalent code it can be assumed that this automorphism is $\sigma = (1, 2, \ldots, n)$. Normally notation is chosen such that this is the case. In the next chapter we develop the theory of linear cyclic codes when n and q are coprime. We start from a different point of view. It will be easy to see that the codes we describe are cyclic. The fact that the inverse is true is not too hard to prove either. Each cyclic linear code satisfying $gcd(q, n) = 1$ is one of the codes described in the following chapter.

1. The **automorphism group** of a code of length n is its stabilizer in the **symmetric group** S_n, where S_n consists of all permutations of coordinates.

2. Two codes are **equivalent** (or **isomorphic**) if one can be obtained from the other by an application of an element of S_n.

3. The basic coding parameters are preserved by the larger **linear monomial group** (permute coordinates and multiply by arbitrary nonzero field elements in each coordinate).

4. Warning: if two codes are linearly monomially equivalent, their subfield codes can be much different.
   The same is true of their trace codes.

5. A linear code of length n is cyclic if it has a cyclic permutation of coordinates consisting of one cycle of length n as an automorphism.

Exercises 12.4

12.4.1. *Find an example of two linear quaternary codes, which are monomially equivalent but have subfield codes and trace codes (down to $\mathbb{F}_2$) of different dimension.*

12.4.2. *Find all cyclic binary codes of length 3 (with respect to the cyclic permutation $\sigma = (1, 2, 3)$).*

12.4.3. *Show that the rule $\sigma_2 \sigma_1 = \sigma_1 \sigma_2$ is* **not** *true in general for permutations σ_1, σ_2.*

12.4.4. *Can a permutation on n objects have order larger than n?*

12.4.5. *Determine the automorphism group of the extended binary Hamming code $[8, 4, 4]_2$, in particular, its order and its degree of transitivity.*

12.4.6. *Consider the extended binary Hamming code $[8, 4, 4]_2$. Count the unordered 4-sets of coordinates with the property that the corresponding columns of a generator matrix sum to 0. Prove that they form a Steiner system $S(3, 4, 8)$.*

Chapter 13

Cyclic codes

13.1 Some primitive cyclic codes of length 15

An example for the general mechanism of cyclic codes.

Our "large field" is $\mathbb{F}_{16}$, so we start with a description of this field. This is an application of the methods from Section 3.2.

The field $\mathbb{F}_{16}$

We have two choices. Either we can represent $\mathbb{F}_{16}$ as an extension of degree 4 of $\mathbb{F}_2$ (this needs an irreducible polynomial of degree 4 over $\mathbb{F}_2$) or we can describe it as an extension of degree 2 of $\mathbb{F}_4$. Let us use the former method. As an irreducible polynomial, choose $f(X) = X^4 + X + 1$. Is it really irreducible? As $f(0) = f(1) = 1$, it has no linear factors. If it is reducible, it will have to be the product of two irreducible quadratic polynomials. However, the only irreducible quadratic polynomial over $\mathbb{F}_2$ is $X^2 + X + 1$. As $f(X)$ is not $(X^2 + X + 1)^2 = X^4 + X^2 + 1$, we conclude that $f(X)$ is irreducible.

It follows that $\mathbb{F}_{16} = \mathbb{F}_2(\alpha)$, where α corresponds to $X \bmod f(X)$. This means that α satisfies the equation

$$\alpha^4 = \alpha + 1.$$

Each element of $\mathbb{F}_{16}$ is a uniquely determined linear combination of $1, \alpha, \alpha^2, \alpha^3$. Here are the elements of $\mathbb{F}_{16}$, written as powers of α and as linear combinations of $1, \alpha, \alpha^2, \alpha^3$ with coefficients in $\mathbb{F}_2$. The following table of the powers of α is obtained:

$\alpha^4 =$	$\alpha + 1$	$\alpha^5 =$	$\alpha^2 + \alpha$	$\alpha^6 =$	$\alpha^3 + \alpha^2$
$\alpha^7 = \alpha^3 + \alpha + 1$		$\alpha^8 =$	$\alpha^2 + 1$	$\alpha^9 =$	$\alpha^3 + \alpha$
$\alpha^{10} = \alpha^2 + \alpha + 1$		$\alpha^{11} = \alpha^3 + \alpha^2 + \alpha$		$\alpha^{12} = \alpha^3 + \alpha^2 + \alpha + 1$	
$\alpha^{13} = \alpha^3 + \alpha^2 + 1$		$\alpha^{14} =$	$\alpha^3 + 1$	$\alpha^{15} =$	1

In particular, we see that 15 is the smallest exponent $i > 0$ such that $\alpha^i = 1$ (15 is the **order** of α). This verifies that $f(X)$ is primitive; see Definition 12.10.

The structure of $\mathbb{F}_{16}$ has been determined. We can add any two field elements, as we have the additive representation in terms of the basis, and we can multiply, as all nonzero field elements are powers of α.

We expect $\mathbb{F}_4$ to be a subfield of $\mathbb{F}_{16}$. Where is it? It is left as Exercise 13.1.1 to show that $\mathbb{F}_4 = \{0, 1, \alpha^5, \alpha^{10}\}$. So we can choose $\omega = \alpha^5$, $\overline{\omega} = \alpha^{10}$.

Let $tr : \mathbb{F}_{16} \to \mathbb{F}_2$ be the trace. The list above shows

$$tr(\alpha) = \alpha + \alpha^2 + \alpha^4 + \alpha^8 = 0.$$

As $tr(u^2) = tr(u)$ for all $u \in \mathbb{F}_{16}$ (applying the Frobenius automorphism does not change the trace) it follows that $tr(\alpha^2) = tr(\alpha^4) = tr(\alpha^8) = 0$. The list shows also that

$$tr(\alpha^5) = 0, \ tr(\alpha^3) = tr(\alpha^7) = 1.$$

Application of the Frobenius automorphism determines the trace of each element of $\mathbb{F}_{16}$.

A code $\mathcal{B}(A)$

We decided to use $F = \mathbb{F}_{16}$ as the big field and to construct codes of length 15. The next ingredient to choose is a set of monomials. Denote by A the corresponding set of exponents. We choose $A = \{0, 1\}$, that is, the only monomials considered are $X^0 = 1$ and $X^1 = X$. A generator matrix of $\mathcal{B}(A)$ is

$j:$	14	13	12	11	10	9	8	7	6	5	4	3	2	1	0
$X^0:$	1	1	1	1	1	1	1	1	1	1	1	1	1	1	1
$X^1:$	α^{14}	α^{13}	α^{12}	α^{11}	α^{10}	α^9	α^8	α^7	α^6	α^5	α^4	α^3	α^2	α^1	1

This means that each nonzero element $u \in F$ determines a column, and each $i \in A$ determines a row of the generator matrix. The corresponding entry is u^i. In our example the columns are α^j, where $j = 0, 1, \ldots, 14$ and we have labelled the columns by j. The entry in row i and column j is α^{ij}.

The words of $\mathcal{B}(A)$ are linear combinations of the rows of the generator matrix. It follows that the codewords are indexed by polynomials $p(X)$ of degree ≤ 1; the coordinates are indexed by elements $0 \neq u \in F$, and the corresponding entry is the evaluation $p(u)$. This shows that $\mathcal{B}(A)$ is a punctured two-dimensional Reed-Solomon code. We talk about a two-dimensional RS-code, as the set of exponents is $A = \{0, 1\}$. It is punctured, as the coordinate corresponding to $u = 0$ is not present. In the general theory, arbitrary sets A of exponents are admitted, so $\mathcal{B}(A)$ will not in general be a punctured Reed-Solomon code.

It is easy to determine the Galois closure of $\mathcal{B}(A)$. We simply have to apply the Frobenius automorphism $x \mapsto x^2$ over and over again. This means we keep doubling in the exponent. As $1 \in A$ is one of the exponents, exponents $2, 4, 8$ will also be present in the Galois closure. However, this is all. Exponent 16 is not needed as $u^{16} = u^1$ for all $u \in F$. In other words, we can calculate mod 15 when working in the exponents. Let $\tilde{A} = \{0, 1, 2, 4, 8\}$ be the set of all exponents obtained from A by doubling repeatedly and working mod 15. It is clear that the Galois closure of $\mathcal{B}(A)$ is $\mathcal{B}(\tilde{A})$. Here is the generator matrix of $\mathcal{B}(\tilde{A})$:

j :	14	13	12	11	10	9	8	7	6	5	4	3	2	1	0
X^0 :	1	1	1	1	1	1	1	1	1	1	1	1	1	1	1
X^1 :	α^{14}	α^{13}	α^{12}	α^{11}	α^{10}	α^9	α^8	α^7	α^6	α^5	α^4	α^3	α^2	α^1	1
X^2 :	α^{13}	α^{11}	α^9	α^7	α^5	α^3	α^1	α^{14}	α^{12}	α^{10}	α^8	α^6	α^4	α^2	1
X^4 :	α^{11}	α^7	α^3	α^{14}	α^{10}	α^6	α^2	α^{13}	α^9	α^5	α^1	α^{12}	α^8	α^4	1
X^8 :	α^7	α^{14}	α^6	α^{13}	α^5	α^{12}	α^4	α^{11}	α^3	α^{10}	α^2	α^9	α^1	α^8	1

Now consider the trace code $tr(\mathcal{B}(\tilde{A}))$, which, by Theorem 12.16, is the same as $tr(\mathcal{B}(A))$. As the F-linear code $\mathcal{B}(\tilde{A})$ has dimension 5, it follows from Theorem 12.17 that $dim_{\mathbb{F}_2}(tr(\mathcal{B}(A))) = 5$. As $\{0, 1, 2\} \subset \tilde{A}$, it follows from Lagrange interpolation that $\mathcal{B}(\tilde{A})$ has strength 3. By Theorem 12.11 the same is true of the binary code $tr(\mathcal{B}(A))$. It follows that we know all the basic parameters of its dual $tr(\mathcal{B}(A))^\perp$. It is a $[15, 10, 4]_2$ code.

In the same way our basic properties will in general suffice to determine the parameters of $\mathcal{C}(A) = tr(\mathcal{B}(A))^\perp$ for an arbitrary set A of exponents. Observe that, while we obtain the precise dimension, we only have a lower bound on the minimum distance. Observe also that, instead of starting with A, we could have started with $\tilde{A}$ right away, as $\mathcal{C}(A) = \mathcal{C}(\tilde{A})$.

By dualization we obtain a more interesting code. Denote by $\overline{A}$ the complement of A in the set of all exponents $\{0, 1, 2, \ldots, 14\}$. In our example $\overline{A} = \{2, 3, \ldots, 14\}$ and $\tilde{\overline{A}} = \{3, 5, 6, 7, 9, 10, \ldots, 14\}$.

Let $B = -\tilde{\overline{A}} = \{1, 2, 3, 4, 5, 6, 8, 9, 10, 12\}$. Consider $\mathcal{B}(B)$. As the set B is closed under doubling, it follows that $\mathcal{B}(B)$ is Galois closed. What can we say about $\mathcal{C}(B)$? The dimension is $15 - 10 = 5$. In order to bound its minimum distance, we need to bound the strength of $\mathcal{B}(B)$. Due to the presence of the interval $1, 2, 3, 4, 5, 6$ of exponents in B, the strength of $\mathcal{B}(B)$ is at least 6. This follows from a slight generalization of Lagrange interpolation; see Exercise 4.1 in Chapter 4. We conclude that $\mathcal{C}(B)$ is a $[15, 5, 7]_2$ code. Moreover, the codes

$\mathcal{C}(A)$ and $\mathcal{C}(B)$ are duals of one another, as we will see in the next section. Observe that we know those code parameters: $[16, 5, 8]_2$ are the parameters of a Reed-Muller code.

Exercises 13.1

13.1.1. *Which powers of the primitive element $\alpha \in \mathbb{F}_{16}$ are in $\mathbb{F}_4$?*

13.1.2. *Find the parameters of the code obtained when $A = \{1, 3\}$.*

13.2   Theory of cyclic codes

Basic concepts: Cyclotomic cosets, dimension formula, BCH bound, BCH codes, the Roos bound, the van Lint-Wilson bound, parametric examples, an application in fingerprinting, generator matrices.

In this section we describe the general mechanism of cyclic codes. It is not much harder than the example given in Section 13.1. We concentrate on the parameters of these codes, leaving the questions of efficient encoding and decoding for later.

The ingredients

We work over a field $\mathbb{F}_q$. Our end products will be linear q-ary codes. $F = \mathbb{F}_{q^r}$ is an extension field of $\mathbb{F}_q$, $tr : F \to \mathbb{F}_q$ the trace, and ϵ a primitive element. This means that the nonzero elements of F are $1, \epsilon, \epsilon^2, \ldots, \epsilon^{q^r-2}$. Let n be a divisor of $q^r - 1$ and W the subgroup of order n of the multiplicative group of F. This means that W consists of all elements $\beta \in F$ satisfying $\beta^n = 1$, equivalently of all ϵ^i such that i is a multiple of $(q^r - 1)/n$. Fix notation by choosing $\alpha = \epsilon^{(q^r-1)/n}$. Then

$$W = \{1, \alpha, \alpha^2, \ldots, \alpha^{n-1}\}.$$

When calculating with exponents of α, we can calculate mod n.

Let $G = \{g_0, \ldots, g_{r-1}\}$ be the Galois group (see Definition 12.6).

13.1 Definition. *Let i be an integer mod n. The set $C_n(i)$ consists of all integers mod n which can be reached by multiplying i with a power of q. In formulas we have $j \in C_n(i)$ if there exists an exponent a such that $j \equiv iq^a \pmod{n}$. The $C_n(i)$ are called* **cyclotomic cosets.**

Observe that the cyclotomic cosets partition Z_n (the integers mod n). In practice they are easy to determine. Just keep multiplying by q and do not forget to calculate mod n. In the case $q = 2, n = 15$ considered in Section 13.1, we obtain the following cyclotomic cosets:

$$\{0\}, \ \{1, 2, 4, 8\}, \ \{3, 6, 12, 9\}, \ \{5, 10\}, \ \{7, 14, 13, 11\}.$$

The concept of cyclotomic cosets is very natural for us. We think of the integers mod n as exponents of α. The mapping $j \mapsto jq$ corresponds to applying the automorphism g_1 to α^j ($\alpha^j \mapsto \alpha^{jq}$).

The final ingredient is $A \subset Z_n$, a set of exponents of α.

13.2 Definition. *Let $A \subseteq Z_n$. The* **Galois closure** *$\tilde{A}$ is the union of all cyclotomic cosets which intersect A nontrivially. Denote by $\overline{A}$ the complement of A in Z_n.*

In the case $q = 2, n = 15, A = \{0, 1\}$, we have $\tilde{A} = C_{15}(0) \cup C_{15}(1) = \{0, 1, 2, 4, 8\}$.

Finally let B be the negative of the complement of the Galois closure of A, in our example $B = C_{15}(1) \cup C_{15}(3) \cup C_{15}(5)$. Then B is of course Galois closed (equivalently a union of cyclotomic cosets), and $|\tilde{A}| + |B| = n$. When considering $\mathcal{C}(A)$, we call A or $\tilde{A}$ the **defining set**, B the **set of zeroes.**

Cyclic codes and their parameters

Start from the F-linear code $\mathcal{B}(A)$ whose generator matrix has

- coordinates indexed by $\beta \in W$ (we will always write
  $\beta = \alpha^j$, $j = 0, 1, \ldots n - 1$ and index the coordinates by j),

- rows indexed by $i \in A$,

- entry $\beta^i = \alpha^{ij}$ in row i, column j.

Because of our preparations and examples, we can be quick on the basic facts, as they are obvious by now.

$\mathcal{B}(A)$ is an F-linear code of length n, dimension $|A|$. The Galois closure of $\mathcal{B}(A)$ is $\mathcal{B}(\tilde{A})$, of dimension $|\tilde{A}|$ and strength at least t, where t is the number of exponents $i \in \tilde{A}$ in the largest interval contained in A. In fact, we can slightly improve on this.

13.3 Definition. *An interval $I \subset Z_n$ is a subset $I = \{a, a+s, a+2s, \ldots, a+(t-1)s\}$, where s is coprime to n. In other words, an interval is an arithmetic progression with stepwidth coprime to n.*

13.4 Proposition. *Let t be the number of elements of the largest interval contained in $\tilde{A}$. Then $\mathcal{B}(\tilde{A})$ has strength at least t.*

PROOF   The Reed-Solomon codes show that $\mathcal{B}(\tilde{A})$ has strength t if $\{0, 1, \ldots, t-1\} \subseteq \tilde{A}$. Exercise 4.1 in Chapter 4 (Lagrange interpolation) shows that the same is true if any t consecutive numbers are contained in $\tilde{A}$.

Let $\beta = \alpha^s$ (recall that s is the stepwidth of the interval). Then β is a primitive element mod n, that is, we can use β in the role of α. Let $\alpha^a = \beta^{a'}$. Then

$$\{\alpha^a, \alpha^{a+s}, \ldots, \alpha^{a+(t-1)s}\} = \{\beta^{a'}, \beta^{a'+1}, \ldots, \beta^{a'+(t-1)}\}.$$

▯

We think of the property that $\mathcal{B}(A)$ has strength $|A|$ as the **interpolation property.** Proposition 13.4 is equivalent to saying that intervals have the interpolation property. It is not easy to find sets of exponents which are not intervals and yet have the interpolation property.

As $\mathcal{B}(\tilde{A})$ is Galois closed of dimension $|\tilde{A}|$ and strength t, it follows that $tr(\mathcal{B}(\tilde{A}))$ has $\mathbb{F}_q$-dimension $|\tilde{A}|$ and strength t. By duality, the $\mathbb{F}_q$-linear code $\mathcal{C}(A) = tr(\mathcal{B}(\tilde{A}))^\perp$, our principal aim, has dimension $n - |\tilde{A}| = |B|$ and minimum distance $\geq t+1$.

13.5 Theorem (BCH bound). *If $\tilde{A}$ contains an interval of size t, then $\mathcal{C}(A) = \mathcal{C}(\tilde{A})$ has minimum distance $\geq t+1$.*

One speaks of BCH codes if A is chosen as an interval.

13.6 Theorem. $\mathcal{B}(\tilde{A})^\perp = \mathcal{B}(B), \ \mathcal{C}(A)^\perp = \mathcal{C}(B)$

PROOF   It suffices to prove the statement for the $\mathcal{B}$ codes. Observe that the dimension is right. It suffices therefore to prove that $\mathcal{B}(\tilde{A})$ and $\mathcal{B}(B)$ are orthogonal. Because of linearity it suffices to show that each of our standard generators of the first code is orthogonal to each standard generator of the second. Let $i \in \tilde{A}, i' \in B, l = i + i'$. By definition of B, we have $l \neq 0 \pmod{n}$. The dot product in question is

$$S = \sum_{\beta \in W} \beta^l.$$

As W consists precisely of the field elements β satisfying $\beta^n = 1$, there must exist some $\beta_0 \in W$ such that $\beta_0^l \neq 1$. We use the same trick as in the proof of Theorem 4.6: compare $\beta_0^l S$ and S. As $\beta_0 \beta$ varies over all elements of W when β does, it follows that $\beta_0^l S = S$, or $S(\beta_0^l - 1) = 0$. As the second factor is $\neq 0$, we must have $S = 0$. □

By Delsarte's theorem and Theorem 12.17, we can write $\mathcal{C}(A)$ in various ways:

$$\mathcal{C}(A) = tr(\mathcal{B}(\tilde{A}))^\perp = \mathcal{B}(B)_{\mathbb{F}_q} = (\mathcal{B}(\tilde{A})_{\mathbb{F}_q})^\perp = tr(\mathcal{B}(B)).$$

In order to illustrate the general mechanism, consider $q = 2, n = 15$ again, with its cyclotomic cosets $C(0), C(1), C(3), C(5),\ C(7)$ (we drop the index $n = 15$ as there is no danger of confusion). As $\mathcal{C}(A) = \mathcal{C}(\tilde{A})$, there is no harm in choosing A to be a union of cyclotomic cosets. As there are five cyclotomic cosets, we have a total of $2^5 = 32$ cyclic codes $\mathcal{C}(A)$ in this case. Some extremal cases are obvious: $A = \emptyset$ yields the whole space $\mathbb{F}_2^{15}$, and $A = Z_{15}$ yields the 0-code. It is equally obvious that $\mathcal{C}(\{0\})$ is the sum zero code $[15, 14, 2]_2$ and its dual $\mathcal{C}(Z_{15} \setminus \{0\})$ is the repetition code $[15, 1, 15]_2$.

The case $A = C(0) \cup C(1) = \{0, 1, 2, 4, 8\}$ has been considered in Section 13.1. Due to the presence of the interval $\{0, 1, 2\}$ (also $\{0, 2, 4\}, \{0, 4, 8\}$, $\{1, 8, 0\}$ are intervals) this is a $[15, 10, 4]_2$ code. Its dual is $\mathcal{C}(B)$, where $B = C(1) \cup C(3) \cup C(5)$. As B contains the interval $\{1, 2, 3, 4, 5, 6\}$, we have a $[15, 5, 7]_2$ code, after parity check a $[16, 5, 8]_2$. It meets the Griesmer bound with equality.

As another example let $A = C(1) \cup C(3)$, containing the interval $\{1, 2, 3, 4\}$. It follows that $\mathcal{C}(A)$ is a $[15, 7, 5]_2$ code. After using a parity check bit, we obtain a $[16, 7, 6]_2$ code.

Parametric examples of cyclic codes

We will simply determine parameters of some cyclic codes $\mathcal{C}(A)$, using the BCH bound Theorem 13.5 as the bound on the minimum distance. Recall that the dimension of $\mathcal{C}(A)$ is $|B|$, the number of elements in Z_n in cyclotomic cosets, which are disjoint from $\tilde{A}$. Each cyclotomic coset has at most r elements.

Start from the conceptually simplest case $q = 2, n = 2^r - 1$ (binary, primitive). $A = \{1\}, \tilde{A} \supset \{1, 2\}$ yields a cyclic code $[2^r - 1, 2^r - 1 - r, 3]_2$. We recognize the parameters of the binary Hamming codes. It is in fact easy to see that we have rediscovered the binary Hamming codes from Section 2.5; see Exercise 13.2.5. The choice $A = \{1, 3\}$ yields, because of $\tilde{A} \supset \{1, 2, 3, 4\}$, a code $[2^r - 1, 2^r - 1 - 2r, 5]_2$, with a parity check bit $[2^r, 2^r - 1 - 2r, 6]_2$. This includes parameters $[16, 7, 6]_2$, $[32, 21, 6]_2$ and $[64, 51, 6]_2$.

Choose $A = \{1, 3, 5\}$. As $\tilde{A} \supset \{1, 2, 3, 4, 5, 6\}$, we obtain, after parity check, codes $[2^r, 2^r - 1 - 3r, 6r + 2]_2$. It is clear how this generalizes:

13.7 Theorem. *For every r and $a < 2^{r-1}$ there is a cyclic binary linear code of length $2^r - 1$, dimension $\geq 2^r - 1 - ar$ and minimum distance $\geq 2a + 1$.*

In the case $r = 4$ the cyclotomic coset $C_{15}(5) = \{5, 10\}$ has only length 2, so we obtain better parameters $[16, 5, 8]_2$.

Next we construct a family of ternary cyclic codes. Let $q = 3$, where r is odd, and $n = (3^r - 1)/2$. As r is odd, it follows that n is odd; see Exercise 13.2.7. Choose $A = \{1, -1\}$. Then $\tilde{A}$ contains $\{-3, -1, 1, 3\}$. As n is odd, this is an interval (see Definition 13.3).

13.8 Theorem. *The cyclic code $C(A)$ in the case $q = 3, n = (3^r - 1)/2$ (r odd), $A = \{1, -1\}$ has parameters $[(3^r - 1)/2, (3^r - 1)/2 - 2r, 5]_3$.*

The first members of this family are $[13, 7, 5]_3$, $[121, 111, 5]_3$.

Let $q = 2$, $n = 2^m + 1$, so $r = 2m$. Choose $A = \{0, 1\}$. As $2^m \equiv -1 \pmod{n}$, the interval $\{-2, -1, 0, 1, 2\}$ is contained in $\tilde{A}$. This produces binary cyclic codes

$$[2^m + 1, 2^m - 2m, 6]_2$$

for all m, in particular, $[17, 8, 6]_2$ and $[33, 22, 6]_2$.

Let $q = 3, n = (3^{2m} + 1)/2$, (so $r = 4m$) and $A = \{1\}$. As n is odd and $\{-3, -1, 1, 3\} \subset \tilde{A}$, we obtain ternary cyclic codes

$$[\frac{3^{2m} + 1}{2}, \frac{3^{2m} + 1}{2} - 4m, 5]_3 \text{ for all } m,$$

in particular, $[41, 33, 5]_3$.

More parametric examples will be given shortly, after having proven another bound on the minimum distance of cyclic codes.

An application in fingerprinting

Binary linear codes of minimum distance $d = 5$ have found an interesting application in the cryptographic problem of **fingerprinting.** For an introduction, see Boneh and Shaw [35]. In the past tables of logarithms were fingerprinted by introducing tiny errors in some randomly chosen values. In the era of electronic documents there is the danger that two owners of fingerprinted copies detect the location of the fingerprints (these are the locations where the documents differ) and make them unreadable. Let $x, y \in \mathbb{F}_2^n$ be the versions of the document. The owners will produce a document $\zeta(x, y) = z \in \{0, 1, \epsilon\}^n$ where $z_i = x_i = y_i$ when $x_i = y_i$ and $z_i = \epsilon$ when $x_i \neq y_i$. These pirates will then distribute the new document $\zeta(x,y)$, hoping these copies cannot be traced back to them. The system designer will choose the fingerprints in such a way that each pirate copy $\zeta(x,y)$ generated by collusion of two owners can be traced back to one of the owners. This leads to a variant of a famous combinatorial-number theoretic problem, as follows: interpret the binary digits $0, 1$ as natural numbers. Then knowledge of $\zeta(x,y) \in \{0, 1, \epsilon\}^n$ is equivalent to knowledge of the integer sum $x + y \in \mathbb{Z}^n$. Each fingerprint is a set of coordinates. We describe it by its characteristic function, an element of $\{0, 1\}^n$. We want to find a family of fingerprints such that each pair of different fingerprints generates a different sum. Such sets of tuples are known as **Sidon sets.**

13.9 Definition. *Let $A = \{0, 1\} \subset \mathbb{Z}$ and S_n be the maximum size of a subset $S \subset A^n$ (a Sidon set) such that $x + y = u + v$ for $x, y, u, v \in S, x \neq y, u \neq v$ implies $\{x, y\} = \{u, v\}$. Let $\sigma = \lim_{n \to \infty} \log_2(S_n)/n$.*

The best known lower bound is $\sigma \geq .5$ The construction is due to B. Lindström [135] and uses cyclic codes of minimum distance 5. Use binary cyclic codes of length $2^r - 1$, with $A = \{0, 1, 3\}$. Then $tr(\mathcal{B}(A))$, the dual of $\mathcal{C}(A)$, has dimension $2r+1$ and strength 5 (we need only strength 4). A generator matrix of $tr(\mathcal{B}(A))$ has $2r + 1$ rows and $2^r - 1$ columns. Any four of the columns are linearly independent. Let $n = 2r + 1$ and let S be the set of columns of the generator matrix, where the entries $0, 1$ are interpreted as integers. Then S is a Sidon set and $\log_2|S|/(2r + 1) \approx r/(2r + 1) \longrightarrow 0.5$ A recent improvement of the upper bound is in Cohen, Litsyn and Zémor [56]: $\sigma \leq 0.5753$.

The Roos bound

The BCH bound Theorem 13.5 is not always sufficient. There are quite a number of good cyclic codes whose minimum distance is larger than the BCH bound guarantees. In this section we prove and then apply an important special case of the Roos bound from [176]. We include it here not only in view of the applications. It is attractive also because the proof makes use of some basic linear algebra in an interesting way.

Let $0 \neq v = (v_u) \in \mathcal{C}(A)$ of weight d, where d is the minimum weight of our cyclic code. The aim is to obtain a lower bound on d. Let S be the support of v. Consider the space F^d with coordinates indexed by the $u \in S$. Define a scalar product (a symmetric bilinear form) $\langle, \rangle$ on F^d by

$$\langle x, y \rangle = \sum_{u \in S} v_u x_u y_u$$

where $x = (x_u), y = (y_u) \in F^d$. As the v_u are all nonzero, this symmetric bilinear form is nondegenerate. The situation is exactly as in the case of the dot product; see Sections 2.4 and 3.4. Bilinear forms will be studied in more generality in Section 17.2.

The idea of the Roos bound is to find two subspaces of large dimension in F^d which are orthogonal to one another with respect to $\langle , \rangle$. The sum of their dimensions will then be a lower bound on d, the dimension of the space we work in.

13.10 Theorem. *Let I_1, I_2 be intervals such that*

$$I_1 + I_2 = \{i_1 + i_2 | i_1 \in I_1, i_2 \in I_2\} \subseteq \tilde{A}.$$

Then the minimum distance d of the cyclic code $\mathcal{C}(A)$ satisfies

$$d \geq |I_1| + |I_2|.$$

PROOF The intervals $I_1 + j, j \in I_2$ and $i + I_2, i \in I_1$ show that $d > |I_1|, |I_2|$. Denote by $\mathcal{P}(I_1)$ the space of polynomials with coefficients in F all of whose terms have degrees in I_1, analogously for $\mathcal{P}(I_2)$. Clearly these spaces have dimension $|I_1|, |I_2|$, respectively. Denote by $\mathcal{B}_S(I_1)$ the evaluation at S, that is,

$$\mathcal{B}_S(I_1) = \{(p(u))_{u \in S} | p(X) \in \mathcal{P}(I_1)\},$$

analogously for I_2. We claim that $\mathcal{B}_S(I_1)$ and $\mathcal{B}_S(I_2)$ are orthogonal under $\langle , \rangle$. In fact, let $i_1 \in I_1, i_2 \in I_2, i_1 + i_2 = a$. As $a \in \tilde{A}$, we have that X^a describes an element of $\mathcal{B}(\tilde{A})$. By the definition of $\mathcal{C}(A)$ as dual of $tr(\mathcal{B}(\tilde{A}))$, we have, for every $\beta \in F$,

$$0 = \sum_{u \in W} v_u tr(\beta u^a) = tr(\beta \sum_{u \in S} v_u u^a).$$

As $\beta \in F$ is arbitrary and tr not the 0 mapping, it follows that $0 = \sum_{u \in S} v_u u^a = \sum_{u \in S} v_u u^{i_1} u^{i_2}$. This means exactly that $\mathcal{B}_S(I_1)$ and $\mathcal{B}_S(I_2)$ are orthogonal. As the dual space has complementary dimension this shows

$$d \geq dim_F(\mathcal{B}_S(I_1)) + dim_F(\mathcal{B}_S(I_2)).$$

Recall that $\mathcal{P}(I_1)$ has dimension $|I_1|$ and that $\mathcal{B}_S(I_1)$ is obtained by projection from $\mathcal{P}(I_1)$. However, this projection (evaluation) map has no kernel, as a polynomial in the kernel must be the 0 polynomial by the defining property of an interval (there is **exactly one** polynomial from $\mathcal{P}(I_1)$ having prescribed values at any $|I_1|$ coordinates). It follows that $dim_F(\mathcal{B}_S(I_1)) = |I_1|$, analogously for I_2. ∎

This was somewhat harder work than in the rest of this text. The examples will show that it was worthwhile making the effort.

Choose $q = 2, n = 2^{2m} + 1$ (so $r = 4m$), $A = \{1\}$. We have

$$\{-4, -2\} + \{0, 3, 6\} = \{-4, -2, -1, 1, 2, 4\} \subset \tilde{A}.$$

As n is coprime to 6, we have that $I_1 = \{-4, -2\}$ and $I_2 = \{0, 3, 6\}$ are intervals, so the Roos bound implies that $\mathcal{C}(A)$ has minimum distance ≥ 5.

13.11 Theorem. *The binary cyclic code* $\mathcal{C}(\{1\})$ *of length* $n = 2^{2m} + 1$ *has parameters*

$$[4^m + 1, 4^m + 1 - 4m, 5]_2.$$

In particular, codes $[18, 9, 6]_2$, $[66, 53, 6]_2$, $[258, 241, 6]_2$ are obtained. A proof from scratch is in [21]. The codes of Theorem 13.11 are known as **Zetterberg codes**.

The next construction is due to C. L. Chen [48]. Let q be a power of 2 and $n = q^2 + 1$, so $r = 4$. Let $A = \{1\}$. We have

$$\tilde{A} = C_{q^2+1}(1) = \{\pm 1, \pm q\} = \{1, q\} + \{-(q+1), 0\}.$$

As q is even, both $\{1, q\}$ and $\{-(q+1), 0\}$ are intervals. It follows from the Roos bound that we have minimum distance $d \geq 4$.

13.12 Theorem. *Let* q *be a power of 2. The* q*-ary cyclic code* $\mathcal{C}(\{1\})$ *of length* $n = q^2 + 1$ *has parameters*

$$[q^2 + 1, q^2 - 3, 4]_q.$$

The Roos theorem as proved in [176] is slightly more general than Theorem 13.10; see Exercise 13.2.10.

The van Lint-Wilson bound

Another method to obtain lower bounds on the minimum distance of cyclic codes is due to van Lint and Wilson [212]. The idea is to build large sets E of exponents such that the projection $\mathcal{B}_S(E)$ still has dimension $|E|$, which then is a lower bound on d. Use the same conventions as in the previous section. In particular, $A \subseteq \mathbb{Z}/n\mathbb{Z}$ is a set of exponents and $S \subseteq W$.

13.13 Definition. *Call* A **independent** *of* S *if* $dim(\mathcal{B}_S(A)) = |A|$ *(equivalently, the projection* $: \mathcal{P}(A) \longrightarrow \mathcal{B}_S(A)$ *has trivial kernel). A is t-**wise** **independent** if it is independent of every set S of size t. We say that A has the **interpolation property** if it is $|A|$-wise independent.*

Observe that A has the interpolation property if and only if $\mathcal{B}(A)$ is an orthogonal array of strength $|A|$ (and index $\lambda = 1$). Each interval has the interpolation property.

13.14 Lemma. *If A is independent of S, then $A + j$ is independent of S, for every j.*

PROOF   Consider the matrix with rows indexed by $i \in A$, columns indexed by $u \in S$ and entries u^i. The fact that A and S are independent is equivalent to this matrix having rank $|A|$ (the rows are independent). Replacing A by $A + j$ has the effect of multiplying column u by u^j. The rank remains unchanged. ⬜

Choose S to be the support of a codeword of $\mathcal{C}(A)$ of minimum weight d, so $|S| = d$. Let $I \subseteq \mathbb{Z}/n\mathbb{Z}$.

13.15 Lemma. *If I is contained in an interval $J \subseteq \tilde{A}$ of length at most d, then I is independent with respect to S.*

PROOF   As J is an interval, the projection from $\mathcal{P}(J)$ to $\mathcal{B}_S(J)$ has trivial kernel. If we restrict to the subspace $\mathcal{P}(I)$, this is still true. ⬜

The following recursive constructions are used to find large independent sets (with respect to any nonzero codeword of the cyclic code). If I is independent with respect to every support S, then $|I|$ is a lower bound on the minimum distance d of $\mathcal{C}(A)$. If, moreover, $I \subseteq \tilde{A}$, then $d \geq |I| + 1$. This follows from the fact that $\mathcal{B}_S(I)$ is contained in a proper subspace of F^S in this case. The recursive construction is based on the following two lemmas.

13.16 Lemma (shifting).

$$dim(\mathcal{B}_S(I)) = dim(\mathcal{B}_S(I + j))$$

for every S, I and j. Here $I + j = \{i + j \mid i \in I\}$.
In particular: if I is independent with respect to S, then $I + j$ has that same property.

PROOF   The mapping $p(X) \mapsto X^j p(X)$ induces an isomorphism between $\mathcal{B}_S(I)$ and $\mathcal{B}_S(I + j)$ for every S. ⬜

13.17 Lemma. *Let $I \subseteq \tilde{A}, j \notin \tilde{A}$. If S is the support of a codeword of $\mathcal{C}(A)$ but not the support of a codeword in $\mathcal{C}(A \cup \{j\})$, then*

$$dim(\mathcal{B}_S(I \cup \{j\})) = 1 + dim(\mathcal{B}_S(I)).$$

In particular, if I is independent with respect to S, then so is $I \cup \{j\}$.

PROOF   Clearly $dim(\mathcal{B}_S(I \cup \{j\})) \leq 1 + dim(\mathcal{B}_S(I))$. The tuple of $\mathcal{B}_S(I \cup \{j\})$ defined by polynomial X^j is not orthogonal (with respect to the dot product) to codeword $v \in \mathcal{C}(A)$ with support S. This follows from the assumption of the lemma. As the elements of $\mathcal{B}_S(I)$ are orthogonal to v, we see that the dimension does increase. $\square$

As an illustration, consider the binary Golay code, the binary cyclic code of length 23 with defining set $A = \{1\}$. The cyclotomic cosets in the case $n = 23$ are $C(0) = \{0\}$,

$$C(1) = \{1, 2, 3, 4, 6, 8, 9, 12, 13, 16, 18\} \text{ and } C(-1) = -C(1).$$

The dimension is $23 - 11 = 12$. We want to show that $\mathcal{C}(A)$ has minimum distance ≥ 7. Observe that, for every $j \neq 0, j \notin A$, we have that $\mathcal{C}(A \cup \{j\})$ is the repetition code of minimum distance 23 (this follows formally from the BCH bound, as this code has as defining set an interval of length 22). It follows that one condition of Lemma 13.17 will always be satisfied.

Let S be the support of a codeword of weight d. Start from the interval $\{1, 2\} \subset C(1)$, which shows $d \geq 3$ by the BCH bound. We have $5 \notin C(1)$. Lemma 13.17 shows that $\{1, 2, 5\}$ is independent of S. The shift lemma shows that $\{8, 9, 12\}$ is independent of S as well. As $\{8, 9, 12\} \subset C(1)$, we have $d \geq 4$. We continue in this way, using alternately Lemma 13.17 and Lemma 13.16:

$$\{8, 9, 12\} \longrightarrow \{8, 9, 12, 14\} \longrightarrow \{12, 13, 16, 18\} \longrightarrow$$

$$\{12, 13, 16, 18, 5\} \longrightarrow \{2, 3, 6, 8, 18\} \longrightarrow \{2, 3, 6, 8, 18, 5\}.$$

This shows $d \geq 6$. Assume $d = 6$. A codeword v of weight 6 is in the sum zero subcode $\mathcal{C}(A \cup \{0\})$. Application of the method to this code yields

$$\{2, 3, 6, 8, 18, 5\} \longrightarrow \{0, 1, 4, 6, 16, 3\} \subset C(0) \cup C(1)$$

which shows that $\mathcal{C}(\{0, 1\})$ has $d > 6$, a contradiction.

We have shown that $\mathcal{C}(\{1\})$ is a $[23, 12, 7]_2$ code. This is the famous binary Golay code, which we constructed in Chapter 7 without any theoretical backing. This application is from the original paper [212].

The following family of cyclic codes of minimum distance 6 was described by Danev and Olsson [63]. Let $q \geq 4$ and $n = q^2 - q + 1$. As $q^3 + 1 = (q+1)(q^2 - q + 1)$ and $(q^3 + 1) \mid q^6 - 1$, we have $F = \mathbb{F}_{q^6}$ and consequently all cyclotomic cosets have as lengths divisors of 6. We use the cyclotomic cosets

$$C(1) = \pm\{1, q - 1, q\} \text{ and } C(q - 2) = \pm\{q - 2, q + 1, 2q - 1\}.$$

Consider

$$\mathcal{C} = \mathcal{C}(\{0\} \cup C(1)) \supset \mathcal{D} = \mathcal{C}(\{0\} \cup C(1) \cup C(q - 2)).$$

As $\{0\} \cup C(1) \cup C(q - 2) \supset \{-1, 0, 1\} + \{-(q - 1), 0, q - 1\}$ contains the sum of two intervals of length 3, it follows from the Roos bound Theorem 13.10

that $\mathcal{D}$ has minimum distance ≥ 6. We want to show that the same is true of the $(n-7)$-dimensional cyclic code $\mathcal{C}$. Let d be the minimum distance of $\mathcal{C}$ and assume $d < 6$. In each application of Lemma 13.17 we choose $j \in C(q-2)$. The condition on j from Lemma 13.17 is then satisfied. We start with $\{q-1, q\} \longrightarrow \{q-1, q, 2q-1\}$ and continue

$$\{0, 1, q\} \longrightarrow \{0, 1, q, q+1\} \longrightarrow \{-1, 0, q-1, q\} \longrightarrow$$

$$\longrightarrow \{-1, 0, q-1, q, 2q-1\} \longrightarrow \{-q, -(q-1), 0, 1, q\}.$$

The resulting independent set of size 5 is contained in $\{0\} \cup C(1)$. We have proved the following:

13.18 Theorem. *Let $q \geq 4$. The cyclic q-ary code of length $q^2 - q + 1$ with $A = \{0, 1\}$ is a $[q^2 - q + 1, q^2 - q - 6, 6]_q$-code.*

Short examples are $[13, 6, 6]_4$, $[21, 14, 6]_5$, $[43, 36, 6]_7$.

Generator matrices of cyclic codes

In order to be able to work with cyclic codes, we have to determine generator matrices for them. It turns out that we can find an almost canonical form.

Recall the situation: we have $n | q^r - 1$, $F = \mathbb{F}_q(\epsilon)$, where ϵ is a primitive element of F. Further, W consists of all elements whose n-th power is 1 and $W = \{1, \alpha, \alpha^2, \ldots, \alpha^{n-1}\}$. Our q-ary cyclic code is $\mathcal{C}(A)$. Its dimension is $|B|$, where $-B$ is the union of the cyclotomic cosets, which are disjoint from $\tilde{A}$. Let C be a cyclotomic coset disjoint from $\tilde{A}$. Then $tr(\mathcal{B}(-C)) \subset \mathcal{C}(A)$ and $\mathcal{C}(A)$ is the direct sum of all those codes. In other words, $\mathcal{C}(A)$ is the direct sum of the codes $tr(\mathcal{B}(C))$, where C varies over the cyclotomic cosets contained in B. Each such code has dimension $|C|$. We have $tr(\mathcal{B}(C)) = \mathcal{B}(C)_{\mathbb{F}_q}$, but the expression using the trace is handier: it may be hard to find a systematic way to write down the words of an F-linear code all of whose entries happen to be in the subfield $\mathbb{F}_q$, whereas there is no problem in applying the trace mapping.

In particular, each cyclotomic coset $C, C \cap \tilde{A} = \emptyset$ contributes $|C|$ rows to a generator matrix of $\mathcal{C}(A)$.

To fix notation, denote $v(l, i) \in \mathbb{F}_q^n$ the vector with entry $tr(\epsilon^l \beta^i)$ in coordinate $\beta \in W$. If $i \in C$, then $v(l, i) \in tr(\mathcal{B}(C))$. Observe that $v(l, i)$ is the codeword in $tr(\mathcal{B}(C))$ corresponding to the monomial $\epsilon^l X^i$. As $1, \epsilon, \ldots, \epsilon^{r-1}$ is a basis of F over $\mathbb{F}_q$, we can choose $0 \leq l \leq r - 1$.

Let C be a cyclotomic coset, $C \cap \tilde{A} = \emptyset$. Choose $i \in C$. In the generic case when $|C| = r$, there is no problem: the words $v(0, i)$, $v(1, i), \ldots v(r-1, i)$ form a basis of $tr(\mathcal{B}(C))$. When $|C| < r$, these vectors still generate $tr(\mathcal{B}(C))$.

We will have to choose $|C|$ linearly independent among these vectors. If this is done for each cyclotomic coset avoiding $\tilde{A}$, the corresponding words $v(l, -i)$ form the rows of a generator matrix.

13.19 Theorem. *Let $v(l, i) \in \mathbb{F}_q^n$ be the word with entry $tr(\epsilon^l \beta^i)$ in coordinate $\beta \in W$.*

For each cyclotomic coset C, choose a representative $i \in C$ and a set $S(i) \subseteq \{0, 1, \ldots, r-1\}$ of $|C|$ indices l such that the $v(l, i)$, $l \in S(i)$ are linearly independent. The $v(l, i)$, $l \in S(i)$ are a basis of $tr(\mathcal{B}(C))$.

The $v(l, -i)$, $l \in S(-i)$ where i varies over representatives of cyclotomic cosets C avoiding $\tilde{A}$, form the rows of a generator matrix of $\mathcal{C}(A)$.

Again, the generator matrices as described in Theorem 13.19 are close to being canonically determined. There is the choice of the representatives of cyclotomic cosets, but this does not really matter. When C has full length r, and this is the generic case, we have $S(i) = \{0, 1, \ldots, r-1\}$. A difficulty arises only when C is shorter.

Check matrices of cyclic codes

It is important to have a concrete representation of a check matrix H of $\mathcal{C}(A)$, equivalently a generator matrix of $tr(\mathcal{B}(A)) = tr(\mathcal{B}(\tilde{A}))$. Strictly speaking, we know this standard form, as the dual of a cyclic code is cyclic, so H is a generator matrix of a cyclic code and we can apply Theorem 13.19. We are redoing the work of the previous section from a slightly different perspective. The main difference is that we view the entries in each section corresponding to a cyclotomic coset of length s not as s-tuples but as elements of $\mathbb{F}_{q^s}$.

The Galois closure $\tilde{A}$ is a union of cyclotomic cosets. Let R be a set of representatives for the cyclotomic cosets contained in $\tilde{A}$. We know that $tr(\mathcal{B}(A))$ is the direct sum of the $tr(\mathcal{B}(C_n(i)))$, where $i \in R$, and $dim(tr(\mathcal{B}(C(i)))) = |C_n(i)|$; in other words, each cyclotomic coset $C_n(i) \subseteq \tilde{A}$ contributes $s = |C_n(i)|$ rows of H. It is proved in Exercise 13.2.9 that s can also be described as the degree of $\mathbb{F}_q(\alpha^i)$ over $\mathbb{F}_q$, equivalently $\mathbb{F}_q(\alpha^i) = \mathbb{F}_{q^s}$. In particular, $\alpha^i \in \mathbb{F}_{q^s}$. It is now obvious how a check matrix H of $\mathcal{C}(A)$ can be described:

1. The columns of H are indexed by $\beta = \alpha^j \in W$, $j = 0, 1, \ldots, n-1$.

2. Choose representatives $i \in R$ for the cyclotomic cosets $C_n(i) \subseteq \tilde{A}$. Let $s = s(i) = |C_n(i)|$. Each $i \in R$ contributes a section of s rows to H.

3. For each $\mathbb{F}_{q^s} \subseteq F$, choose a basis over $\mathbb{F}_q$.

4. The section of column $\beta \in W$ corresponding to $i \in R$ is β^i, where $\beta^i \in \mathbb{F}_{q^s}$ is represented by a column vector of length s, whose entries are the coefficients of the representation of β^i in terms of the fixed basis.

A binary example

Consider the primitive binary BCH-code $\mathcal{C}(A)$ of length 15 and $A = \{1, 2, \ldots, 6\}$ again (of designed distance 7). This is one of the pair of dual codes considered in Section 13.1. The cyclotomic cosets avoiding $\tilde{A}$ are $\{0\}$ and $C(7) = \{7, 14, 13, 11\}$. We have $-C(7) = C(1)$, and clearly we choose representatives 0 from $C(0)$ (no choice here) and 1 from $C(1)$. The rows of a generator matrix of $\mathcal{C}(A)$ are $v(3, 0), v(0, 1), v(1, 1), v(2, 1), v(3, 1)$. There is no choice for the last four of these rows, as $C(1)$ has length $r = 4$, hence $S(1) = \{0, 1, 2, 3\}$ in the terminology of Theorem 13.19. Also $\{0\}$ contributes only one row. We will always choose it to be the all-1 word. The following generator matrix is obtained:

$index$:	14	13	12	11	10	9	8	7	6	5	4	3	2	1	0
$v(3, 0)$:	1	1	1	1	1	1	1	1	1	1	1	1	1	1	1
$v(0, 1)$:	1	1	1	1	0	1	0	1	1	0	0	1	0	0	0
$v(1, 1)$:	0	1	1	1	1	0	1	0	1	1	0	0	1	0	0
$v(2, 1)$:	0	0	1	1	1	1	0	1	0	1	1	0	0	1	0
$v(3, 1)$:	0	0	0	1	1	1	1	1	0	1	0	1	1	0	0	1

Here the column index j corresponds to $\alpha^j \in W = \mathbb{F}_{16}^*$. Let us check that we understand how this matrix was calculated. Rows $v(l, 1), l = 0, 1, 2, 3$ correspond to the monomials $\alpha^l X$. Consider the column indexed by 6, say. The corresponding entries are $tr(\alpha^l \alpha^6), l = 0, 1, 2, 3$, concretely $tr(\alpha^6) = 1, tr(\alpha^7) = 1, tr(\alpha^8) = 0$ and $tr(\alpha^9) = 1$. This explains the entries in column $j = 6$. We know that $\mathcal{C}(A)$ is a $[15, 5, 7]_2$ code.

A ternary example

Consider the case $q = 3, r = 4, n = 20$. As 20 divides $3^4 - 1$, we have $r = 4$, so the field $F = \mathbb{F}_{81}$ needs to be described before we can work with ternary cyclic codes of length 20.

We choose $X^4 - X - 1$ as an irreducible polynomial over $\mathbb{F}_3$. This leads to a description $\mathbb{F}_{81} = \mathbb{F}_3(\epsilon)$, where $\epsilon^4 = \epsilon + 1$. Here are the elements of the field:

$1 =$		1	$\epsilon^{21} = -\epsilon^3 -\epsilon^2 +\epsilon +1$
$\epsilon =$		ϵ	$\epsilon^{22} = -\epsilon^3 +\epsilon^2 \quad -1$
$\epsilon^2 =$	ϵ^2		$\epsilon^{23} = \epsilon^3 \quad +\epsilon -1$
$\epsilon^3 = \epsilon^3$			$\epsilon^{24} = \quad \epsilon^2 \quad +1$
$\epsilon^4 =$		$\epsilon +1$	$\epsilon^{25} = \epsilon^3 \quad +\epsilon$
$\epsilon^5 =$	$\epsilon^2 +\epsilon$		$\epsilon^{26} = \quad \epsilon^2 +\epsilon +1$
$\epsilon^6 = \epsilon^3 +\epsilon^2$			$\epsilon^{27} = \epsilon^3 +\epsilon^2 +\epsilon$
$\epsilon^7 = \epsilon^3$		$+\epsilon +1$	$\epsilon^{28} = \epsilon^3 +\epsilon^2 +\epsilon +1$
$\epsilon^8 =$	$\epsilon^2 -\epsilon +1$		$\epsilon^{29} = \epsilon^3 +\epsilon^2 -\epsilon +1$
$\epsilon^9 = \epsilon^3 -\epsilon^2 +\epsilon$			$\epsilon^{30} = \epsilon^3 -\epsilon^2 -\epsilon +1$
$\epsilon^{10} = -\epsilon^3 +\epsilon^2 +\epsilon +1$			$\epsilon^{31} = -\epsilon^3 -\epsilon^2 -\epsilon +1$
$\epsilon^{11} = \epsilon^3 +\epsilon^2 \quad -1$			$\epsilon^{32} = -\epsilon^3 -\epsilon^2 \quad -1$
$\epsilon^{12} = -\epsilon^3 \quad +1$			$\epsilon^{33} = -\epsilon^3 \quad +\epsilon -1$
$\epsilon^{13} =$	$-\epsilon +1$		$\epsilon^{34} = \quad \epsilon^2 +\epsilon -1$
$\epsilon^{14} =$	$-\epsilon^2 +\epsilon$		$\epsilon^{35} = \epsilon^3 +\epsilon^2 -\epsilon$
$\epsilon^{15} = -\epsilon^3 +\epsilon^2$			$\epsilon^{36} = \epsilon^3 -\epsilon^2 +\epsilon +1$
$\epsilon^{16} = -\epsilon^3$		$-\epsilon -1$	$\epsilon^{37} = -\epsilon^3 +\epsilon^2 -\epsilon +1$
$\epsilon^{17} =$	$-\epsilon^2 \quad +1$		$\epsilon^{38} = \epsilon^3 -\epsilon^2 \quad -1$
$\epsilon^{18} = -\epsilon^3$	$+\epsilon$		$\epsilon^{39} = -\epsilon^3 \quad +1$
$\epsilon^{19} =$	$\epsilon^2 -\epsilon -1$		$\epsilon^{40} = \quad -1$
$\epsilon^{20} = \epsilon^3 -\epsilon^2 -\epsilon$			

For exponents higher than 40, use $\epsilon^{40+i} = -\epsilon^i$. In particular, ϵ is a primitive element.

The cyclotomic cosets in case $q = 3, n = 20$ are $\{0\}$, $\{10\}$ and

$$\{1, 3, 9, 7\}, \quad \{2, 6, 18, 14\}, \quad \{4, 12, 16, 8\}, \quad \{5, 15\}, \quad \{11, 13, 19, 17\}.$$

Choose $A = \{1, \ldots, 10\}$. The cyclotomic cosets avoiding $\tilde{A}$ are $C(0)$ and $C(19)$, so $\mathcal{C}(A)$ is a $[20, 5, 11]_3$ code. We can choose as rows of a generator matrix G the words $v(0, 0)$ and $v(j, 1), j = 0, 1, 2, 3$. In order to write out the generator matrix, we just need to determine the trace function. As $tr(x) = tr(x^3)$ and $tr(-x) = -tr(x)$, it suffices to calculate $tr(\epsilon^i)$, where i varies over representatives < 40 of the cyclotomic cosets $C_{80}(i)$. We have

$tr(0) = tr(\epsilon) = tr(\epsilon^2) = tr(\epsilon^5) = tr(\epsilon^{14}) = tr(\epsilon^{20}) = tr(\epsilon^{25}) = 0$
$tr(1) = tr(\epsilon^4) = tr(\epsilon^7) = tr(\epsilon^8) = tr(\epsilon^{10})$
$= tr(\epsilon^{13}) = tr(\epsilon^{17}) = tr(\epsilon^{26}) = 1$
$tr(\epsilon^{11}) = tr(\epsilon^{16}) = tr(\epsilon^{22}) = tr(\epsilon^{23}) = -1$

Choose $\alpha = \epsilon^4$ as the generator of W. The following generator matrix for $\mathcal{C}(A)$ is obtained:

index :	19	18	17	16	15	14	13	12	11	10
$v(0,0)$:	1	1	1	1	1	1	1	1	1	1
$v(0,1)$:	2	1	2	2	0	1	2	2	2	2
$v(1,1)$:	2	1	2	0	2	2	2	0	0	0
$v(2,1)$:	1	1	2	2	1	0	0	2	0	0
$v(3,1)$:	2	0	1	0	1	1	0	1	2	0

index :	9	8	7	6	5	4	3	2	1	0
$v(0,0)$:	1	1	1	1	1	1	1	1	1	1
$v(0,1)$:	1	2	1	1	0	2	1	1	1	1
$v(1,1)$:	1	2	1	0	1	1	1	0	0	0
$v(2,1)$:	2	2	1	1	2	0	0	1	0	0
$v(3,1)$:	1	0	2	0	2	2	0	2	1	0

The columns are coordinatized by the numbers $j \in \{0 \ldots, 19\}$ corresponding to the elements α^j of W. The entry of $v(l,1)$ in coordinate j is $tr(\epsilon^{4j+l})$.

The code generated by the four last rows is $\mathcal{C}(\{0 \ldots, 10\})$, of designed distance 12. Each $(v_j) \in \mathcal{C}(\{0, \ldots, 10\})$ satisfies the symmetry $v_{i+10} = -v_i$. In particular, all the weights of $\mathcal{C}(\{0, \ldots, 10\})$ are even. An explanation for this phenomenon will be given in Section 13.4.

1. The ingredients: ground field $\mathbb{F}_q$, extension field $F = \mathbb{F}_{q^r}$, divisor $n \mid (q^r - 1)$, the multiplicative subgroup $W = \langle \alpha \rangle$ of order n in F^*, defining set A of exponents (mod n).

2. The Galois group G acts on the exponents (integers mod n) by multiplication with q. The orbits are cyclotomic cosets.

3. The Galois closure $\tilde{A}$ is the union of all cyclotomic cosets nontrivially intersected by A.

4. $\mathcal{B}(A)$ is an $[n, |A|]_F$-code with basis: the monomials $X^i, i \in A$, evaluated at the elements of W.

5. The cyclic code $\mathcal{C}(A) = \mathcal{C}(\tilde{A}) = tr(\mathcal{B}(A))^\perp$ is an $[n, n - |\tilde{A}|, t + 1]_q$ code, where t is the size of the largest interval contained in $\tilde{A}$ (BCH bound).

6. The dual of $\mathcal{C}(A)$ has defining set $B = -\overline{\tilde{A}}$.

7. A special case of the Roos bound: if I_1, I_2 are intervals, $I_1 + I_2 \subseteq \tilde{A}$, then $\mathcal{C}(A)$ has $d \geq |I_1| + |I_2|$.

8. There is an essentially canonically determined generator matrix of $\mathcal{C}(A)$.

9. There is an application of binary primitive cyclic codes of minimum distance 5 to fingerprinting.

Exercises 13.2

13.2.1. *Determine the cyclotomic cosets in the case $q = 2, n = 31$.*

13.2.2. *Determine the parameters of binary cyclic codes of length 31 in cases $A = \{1, 3, 5\}$ and $A = \{-1, 0, 1\}$.*

13.2.3. *Determine the parameters of the binary cyclic code of length 63 with $A = \{1, 3, 5, 7, 9\}$.*

13.2.4. *Determine the cyclotomic cosets in the case $q = 3, n = 26$.*

13.2.5. *Show that the binary primitive cyclic code (length $2^r - 1$) with defining set $A = \{1\}$ is equivalent to the binary Hamming code.*

13.2.6. *Find a cyclic code C with parameters $[26, 8, 13]_3$, containing two cyclic codes D_1, D_2 of parameters $[26, 7, 14]_3$ such that $D_1 \cap D_2$ is a cyclic $[26, 6, 15]_3$-code.*

13.2.7. *Prove that, if r is odd, then $(3^r - 1)/2$ is odd.*

13.2.8. *Construct a generator matrix of a cyclic $[13, 7, 5]_3$ code.*

13.2.9. *Let $W = \langle \alpha \rangle$ the subgroup of order n of F^* and $C(i)$ a cyclotomic coset mod n. Prove that the length $|C(i)| = s$ is determined by $\mathbb{F}_q(\alpha^i) = \mathbb{F}_{q^s}$, the degree of the field generated by α^i.*

13.2.10. *Prove the Roos bound for cyclic codes in its general form: Let I be an interval and J contained in an interval J_0 such that $|J_0| - |J| < |I|$ and $I + J \subseteq \tilde{A}$. Then $d(\mathcal{C}(A)) \geq |I| + |J|$.*

13.2.11. *Determine the binary cyclic codes of length 9 and their parameters.*

13.2.12. *Show that the maximum minimum distance of a $[17, 9, d]_2$ code is $d = 5$.*

13.2.13. *Determine the binary cyclic codes of length 17 and their parameters.*

13.2.14. *Show that all three parameters of $[13, 7, 5]_3$ are optimal.*

13.2.15. *Show that the $[q^2 - q + 1, q^2 - q - 6, 6]_q$ codes of Theorem 13.18 have optimal d for $q \geq 5$. In fact this is true also for $q = 4$. We will prove the nonexistence of $[13, 6, 7]_4$ in Exercise 17.1.9, using geometric methods.*

13.3   Decoding BCH codes

---

using the Euclidean algorithm.

The Euclidean algorithm

As everybody knows the Euclidean algorithm is a fast method to compute the **greatest common divisor** (a, b) of two elements a, b of a Euclidean domain. The Euclidean domain R comes with a norm N and the basis of the Euclidean algorithm is division with remainder

$$a = q \cdot b + r,$$

where either $r = 0$ or $N(r) < N(b)$. The classical case is $R = \mathbb{Z}$ with norm $N(x) = |x|$. As an example, we determine $(910, 143)$

$$
\begin{array}{|l}
910 = 6 \cdot 143 + 52 \\
143 = 2 \cdot 52 + 39 \\
52 = 1 \cdot 39 + 13 \\
39 = 3 \cdot 13 + 0
\end{array}
$$

As soon as the remainder becomes 0, we read off the gcd in the preceding row, in the example

$$(910, 143) = (143, 52) = (52, 39) = (39, 13) = (13, 0) = 13.$$

Row k reads $r_k = q_k \cdot r_{k-1} + r_{k-2}$, with r_k in the first column ($r_{-1} = a = 910$, $r_0 = b = 143$, $r_1 = 52$, $r_2 = 39$). The second column contains the quotients: $q_{-1} = 6$, $q_0 = 2$, $q_1 = 1$, $q_2 = 3$.

An important consequence of the Euclidean algorithm, from both a practical and a theoretical point of view, is the fact that (a, b) can be written as a linear combination of a and b :

$$(a, b) = u \cdot a + v \cdot b.$$

In our example we start from the penultimate row. It has the gcd as the last entry. Solving for it yields $13 = 1 \cdot 52 - 1 \cdot 39$, an expression of 13 as a linear combination of 52 and 39. Then we work our way up. The previous row expresses 39 as a linear combination of 143 and 52. Substituting this into our expression, we obtain $13 = 3 \cdot 52 - 1 \cdot 143$. Proceeding like this once more: $13 = 3 \cdot 910 - 19 \cdot 143$. This is the desired expression $(910, 143) = u \cdot 910 + v \cdot 143$, where $u = 3$, $v = -19$.

We want to generate these data by the algorithm, without having to go back. That is why we introduce two more columns, with entries u_k, v_k such that $r_k = u_k \cdot a + v_k \cdot b$. The initialization is easy. The first two rows with indices -1 and 0 are

k	Q	R	U	V
-1		a	1	0
0		b	0	1

Row $k + 1$ will be calculated from the two preceding rows. So assume rows $k - 1$ and k are given, as follows:

Q	R	U	V
q_{k-1}	r_{k-1}	u_{k-1}	v_{k-1}
q_k	r_k	u_k	v_k

We apply long division to the two preceding remainders:
$r_{k-1} = q_{k+1} \cdot r_k + r_{k+1}$. This gives us the new remainder r_{k+1} and the new quotient q_{k+1}. Moreover, we have $r_k = u_k \cdot a + v_k \cdot b$ and $r_{k-1} = u_{k-1} \cdot a + v_{k-1} \cdot b$. This yields $r_{k+1} = r_{k-1} - q_{k+1} \cdot r_k = (u_{k-1} - q_{k+1} u_k) \cdot a + (v_{k-1} - q_{k+1} v_k) \cdot b$. We obtain the recursions

$$u_{k+1} = u_{k-1} - q_{k+1} u_k \text{ and } v_{k+1} = v_{k-1} - q_{k+1} v_k.$$

Here is the scheme for our example:

Q	R	U	V
	910	1	0
	143	0	1
6	52	1	-6
2	39	-2	13
1	13	3	-19
3	0	-11	70

As soon as 0 appears in column R, we know that the gcd is in the preceding row. There is more information in the complete scheme:

13.20 Theorem (Cross-product theorem). *In the complete Euclidean scheme for the calculation of the greatest common divisor (a, b), the following relations hold:*

- $r_{k-1} u_k - r_k u_{k-1} = \pm b$

- $r_{k-1} v_k - r_k v_{k-1} = \pm a$

- $u_{k-1} v_k - u_k v_{k-1} = \pm 1$

PROOF   These elementary statements are proved by induction. We consider only the first case: $r_k u_{k+1} - r_{k+1} u_k = r_k(u_{k-1} - q_{k+1} u_k) - (r_{k-1} - q_{k+1} r_k) u_k = r_k u_{k-1} - r_{k-1} u_k$, which is the negative of the cross-product one step earlier. It suffices to check the cross-product for the two initial rows, and this is $-b$. □

The third statement of Theorem 13.20 shows that $(u_k, v_k) = 1$ for all k. Consider the last row (row j, say) of the complete scheme, and put $u = u_j$, $v = v_j$. We have $r_j = 0 = u \cdot a + v \cdot b$. It follows that $v/u = -a/b$. As $(u, v) = 1$, we have that v/u is the reduced form of $-a/b$.

13.21 Lemma. *The norms of the elements in the R-column are strictly decreasing starting from row 0.*
The norms of the elements in the U column and in the V column are strictly increasing starting from row 1.

PROOF The statement concerning row R is obvious from the way the algorithm was defined (division with remainder). Consider the recursion $u_{k+1} = u_{k-1} - q_{k+1}u_k$. Proceed by induction. Assume $N(u_{k-1}) \leq N(u_k)$. As $N(q_{k+1}) > 0$, it follows that $N(u_{k+1}) = N(q_{k+1}u_k) > N(u_k)$. The same argument holds for the V column. □

Decoding BCH codes

We consider the q-ary BCH code of length n, where n divides $q^r - 1$, with designed distance $2e + 1$ (although the term *designed distance* may not be appropriate in cases when one of our lower bounds gives a larger value), with defining set $A = \{l, l + 1, \ldots, l + 2e - 1\} = [l, l + 2e - 1]$. Decoding is based on a check matrix for $\mathcal{C}(A)$.

13.22 Definition. *H is a $(2e, n)$ matrix with rows indexed by the elements of A (the numbers from l to $l + 2e - 1$), columns indexed by the elements $\beta \in W$ and entry $\beta^i \in F$ in row i, column β.*

Clearly H is a generator matrix of $\mathcal{B}(A)$, possibly with some redundant rows (compare Theorem 13.19 and the section following it). In particular, an n-tuple $c = (c_\beta) \in \mathbb{F}_q^n$ is contained in $\mathcal{C}(A)$ if and only if its dot products with the rows of H vanish. These products are the syndromes.

13.23 Definition. *Let $i \in [l, l + 2e - 1] = A$ and $c = (c_\beta), \beta \in W$. The syndrome $S_i \in F$ is defined as*

$$S_i = S_i(c) = \sum_{\beta \in W} c_\beta \cdot \beta^i.$$

*The **syndrome polynomial** is $s(z) = \sum_{i=0}^{2e-1} S_{i+l} z^i \in F[z]$.*

In particular, we have $c \in \mathcal{C}(A)$ if and only if $S_l = S_{l+1} = \cdots = S_{l+2e-1} = 0$. Definition 13.23 is a slight generalization of the notion of syndrome from Section 3.7, as the syndromes are in F, not in the field over which $\mathcal{C}(A)$ is defined.

The fundamental equation

Let $x = (x_\beta) \in \mathcal{C}(A)$ be the codeword which was sent. The received word is $y = (y_\beta) \in \mathbb{F}_q^n$. Define the error-vector as $err = y - x = (e_\beta)$. We have $S_i(err) = S_i(y)$ for all i. As the receiver knows y, he can calculate the syndromes. We work under the assumption that no more than e errors occurred, equivalently that the vector err has weight $\leq e$. Let $M = \{\beta \mid e_\beta \neq 0\}$

be the set of coordinates where errors occurred. We express the syndrome polynomial in a different way, based on the error vector:

$$s(z) = \sum_{i=0}^{2e-1} S_{i+l} z^i = \sum_{i=0}^{2e-1} \sum_{\beta \in M} e_\beta \beta^{i+l} z^i = \sum_{\beta \in M} e_\beta \beta^l \sum_{i=0}^{2e-1} (\beta z)^i.$$

We use the geometric series and obtain

$$s(z) = \sum_{\beta \in M} e_\beta \beta^l (1 - (\beta z)^{2e})/(1 - \beta z) =$$

$$\sum_{\beta \in M} e_\beta \beta^l /(1 - \beta z) - \sum_{\beta \in M} e_\beta \beta^{2e+l} z^{2e}/(1 - \beta z).$$

The common denominator for these rational functions is

$$l(z) = \prod_{\beta \in M} (1 - \beta z).$$

The polynomial $l(z)$ is known as the **error locator**. The roots of $l(z)$ are precisely the inverses of the coordinates where errors occurred. With $l(z)$ as denominator, we have

$$s(z) = \frac{w(z)}{l(z)} - \frac{u(z)}{l(z)} z^{2e}$$

and it is clear what the polynomials $w(z), u(z)$ look like. After multiplication with $l(z)$, we obtain the fundamental equation:

13.24 Theorem (fundamental equation). *We have*

$$s(z) \cdot l(z) = w(z) - u(z) \cdot z^{2e}.$$

Here $l(z)$ is the error locator, as introduced above, and we have

$$w(z) = \sum_{\beta \in M} e_\beta \beta^l \prod_{\gamma \in M, \gamma \neq \beta} (1 - \gamma z),$$

$$u(z) = \sum_{\beta \in M} e_\beta \beta^{2e+l} \prod_{\gamma \in M, \gamma \neq \beta} (1 - \gamma z).$$

*Polynomial $w(z)$ is known as the **error evaluator**.*

Observe that the receiver can calculate $s(z)$ and needs to know $l(z)$ and $w(z)$. In the binary case $q = 2$ it suffices to know $l(z)$.

Using the Euclidean algorithm

The idea which leads to the desired decoding algorithm is based on the Euclidean algorithm and goes as follows: let $a = z^{2e}$ and $b = s(z)$. Apply the Euclidean algorithm to a and b. We work in the Euclidean domain $F[z]$ (a polynomial ring over a field), with the polynomial degree as Euclidean norm. Use the complete scheme of Section 13.3. We are not interested in the final result, the gcd of a and b. Rather, we stop the algorithm at row j, which is defined by $deg(r_j) < e$, $deg(r_{j-1}) \geq e$. The basic observation is that $v_j(z)$ (in the terminology of Section 13.3) is, up to a constant factor, identical with $l(z)$. The details are in the following theorem. Once it is proved, it is clear how to determine the set M of coordinates which are in error: find the roots of $v_j(z)$. The inverses of these roots are the coordinates β, where $y_\beta \neq x_\beta$. In the binary case $q = 2$, this suffices to decode.

13.25 Theorem (the algorithm). *In the current terminology the following holds: apply the Euclidean algorithm to $a = z^{2e}$ and $b = s(z)$ in the form of the complete scheme. Determine row j by $deg(r_j) < e$, $deg(r_{j-1}) \geq e$. Assume not more than e errors have occurred. Then there is a constant $\lambda = v_j(0) \in \mathbb{F}_q^*$ such that*

$$\boxed{\begin{aligned} v_j(z) &= \lambda \cdot l(z) \\ r_j(z) &= \lambda \cdot w(z) \\ u_j(z) &= \lambda \cdot u(z) \end{aligned}}$$

The proof of Theorem 13.25 is a routine algebraic argument. It may as well be skipped at first reading.

The following easy lemma will be needed in the proof.

13.26 Lemma. $(l(z), u(z)) = 1$.

PROOF As $l(z)$ is a product of linear factors, we have to show that roots of $l(z)$ cannot be roots of $u(z)$. Let $\beta \in M$. Then

$$u(1/\beta) = e_\beta \beta^{2e+l} \prod_{\gamma \in M, \gamma \neq \beta} (1 - \gamma/\beta) \neq 0. \qquad \square$$

Consider the last rows before the algorithm stops:

q_{j-1}	$r_{j-1}(z)$	$u_{j-1}(z)$	$v_{j-1}(z)$
q_j	$r_j(z) = w^*(z)$	$u_j(z) = u^*(z)$	$v_j(z) = l^*(z)$

Compare the starred polynomials with their nonstarred counterparts. The defining property of the entries in columns U and V shows

$$w^*(z) = u^*(z)z^{2e} + l^*(z)s(z).$$

This means that the starred polynomials satisfy the fundamental equation of Theorem 13.24.

13.27 Lemma. *We have*

$$deg(l^*(z)) \le e, \ deg(u^*(z)) < e \ and \ deg(w^*(z)) < e.$$

PROOF In the case of $w^*(z)$, this belongs to the stop condition. Use Theorem 13.20, the cross-product theorem. One of its statements is

$$l^*(z)r_{j-1}(z) - v_{j-1}(z)w^*(z) = \pm z^{2e}.$$

The monotonicity of the degrees (Lemma 13.21) shows that the first summand is dominating on the left side. This shows $deg(l^*(z)r_{j-1}(z)) = 2e$. It follows from the choice of j that $deg(r_{j-1}(z)) \ge e$, hence $deg(l^*(z)) \le e$.

In the case of $u^*(z)$ we proceed in an analogous way, starting from another statement of Theorem 13.20:

$$u^*(z)r_{j-1}(z) - u_{j-1}(z)w^*(z) = \pm s(z).$$

We obtain $deg(u^*(z)r_{j-1}(z)) = deg(s(z)) < 2e$ and finally $deg(u^*(z)) < e$, as required. ⬜

Start from the fundamental equations, for the original polynomials and for the starred polynomials:

$$\boxed{\begin{aligned} l(z)s(z) + u(z)z^{2e} &= w(z) \\ l^*(z)s(z) + u^*(z)z^{2e} &= w^*(z) \end{aligned}}$$

Multiply the first equation by $l^*(z)$, the second by $l(z)$ and subtract. This yields

$$(l^*(z)u(z) - l(z)u^*(z))z^{2e} = l^*(z)w(z) - l(z)w^*(z).$$

The right side of this equation has degree $< 2e$. It follows that both sides must vanish:

$$\boxed{\begin{aligned} l^*(z)u(z) &= l(z)u^*(z) \\ l^*(z)w(z) &= l(z)w^*(z) \end{aligned}}$$

Next use Lemma 13.26. The Euclidean algorithm shows that there are polynomials $f(z), g(z)$ such that $1 = f(z)l(z) + g(z)u(z)$. Multiply both sides by $l^*(z) : l^*(z) = f(z)l(z)l^*(z) + g(z)l^*(z)u(z)$. Use the first equation above to substitute $l^*(z)u(z)$. We obtain

$$l^*(z) = (f(z)l^*(z) + g(z)u^*(z))l(z).$$

We have found a polynomial $k(z)$ such that $l^*(z) = k(z)l(z)$. The equations above yield $w^*(z) = k(z)w(z)$ and $u^*(z) = k(z)u(z)$. As $u_j(z) = u^*(z)$ and $v_j(z) = l^*(z)$ and ± 1 can be written as a linear combination of $u_j(z)$ and $v_j(z)$ by Theorem 13.20, we have that $(u^*(z), v^*(z)) = 1$. This shows that the polynomial $k(z)$ is indeed a nonzero constant λ. As $l(0) = 1$, we have $\lambda = l^*(0)$. The proof of Theorem 13.25 is complete.

The binary example continued

Consider the primitive binary BCH code $C(A)$ of length 15 and designed distance 7 of Subsection 13.2. We organize the check matrix H such that the columns are indexed by the powers of α, in descending order, the first column corresponding to α^{14}, the last to $\alpha^0 = 1$. The rows correspond to $X, X^2, \ldots, X^6$. The entries are powers of α again. We use the convention to write out the exponents only. The following matrix H is obtained. The first column is a numbering of the rows.

1	14	13	12	11	10	9	8	7	6	5	4	3	2	1	0
2	13	11	9	7	5	3	1	14	12	10	8	6	4	2	0
3	12	9	6	3	0	12	9	6	3	0	12	9	6	3	0
4	11	7	3	14	10	6	2	13	9	5	1	12	8	4	0
5	10	5	0	10	5	0	10	5	0	10	5	0	10	5	0
6	9	3	12	6	0	9	3	12	6	0	9	3	12	6	0

The first row also serves to show the parametrization. Observe that the row of the generator matrix $G(A)$ as given in Section 13.2 is indeed orthogonal to the rows of H. It follows from the definition that $S_2 = S_1^2$, $S_4 = S_2^2 = S_1^4$, $S_6 = S_3^2$. In general, we have $S_{2i} = S_i^2$. This shows that we do not need to write out all the rows of the check matrix but rather only rows forming a set of representatives of the cyclotomic cosets intersecting A nontrivially. In our cases it suffices to know rows number $1, 3$ and 5. Let

$$y = (1, 0, 0, 0, 1, 1, 1, 0, 0, 1, 0, 0, 1, 1, 1)$$

be the received vector. We calculate $S_1 = \alpha^{14} + \alpha^{10} + \alpha^9 + \alpha^8 + \alpha^5 + \alpha^2 + \alpha + 1 = 0$. It follows that $S_2 = S_4 = 0$. Likewise $S_5 = 0$. However, we have $y \notin C(A)$ as $S_3 = \alpha^{12} + 1 + \alpha^{12} + \alpha^9 + 1 + \alpha^6 + \alpha^3 + 1 = \alpha^{12}$. The second nonzero syndrome is therefore $S_6 = S_3^2 = \alpha^9$. We obtain

$$s(z) = \alpha^9 z^5 + \alpha^{12} z^2.$$

Use the Euclidean algorithm:

k	Q	R	U	V
-1		z^6	1	0
0		$\alpha^9 z^5 + \alpha^{12} z^2$	0	1
1	$\alpha^6 z$	$\alpha^3 z^3$	1	$\alpha^6 z$
$j = 2$	$\alpha^6 z^2$	$\alpha^{12} z^2$	$\alpha^6 z^2$	$\alpha^{12} z^3 + 1$

Only two divisions with remainder are needed. We have to factor $l^*(z) = \alpha^{12} z^3 + 1$. Its roots correspond to $z^3 = \alpha^3$ and are therefore $\alpha, \alpha^6, \alpha^{11}$. The errors occurred in coordinates α^{14}, α^9 and α^4. Correcting in these coordinates yields the vector

$$x = (0, 0, 0, 0, 1, 0, 1, 0, 0, 1, 1, 0, 1, 1, 1).$$

It is easy to see that this is indeed a codeword. In the notation of Section 13.2, we have $x = v(3,0) + v(0,1)$.

Let us see what happens if fewer than e errors occur. Let the received vector be

$$y = (0,0,0,0,1,0,1,1,0,1,1,0,1,1,1).$$

The syndrome vector is

$$s(z) = \alpha^{12}z^5 + \alpha^5 z^4 + \alpha^{13}z^3 + \alpha^6 z^2 + \alpha^{14}z + \alpha^7.$$

The Euclidean algorithm terminates earlier:

k	Q	R	U	V
-1		z^6	1	0
0		$s(z)$	0	1
$j=1$	$\alpha^3 z + \alpha^{11}$	α^3	1	$\alpha^3 z + \alpha^{11}$

As $v_1(z)$ has degree 1, only one error occurred. The root of $v_1(z)$ is α^8. It follows that the error occurred in coordinate $\alpha^{-8} = \alpha^7$. The vector x, which was sent, is

$$x = (0,0,0,0,1,0,1,0,0,1,1,0,1,1,1),$$

the same word as before.

The ternary example continued

We are in the case $q = 3, r = 4, n = 20, A = [1,10]$. The first step is the calculation of the syndromes. The cyclotomic cosets (mod 20) intersecting A nontrivially have representatives $1,2,4,5,10$. It suffices to calculate the corresponding five rows of the check matrix of $\mathcal{C}(A)$. Here they are:

index	19	18	17	16	15	14	13	12	11	10
1	76	72	68	64	60	56	52	48	44	40
2	72	64	56	48	40	32	24	16	8	0
4	64	48	32	16	0	64	48	32	16	0
5	60	40	20	0	60	40	20	0	60	40
10	40	0	40	0	40	0	40	0	40	0

index	9	8	7	6	5	4	3	2	1	0
1	36	32	28	24	20	16	12	8	4	0
2	72	64	56	48	40	32	24	16	8	0
4	64	48	32	16	0	64	48	32	16	0
5	20	0	60	40	20	0	60	40	20	0
10	40	0	40	0	40	0	40	0	40	0

Recall that entry i in the check matrix stands for ϵ^i. Let the received vector be

$$y = (2, 0, 1, 0, 1, 1, 0, 1, 2, 0, 2, 1, 0, 0, 2, 2, 0, 2, 1, 2).$$

The dot products of y with the rows of our check matrix yield

$$S_1 = \epsilon^{20}, \; S_2 = \epsilon, \; S_4 = \epsilon^{33}, \; S_5 = 0, \; S_{10} = 1.$$

The remaining syndromes are obtained from the rule $S_{3i} = S_i^3$, where the subscripts are mod 20. The syndrome polynomial is

$$s(z) = z^9 + \epsilon^{20} z^8 + \epsilon^{11} z^7 - \epsilon^{20} z^6 + \epsilon^3 z^5 + \epsilon^{33} z^3 - \epsilon^{20} z^2 + \epsilon z + \epsilon^{20}.$$

The next step is the application of the complete Euclidean algorithm to $a = z^{10}$ and $b = s(z)$. This involves a series of long divisions of polynomials. We use the algorithm of Section 13.3. As $deg(v_k(z)) = k$ equals the number of errors, the algorithm must terminate when $k = 5$ or earlier.

k	Q	R	U	V
-1		z^{10}	1	0
0		$z^9 + \epsilon^{20} z^8 + \epsilon^{11} z^7$ $-\epsilon^{20} z^6 + \epsilon^3 z^5 + \epsilon^{33} z^3$ $-\epsilon^{20} z^2 + \epsilon z + \epsilon^{20}$	0	1
1	$z - \epsilon^{20}$	$-\epsilon^6 z^8 + \epsilon^{26} z^7 + \epsilon^{39} z^6$ $+\epsilon^{23} z^5 - \epsilon^{33} z^4 + \epsilon^{38} z^3$ $+\epsilon^{13} z^2 - \epsilon^{33} z - 1$	1	$-z + \epsilon^{20}$
2	$\epsilon^{34} z + \epsilon^{14}$	$\epsilon^{34} z^7 - \epsilon^{12} z^6 + \epsilon^{11} z^5$ $+\epsilon^{30} z^4 - \epsilon^{33} z^3 - \epsilon^5 z^2$ $+\epsilon^{10} z + \epsilon^6$	$-\epsilon^{34} z - \epsilon^{14}$	$\epsilon^{34} z^2 - \epsilon^{14} z$ $-\epsilon^{26}$
3	$\epsilon^{12} z + \epsilon^{14}$	$\epsilon^6 z^6 - \epsilon^{17} z^5$ $-\epsilon^{10} z^4 + \epsilon^{19} z^3$ $+\epsilon^7 z^2 - \epsilon^{35} z - \epsilon^{30}$	$-\epsilon^6 z^2$ $-\epsilon z - \epsilon^{31}$	$\epsilon^6 z^3 - \epsilon^{24} z^2$ $+\epsilon^{33} z - \epsilon^{10}$
4	$\epsilon^{28} z - \epsilon^{22}$	$-\epsilon^{34} z^3 - z^2$ $-z - \epsilon^{38}$	$\epsilon^{34} z^3 + \epsilon z^2$ $-\epsilon^9 z + \epsilon^{26}$	$-\epsilon^{34} z^4 + \epsilon^5 z^3$ $-\epsilon^{14} z^2 - \epsilon^{28} z$ $-\epsilon^{18}$

The algorithm ends at $k = 4$, so four errors occurred. We have $v_4(z) = \lambda \cdot l(z)$, where $\lambda = v_4(0) = -\epsilon^{18}$. The roots of $v_4(z)$ are $1, -\epsilon^4, -\epsilon^8, -\epsilon^{12}$. The reciprocals of these roots are $1 = \alpha^0, \epsilon^{36} = \alpha^9, \epsilon^{32} = \alpha^8$ and $\epsilon^{28} = \alpha^7$. We conclude that the errors occurred in coordinates $0, 7, 8, 9$; in other words $M = \{0, 7, 8, 9\}$. In order to determine these errors, we use $r_4(z) = \lambda \cdot w(z)$, where $w(z)$ is the error evaluator. Let $\beta \in M$. It follows from the definition of $w(z)$ that $w(1/\beta) = e_\beta \beta^l \prod_{\gamma \in M, \gamma \neq \beta} (1 - \gamma/\beta)$. The only unknown in this expression is e_β. Let us work out the case $\beta = 1$. We have $r_4(1) = -\epsilon^{34} - 1 - 1 - \epsilon^{38} = -\epsilon^{25}$. As $\prod_{\gamma \in M, \gamma \neq 1} (1 - \gamma) = -\epsilon^7$, we obtain

$$-\epsilon^{25} = -\epsilon^{18} e_0 \epsilon^7.$$

It follows that $e_0 = -1$. Similar calculations in the three remaining cases show $e_7 = e_8 = e_9 = 1$. We decode:

$$x = (2, 0, 1, 0, 1, 1, 0, 1, 2, 0, 1, 0, 2, 0, 2, 2, 0, 2, 1, 0).$$

In fact, $x = v(3, 1) \in \mathcal{C}([1, 10])$.

1. $\mathcal{C}(A)$ is a cyclic code, $A = [l, l + 2e - 1]$
   (BCH code, designed distance $2e + 1$).

2. H is the standard $(2e, n)$ generator matrix of $\mathcal{B}(A)$
   (seen as a check matrix for $\mathcal{C}(A)$).

3. $x \in \mathcal{C}(A)$ is the sent vector, y the received vector.
   Error correction is guaranteed if $d(x, y) \le e$.
   Assume this is the case.

4. Compute the syndromes $S_i \in F$ and the syndrome polynomial
   $$s(z) = \sum_{i=0}^{2e-1} S_{i+l} z^i \in F[z].$$

5. Apply the complete Euclidean algorithm to the
   polynomials $a = z^{2e}$ and $b = s(z)$.

6. In step k $(k = -1, 0, 1, \dots)$ this produces
   $r_k(z) = u_k(z) \cdot z^{2e} + v_k(z) \cdot s(z)$.

7. Stop the algorithm in step j defined by
   $deg(r_{j-1}) \ge e$, $deg(r_j) < e$.

8. Determine the roots of $v_j(z)$. The reciprocals of these roots are
   the coordinates where errors occurred.
   This does it in the binary case.

9. Let M be the set of coordinates where errors occurred,
   e_β the error in coordinate $\beta \in M$.

10. Compute $r_j(\beta^{-1}) = e_\beta \cdot v_j(0)\beta^l \prod_{\gamma \in M, \gamma \ne \beta}(1 - \gamma/\beta)$.
    Solve for e_β.

13.4   Constacyclic codes

In this section we give a general explanation for a phenomenon which we observed in the ternary example from Section 13.2. All we need is the explicit description of the generator matrix of cyclic codes given in Theorem 13.19. Recall the situation:

$\mathcal{C}(A)$ is a q-ary cyclic code. The length N divides $q^r - 1$. Assume there is a divisor u of $q - 1$ such that $N = nu$. As usual, let B be the negative of the union of cyclotomic cosets (mod N), which avoid A. Write $q^r - 1 = Nw$, let ϵ be a primitive element of $F = \mathbb{F}_{q^r}$ and $\alpha = \epsilon^w$ a generator of the group W of N-th roots of unity. Let $\omega = \epsilon^{nw} \in \mathbb{F}_q$ an element of order u.

Our main assumption is

- All elements $i \in B$ have the same modulus mod u.

Let us check the consequences upon the generator matrix of $\mathcal{C}(A)$ as described in Theorem 13.19. The coordinates (columns) are described by $\beta = \alpha^j = \epsilon^{wj}$, where $j = 0, 1, \ldots N - 1$. The rows of the generator matrix have the form $v(l, i)$, where the entry of $v(l, i)$ in coordinate j is

$$tr(\epsilon^l \beta^i) = tr(\epsilon^l \alpha^{ij}) = tr(\epsilon^{l + ijw}).$$

Here i varies over representatives of cyclotomic cosets contained in B.

The main observation is the following: the substitution $j \mapsto j+n$ introduces a factor $\epsilon^{inw} = \omega^i$ under the trace. As $\omega \in \mathbb{F}_q$ and the trace is $\mathbb{F}_q$-linear, this leads to a factor ω^i in the entry. Because of the main assumption this factor is independent of i. It follows that, if we project onto the first n coordinates corresponding to $j = 0, 1, \ldots, n-1$, we obtain a code with the same dimension k as $\mathcal{C}(A)$, and the subsequent blocks of n coordinates are obtained from the first one by multiplication by powers of ω. We call the projection onto the n first coordinates a **contraction** of the cyclic code $\mathcal{C}(A)$. The contractions of cyclic codes are the **constacyclic** codes. Cyclic codes can be considered as special cases of constacyclic codes (case $u = 1$).

13.28 Theorem. *Let $u \mid q-1$ and $N \mid q^r - 1$ be a multiple of u. Put $N = n \cdot u$. Consider the cyclic code $C = \mathcal{C}(A)$ of length N with defining set A, such that all elements of $B = -\overline{A}$ are in the same modulus mod u. Let ω be an element of order u in $\mathbb{F}_q$.*

Then all weights of C are multiples of u. More precisely, there is a code $\mathcal{K}$ of length n which has the same dimension k as C, and there is some $\gamma \in \mathbb{F}_q, \gamma^u = 1$ such that C is the code whose codewords are the following:

$$(v, \gamma v, \ldots, \gamma^{u-1} v), \text{ where } v \in \mathcal{K}.$$

*If C has parameters $[N, k, d]_q$, then $\mathcal{K}$ (a **contraction** of C) has parameters $[n, k, d/u]_q$.*

For an obvious converse, see Exercise 13.4.6. As an example, consider the case $q = 3, r = 4, N = 20, u = 2$. If we pick $A = [0, 10]$, then $B = \{1, 3, 7, 9\}$. It follows that $\mathcal{C}(A)$ has parameters $[20, 4, 12]_3$. This is the ternary example from Section 13.2. The contraction $\mathcal{K}$ has parameters $[10, 4, 6]_3$. We remark that $\mathcal{K}$ can be extended to the ternary Golay code. It is a natural question when contractions of cyclic codes are cyclic codes. Our example shows that this is not always the case. In fact, there is no cyclic $[10, 4, 6]_3$ code; see Exercise 13.4.8.

Let $q = 3, N = 56$. We have $56 \mid 3^6 - 1$. Use the following cyclotomic cosets:

$$\{1, 3, 9, 27, 25, 19\}, \ \{5, 13, 15, 23, 39, 45\}, \ \{7, 21\}.$$

Consider at first the cyclic code with all cyclotomic cosets except $C(1)$ and $C(7)$ as the defining set. The interval $[28, 0]$ shows that the code has parameters $[56, 8, 30]_3$. As the elements of $C(1) \cup C(7)$ are odd, Theorem 13.28 applies and we obtain a contraction with parameters $[28, 8, 15]_3$. This constacyclic code has optimal parameters and was first constructed in Kschischang and Pasupathy [129]. By our theory it is no problem to construct a generator matrix of this code. It simply consists of the "first half" of the generator matrix of the corresponding cyclic code of length 56.

In the same situation, let $\tilde{A}$ consist of all cyclotomic cosets except $C(1)$, $C(5)$, $C(7)$. The interval $[46, 0]$ and also interval $[28, 38]$ show that $\mathcal{C}(A)$ has minimum distance $d \geq 12$. Assume $d = 12$ and apply the method of the Roos bound. Let $I = [28, 38]$, $J = \{0, 18\}$ (observe that J is not an interval). Let S be the support of a codeword of weight 12. We have $dim(\mathcal{B}_S(I)) = 11$ and the one-dimensional spaces $\mathcal{B}_S(\{0\})$ (the repetition code) and $\mathcal{B}_S(\{18\})$ are contained in its dual. As all this happens in a 12-dimensional space, it follows that $\mathcal{B}_S(\{18\}) = \mathcal{B}_S(\{0\})$. This implies that polynomial X^{18} has constant values on S. As raising to the ninth power is a field automorphism, we must have that X^2 has constant values on S, which is impossible, as $|S| = 12 > 2$.

We have shown $d > 12$. It follows from Theorem 13.6 that $\mathcal{C}(A)^\perp = \mathcal{C}(B)$, where B is the negative of the complement of A. In our case, $B \subset A$; consequently $\mathcal{C}(A)$ is contained in its dual. This implies that all weights in $\mathcal{C}(A)$ are multiples of 3. Theorem 13.28 shows that $\mathcal{C}(A)$ can be contracted. It follows that all weights are even. As $d > 12$ and all weights are multiples of 6, we conclude that our code has parameters $[56, 14, 18]_3$. Its contraction is then a constacyclic $[28, 14, 9]_3$ code, another code from [129].

Two families of constacyclic quaternary codes

Let $Q = 2^f$, f odd, $F = \mathbb{F}_{Q^2}$ and W the subgroup of order $Q+1$ in F^*. Choose a set $Z \subset W$ of representatives of $\mathbb{F}_4^*$ in W. Equivalently Z is a set of $n = (Q+1)/3$ elements of W $((Q+1)$-st roots of unity) such that $z_1/z_2 \notin \mathbb{F}_4$ whenever $z_1, z_2 \in Z, z_1 \neq z_2$. Use the $z \in Z$ as columns of a check matrix H of a quaternary code S_f. Clearly S_f has length n. Interpret the columns of H as quaternary f-tuples (use a basis of F over $\mathbb{F}_4$). The rank of H is f, as otherwise W would be contained in a proper $\mathbb{F}_4$ subspace of F (see Exercise 13.4.5).

By definition H is 2-independent. We claim it is 4-independent. Assume there is an $\mathbb{F}_4$ linear combination involving three columns of H. We can choose notation such that $\alpha_1 z_1 = \alpha_2 z_2 + z_3$, where $0 \neq \alpha_i \in \mathbb{F}_4$ and the z_i are different elements in Z. Raise this equation to the $(Q+1)$-st power (this is the norm down to $\mathbb{F}_Q$), observing that raising to power Q is a field automorphism and that $w^Q = 1/w$ for $w \in W$. This yields $1 = (\alpha_2 z_2 + z_3)(\alpha_2^2/z_2 + 1/z_3) = 1 + 1 + \alpha_2 z_2/z_3 + \alpha_2^2 z_3/z_2$. Let $x = \alpha_2 z_2/z_3$. We have $x + 1/x + 1 = 0$, equivalently $x^2 + x + 1 = 0$. As x satisfies a quadratic equation with coefficients in $\mathbb{F}_2$, it follows that $x \in \mathbb{F}_4$, consequently $z_2/z_3 \in \mathbb{F}_4$. This contradicts the definition of Z.

We have seen that H is 3-independent. Assume there is a linear combination involving four columns of H. We can write

$$\alpha_1 z_1 + \alpha_2 z_2 = \alpha_3 z_3 + \alpha_4 z_4$$

with obvious notation (in particular $0 \neq \alpha_i \in \mathbb{F}_4$). Raise both sides to power $Q + 1$. Write $x = \alpha_1 \alpha_2^2 z_1/z_2$, $y = \alpha_3 \alpha_4^2 z_3/z_4$. We obtain $x + 1/x = y + 1/y$, equivalently $(x + y)(xy + 1) = 0$. It follows that either $y = x$ or $y = 1/x$.

Write $a \sim b$ if $a, b \in W$ and $a/b \in \mathbb{F}_4$. Assume at first $y = x$. Then $z_1/z_2 \sim z_3/z_4$. Proceed as before, grouping indices $1, 4$ on one side, $2, 3$ on the other. By what we have shown, either $z_1/z_4 \sim z_2/z_3$ or $z_1/z_4 \sim z_3/z_2$. In the first case we obtain $z_3^2 \sim z_4^2$, hence $z_3 \sim z_4$, and, by definition of Z, finally $z_3 = z_4$, a contradiction. The second case leads to the analogous contradiction $z_2 = z_4$.

Case $y = 1/x$ is analogous. It follows that H is 4-independent. Equivalently S_f is a $[(2^f + 1)/3, (2^f + 1)/3 - f, 5]_4$ code for every odd $f \geq 5$. In the case $f = 5$ this yields an $[11, 6, 5]_4$ code, S_7 is an $[43, 36, 5]_4$ code.

As the subgroup W is the disjoint union $W = Z \cup \omega Z \cup \overline{\omega} Z$, it follows from the converse of Theorem 13.28 that S_f is constacyclic. In fact, S_f can be chosen to be cyclic rather frequently. Assume $Q + 1$ is not divisible by 9. This is the case if $f \equiv 3, 5 \pmod{6}$. Then we can choose Z as the subgroup of order $(Q+1)/3$ and S_f is a cyclic code.

13.29 Theorem. *Let* $Q = 2^f$, *f odd, $F = \mathbb{F}_{Q^2}$ and W the subgroup of order $N = Q + 1$ in F^*. The contraction S_f of length $n = (Q + 1)/3$ of the quaternary cyclic code of length N with defining set $A = \{1\}$ is a constacyclic $[(Q + 1)/3, \ (Q + 1)/3 - f, 5]_4$ code.*

If $f \equiv 3, 5 \pmod 6$, then the cyclic quaternary code $\mathcal{C}(\{1\})$ of length $n = (Q + 1)/3$ is a $[(Q + 1)/3, \ (Q + 1)/3 - f, 5]_4$ code.

In particular, we obtain a cyclic $[11, 6, 5]_4$ code and a cyclic $[43, 36, 5]_4$ code. We will see later on that the first of these codes can be used for the construction of a $[12, 6, 6]_4$-code.

A second family of constacyclic quaternary codes arises in the following situation: let $Q = 2^f, F = \mathbb{F}_{Q^2}$ and W a set of coset representatives of $\mathbb{F}_4^*$ in F^*, $n = |W| = (Q^2 - 1)/3$. Define a quaternary code R_f of length n whose check matrix H has as columns the $(w, 1/w^2)^t$, where $w \in W$. Here we imagine w and w^{-2} expanded in terms of an $\mathbb{F}_4$ basis of F. It is clear that H has rank $2f$. We claim that H is 4-independent. The proof is very similar to the first family. It is left as an exercise.

13.30 Theorem. *Let $f \geq 3$, $Q = 2^f$, $F = \mathbb{F}_{Q^2}$ and W a set of coset representatives of $\mathbb{F}_4^*$ in F^*. The contraction R_f of length $n = (Q^2 - 1)/3$ of the quaternary cyclic code of length N with defining set $A = \{1, -2\}$ is a constacyclic $[(Q^2 - 1)/3, \ (Q^2 - 1)/3 - 2f, 5]_4$ code.*

If f is not a multiple of 3, then the cyclic quaternary code $\mathcal{C}(\{1, -2\})$ of length $n = (Q^2 - 1)/3$ is a $[(Q^2 - 1)/3, \ (Q^2 - 1)/3 - 2f, 5]_4$ code.

The smallest examples of Theorem 13.30 are R_3, a constacyclic $[21, 15, 5]_4$ code, and the cyclic $[85, 77, 5]_4$ code R_4.

These families of quaternary constacyclic codes have been introduced by Gevorkyan, Avetisyan and Tigranyan [90]; see also Dumer and Zinoviev [73].

Exercises 13.4

13.4.1. *Let $\mathcal{K}$ be an $[n, k, d]_q$ code with the property that for every $(a_0, a_1, \ldots, a_{n-1}) \in \mathcal{K}$ we also have $(a_1, a_2, \ldots, a_{n-1}, \gamma a_0) \in \mathcal{C}$, where $\gamma \in \mathbb{F}_q$ such that $\gamma^u = 1$.*

*Prove the following: there is a cyclic $[un, k, ud]_q$-code $\mathcal{C}$ which possesses $\mathcal{K}$ as a projection (contraction). Here the term **cyclic** is to be interpreted as follows: there is a cyclic permutation on the coordinates which is an automorphism of $\mathcal{C}$.*

13.4.2. *Let $\mathcal{K}$ be a constacyclic code constructed according to Theorem 13.28 from a cyclic code with the set of zeroes B.*

Prove: $\mathcal{K}^\perp$ *is the contraction of the cyclic code with the set of zeroes* B', *where* B' *is the complement of* $-B$ *in the set of numbers having modulus (mod* u*) the negative of the elements of* B.

13.4.3. When is a constacyclic code self-dual (equal to its dual)?

13.4.4. Prove that the constacyclic $[28, 14, 9]_3$ code constructed in Section 13.4 is self-dual.

13.4.5. Let $F = \mathbb{F}_{p^{2f}}$ and W the subgroup of order $p^f + 1$ in F^*. Show that W is not contained in a proper subspace of F when F is considered as a $2f$-dimensional vector space over $\mathbb{F}_p$.

13.4.6. Prove the following converse of Theorem 13.28: Let $\mathcal{C}$ be a cyclic q-ary code of length $N = nu$ and $\gamma \in \mathbb{F}_q$ such that $\gamma^u = 1$, let $\mathcal{K}$ be a code of length n and $dim(\mathcal{K}) = dim(\mathcal{C})$ such that

$$\mathcal{C} = \{(v, \gamma v, \ldots, \gamma^{u-1} v) \mid v \in \mathcal{K}\}.$$

Then $\mathcal{K}$ *is constacyclic.*

13.4.7. Prove that the check matrix H of R_f is 4-independent.

13.4.8. Prove that there is no cyclic $[10, 4, 6]_3$ code.

13.4.9. Does a $[10, 4, 6]_3$ code exist?

13.5   Remarks

In the light of the automorphism group (see Section 12.4), it is most natural to define a code of length n to be cyclic if it is closed under a cyclic permutation which consists of one cycle of length n (see our treatment of the binary Golay code in Chapter 7). It is easy to see that the codes we called cyclic in the present chapter are cyclic in this sense as well. It suffices to check that $\mathcal{B}(A)$ is cyclic. A typical generator, corresponding to the monomial λX^i, where $i \in \tilde{A}$, has entry $\lambda \alpha^{ij}$ in coordinate j, where α is a generator of the group W of order n, $\lambda \in F$ and $j = 0, 1, \ldots, n - 1$. The cyclic shift $j \mapsto j + 1$ maps this into the vector with entry $\lambda \alpha^{i(j-1)}$ in coordinate j. This corresponds to the monomial $\lambda \alpha^{-i} X^i$ and therefore is a codeword of $\mathcal{B}(A)$ as well.

The traditional theory of linear cyclic codes whose length n is coprime to the characteristic p proceeds as follows: each codeword $(a_0, a_1, \ldots, a_{n-1})$ of such a code $\mathcal{C}$ is identified with the polynomial $p(X) = a_0 + a_1 X + \cdots + a_{n-1} X^{n-1}$. Because of linearity $\mathcal{C}$ is a subspace of the polynomial ring $\mathbb{F}_q[X]$. The cyclic shift maps this codeword (polynomial) into $a_0 X + a_1 X^2 + \cdots + a_{n-2} X^{n-1} + a_{n-1}$. This is almost the same as $X p(X)$, in fact, it is the same if we identify

X^n with 1. In other words, if we identify $\mathcal{C}$ with a subspace of the factor ring $R = \mathbb{F}_q[X]/(X^n - 1)$, then it is not only a vector subspace but it is also closed under multiplication by X. We see that the cyclic code $\mathcal{C}$ can be identified with an **ideal** of this ring R. The theory goes on to prove that R is a principal ideal ring and that each ideal (cyclic code) is the sum of some minimal ideals. The minimal ideals are in bijection with the cyclotomic cosets. If there are c cyclotomic cosets, there are altogether 2^c ideals (cyclic codes) in R. We obtained the same total number 2^c of cyclic codes $\mathcal{C}(A)$. This shows that the theory of codes $\mathcal{C}(A)$ as developed in this chapter is precisely the theory of linear cyclic codes, at least in the case when p and n are coprime. It may be noted that the ring-theoretic formulation of the traditional theory did serve a purpose, at least in a political sense. It contributed toward giving some dignity to coding theory in the eyes of the algebraic-number theoretic community and it provided algebraists with a (hypothetical) area of applications.

Cyclic codes constitute the most popular and most widely studied family of error-correcting codes. This is why several special situations are known by historically established names. **BCH** codes are the cyclic codes $\mathcal{C}(A)$, where the defining set A is an interval. This is in fact a very natural idea. If we want to construct a cyclic code with minimum distance 7, simply choose A to be an interval of six exponents. In order to maximize the dimension, A should be chosen such that $|\tilde{A}|$ is minimal. BCH codes are named after their inventors A. Hocquenghem [115], R.C. Bose and D.K. Ray-Chaudhuri [37]. Theorem 13.5 is the **BCH bound.** The value of this bound is also known as the **designed distance.** One speaks of BCH codes in the **narrow sense** if $A = \{1, 2, \ldots, t\}$. A q-ary cyclic code is **primitive** if $n = q^r - 1$.

Linear **constacyclic codes** are by definition linear codes which admit the action of an element of the monomial group, inducing a cyclic permutation on the n coordinates. After passing to equivalent codes and choosing bases accordingly, this means that there is some $0 \neq \gamma \in \mathbb{F}_q$ such that, for all $(a_0, a_1, \ldots, a_{n-1}) \in \mathcal{C}$, we also have $(a_1, a_2, \ldots, a_{n-1}, \gamma a_0) \in \mathcal{C}$. The notion of a constacyclic code generalizes the notion of a linear cyclic code. A constacyclic code with $\gamma = 1$ is nothing but a cyclic code. The traditional ring-theoretic approach to cyclic codes can be generalized to cover constacyclic codes, but once again this is not needed.

In fact, Theorem 13.28 provides a construction of constacyclic codes of length n out of cyclic codes of length un, where the factor γ satisfies $\gamma^u = 1$. The construction in the inverse direction is much easier: given a constacyclic code of length n with factor γ of order u, it is easy to construct a cyclic linear code of length un which projects to the original code (see Exercise 13.4.1). Moreover, these constructions are inverses of each other. This shows that constacyclic codes are precisely the contractions of cyclic codes in the sense of Theorem 13.28. The interesting conclusion is that the theory of constacyclic codes, although formally a generalization of the theory of linear cyclic codes, can in fact be found within the theory of cyclic codes.

Our approach to the decoding of BCH codes follows O. Pretzel [170]. The first version of this algorithm is due to Berlekamp [12]. Massey modified it in [149] and showed a connection with shift-register sequences. Pretzel's presentation is based on a version introduced by Sugiyama et. al. [203] which uses the Euclidean algorithm. All these decoding methods, including another one based on continuous fractions, are essentially equivalent.

Chapter 14

Recursive constructions and the covering radius

14.1 Construction X

and its variants. Applications to cyclic codes. Reed-Muller codes.

The following fundamental construction yields a lengthening (in the sense of Section 5.1) of a given code provided a subcode with a larger minimum distance is known. This has been described as **construction X** in MacWilliams and Sloane [142]. We start from a chain $\mathcal{C} \supset \mathcal{D}$ of linear q-ary codes and produce a code which has the same dimension as $\mathcal{C}$, is longer than $\mathcal{C}$ (this is bad news) and has the same minimum distance as $\mathcal{D}$ (this is very good news).

14.1 Theorem (construction X). *Let $\mathcal{C} \supset \mathcal{D}$ be q-ary linear codes, with parameters $[n, K, d]_q \supset [n, k, D]_q$, where $D > d$.*
Assume further there is an $[l, K - k, \delta]_q$ code, the auxiliary code $\mathcal{E}$. Then there is an l-step lengthening of $\mathcal{C}$ with parameters

$$[n + l, K, min(d + \delta, D)]_q.$$

PROOF The factor space $\mathcal{C}/\mathcal{D}$ and the auxiliary code are $\mathbb{F}_q$ vector spaces of the same dimension. We can therefore find a bijective $\mathbb{F}_q$ linear mapping $\phi : \mathcal{C}/\mathcal{D} \longrightarrow \mathcal{E}$. The lengthening of $\mathcal{C}$ is constructed by appending $\phi(x)$ to each word $x \in \mathcal{C}$. This means that the codewords of the lengthened code are

$$\tilde{x} = (x|\phi(x)) \in \mathbb{F}_q^{n+l} \text{ for } x \in \mathcal{C}.$$

As $\mathcal{D}$ is the kernel of ϕ (the elements of $\mathcal{D}$ are those which are mapped to the 0-tuple under ϕ,) we have $wt(\tilde{x}) = wt(x)$ for $x \in \mathcal{D}$. Let $x \in \mathcal{C} \setminus \mathcal{D}$.

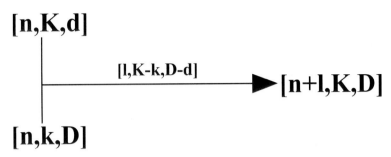

FIGURE 14.1: Construction X

Then $\phi(x)$ is a nonzero codeword of $\mathcal{E}$, therefore of weight $\geq \delta$. It follows that $wt(\tilde{x}) \geq wt(x) + \delta \geq d + \delta$ in this case. ▯

The best-known application of construction X is the parity check bit for binary codes, Theorem 5.3. It is easy to see that this is indeed an application of construction X; see Exercise 14.1.1. The auxiliary code is the truly trivial code $[1, 1, 1]_2$.

The following construction, essentially due to Alltop [1], uses knowledge of a code, two of its subcodes and their intersection:

14.2 Theorem (construction XX). *Let the following be given:*

- *A code $\mathcal{C}$ with parameters $[n, k, d]_q$,*

- *two subcodes $\mathcal{C}_i \subset \mathcal{C}$, $i = 1, 2$ with parameters $[n, k - \kappa_i, d_i]_q$,*

- *with $\mathcal{C}_1 \cap \mathcal{C}_2$ of minimum distance $\geq D$,*

- **auxiliary codes** $\mathcal{E}_i$, $i = 1, 2$ *with parameters $[l_i, \kappa_i, \delta_i]_q$.*

Then there is a code with parameters

$$[n + l_1 + l_2, k, min\{D, d_2 + \delta_1, d_1 + \delta_2, d + \delta_1 + \delta_2\}]_q.$$

PROOF   Apply Theorem 14.1 to the pair $\mathcal{C} \supset \mathcal{C}_1$. The lengthened codes are $\tilde{\mathcal{C}} \supset \tilde{\mathcal{C}}_2$. The larger of these has minimum distance $\geq min\{d_1, d + \delta_1\}$. A codeword in $\mathcal{C}_2 \setminus (\mathcal{C}_1 \cap \mathcal{C}_2)$ obtains a nontrivial lengthening. The corresponding word in the lengthened code has then weight $\geq d_2 + \delta_1$. It follows that the minimum weight of $\tilde{\mathcal{C}}_2$ is $\geq min\{D, d_2 + \delta_1\}$. We apply Theorem 14.1 again, this time to the pair $\tilde{\mathcal{C}} \supset \tilde{\mathcal{C}}_2$, with $\mathcal{E}_2$ as auxiliary code. This yields the desired code. ▯

Applications to cyclic codes

Cyclic codes are ideally suited for applications of constructions X and XX. Consider q-ary cyclic codes of length n. Let $A_1 \subset A_2$. Then $\mathcal{C}(A_1) \supset \mathcal{C}(A_2)$. This is a situation where construction X can be applied. It remains to find a suitable auxiliary code.

In the case $q = 2, n = 15$ the cyclotomic cosets have been given in Section 13.2:

$$\{0\}, \ \{1,2,4,8\}, \ \{3,6,12,9\}, \ \{5,10\}, \{7,14,13,11\}.$$

Consider $A_1 = \{1,2,3,4\} \subset A_2 = \{1,2,3,4,5,6\}$. The corresponding cyclic codes form a chain $[15,7,5]_2 \supset [15,5,7]_2$. In order to apply construction X, we need an auxiliary code $[l,2,2]_2$ with l as small as possible. Clearly $l = 3$. The result is a code $[18,7,7]_2$, after parity check $[19,7,8]_2$. Observe that such a code can also be derived from the extended binary Golay code $[24,12,8]_2$ by repeated shortening.

$\mathcal{C}(A_2)$ contains the repetition code $\mathcal{C}(\{1,2,\ldots,14\})$. Apply construction X to this chain

$$[15,5,7]_2 \supset [15,1,15]_2.$$

Observe that we are not forced to use the exact minimum distance of the smaller code. For example, using the sum-zero code $[5,4,2]_2$ as auxiliary code we obtain $[20,5,9]_2$, after a parity check bit $[21,5,10]_2$. Another choice is the extended Hamming code $[8,4,4]_2$ as auxiliary code. After adding the parity check bit, this yields a $[24,5,12]_2$ code. The Griesmer bound shows that a code $[l,4,6]_2$ has $l \geq 12$. An $[12,4,6]_2$ can be constructed by concatenation (see Exercise 5.2.8). Another possibility is to use construction X. Using $[12,4,6]_2$ as auxiliary code produces a lengthened code $[27,5,13]_2$, equivalently $[28,5,14]_2$, once again a code meeting the Griesmer bound with equality. Finally, $\mathcal{C}(\{0,1,3,5\})$ is a $[15,4,8]_2$ code (it meets the Griesmer bound with equality). Used as an auxiliary code in the same situation, we obtain $[32,6,16]_2$. A code with these parameters was constructed in Section 5.2 by way of the $(u, u + v)$ construction.

It is obvious that there is a multitude of situations where construction X can be applied to chains of cyclic codes.

Standard lengthening

Here is a generic situation. Consider a BCH code $\mathcal{C}(A_1)$, where $A_1 = [l, l + t - 1] = \{l, l + 1, \ldots, l + t - 1\}$ (using obvious notation for intervals) and its subcode $\mathcal{C}(A_2)$, where $A_2 = [l - 1, l + t - 1]$ and $l - 1 \notin \tilde{A}_1$. By the BCH bound, $\mathcal{C}(A_1)$ has minimum distance $\geq t + 1$, and $\mathcal{C}(A_2)$ has minimum distance $\geq t + 2$. The codimension of $\mathcal{C}(A_2)$ in $\mathcal{C}(A_1)$ is $\kappa = |C(l - 1)|$, the length of the cyclotomic coset containing $l - 1$. As auxiliary code we use the trivial code $[\kappa, \kappa, 1]_q$. The result is a code of the same dimension as $\mathcal{C}(A_1)$, minimum distance $\geq t + 2$ and length $n + \kappa$. Clearly we could have used $[l - 1, l + t]$ in the role of A_2 as well.

A typical situation where construction XX can be applied is formed by quadrangles of BCH codes with defining sets $A_1, A_2, A_1 \cup A_2, A_1 \cap A_2$. We have $\mathcal{C}(A_1) \cap \mathcal{C}(A_2) = \mathcal{C}(A_1 \cup A_2)$ (see Exercise 14.1.3). As largest code we can use $\mathcal{C}(A_1 \cap A_2)$. In the case $A_1 = [l - 1, l + t - 1]$, $A_2 = [l, l + t]$, we can apply construction XX with trivial auxiliary codes and obtain a lengthening of $\mathcal{C}(A_1 \cap A_2)$ with minimum distance $\geq t + 3$, two more than the distance of $\mathcal{C}(A_1 \cap A_2)$ guaranteed by the BCH bound.

As an example, consider the case $q = 3, n = 26$, where the interval $[1, 12]$ is disjoint from the cyclotomic cosets

$$\{0\}, \{13\}, \{14, 16, 22\} \text{ and } \{17, 23, 25\}.$$

It follows that $\mathcal{C}([1, 12])$ is a $[26, 8, 13]_3$ code. Subcodes $\mathcal{C}([0, 12])$ and $\mathcal{C}([1, 13])$ are $[26, 7, 14]_3$ codes intersecting in $\mathcal{C}([0, 13])$, a $[26, 6, 15]_3$ code. By standard lengthening we obtain an (optimal) $[28, 8, 15]_3$ code.

A four-dimensional family

For arbitrary q let $n = q^2 - 1$. The interval $A_1 = [1, q^2 - q - 3]$ avoids the cyclotomic cosets

$$\{q^2 - q - 1, q^2 - 2\} = \{-q, -1\} \text{ and } \{0\}, \{q^2 - q - 2\} = \{-(q-1)\}.$$

It follows that $\mathcal{C}(A_1)$ is a $[q - 1, 4, q^2 - q - 2]_q$ code. Consider the intervals $A_1 \subset A_2 = [1, q^2 - q - 2] \subset [1, n - 1]$. The corresponding BCH codes form a chain

$$[q^2 - 1, 4, q^2 - q - 2] \supset [q^2 - 1, 3, q^2 - q - 1] \supset [q^2 - 1, 1, q^2 - 1].$$

Using the intervals $\{0\} \cup A_i$ instead yields a chain of codes of codimension 1 in $\mathcal{C}(A_i)$ and of minimum distance one larger than that of $\mathcal{C}(A_i)$. Application of construction X with $[1, 1, 1]_q$ as auxiliary code yields the following chain of codes:

14.3 Proposition. *There is a chain $C_1 \supset C_2 \supset C_3$ of lengthened BCH codes with parameters*

$$[q^2, 4, q^2 - q - 1]_q \supset [q^2, 3, q^2 - q]_q \supset [q^2, 1, q^2]_q.$$

Application of construction X to the chain $C_1 \supset C_2$ yields a $[q^2 + 1, 4, q^2 - q]_q$ code. We call codes with such parameters **ovoid codes**.

14.4 Lemma. *The weight distribution of a $[q^2 + 1, 4, q^2 - q]_q$ code is* $A_0 = 1$, $A_{q^2 - q} = (q - 1)(q^3 - q)$, $A_{q^2} = (q - 1)(q^2 + 1)$.

PROOF   Assume $w = q^2 - q + i$ is the weight of a codeword, where $1 \leq i \leq q-1$. Use the generalization of the residual code construction proved in Exercise 5.1.12. As $\lceil w/q \rceil = q$, we obtain a $[q+1, i, 3, q-i]_q$ code, which is impossible because of the Singleton bound. The only possible weights of an ovoid code are 0, $q^2 - q$, q^2, $q^2 + 1$.

Assume $q^2 + 1$ is a weight. We can assume the all-1-word **1** is a codeword. Let $x \notin \langle \mathbf{1} \rangle$ be a codeword and δ_i the frequency of $i \in \mathbb{F}_q$ as an entry of x. As $wt(x - i\mathbf{1}) = q^2 + 1 - \delta_i$, we have $\delta_i \in \{0, 1, q+1\}$. However, the equation $\sum_i \delta_i = q^2 + 1$ cannot be satisfied, a contradiction. It follows that $q^2 - q$, q^2 are the only nonzero weights. The weight numbers satisfy the trivial equation $1 + A_{q^2-q} + A_{q^2} = q^4$. Another equation is obtained by counting the zeroes in codewords:

$$q^2 + 1 + (q + 1)A_{q^2-q} + A_{q^2} = (q^2 + 1)q^3.$$

The resulting system of two equations has a unique solution.   ▯

All codewords in $\mathcal{C}_2 \setminus \mathcal{C}_3$ have weight $q^2 - q$ (see Exercise 14.1.6). The minimum distance of $\mathcal{C}_1$ is indeed $q^2 - q - 1$, as the Griesmer bound excludes parameters $[q^2, 4, q^2 - q]_q$. As the ovoid code $[q^2 + 1, 4, q^2 - q]_q$ is a lengthening of $\mathcal{C}_1$, it follows from Lemma 14.4 that the only possible nonzero weights of $\mathcal{C}_1$ are $q^2 - q - 1, q^2 - q, q^2 - 1, q^2$. It is easy to see that all these weights really occur (see Exercise 14.1.7).

In the octal case ($q = 8$) the chain of Proposition 14.3 has parameters

$$[64, 4, 55]_8 \supset [64, 3, 56]_8 \supset [64, 1, 64]_8.$$

Concatenation with the Simplex code $[7, 3, 4]_2$ (of constant weight 4) yields a chain

$$[448, 12, 220]_2 \supset [448, 9, 224]_2 \supset [448, 3, 256]_2$$

of binary linear codes. The nine-dimensional code meets the Griesmer bound; the 12-dimensional code is d-optimal (a $[448, 12, 221]_2$ code cannot exist). In fact, the residual code, a $[228, 11, 110]_2$ code, is known to be d-optimal.

The 12-dimensional code contains a codeword of weight 224. The variant of the residual code construction from Exercise 5.1.12 produces a $[224, 11, 108]_2$ code which is known to be d-optimal.

Apply construction X to the chain $[448, 12, 220]_2 \supset [448, 9, 224]_2$, using $[4, 3, 2]_2$ or $[7, 3, 4]_2$ as auxiliary code. The resulting parameters $[452, 12, 222]_2$ and $[455, 12, 224]_2$ are d-optimal, as is shown by the Griesmer bound. Many more good codes can be constructed in this situation. For details see [22].

A six-dimensional family

For arbitrary q let $n = (q-1)(q^2+1)|(q^4-1)$. Elements $i(q^2+1)$ form cyclotomic cosets of length 1. Interval $A = [1, (q-2)(q^2+1)-1]$ is disjoint from $\{0\}, \{(q-2)(q^2+1)\}$ and from $C(-1)$. It follows that $\mathcal{C}(A)$ is an $[n, 6, (q-2)(q^2+1)]_q$ code. Standard lengthening (construction XX) yields a code with parameters

$$[(q-1)(q^2+1)+2, 6, (q-2)(q^2+1)+2]_q.$$

In the ternary case the resulting $[22, 6, 12]_3$-code is optimal. Apparently this code was first constructed by Kschischang and Pasupathy [129]. In the quaternary case $[53, 6, 36]_4$ is obtained, another optimal code which, after concatenation with $[3, 2, 2]_2$, yields a binary $[159, 12, 72]_2$ code. The $[106, 6, 80]_5$ is a good code as well.

A 10-dimensional family

A similar situation occurs when $n = (q-1)(q^3+1) = (q-1)(q+1)(q^2-q+1)$. The multiples of q^3+1 form cyclotomic cosets of length 1; multiples of q^2-q+1 are in cyclotomic cosets of length either 1 or 2. The cyclotomic cosets

$$\{1, q, q^2, q^3, q^3-q+1, q^3-q^2+1\}, \ \{q^2-q+1, q^3-q^2+q\}$$

are both contained in the interval $[1, q^3]$ and therefore avoid an interval of length $n - q^3 = q^4 - 2q^3 + q - 1$, whose endpoints 0 and q^3+1 each form a cyclotomic coset of length 1.

Application of construction XX as before yields codes

$$[(q-1)(q^3+1)+2, 10, (q-2)(q^3+1)+2]_q.$$

Both $[58, 10, 30]_3$ and $[197, 10, 132]_4$ are good codes, realizing the highest minimal distance currently known.

More applications of X and XX

Let $r = 3, n = q^3 - 1$. Denote by $[i, j]'$ the complement of interval $[i, j]$. Consider the following chains of BCH codes:

- $\mathcal{C}_1 \supset \mathcal{C}_2 \supset \mathcal{C}_3$, corresponding to
  $[0, q^2+q+1]' \supset [0, q^2+q]' \supset [0, q^2]'$

- $\mathcal{D}_1 \supset \mathcal{D}_2 \supset \mathcal{D}_3$ corresponding to
  $[1, q^2+q+1]' \supset [1, q^2+q]' \supset [1, q^2]'.$

The cyclotomic cosets

$$\{0\}, \ \{1, q, q^2\}, \ \{q+1, q^2+q, q^2+1\}, \ \{q^2+q+1\}$$

give lower bounds on the dimensions of these codes (in fact, these are the precise dimensions). The $\mathcal{C}$-chain has parameters

$$[n, 8, q^3 - q^2 - q - 2] \supset [n, 7, q^3 - q^2 - q - 1] \supset [n, 4, q^3 - q^2 - 1],$$

the $\mathcal{D}$-chain has parameters

$$[n, 7, q^3 - q^2 - q - 1] \supset [n, 6, q^3 - q^2 - q] \supset [n, 3, q^3 - q^2].$$

Moreover, $\mathcal{C}_2 \cap \mathcal{D}_1 = \mathcal{D}_2$, $\mathcal{C}_3 \cap \mathcal{D}_2 = \mathcal{D}_3$.

Apply construction X to $\mathcal{C}_1 \supset \mathcal{D}_1$ with $[1, 1, 1]$ as auxiliary code (standard lengthening). The images of $\mathcal{C}_2$ and $\mathcal{C}_3$ form a chain $[q^3, 7, q^3 - q^2 - q] \supset [q^3, 4, q^3 - q^2]$. Apply construction X to this chain. The auxiliary codes $[q + 1, 3, q - 1]_q$ (extended Reed-Solomon codes, to be introduced in Section 17.1) produce $[q^3 + q + 1, 7, q^3 - q^2 - 1]_q$. In the ternary case this $[31, 7, 17]_3$ code is optimal. In the quaternary case it is better to use the hexacode $[6, 3, 4]_4$ as auxiliary code (we encountered it in Exercises 3.4.5 and 3.4.6) An optimal code $[70, 7, 48]_4$ is obtained. Auxiliary code $[6, 3, 3]_3$ (obtainable by shortening from a Hamming code) yields $[33, 7, 18]_3$, another optimal code.

If aiming at eight-dimensional codes, one should instead apply standard lengthening to the images of $\mathcal{C}_1 \supset \mathcal{C}_2$ obtained in the first step. We obtain a chain

$$[q^3 + 1, 8, q(q^2 - q - 1)] \supset [q^3 + 1, 4, q^2(q - 1)].$$

The larger of these codes is interesting in itself. In the ternary case we obtain the by now familiar parameters $[28, 8, 15]_3$; the quaternary case yields the optimal $[65, 8, 44]_4$, which apparently was first considered in Groneick and Grosse [100]. In the ternary case application of construction X with the sum-zero code $[5, 4, 2]_3$ or with $[7, 4, 3]_3$ as auxiliary code yields the optimal codes $[33, 8, 17]_3$ and $[35, 8, 18]_3$. For $q \geq 4$, use of the extended Reed-Solomon code $[q + 1, 4, q - 2]_q$ yields the following:

14.5 Theorem. *There is a code* $[q^3 + q + 2, 8, q^3 - q^2 - 2]_q$ *$(q \geq 4)$.*

In the case $q = 4$, the optimal parameters $[70, 8, 46]_4$ are obtained.

Reed-Muller codes

Consider the binary primitive case $q = 2$, $n = 2^r - 1$. The numbers between 0 and $2^r - 2$ can be identified with their duadic representation: if $x = \sum_{i=0}^{r-1} x_i 2^i$, then x is described by the binary r-tuple $(x_{r-1}, \ldots, x_1, x_0)$. The number $2x$ is represented by the cyclic shift $(x_{r-2}, \ldots, x_0, x_{r-1})$. It follows that a set of exponents is Galois closed if the corresponding duadic representations are closed under cyclic shifts. Choose A to consist of the exponents whose duadic representations have weight at most b, for some $b < r$. Then A is Galois closed and the cyclic code $\mathcal{C}(A)$ therefore has dimension

$$n - |A| = 2^r - 1 - \sum_{i=0}^{b} \binom{r}{i} = \left(\sum_{i=0}^{r-b-1} \binom{r}{b} \right) - 1.$$

The smallest number having a duadic representation of weight $> b$ is $2^{b+1} - 1$. It follows that A contains the interval $[0, 2^{b+1} - 2]$ of length $2^{b+1} - 1$, and $\mathcal{C}(A)$ therefore has minimum distance $\geq 2^{b+1}$. Applying construction X to the chain $\mathcal{C}(A) \subset \mathcal{C}(A \setminus \{0\})$, we obtain a code with parameters $[2^r, \sum_{i=0}^{r-b-1} \binom{r}{b}), 2^{b+1}]_2$. It is more customary to use parameter a, where $r = a+b+1$. The corresponding code parameters are

$$[2^r, \sum_{i=0}^{a} \binom{r}{i}, 2^{r-a}]_2.$$

These are the **Reed-Muller codes** $\mathcal{RM}(a, r)$, which were constructed in Exercise 5.2.9 using the $(u, u + v)$-method. The Reed-Muller codes are a classical family of codes, first studied by D. E. Muller [156] and I. S. Reed [172], with close ties to affine geometry. They are also popular because they possess an effective decoding algorithm.

Generalized concatenation

The method of concatenation, as discussed in Chapter 5, can be generalized in various ways. A far-reaching generalization is the Zinoviev codes [227]. As an illustration we give here just one example of a Zinoviev code.

The cyclic theory provides us with a chain of binary cyclic codes $B_1 = [15, 4, 8]_2 \subset B_2 = [15, 5, 7]$. Let $A_1 = [16, 1, 16]_2$ be the repetition code and $A_2 = [16, 3, 14]_{16}$ an $\mathbb{F}_{16}$ linear Reed-Solomon code. Consider the space of

$$\text{matrices } M = \begin{pmatrix} u_1 & v_1 \\ u_2 & v_2 \\ \cdots & \cdots \\ u_{16} & v_{16} \end{pmatrix}$$

where $u_i \in \mathbb{F}_2, v_i \in \mathbb{F}_{16}$ and $u = (u_1, \ldots, u_{16}) \in A_1, v = (v_1, \ldots, v_{16}) \in A_2$. The number of matrices obtained is $2 \times 16^3 = 2^{13}$ and they form a binary vector space of dimension 13. Let $\alpha : \mathbb{F}_2 \times \mathbb{F}_{16} \longrightarrow B_2$ such that $\alpha(u, v) \in B_1$ if and only if $u = 0$. The binary code C we aim at consists of the codewords $(\alpha(u_1, v_1), \ldots, \alpha(u_{16}, v_{16}))$ where M varies over our space of matrices. Clearly C is a linear code $[16 \times 15, 13, d]_2$. What can we say about d? Assume at first $u_1 = \cdots = u_{16} = 1$. Then $\alpha(u_i, v_i) \neq 0$ for all i and the weight of our codeword is $\geq 16 \times 7 = 112$. The remaining case is $u_1 = \cdots = u_{16} = 0$. If the codeword is nonzero, then at least 14 of the v_i are nonzero. As $\alpha(0, v_i) \in B_1$, we obtain weight $\geq 14 \times 8 = 112$. We conclude that we obtained a $[240, 13, 112]_2$ code.

---

1. **construction X:** given a chain $[n, K, d]_q \supset [n, k, D]_q$ and an $[l, K - k, D - d]_q$ code, an $[n + l, K, D]_q$ code is constructed.

2. Construction XX is a repeated application of construction X.

Exercises 14.1

14.1.1. *Show that Theorem 5.3 follows from construction X.*

14.1.2. *Use construction X to obtain a $[12, 4, 6]_2$ code.*

14.1.3. *Prove the following rule for cyclic codes: $\mathcal{C}(A_1) \cap \mathcal{C}(A_2) = \mathcal{C}(A_1 \cup A_2)$.*

14.1.4. *Show that $[8, 4, 4]_4$ exists.*

14.1.5. *Construct a $[73, 8, 48]_4$ code using construction X.*

14.1.6. *Show that all codewords of $\mathcal{C}_2 \setminus \mathcal{C}_3$ (see Proposition 14.3) have weight $q^2 - q$.*

14.1.7. *Show that the nonzero weights of $\mathcal{C}_1$ from Proposition 14.3 are $q^2 - q - 1$, $q^2 - q$, $q^2 - 1$, q^2.*

14.1.8. *Show that all three parameters of $[28, 8, 15]_3$ are optimal.*

14.1.9. *Combine construction X with the Varshamov bound to prove the following construction from [75]:*
Assume $V_q(n - 1, d - 2) < q^{n-k}$ and the existence of an $[n - 1, k - 1, d + \delta]_q$ code. Then an $[n + \delta, k, d + \delta]_q$ code exists.

14.1.10. *A $[126, 36, 34]_2$ code is known to exist. Use the preceding exercise and the fact that $V_2(26, 126) < 2^{90}$ to show the existence of an $[133, 37, 34]_2$ code.*

14.2 Covering radius

and its relation to lengthening. Applications of covering codes.

Lengthening and the covering radius

The following parameter is important in its own right:

14.6 Definition. *The **covering radius** $\rho(C)$ of the code C is the smallest number ρ such that every n-tuple in the ambient space $\mathbb{F}_q^n$ has distance $\leq \rho$ from some codeword.*

Expressed in different words: the covering radius is the smallest number ρ such that the balls of radius ρ centered at the codewords cover all $\mathbb{F}_q^n$. Yet another way of expressing this: each coset $C + x$ of C contains a vector of weight $\leq \rho$. It follows that ρ is the maximum among the weights of coset leaders (compare Section 3.7).

Yet another description of the covering radius involves the check matrix H of C, a matrix with $n - k$ rows and n columns. Recall that the syndrome vector σ is defined by $\sigma(x) = xH^t \in \mathbb{F}_q^{n-k}$, where we think of $x \in \mathbb{F}_q^n$ as a row vector. The coordinates of $\sigma(x)$ are the dot products with the rows of H. All these dot products vanish if and only if $x \in C$. We have $\sigma(x) = \sigma(x')$ if and only if x, x' are in the same coset (we know all this from Section 3.7). How can we read off the covering radius after applying σ? Let $z \in \mathbb{F}_q^{n-k}$; write $z = \sigma(x)$. We can find x' of weight $\leq \rho$ such that $z = \sigma(x')$. This shows that z is a linear combination of at most ρ columns of H. It is clear that the reverse is also true.

14.7 Theorem. *For an $[n, k]_q$ code C, the following are equivalent:*

- *The balls of radius i cover all $\mathbb{F}_q^n$.*

- *Each coset of C contains a vector of weight $\leq i$.*

- *Each vector in $\mathbb{F}_q^{n-k}$ can be written as a linear combination of at most i columns of a check matrix of C.*

The covering radius ρ is the smallest number i satisfying the equivalent conditions in Theorem 14.7. The concept of the covering radius represents an idea which is different from what we have seen thus far. It is not an intrinsic parameter of the code, like the dimension, minimum distance and strength.

Rather it describes how the code is embedded in its ambient space $\mathbb{F}_q^n$. If we are interested primarily not in the distance or strength of a code, but in its covering radius, we also speak of a **covering code.** Covering codes are the subject of Cohen, Honkala, Litsyn, Lobstein [55].

Fix the field $\mathbb{F}_q$, length n and dimension k. In order to maximize the minimum distance, the codewords have to be spread out such that any two of them are fairly far from each other. In order to minimize the covering radius, the codewords have to be placed such that no vector from the ambient space is very far from a codeword. So the aims are slightly different. In the case of the minimum distance, we do not worry about the vectors outside the code at all, whereas in the case of the covering radius, we are not allowed to leave a vector behind (at a large distance from all codewords). However, it can be expected that these two optimization problems will be closely related. In both cases the codewords have to be spread out in a uniform way. For example, in order to construct a good covering code, how could it be helpful to use codewords which are very close to one another? So we suspect that a good covering code will typically have large minimum distance and vice versa. In one instance we have had occasion to observe this phenomenon already.

Recall the sphere-packing bound from Section 1.6. It relies on the observation that, for a code of minimum distance d, the balls of radius $e = \lfloor (d-1)/2 \rfloor$ are pairwise disjoint. This shows that the covering radius certainly satisfies $\rho \geq e$. Equality will happen precisely when the sphere-packing bound is met with equality, that is, when the code is a perfect code.

14.8 Proposition. *The covering radius ρ of a code with minimum distance d satisfies $\rho \geq e = \lfloor (d-1)/2 \rfloor$, with equality if and only if the code is perfect.*

In particular, the Hamming codes have covering radius $\rho = 1$, the ternary Golay code $[11, 6, 5]_3$ has $\rho = 2$ and the binary Golay code $[23, 12, 7]_2$ satisfies $\rho = 3$.

Proposition 14.8 shows that in some sense the covering radius is extremely small if and only if the minimum distance is extremely large. What happens at the other end of the spectrum? If $\rho \geq d$, then $\mathcal{C}$ is not maximal. This means that we can pick an element $x \in \mathbb{F}_q^n \setminus \mathcal{C}$ at distance $\geq d$ from $\mathcal{C}$. If the parameters of $\mathcal{C}$ are $[n, k, d]_q$, then $\mathcal{C} + \mathbb{F}_q x$ has parameters $[n, k+1, d]_q$. We have seen the following trivial fact:

14.9 Lemma. *Let $\mathcal{C}$ be an $[n, k, d]_q$ code. If $\rho(\mathcal{C}) \geq d$, then $\mathcal{C}$ can be embedded in an $[n, k+1, d]_q$ code.*

14.10 Proposition. *Let $\mathcal{C}$ be an $[n, k, d]_q$-code.*
Then the following are equivalent:

- $\rho(\mathcal{C}) \geq d - 1$.

- $\mathcal{C}$ can be extended to an $[n+1, k+1, d]_q$ code.

PROOF   Consider a check matrix H for $\mathcal{C}$. The existence of a one-step extension is equivalent to the existence of a vector $x \in \mathbb{F}_q^{n-k}$ (to be used as the $(n+1)$-st column), which is not linearly dependent on any set of $d-2$ columns of H. Theorem 14.7 shows that this is equivalent to $\rho(\mathcal{C}) \geq d-1$. ▯

The proof of Proposition 14.10 shows that the existence of one-step extensions can be expressed in terms of the covering radius.

14.11 Theorem. *Let $\mathcal{C}$ be a linear $[n, k, d]_q$ code with covering radius $\rho < d$. The maximum minimum distance of a one-step extension $[n+1, k+1]_q$ of $\mathcal{C}$ is $\rho + 1$.*

In a way Theorem 14.11 solves a problem from Section 5.1 concerning the existence of one-step extensions.

14.12 Corollary. *Let $\mathcal{C}$ be a linear $[n, k, d]_q$-code with covering radius $\rho < d-1$. Then $\mathcal{C}$ can be extended to an $[n+1, k+1, \rho+1]_q$ code, which can be lengthened to yield an $[n+d-\rho, k+1, d]_q$ code.*

PROOF   The first statement follows from Theorem 14.11. The second statement is then an application of construction X with the repetition code as auxiliary code. ▯

Covering codes and steganography

In this section we sketch an amusing link between covering codes and **steganography.** It is the aim of steganographic techniques to obscure the existence of a message from the adversary (think of invisible ink). Let us try to hide a small picture in a large picture in such a way that no third person will suspect the existence of the hidden picture. The big picture is given to us. It consists of a lot of pixels. Assume we wish to "extract" one bit from each pixel. We use a natural idea, block coding. The pixels are divided into blocks of N. Let $x = (x_1, x_2, \ldots, x_N) \in \mathbb{F}_2^N$ be the N-tuple of bits extracted from the given block of N pixels. We want to use this block to build n bits of the image to be hidden. Let $y = (y_1, y_2, \ldots, y_n)$ be the part of the hidden image to be constructed from x. Observe that we have no control over x and that the tuple y is given to us. We need to devise a function $f : \mathbb{F}_2^N \longrightarrow \mathbb{F}_2^n$. Assume f is given. We are in good shape when $f(x) = y$. Typically this is not the case. We have to change x, that is, to replace x by x' such that $f(x') = y$. This means that we have to change the pixels, thus doing damage to the original picture. As the changes made within the picture should not be evident, we are interested in changing a minimal amount of pixels. The number of pixels that have to be changed is the Hamming distance $d(x, x')$. This leads to the following conflicting aims in the construction of f : when N is given, then we want n to be large (our block of pixels should give a nonnegligible part of the hidden image), and we want the maximum possible $d(x, x')$ to be small (the number of pixels that have to be changed within each block should be small), say $d(x, x') \leq \rho$. This leads to the binary case of the following definition. There is in fact no reason not to consider arbitrary alphabets.

14.13 Definition. *Let $|\mathcal{A}| = q$. A function $f : \mathcal{A}^N \longrightarrow \mathcal{A}^n$ is a **covering function** $COV(\rho, N, n)_q$ if, for every $y \in \mathcal{A}^n$ and $x \in \mathcal{A}^N$, there is $x' \in \mathcal{A}^N$ such that $f(x') = y$ and $d(x, x') \leq \rho$.*

For our application we are interested in constructing binary covering functions $COV(\rho, N, n) = COV(\rho, N, n)_2$ such that the **change rate** ρ/N is small and the **information rate** n/N is large. Clearly, both rates are bounded by 1 and the aims are in conflict. Fix $y \in \mathcal{A}^n$. The defining property of Definition 14.13 says that every vector $x \in \mathcal{A}^N$ is at Hamming distance at most ρ from some word of $f^{-1}(y)$. In other words, $f^{-1}(y)$ is a code with **covering radius** $\leq \rho$ for every $y \in \mathcal{A}^n$. Equivalently we may say that $\mathcal{A}^N$ is partitioned into q^n codes, where each of these codes has covering radius $\leq \rho$.

14.14 Definition. *A **large set** of m covering codes $LCOV[m](N, \rho)_q$ is a partition of $\mathcal{A}^N$ into m subcodes, where each subcode has covering radius $\leq \rho$ (and $|\mathcal{A}| = q$).*

14.15 Theorem. *The following are equivalent:*

- *A covering function $COV(\rho, N, n)_q$.*

- *A large set $LCOV[q^n](N, \rho)_q$.*

It is natural to consider the special case of **linear** covering functions. In that case $f^{-1}(0)$ will be a linear space, and hence a linear covering code. As $f^{-1}(y)$ is a coset of $f^{-1}(0)$ for every y and these cosets have the same properties as the linear code, we see that the large set will exist automatically.

14.16 Theorem. *The following are equivalent:*

- *A linear covering function $COV(\rho, N, n)_q$.*

- *A linear code $[N, N - n]_q$ (q-ary, length N, codimension n) with covering radius $\leq \rho$.*

Let H be a check matrix of the linear code in Theorem 14.16. Then H is an (n, N) matrix. We may describe the corresponding linear covering function f by $f(x) = Hx$. Here we write our vectors as column vectors.

14.17 Definition. *Let $l(n, \rho; q)$ be the smallest number N such that there exists a linear q-ary code $\mathcal{C}$ of length N, dimension $N - n$ and covering radius $\leq \rho$. Analogously, for arbitrary q, let $l^*(n, \rho; q)$ be the smallest N such that $COV(\rho, N, n)_q$ exists.*

The function $l(n, \rho; q)$ has been studied by coding theorists for a long time, especially in the binary case. Theorem 14.16 shows that $l(n, \rho; q)$ is the smallest N such that a linear covering function $COV(\rho, N, n)_q$ exists.

In practice one would probably fix an upper bound x on the change rate ρ/N. For example, $x = 0.1$ guarantees that not more than one tenth of all pixels need to be changed. The problem is then to construct $COV(\rho, N, n)_2$ (in the binary case) such that $\rho/N \leq 0.1$ and $y = n/N$ as large as possible. The asymptotic sphere-packing bound (see Exercise 14.2.3) implies an upper bound $y \leq H_q(x)$, in the case at hand $y \leq h(0.1) \approx 0.469$. The Hamming code $[7, 4, 3]_2$ yields $COV(1, 7, 3)_2$, which implies a $COV(1, 10, 3)_2$ (see Exercise 14.2.4). This covering function has $x = 0.1$ and $y = 0.3$, which already is rather close to the upper bound of 0.469. The binary Golay code $[23, 12, 7]_2$ yields $COV(3, 23, 11)_2$ and $COV(3, 30, 11)_2$, achieving $y = 11/30$, a slight improvement upon the Hamming code. There is a $COV(6, 59, 23)_2$ in [55], implying $COV(6, 60, 23)_2$ and therefore $y = 23/60 \approx 0.383$.

Covering codes and switching noise

Switching noise is generated when the values of bits change (from 0 to 1 or from 1 to 0). It may be desirable, for technical reasons, to reduce this noise, and hence to devise a system such that, in each step, we have an upper bound on the number of bits which change their values. More precisely, the problem can be formalized as follows: A stream of bitstrings $x(t) \in \mathbb{F}_2^k, t = 1, 2, \ldots$ is presented to us. We want to encode these in bitstrings of length n

$$x(t) \longrightarrow y(t) \in \mathbb{F}_2^n$$

in such a way that for every t we have $d(y(t-1), y(t)) \leq R$ (at each step the number of bit switches is bounded by R) and such that we have an easy decoding algorithm, which allows us to recover $x(t)$ from $y(t)$. This problem is described in Chen and Curran [50] and the following construction is given:

Let $\mathcal{C}$ be a code $[n, n - k]_2$ with covering radius $\rho(\mathcal{C}) = R$. Define $x(0) = 0 \in \mathbb{F}_2^k$ and $y(0) = 0 \in \mathbb{F}_2^n$. The definition of $y(t)$ depends on $x(t)$ and on $y(t - 1)$ (of course). Let H be a check matrix of $\mathcal{C}$ and H^t its transpose. The procedure is as follows:

1. Compute $x(t) + x(t - 1)$.

2. Find $u(t) \in \mathbb{F}_2^n$ of weight $\leq R$ such that $x(t) + x(t - 1) = u(t)H^t$ (this is possible because of the definition of the covering radius; see Theorem 14.7).

3. Define $y(t) = y(t - 1) + u(t)$.

How does decoding work? Compute $y(t)H^t = y(t-1)H^t + x(t-1) + x(t)$. We see by induction that $y(t)H^t = x(t)$. This describes a very easy decoding method. We have described an (n, k) switching code. The price we pay is that we produce longer bit strings (length n instead of k). The benefit is the upper bound of $R = \rho(\mathcal{C})$ on the number of bit switches.

As an example, consider the repetition code $[3, 1, 3]_2$ of covering radius 1 (case $k = 2, n = 3, \rho = 1$). The length of the bitstrings increases by 50%; the encoded bitstrings change in at most one coordinate. This is better than using the Hamming code $[7, 4, 3]_2$ (case $k = 3, n = 7, \rho = 1$), where the length of the bitstrings increases by more than 100%.

---

1. The **covering radius** $\rho(\mathcal{C})$ is the smallest i such that the spheres of radius i centered at the codewords cover the ambient space.

2. Let $\mathcal{C}$ be an $[n, k, d]_q$ code, H a check matrix, $\rho = \rho(\mathcal{C})$. Then ρ is the smallest i such that each $y \in \mathbb{F}_q^{n-k}$ is a linear combination of at most i columns of H.

3. ρ is the maximum weight of a coset leader.

4. $\rho \geq d$ if and only if $\mathcal{C}$ can be embedded in an $[n, k+1, d]_q$.

5. $\rho < d$ is the maximum i such that $\mathcal{C}$ possesses an extension $[n+1, k+1, i+1]_q$.

6. A problem in **steganography** leads to covering functions, equivalently partitions of the ambient space into covering codes.

7. We saw an application of covering codes to switching noise.

Exercises 14.2

14.2.1. *Formulate and prove a sphere-packing bound for covering codes.*

14.2.2. *Characterize the case of equality in the sphere-packing bound for covering codes.*

14.2.3. *Prove the asymptotic sphere-packing bound for covering functions: Let $\alpha(x)$ be the lim sup of the information rate n/N for families of covering functions $COV(\rho, N, n)_q$ whose change rate is $\rho/N \le x$. Then $\alpha(x) \le H_q(x)$.*
  Here $H_q(x) = x\log_q(q-1) + \log_q(2)h(x)$ is the q-ary entropy function. It was introduced in Exercise 9.7.

14.2.4. *Show that the existence of $COV(\rho, N, n)_q$ implies the existence of $COV(\rho, N', n)_q$ for all $N' \ge N$.*

14.2.5. *Show that the existence of $COV(\rho, N, n)_q$ implies the existence of $COV(c\rho, cN, cn)_q$ for every natural number c.*

14.2.6. *Determine the covering radius of the repetition code $[n, 1, n]_q$.*

14.2.7. *Determine the covering radius of the extended Hamming code $[8, 4, 4]_2$.*

Chapter 15

The linear programming method

We are primarily interested in the powerful bounds on codes, orthogonal arrays and related objects which can be derived by the **LP** method. In the first section we start with a brief self-contained introduction to linear programming.

15.1 Introduction to linear programming

Basic concepts: Linear programming problems, duality.

After a brief introduction to linear programming, centered around the duality theorem, we show how the *LP* method, as applied by Delsarte [66] to obtain upper bounds on the number of code words, yields lower bounds on the number of rows of orthogonal arrays as well. A key result is that every upper bound on codes which can be derived from the LP bound, automatically yields a lower bound on orthogonal arrays.

In this first section we follow essentially Collatz and Wetterling [58]. Here is the basic **LP** problem in a standard form:

15.1 Definition. *Let the following be given:*

- *A real (m, n) matrix A,*

- *a column vector $\mathbf{p}$ of length n,*

- *a column vector $\mathbf{b}$ of length m.*

Assume $\mathbf{b} \geq 0$ *(meaning that the entries of* $\mathbf{b}$ *are nonnegative). Consider column vectors* $\mathbf{x}$ *of length* n *satisfying*

$$\mathbf{x} \geq 0 \text{ and } A \cdot \mathbf{x} = \mathbf{b}.$$

Under these conditions **minimize** $Q(\mathbf{x}) = \mathbf{p} \cdot \mathbf{x}$.

Let

$$\mathcal{M} = \{\mathbf{x} \mid \mathbf{x} \geq 0, A \cdot \mathbf{x} = \mathbf{b}\}$$

be the set of admissible points in n-dimensional space $\mathbb{R}^n$.

A toy example

Consider the case $m = 2, n = 3$, matrix $A = \begin{pmatrix} 2 & 0 & 1/2 \\ -5 & 1 & 0 \end{pmatrix}$ and $\mathbf{b} = (3, 5/3), \mathbf{p} = (3, 8, -11)$. What does $\mathcal{M}$ look like? We have $\mathbf{x} = (x_1, x_2, x_3)$ and the conditions are $x_1, x_2, x_3 \geq 0$ and

$$\begin{array}{|lll|} \hline 2x_1 & + 0.5x_3 & = 3 \\ -5x_1 + x_2 & & = 5/3 \\ \hline \end{array}$$

Use x_1 as parameter. Then x_2, x_3 can be expressed in terms of x_1 :

$$x_2 = 5/3 + 5x_1, x_3 = 6 - 4x_1.$$

The conditions are $x_1 \geq 0, x_1 \geq -1/3, x_1 \leq 1.5$, equivalently $0 \leq x_1 \leq 1.5$. It follows that $\mathcal{M}$ is a line segment:

$$\mathcal{M} = \{(x_1, 5/3 - 5x_1, 6 - 4x_1) \mid 0 \leq x_1 \leq 1.5\}$$

with end points $A = (0, 5/3, 6)$ and $B = (3/2, 55/6, 0)$. The minimum and maximum must be achieved in one of the endpoints. It suffices to compute $Q(A)$ and $Q(B)$. One is the maximum, the other is the minimum.

Obtaining standard form

Let us convince ourselves that any problem involving nonnegative unknowns and linear equations and inequalities can be transformed into the standard form of Definition 15.1. If the maximum is searched instead of the minimum, it suffices to replace $\mathbf{p}$ by $-\mathbf{p}$. If $\mathbf{b}$ should not satisfy the non-negativity condition, we should change the sign of the negative entries, at the same time replacing the corresponding row of A by its negative. A little trickier is the case when some of the main conditions are given in the form of inequalities instead of the equalities in our basic problem. Assume row i of A yields the condition $\sum_j a_{ij}x_j \leq b_i$. In that case one introduces an additional variable $y_i \geq 0$ and replaces the inequality above by $y_i + \sum_j a_{ij}x_j = b_i$. In the case of a lower inequality, one has to subtract y_i. Clearly $\mathbf{p}$ has to be replaced by a vector with entries 0 in the additional coordinates.

The vertices

It can be assumed that the rows of A are linearly independent. In fact, if a row is a linear combination of others, then the corresponding condition is either automatically satisfied or it contradicts the others. In the first case we can forget about that condition, reducing m by 1; in the second case $\mathcal{M}$ is empty.

15.2 Definition. *A* **convex combination** *of two points $P, Q \in \mathbb{R}^n$ is $x(t) = tP + (1-t)Q$ where $0 \leq t \leq 1$.*

A point set $\mathcal{M} \subseteq \mathbb{R}^n$ is **convex** *if it is closed under convex combinations: any convex combination of two points in $\mathcal{M}$ is in $\mathcal{M}$.*

The geometric intuition is: $x(t)$ is a point on the line segment between P and Q. In fact, $x(0) = Q, x(1) = P$. Think of $x(t)$ as a parametric curve, a point moving in space as time t increases from 0 to 1.

The set $\mathcal{M}$ of admissible points is convex. Formally this follows directly from the conditions

$$\mathbf{x} \geq 0 \text{ and } A \cdot \mathbf{x} = \mathbf{b}$$

describing $\mathcal{M}$: if points $\mathbf{x}, \mathbf{x}'$ satisfy them, then also convex combinations do.

This suggests how to define those special points, the vertices, where we think max and min will be reached:

15.3 Definition. *In the situation of Definition 15.1 a point $\mathbf{x} \in \mathcal{M}$ is a* **vertex** *of $\mathcal{M}$ if $\mathbf{x}$ cannot be written as a convex combination of two different points in $\mathcal{M}$.*

This means that there is no line segment which is completely contained in $\mathcal{M}$ and has $\mathbf{x}$ in its interior.

15.4 Lemma. *Let the columns of A be $\mathbf{c}_1, \ldots \mathbf{c}_n$, and let*

$$\mathbf{x} = (x_1, x_2, \ldots x_n)^t \in \mathcal{M}.$$

Then $\mathbf{x}$ is a vertex of $\mathcal{M}$ if and only if the columns $\mathbf{c}_k$ where $x_k > 0$ are linearly independent.

PROOF Let $\mathbf{x} \in \mathcal{M}$, and assume without restriction that $x_1, x_2, \ldots, x_r$ are the positive coordinates of $\mathbf{x}$.

Let $\mathbf{x}$ be a vertex. Assume the first r columns of A are dependent: $\sum_{k=1}^r d_k \mathbf{c}_k = 0$ for some $(d_1, \ldots, d_r) \neq 0$. Choose δ small enough such that $x_k \pm \delta \cdot d_k > 0, k = 1, 2, \ldots r$. Then $\sum_{k=1}^r \mathbf{c}_k(x_k \pm \delta \cdot d_k) = \mathbf{b}$. It follows that we can write $\mathbf{x} = \frac{1}{2}(\mathbf{x}_1 + \mathbf{x}_2)$ as a convex combination of vectors $\mathbf{x}_1$ and $\mathbf{x}_2$, where

$$\mathbf{x}_1 = \begin{cases} x_k + \delta \cdot d_k & k \leq r \\ 0 & k > r \end{cases}, \mathbf{x}_2 = \begin{cases} x_k - \delta \cdot d_k & k \leq r \\ 0 & k > r. \end{cases}$$

Thus $\mathbf{x}$ is not a vertex, a contradiction. This proves one direction of the claim.

Assume now the first r columns of A are independent. Write $\mathbf{x} = t \cdot \mathbf{x}_1 + (1 - t) \cdot \mathbf{x}_2$, where $0 < t < 1, \mathbf{x}_1, \mathbf{x}_2 \in \mathcal{M}$. Because of the positivity of t and $1 - t$, the entries of $\mathbf{x}_1, \mathbf{x}_2$ vanish at coordinates $> r$. It follows that $\sum_{k=1}^{r} c_k(a_k - b_k) = 0$, where a_k, b_k are the coordinates of $\mathbf{x}_1, \mathbf{x}_2$, respectively. As these columns are independent, we get $a_k = b_k$, hence $\mathbf{x}_1 = \mathbf{x}_2$ and $\mathbf{x}$ is a vertex. □

15.5 Corollary. *If $\mathbf{x}$ is a vertex of $\mathcal{M}$, then the number of nonzero entries, the* **weight** *$wt(\mathbf{x})$ of $\mathbf{x}$, is at most m. If $wt(\mathbf{x}) < m$, then the vertex $\mathbf{x}$ is called* **degenerate.**

Note the similarity between this notion and the basic notion of coding theory. We also see that the vertex $\mathbf{x}$ is uniquely determined by the corresponding set of linearly independent columns of A. In particular, $\mathcal{M}$ has only a finite number of vertices. It is not clear yet when a linearly independent set of columns of A does yield a vertex of $\mathcal{M}$.

The cows/sheep example

This example is from [58]. The variables x_1, x_2 are the numbers of cows and sheep on a farm. There are stables for at most 50 cows and 200 sheep. Seventy-two acres are available, and each cow needs one acre, each sheep needs 0.2 acres. A cow needs 150 yearly hours of work, a sheep needs 25. At most 10,000 yearly hours are available. The net gain per year is 250 per cow, 45 per sheep. The problem is to maximize the profit. The conditions are

$$
\begin{array}{rrcl}
x_1, & x_2 & \geq & 0 \\
x_1 & & \leq & 50 \\
& x_2 & \leq & 200 \\
x_1 & + 0.2x_2 & \leq & 72 \\
150x_1 & + 25x_2 & \leq & 10,000
\end{array}
$$

Maximize $Q(x) = 250x_1 + 45x_2$.

This is not in standard form. When the region of admissible points is sketched in the plane, the vertices become visible:

$P_0 = (0, 0), P_1 = (0, 200), P_2 = (32, 200), P_3 = (40, 160), P_4 = (50, 100), P_5 = (50, 0).$

The maximum occurs in P_3 : 40 cows and 160 sheep is the optimal choice.

Continuing with the vertices

15.6 Lemma. *If $\mathcal{M} \neq \emptyset$, then $\mathcal{M}$ does have vertices. More precisely, every vector in $\mathcal{M}$ of minimal weight is a vertex.*

PROOF If $0 \in \mathcal{M}$, then it certainly is a vertex. So now let $wt(\mathbf{x}) = r > 0$ and assume r is the minimum weight among elements of $\mathcal{M}$. Without restriction $x_1 \cdot x_2 \cdot \ldots x_r \neq 0$. If $\mathbf{x}$ is not a vertex, then $\sum_{i=1}^{r} d_i \mathbf{c}_i = 0$ for some $(d_1, \ldots, d_r) \neq 0$.
Let $\mathbf{x}(\lambda) = \mathbf{x} + \lambda \cdot (d_1, d_2, \ldots, d_r, 0, \ldots, 0)$. Then $\mathbf{x}(\lambda)$ still satisfies the equality in Definition 15.1. If λ is small enough, then $\mathbf{x}(\lambda)$ will still be a member of $\mathcal{M}$. We can choose λ such that the weight of $\mathbf{x}(\lambda)$ is smaller than r, a contradiction.
□

Three cases arise:

- If $\mathcal{M}$ is empty, then the conditions of our problem can never be satisfied.

- If $\mathcal{M}$ is bounded, then it is called a **convex polyhedron.** We shall see later on that in this case the problem possesses a minimum and that this minimum is attained in one of the vertices.

- Finally, it can happen that $\mathcal{M}$ is unbounded. Then either the objective function Q attains arbitrarily small values or else Q does again attain its minimum.

15.7 Definition. $\mathbf{x} \in \mathcal{M}$ *is a **minimal point** if $Q(\mathbf{x}) \leq Q(\mathbf{y})$ for all $\mathbf{y} \in \mathcal{M}$.*

The set of minimal points is convex. It is in fact the intersection of a hyperplane with $\mathcal{M}$.

15.8 Theorem. *If $\mathcal{M}$ is a convex polyhedron (bounded, nonempty), then every point in $\mathcal{M}$ is a convex combination of vertices.*

PROOF Let $\mathbf{x} \in \mathcal{M}, wt(\mathbf{x}) = r$ and Z the support of $\mathbf{x}$. If $r = 0$, there is no problem. Let $r > 0$. If the columns indexed by Z are independent, then $\mathbf{x}$ is itself a vertex. So assume there is a nontrivial linear combination of the columns of A, with coefficients $d_1, \ldots, d_n$, which vanish outside Z. As before, consider $\mathbf{x}(\lambda) = \mathbf{x} + \lambda \cdot (d_1, \ldots, d_n)$. Then $\mathbf{x}(\lambda) \in \mathcal{M}$ if and only if $\mathbf{x}(\lambda) \geq 0$, if and only if $\lambda_1 \leq \lambda \leq \lambda_2$, where $\lambda_1 < 0$ is the smallest negative such value and $\lambda_2 > 0$ is the largest such value. Clearly the weights of $\mathbf{x}(\lambda_1)$ and of $\mathbf{x}(\lambda_2)$ are smaller than r. By induction they are convex combinations of vertices. It follows that the same holds for $\mathbf{x}$. □

15.9 Theorem. *If Q has a minimum in $\mathcal{M}$, then this minimum is attained in a vertex.*

PROOF   There is a difficulty only when $\mathcal{M}$ is unbounded. Let $\mathbf{x}_0$ be a minimal point which is not a vertex, $\mathbf{x}_1, \mathbf{x}_2, \ldots, \mathbf{x}_p$ the vertices of $\mathcal{M}$. The trick is to consider the sum C_i of the coordinates of each $\mathbf{x}_i, i = 0, 1, \ldots, p$. Let C be the maximum of these numbers. If $C = 0$, then 0 is the only vertex and minimal point. So assume $C > 0$. Consider a new LP in standard form: add a new indeterminate x_{n+1}, do not change the old conditions, add a new row to A which corresponds to the condition $x_1 + x_2 + \ldots x_{n+1} = 2C$, do not change Q. The corresponding space $\tilde{\mathcal{M}}$ of admissible points is certainly bounded because of the last condition. The projection onto the first n coordinates maps $\tilde{\mathcal{M}}$ onto a corresponding bounded part of $\mathcal{M}$, while preserving the objective function Q. In particular, $\tilde{\mathbf{x}}_0$ ($\mathbf{x}_0$ with last coordinate chosen such that the sum of the coordinates is $2C$) is a minimal point. We distinguish two classes of vertices in $\tilde{\mathcal{M}}$. The $\tilde{\mathbf{x}}_i, i = 1, 2, \ldots, p$ are those with last coordinate > 0. The remaining vertices $\tilde{\mathbf{x}}_i, i = p+1, \ldots, r$ have vanishing last coordinate. Express $\tilde{\mathbf{x}}_0 = \sum_j \alpha_j \tilde{\mathbf{x}}_j$ as a convex combination of the vertices. The last coordinate shows that there must be $j \leq p$ such that $\alpha_j > 0$. It follows that $\tilde{\mathbf{x}}_j$ is a minimal point of $\tilde{\mathcal{M}}$ and vertex $\mathbf{x}_j$ is a minimal point of the original LP.   ☐

15.10 Definition. *Let $\mathbf{x}$ be a vertex (we know $wt(\mathbf{x}) \leq m$). A **basis** of $\mathbf{x}$ is a set of m independent columns of A containing the support of $\mathbf{x}$. Observe that the basis is uniquely determined if $\mathbf{x}$ is nondegenerate.*

The vertex exchange

Let $\mathbf{x}_0$ be a vertex with coordinates $x_k^{(0)}, Z$ the set of indices of a basis of $\mathbf{x}_0$. We have $\sum_{k \in Z} x_k^{(0)} \mathbf{c}_k = \mathbf{b}$. Each column-vector of A can be represented in terms of this basis. There must therefore exist constants α_{ki} such that

$$\mathbf{c}_i = \sum_{k \in Z} \alpha_{ki} \cdot \mathbf{c}_k, i = 1, 2, \ldots, n. \tag{15.1}$$

If $j \in Z$, then certainly $\alpha_{kj} = \delta_{kj}$. Assume there is an $\alpha_{\hat{k}j} > 0$, where $\hat{k} \in Z, j \notin Z$. For every $\delta \geq 0$, consider $\mathbf{x}(\delta)$ with the following coordinates:

$$x_k(\delta) = \begin{cases} x_k^{(0)} - \delta \cdot \alpha_{kj} & \text{if } k \in Z \\ \delta & \text{if } k = j \\ 0 & \text{otherwise.} \end{cases} \tag{15.2}$$

Then $A \cdot \mathbf{x}(\delta) = \mathbf{b} - \delta \cdot \mathbf{c}_j + \delta \cdot \mathbf{c}_j = \mathbf{b}$. Put

$$\delta_1 = Min_{k,\alpha_{kj}>0}(x_k^{(0)}/\alpha_{kj}). \tag{15.3}$$

If $\mathbf{x}_0$ is degenerate, then it is possible that $\delta_1 = 0$. In the nondegenerate case, $\delta_1 > 0$. The next vertex is $\mathbf{x}_1 = \mathbf{x}(\delta_1)$ (see Definition 15.2). First of all it is clear that we have an admissible point, by the choice of δ_1. Is $\mathbf{x}_1$ really a vertex? Certainly $wt(\mathbf{x}_1) \le m$, as we have only one new nonzero coordinate when compared with $\mathbf{x}_0$, but at least one nonzero coordinate l of $\mathbf{x}_0$ leads to a vanishing coordinate in $\mathbf{x}_1$.

The columns indexed by $Z' = Z - \{l\} \cup \{j\}$ form a basis of $\mathbf{x}_1$. In fact, the cardinality is right. Each vector is expressible in terms of the c_k, where $k \in Z$ (because Z is a basis). As c_l is expressible in terms of c_j and the $c_k, k \in Z \setminus \{l\}$, it follows that we have a generating set, and thus a basis.

If $\mathbf{x}_0$ is degenerate and j is chosen appropriately ($\delta_1 = 0$), then $\mathbf{x}_1 = \mathbf{x}_0$ with a new basis. Now express now the columns of A with respect to the new basis given by Z' in analogy with Equation 15.1:

$$\mathbf{c}_i = \sum_{k \in Z'} \alpha'_{ki} \cdot \mathbf{c}_k, i = 1, 2, \ldots, n.$$

Comparing the two expressions, the new structure constants are obtained from the old ones:

$$\begin{aligned}
\alpha'_{jl} &= 1/\alpha_{lj} \\
\alpha'_{ji} &= \alpha_{li}/\alpha_{lj} \\
\alpha'_{kl} &= -\alpha_{kj}/\alpha_{lj} \quad (k \in Z, k \ne l) \\
\alpha'_{ki} &= \alpha_{ki} - \frac{\alpha_{li}\alpha_{kj}}{\alpha_{lj}} \quad (k \in Z, k \ne l).
\end{aligned} \tag{15.4}$$

Let $\mathbf{x} \in \mathcal{M}$. By definition $\sum_{i=1}^{n} x_i \cdot \mathbf{c}_i = \mathbf{b}$. The same thing holds for the vertex $\mathbf{x}_0$. Compare:

$$\sum_{k \in Z} x_k^{(0)} \mathbf{c}_k = \mathbf{b} = \sum_{i=1}^{n} x_i \sum_{k \in Z} \alpha_{ki} \mathbf{c}_k.$$

Comparing coefficients and observing that, for $k, i \in Z$, the coefficient α_{ki} is nonzero only if $k = i$, we obtain

$$x_k = x_k^{(0)} - \sum_{i \notin Z} \alpha_{ki} x_i \quad (k \in Z). \tag{15.5}$$

The Simplex algorithm

The idea is to perform a series of vertex exchanges with decreasing value of the objective function Q. So again let a vertex $\mathbf{x}_0$ with basis indexed by Z be given; the corresponding value of the objective function is $Q_0 = \sum_{k \in Z} p_k \cdot x_k^{(0)}$. Put

$$t_i = \sum_{k \in Z} \alpha_{ki} p_k.$$

For an arbitrary point $\mathbf{x} \in \mathcal{M}$, we have $Q(\mathbf{x}) = \sum_{k \in Z} p_k x_k + \sum_{i \notin Z} p_i x_i = \sum_{k \in Z} p_k x_k^{(0)} + \sum_{i \notin Z}(p_i - \sum_{k \in Z} p_k \alpha_{ki}) x_i$, hence

$$Q(\mathbf{x}) = Q_0 - \sum_{i \notin Z}(t_i - p_i) x_i.$$

Observe the following facts:

1. If $\mathbf{x}_0$ is nondegenerate, $\hat{k} \in Z, j \notin Z, t_j > p_j, \alpha_{\hat{k}j} > 0$, then the vertex exchange yields a vertex $\mathbf{x}_1$ with a smaller value of the objective function: $Q_1 < Q_0$ (we have $x_i^{(1)} = 0$ if $i \notin Z, i \neq j$, and $x_j^{(1)} = \delta_1 > 0$. It follows $Q_1 = Q_0 - \delta_1(t_j - p_j) < Q_0$).

2. If, for some $j \notin Z$, we have $t_j > p_j, \alpha_{kj} \leq 0$ for all $k \in Z$, then the LP attains arbitrarily small values $(\mathbf{x}(\delta) \in \mathcal{M}$ all the time and $Q(\mathbf{x}(\delta)) = Q_0 - \delta(t_j - p_j)$.

3. If $t_j \leq p_j$ for all $j \notin Z$, then $\mathbf{x}_0$ is a minimal point $(Q(\mathbf{x}) = Q_0 - \sum_{j \notin Z}(t_j - p_j) x_j \geq Q_0)$.

We can describe the **simplex algorithm** now. To a vertex $\mathbf{x}$ and basis given by Z belongs a rectangular array: the columns are indexed by $i \notin Z$, the rows are indexed by $k \in Z$, the entry in row k and column i is α_{ki}. There is a bottom row with entries $d_i = t_i - p_i = \sum_{k \in Z} \alpha_{ki} p_k - p_i$ and there is a right column containing the coefficients of $\mathbf{x}$ with respect to Z. Write the value of the objective function in the lower right cell. The following steps have to be taken:

1. If $d_i \leq 0$ for all i, then the minimum is reached.

2. Choose $d_j > 0$.

3. If all $\alpha_{kj} \leq 0$, then the LP attains arbitrarily small values: there is no minimum.

4. If there is $k \in Z$ such that $\alpha_{kj} > 0$, insert x_k/α_{kj} in an additional column, for all the corresponding k, let δ_1 be the minimum of all these fractions, l a value of k for which this minimum is attained. The **pivot column** is column j, the **pivot row** is row l. The intersection, with entry α_{lj}, is the **pivot element**. A new array is constructed as follows: Indices l and j change places, the pivot element α_{lj} is replaced by its reciprocal $1/\alpha_{lj}$, the remaining entries of the pivot row are divided by α_{lj}, the remaining entries of the pivot column are divided by $-\alpha_{lj}$, all remaining entries are transformed according to the **rectangle rule:** if d is the entry in question in the old array, consider the submatrix $\begin{pmatrix} a & b \\ c & d \end{pmatrix}$, where b and c are the projections to the pivot column and the pivot row, respectively, a is the pivot element (all this in the old array). Then $d \longmapsto d - \dfrac{b \cdot c}{a}$. These rules are also valid on the border of the array (last row and column). The new value of the objective function is $Q_1 = Q_0 - x_l d_j/\alpha_{lj}$.

An example

Let

$$A = \begin{pmatrix} 2 & 4 & -1 & 1 & 0 & 0 \\ -3 & 2 & -2 & 0 & 1 & 0 \\ 0 & -1 & -3 & 0 & 0 & 1 \end{pmatrix}, b = (9, 4, 5).$$

Then a vertex is clearly visible:

$$x^{(0)} = (0, 0, 0, 9, 4, 5), \text{ with basis } Z = \{4, 5, 6\}.$$

Because of the unit matrix in the right half of A, the coefficients α are the entries in the left half, for example,

$$\alpha_{4,2} = 4, \alpha_{5,2} = 2, \alpha_{6,2} = -1.$$

As positive values occur, we can perform a vertex exchange with $j = 2$. Which of the columns of Z will be kicked out? Compute $\delta_1 = Min(9/4, 2) = 2$. It follows that $Z' = \{2, 4, 6\}$. The new vertex is

$$x^{(1)} = x(2) = x^{(0)} + (0, 2, 0, 0, 0, 0) - 2(0, 0, 0, 4, 2, -1) = (0, 2, 0, 1, 0, 7).$$

Finding a starting vertex

If new variables have been introduced, as described in the beginning, then $\mathbf{x} = 0, \mathbf{y} = \mathbf{b}$ is a vertex. In general, it is not obvious if the conditions are satisfiable at all (or $\mathcal{M} = \emptyset$). In order to decide this and to find a first vertex, one has to solve a related problem beforehand:

$$A \cdot \mathbf{x} + \mathbf{y} = \mathbf{b}, \quad \mathbf{x} \geq 0, \mathbf{y} \geq 0, \quad \sum y_i = Min.$$

A vertex of this problem is $(0, \mathbf{b})$. Let $(\mathbf{x}^*, \mathbf{y}^*)$ be an optimal solution of this problem. If $\mathbf{y}^* \neq 0$, then our primary problem has no solution, for each solution of the primary problem yields $(\mathbf{x}^*, 0)$. If, however, $\mathbf{y}^* = 0$, then $\mathbf{x}^*$ is a vertex of the primary problem. Algorithmically:

1. Let an LP in standard form be given, where $b \geq 0$.

2. Set up the preliminary problem above.

3. Solve it with the Simplex algorithm, using $(0, b)$ as starting vertex.

4. If the minimum value is > 0, then $\mathcal{M} = \emptyset$.

5. If the minimum value $= 0$, with minimum point $(x, 0)$, then x is a vertex for the original problem.

6. Start the Simplex algorithm there.

Examples

Here is an LP: maximize $x + y + z$ subject to $x, y, z \geq 0$ and

$$
\begin{array}{rl}
-x +3y +7z \leq & 9 \\
x \qquad\quad -5z \leq & 9 \\
x \;-2y +7z \leq & 21 \\
-3x +3y -7z \leq & 63 \\
z \leq & 1
\end{array}
$$

In order to obtain a basic LP problem, introduce five additional variables y_1, y_2, y_3, y_4, y_5, as described after Definition 15.1. The starting vertex is then $\mathbf{x}_0 = (0, 0, 0, 9, 9, 21, 63, 1)$, and the basis is $Z = \{4, 5, 6, 7, 8\}$. Use objective function $Q = -x - y - z$. The starting array is

	1	2	3	$\mathbf{x}_0$
4	-1	3	7	9
5	1	0	-5	9
6	1	-2	7	21
7	-3	3	-7	63
8	0	0	1	1
	1	1	1	$Q_0 = 0$

As the bottom entry in the first column is positive, choose this column as the pivot column. The pivot element is printed in **bold.** Proceeding in this fashion, the following arrays are obtained:

	5	2	3	$\mathbf{x}_1$
4	1	3	2	18
1	1	0	-5	9
6	-1	-2	12	12
7	3	3	-22	90
8	0	0	1	1
	-1	1	6	$Q_1 = -9$

	5	4	3	$\mathbf{x}_2$
2	1/3	1/3	2/3	6
1	1	0	-5	9
6	-1/3	2/3	40/3	24
7	2	-1	-24	72
8	0	0	1	1
	-4/3	-1/3	16/3	$Q_2 = -15$

	5	4	8	$\mathbf{x}_3$
2			-2/3	16/3
1			5	14
6			-40/3	
7			24	
3			1	1
	-4/3	-1/3	-16/3	$Q_3 = -20\frac{1}{3}$

As the entries d_i in the last row are negative, we have reached the minimum. Observe that there is no need to calculate all the entries in this last array.

The cows/sheep example again

This was the following problem:

$$
\begin{array}{rrcl}
x_1, & x_2 & \geq & 0 \\
x_1 & & \leq & 50 \\
 & x_2 & \leq & 200 \\
x_1 & + 0.2x_2 & \leq & 72 \\
150x_1 & + 25x_2 & \leq & 10,000
\end{array}
$$

Minimize $Q(x) = -250x_1 - 45x_2$. In standard form we work with equalities, using four new nonnegative variables $y_1, \ldots, y_4$. The starting vertex is $(0, 0, 50, 200, 72, 10000)$. The first tableau is

	1	2	$\mathbf{x}_0$
3	1	0	50
4	0	1	200
5	1	0.2	72
6	150	25	10000
	250	45	$Q_0 = 0$

The rest is by now automatic:

	3	2	$\mathbf{x}_1$
1	1	0	50
4	0	1	200
5	-1	0.2	22
6	-150	**25**	2500
	-250	45	$Q_1 = -12500$

	3	6	$\mathbf{x}_2$
1	1	0	50
4	6	$-.04$	100
5	**0.2**	$-.008$	2
2	-6	0.04	100
	20	-1.8	$Q_2 = -17000$

	5	6	$\mathbf{x}_3$
1	-5	0.04	40
4	-30	.2	40
3	5	$-.04$	10
2	30	-0.2	100
	-100	-1	$Q_3 = -17200$

Duality theorems

15.11 Theorem (duality, first form). *Consider the following LP:*

(G) : *A a real (m,n) matrix of rank m, column vectors $\mathbf{p}$ of length n and $\mathbf{b}$ of length m. Search for $\mathbf{x}$ of length n satisfying*

$$\mathbf{x} \geq 0, \quad A\mathbf{x} = \mathbf{b}.$$

Minimize $Q(\mathbf{x}) = \mathbf{p} \cdot \mathbf{x}$.

(G^*) : *Search for a column vector $\mathbf{w}$ of length m satisfying*

$$A^t\mathbf{w} \leq \mathbf{p}.$$

Maximize $Q^*(\mathbf{w}) = \mathbf{w} \cdot \mathbf{b}$.

Then the following hold:
If $\mathbf{x}$ is admissible for (G), and $\mathbf{w}$ is admissible for (G^*), then $Q(\mathbf{x}) \geq Q^*(\mathbf{w})$ (this is clear: $Q(\mathbf{x}) = \mathbf{p} \cdot \mathbf{x} \geq \mathbf{w}^t A\mathbf{x} = \mathbf{w} \cdot \mathbf{b} = Q^*(\mathbf{w})$).
Exactly then does (G) possess an optimal solution if (G^*) does. In that case these optimal values are equal.

PROOF

first part: Let (G) attain its minimum value at vertex
$\mathbf{x}_0 = (x_1^{(0)}, x_2^{(0)}, \ldots, x_m^{(0)}, 0, \ldots, 0)^t$, basis $Z = \{1, 2, \ldots m\}$. Put $C = (\alpha_{ki})_{m,n}, T = (t_1, t_2, \ldots, t_n)^t$. We know $t_i \leq p_i$. Restrict everything to the first m coordinates: $\tilde{A} = (\mathbf{c}_1, \ldots, \mathbf{c}_m), \tilde{\mathbf{x}}_0 = (x_1^{(0)}, \ldots, x_m^{(0)})^t, \tilde{\mathbf{p}} = (p_1, p_2, \ldots, p_m)^t$. Then $\tilde{A}\tilde{\mathbf{x}}_0 = \mathbf{b}, \tilde{A}C = A$. Now $\tilde{A}$ is a regular quadratic matrix. It follows that $\tilde{\mathbf{x}}_0 = \tilde{A}^{-1}\mathbf{b}, C = \tilde{A}^{-1}A$. By definition of the entries of T, $\tilde{\mathbf{p}}^t C = T \leq \mathbf{p}^t$. We want an admissible point of the dual problem yielding the same value of the objective function:
Put $\mathbf{w}_0 = (\tilde{A}^{-1})^t \cdot \tilde{\mathbf{p}}$. First of all, this is admissible for (G^*) : $\mathbf{w}_0^t A = \tilde{\mathbf{p}}^t \cdot (\tilde{A}^{-1}A) = \tilde{\mathbf{p}}^t C \leq \mathbf{p}^t$. Its value is
$Q^*(\mathbf{w}_0) = \mathbf{w}_0^t \mathbf{b} = \tilde{\mathbf{p}}^t \tilde{A}^{-1}\mathbf{b} = \tilde{\mathbf{p}}^t \tilde{\mathbf{x}}_0 = Q(\mathbf{x}_0)$.

second part: Now let $\mathbf{w}$ be a maximal solution of (G^*). Remember that there was no restriction on the sign of the entries. Split $\mathbf{w}$ in a nonnegative and a nonpositive part: $\mathbf{w} = \mathbf{w}_1 - \mathbf{w}_2$, where $\mathbf{w}_1, \mathbf{w}_2 \geq 0$. Further, $\mathbf{w}_3 = \mathbf{p} - A^t \cdot \mathbf{w}$. Then (G^*) is equivalent to

$$A^t(-\mathbf{w}_1 + \mathbf{w}_2) - \mathbf{w}_3 = -\mathbf{p}, \quad \mathbf{w}_i \geq 0, \quad \mathbf{b}^t(-\mathbf{w}_1 + \mathbf{w}_2) = Min.$$

Put $S = (-A^t \mid A^t \mid -I_n), \mathbf{r}^t = (-\mathbf{b}^t \mid \mathbf{b}^t \mid 0)$. We have the LP

$$S \cdot \mathbf{v} = -\mathbf{p}, \quad \mathbf{v} \geq 0, \quad \mathbf{r}^t \cdot \mathbf{v} = Min.$$

By the first part, the LP which is dual to that one has a finite maximum whose value is $-maximum(G^*)$. This dual is

$$S^t \cdot \mathbf{x} \leq \mathbf{r}, \quad \mathbf{x}^t(-\mathbf{p}) = Max,$$

meaning $-A\mathbf{x} \leq -\mathbf{b}, A\mathbf{x} \leq \mathbf{b}, -\mathbf{x} \leq 0$. We see that this is (G) again.

$\square$

Problems (G) and (G^*) of Theorem 15.11 are called **duals** of one another. This was the hard part. Another form of duality follows easily:

15.12 Theorem (duality, symmetric form). *Consider the following LP:*

(U) : *Let a real (m, q) matrix A, column vectors $\mathbf{p}$ of length q and $\mathbf{b}$ of length m be given. Assume $rk(A) = m$. Search for a column-vector $\mathbf{x}$ of length q satisfying*

$$\mathbf{x} \geq 0, A\mathbf{x} \geq \mathbf{b}.$$

Minimize $Q(\mathbf{x}) = \mathbf{p} \cdot \mathbf{x}.$

(U^*) : *Search for a column vector* $\mathbf{w}$ *of length* m *satisfying*

$$A^t\mathbf{w} \leq \mathbf{p}, \ \mathbf{w} \geq 0.$$

Maximize $Q^*(\mathbf{w}) = \mathbf{w} \cdot \mathbf{b}$.

Then (U) *and* (U^*) *are dual problems in the same sense as in the preceding theorem.*

PROOF   We can simply reduce this to the first form of duality. In fact, (U) is equivalent to

$$A\mathbf{x} - \mathbf{y} = b, \ \mathbf{x}, \ \mathbf{y} \geq 0, \ \mathbf{p} \cdot \mathbf{x} = Min.$$

The corresponding dual problem is (U^*).   ▯

A consequence of the duality theorem has to be mentioned:

15.13 Theorem (Farkas-alternative). *Given a real matrix A and a real vector $\mathbf{b}$, exactly one of the following equations possesses a solution:*

- $A\mathbf{x} = \mathbf{b}, \ \mathbf{x} \geq 0.$

- $\mathbf{y}^t A \geq 0, \ \mathbf{y} \cdot \mathbf{b} < 0.$

PROOF   Consider the duality theorem in its first form, with $\mathbf{p} = 0$. The corresponding primal problem (G) has 0 as the only possible value. Assume both Farkas alternatives have a solution. Then $(\mathbf{y}^t A)\mathbf{x} \geq 0$ for one reason and $\mathbf{y}^t(A\mathbf{x}) = \mathbf{y} \cdot \mathbf{b} < 0$ for another reason, a contradiction. It remains to show that at least one of the alternatives has a solution.

Assume the first alternative has no solution. The dual problem certainly has admissible values ($\mathbf{y} = 0$ gives one). By the duality theorem, the dual problem takes on arbitrarily large values. Working with $-\mathbf{y}$ instead, this shows that the second alternative has solutions.

Assume the second alternative has no solution: whenever $\mathbf{y}^t A \geq 0$, then $\mathbf{y} \cdot \mathbf{b} \geq 0$. Equivalently, whenever $\mathbf{y}^t A \leq 0$ then $\mathbf{y} \cdot \mathbf{b} \leq 0$. This means that G^* has a maximum. By the duality theorem (G) has a finite minimum. In particular, the first Farkas alternative has solutions.   ▯

It has become apparent how the dual problem looks like in general. We want to describe this in detail.

The dual problem in general

Let the primal problem be given by

- a real (m, n) matrix A, a column vector $\mathbf{p}$ of length n and another column vector $\mathbf{b} \geq 0$ of length n.

- Each row $\mathbf{z}_i$ of A yields an equation or an inequality ($\mathbf{z}_i\mathbf{x} = b_i$ or $\mathbf{z}_i\mathbf{x} \geq b_i$ or $\mathbf{z}_i\mathbf{x} \leq b_i$).

- The objective function is $Q(\mathbf{x}) = \mathbf{x} \cdot \mathbf{p} = Min$.

Then the dual problem looks as follows:

- Each row $\mathbf{z}_i$ of A leads to an indeterminate w_i. The main condition is $A^t \cdot \mathbf{w} \leq \mathbf{p}$.

- If the corresponding row of A yields an equation in the primal problem, then there is no restriction on w_i. If $\mathbf{z}_i\mathbf{x} \geq b_i$ in the primal problem, then $w_i \geq 0$; if $\mathbf{z}_i\mathbf{x} \leq b_i$, then $w_i \leq 0$.

- The objective function is $Q^*(\mathbf{w}) = \mathbf{w} \cdot \mathbf{b} = Max$.

These conditions include everything, also the nonnegativity conditions of the variables. As an example, consider type (G), which we wrote as $x \geq 0, Ax = b$ with objective function $x \cdot p$. In the general terminology we would use an $(n + m, n)$ matrix with I in the first n rows and A under it. We have lower inequalities as the first n conditions, equalities in the remaining m rows. The dual problem has unknowns $y_1, \ldots, y_n \geq 0$ and unrestricted $w_1, \ldots, w_m$. The main conditions are $y_i + (A^t w)_i \leq p_i$ for $i = 1, \ldots, n$. As $y_i \geq 0$, this is equivalent to $A^t w \leq p$. The objective function is $w \cdot b$. This is our dual problem (G^*).

Remark

One might think that the primal problem has no solution if and only if the dual problem admits all values. However, this does not follow from the duality theorems and it is in fact not true, as the following example shows.

Maximize $2x_1 - x_2$ subject to the conditions

$$x_1, x_2 \geq 0, \ x_1 - x_2 \leq 1, \ -x_1 + x_2 \leq -2.$$

Clearly there is no solution. Consider the dual problem. The variables are w_1, w_2, w_3, w_4, where $w_1, w_2 \geq 0$ and $w_3, w_4 \leq 0$. The inequalities are

$$w_1 + w_3 + w_4 \leq -2 \text{ and } w_2 - w_3 - w_4 \leq 1.$$

Addition yields $w_1 + w_2 \leq -1$, a contradiction. This shows that the dual problem has no solutions either.

The dual of the cows/sheep example

The cows/sheep example in standard form was

$$A = \begin{pmatrix} 1 & 0 & 1 & 0 & 0 & 0 \\ 0 & 1 & 0 & 1 & 0 & 0 \\ 1 & 0.2 & 0 & 0 & 1 & 0 \\ 150 & 25 & 0 & 0 & 0 & 1 \end{pmatrix}, b = (50, 200, 72, 10000)^t, \mathbf{p} = (-250, -45, 0, 0, 0, 0)^t.$$

Its minimal value was -17200.

The dual problem has four unknowns and six conditions. The vector of unknowns is $\mathbf{w} = (w_1, w_2, w_3, w_4)^t$, the condition is $A^t \mathbf{w} \le \mathbf{p}$ and the objective function $\mathbf{b} \cdot \mathbf{w}$ has to be maximized. The last four inequalities are $w_i \le 0$. We should work with $z_i = -w_i \ge 0$ instead. Write the two remaining as inequalities, using two new variables z_5, z_6 :

$$\boxed{\begin{array}{llll} z_1 & + \;\; z_3 & + \; 150 z_4 - z_5 & = 250 \\ z_2 & + \; 0.2 z_3 & + \;\; 25 z_4 \quad\quad -z_6 & = 45 \end{array}}$$

This has standard form, with six variables and two conditions. The matrix is

$$B = \begin{pmatrix} 1 & 0 & 1 & 150 & -1 & 0 \\ 0 & 1 & 0.2 & 25 & 0 & -1 \end{pmatrix},$$

the right side is $(250, 45)^t$, the objective function is described by the vector $(50, 200, 72, 10000, 0, 0)$.

As a starting vertex we can choose $\mathbf{w}_0 = (250, 45, 0, 0, 0, 0)^t$. The starting tableau is

	3	4	5	6	$\mathbf{w}_0$
1	1	150	−1	0	250
2	**0.2**	25	0	−1	45
	18	2500	−50	−200	21500

The rest is automatic again:

	2	4	5	6	$\mathbf{w}_1$
1	−5	**25**	−1	5	25
3	5	125	0	−5	225
	−90	250	−50	−110	17450

	2	1	5	6	$\mathbf{w}_2$
4	−1/5	1/25	−1/25	1/5	1
3		−5			100
	−80	−10	−40	−160	17200

Principle of complementary slackness

This follows directly from the proof of Theorem 15.11. The basic inequality was

$$Q(\mathbf{x}) = \mathbf{p} \cdot \mathbf{x} \geq \mathbf{w}^t A \mathbf{x} = \mathbf{w} \cdot \mathbf{b} = Q^*(\mathbf{w}).$$

When will equality hold? The only inequality used is $(\mathbf{p} - A^t \mathbf{w}) \cdot \mathbf{x} \geq 0$. In order to have $Q(\mathbf{x}) = Q^*(\mathbf{w})$, it must be that $p_i = (A^t \mathbf{w})_i$ whenever $x_i > 0$.

15.14 Proposition (complementary slackness). *Consider the dual problems (G) and $(G)^*$. Let $\mathbf{x} \in \mathcal{M}, \mathbf{w} \in \mathcal{M}^*$. Then the following are equivalent:*

- $Q(\mathbf{x}) = Q^*(\mathbf{w})$.

- *Whenever $x_i > 0$, we must have $p_i = \mathbf{c}_i \cdot \mathbf{w}$.*

In the case of the symmetric form of duality:

15.15 Corollary. *An admissible point $\mathbf{x} \in \mathcal{M}(U)$ is a minimal solution of the linear program (U) if and only if there is $\mathbf{w} \in \mathcal{M}(U^*)$ such that the following conditions are satisfied:*

- $x_k(\mathbf{c}_k \cdot \mathbf{w} - p_k) = 0, \ k = 1, 2, \ldots q.$

- $w_i(\mathbf{z}_i \cdot \mathbf{x} - b_i) = 0, \ i = 1, 2, \ldots, m.$

Here $\mathbf{c}_k$ is column number k, $\mathbf{z}_i$ is row number i of A.

We will use linear programming to derive explicit bounds for codes and orthogonal arrays.

15.2 The Fourier transform

We aim at bounds on not necessarily linear codes and orthogonal arrays. One tool is the Fourier transform: impose the structure of a cyclic group on the set of entries:

15.16 Definition. *Let $E = \mathbb{Z}/v\mathbb{Z}, V = (\mathbb{Z}/v\mathbb{Z})^n, \ g = |V| = v^n$. Denote by $L^2(V)$ the set of complex-valued functions $f : V \longrightarrow \mathbb{C}$, with scalar product (bilinear form) $\langle, \rangle$ defined by*

$$\langle f, g \rangle = \frac{1}{g} \sum_{x \in V} f(x) \cdot \overline{g(x)}.$$

Here $x = (x_1, x_2, \ldots, x_n) \in V$.

15.17 Definition. *For every element* $z = (z_1, z_2, \ldots, z_n) \in V$, *consider the function* $\phi_z \in L^2(V)$ *defined by*

$$\phi_z(x) = \zeta^{x \cdot z}.$$

The ϕ_z *are known as the* **characters** *of the group* V. *Here* ζ *is a primitive complex v-th root of unity, which will be fixed throughout the discussion, and* $x \cdot z$ *denotes the standard dot product.*

The characters form an orthonormal basis of the g-dimensional complex vector space $L^2(V)$.

15.18 Definition. *Let* $f \in L^2(V)$. *Express* f *in terms of the characters in the form*

$$f = \sum_{z \in V} \hat{f}(z)\phi_z.$$

The complex numbers $\hat{f}(z)$ *are known as the* **Fourier coefficients** *of* f. *The* **Fourier transform** *of* f *is the function* $\hat{f}$, *which has value* $\hat{f}(z)$ *at* z.

Because of the orthonormality of the characters, the Fourier coefficients are given by

$$\hat{f}(z) = \langle f, \phi_z \rangle = \frac{1}{g} \sum_{x \in V} f(x)\zeta^{-x \cdot z}. \tag{15.6}$$

The objects we are primarily interested in, codes or orthogonal arrays, consist of n-tuples with entries in the alphabet E. They can therefore be identified with their characteristic functions:

15.19 Definition. *Let* A *be an array with* n *columns and entries in the v-set* E. *The characteristic function* $f(x) = \chi_A(x)$ *counts with which multiplicity* $x \in E^n$ *occurs as a row of* A. *We define the Fourier transform* $\hat{A}(z)$ *of* A *by identifying* A *with its characteristic function and applying Definition 15.18.*

We will prove a close link between the Fourier coefficients and the strength of an array. If A has strength t, then $\hat{A}(z) = 0$ for every $0 \neq z \in G$ of weight $wt(z) \leq t$. This follows from the definition of an orthogonal array and the fact that the sum of all powers of a root of unity vanishes. The converse is also true, as the condition $\hat{A}(z) = 0$ for all $0 \neq z \in G$ defines a one-dimensional subspace of $L^2(G)$, the constant functions.

15.20 Theorem. *The following are equivalent for each array* $A \subseteq G = E^n$ *and each subset* D *of coordinate positions:*

- $\hat{A}(z) = 0$ *for all* $0 \neq z \in V$ *such that* $\text{supp}(z) \subseteq D$.

- *The projection of* A *to* D *is uniformly distributed.*

In particular, A has strength t if and only if the Fourier coefficients with respect to all nonzero vectors z of weight at most t vanish.

LP bounds for codes and OA

Let $C \subseteq E^n$ be a multisubset, which we identify with its characteristic function $\chi_C \in L^2(V)$. As a tool to derive bounds, we use the distance distribution (A_i) of C :

15.21 Definition. *Let C be a q-ary code of length n.*
*The **distance distribution** of C is the $(n+1)$-tuple of rational numbers $A_i, i = 0, 1, \ldots, n$, where*

$$A_i = \frac{1}{|C|}(\#(x, y) \mid x, y \in C, d(x, y) = i).$$

*The **distance polynomial** of C is $W_C(X, Y) = \sum_{i=0}^{n} A_i X^{n-i} Y^i.$*

We encountered this earlier, in Section 3.5, where we derived the MacWilliams transform for linear codes. In the case of a linear or additive code, A_i is the number of codewords of weight i. Clearly always $A_0 = 1$. By definition the A_i are nonnegative rational numbers and the minimum distance is the smallest $i > 0$ such that $A_i > 0$. Also $\sum A_i = M$, the number of codewords.

Here comes the key idea: We wish to define a transform (A'_k) of the weight distribution which satisfies the following conditions:

- $A'_k \geq 0$ for all k, and

- C is an OA of strength t if and only if $A'_1 = \cdots = A'_t = 0$.

The fact that (A'_k) is a transform of (A_i) means that we can write

$$A'_k = \frac{1}{M} \sum_i A_i K_k(i)$$

for suitable coefficients $K_k(i)$. How are we going to choose those coefficients? The following definition will be justified by two lines of computation which suffice to check that the conditions above are satisfied. Here is the definition.

15.22 Definition.
$$K_k(i) = \sum_{z:wt(z)=k} \zeta^{x \cdot z},$$

where x is a vector of weight i.

It is rather obvious that this expression is indeed independent of the choice of x. This follows from Lemma 15.24 below. Here comes the justification:

$$A'_k = \frac{1}{M^2} \sum_i \sum_{a,b\in C, d(a,b)=i} \sum_{z:wt(z)=k} \zeta^{(a-b)\cdot z} =$$

$$= \frac{1}{M^2} \sum_{z:wt(z)=k} \left(\sum_{a\in C} \zeta^{a\cdot z}\right)\left(\sum_{b\in C} \zeta^{-b\cdot z}\right) = \frac{g^2}{M^2} \sum_{z:wt(z)=k} |\hat{C}(z)|^2,$$

where $\hat{C}(z)$ is the Fourier coefficient of C when we identify it with its characteristic function. This shows that the A'_k have the desired properties. All we needed was the idea of reversing the order of summation and the basic fact that the sum of all powers of ζ is 0.

To sum up: given a code C, we defined the distance distribution (A_i) and its transform (A'_k). The minimum distance d of C is the largest number such that $A_1 = A_2 = \cdots = A_{d-1} = 0$; the strength of C is the largest number t such that $A'_1 = \cdots = A'_t = 0$. For future reference:

15.23 Theorem. *Let C be an $(n, M, d)_q$ code, t its maximal strength as an orthogonal array and (A_i) its distance distribution. Then the following hold:*

1. *The A'_k are nonnegative rational numbers.*

2. *$t = d' - 1$, where the dual distance d' is defined as the smallest positive number i such that $A'_i > 0$.*

Theorem 15.23 shows that the principle of duality is in a way also valid for general nonlinear codes. Although the dual distance d' is defined in a purely algebraic way, using the Kravchouk transform, and although there is no dual code in general, still d' is just one more than the strength of the code.

Definition 15.22 is not satisfactory yet. It would be nice to have a more combinatorial interpretation of the coefficients $K_k(i)$:

15.24 Lemma. *Let $x \in V$ of weight i. Then*

$$\sum_{z\in V, wt(z)=k} \zeta^{x\cdot z} = \sum_{j=0}^{k} (-1)^j (v-1)^{k-j} \binom{i}{j}\binom{n-i}{k-j}.$$

PROOF   Without restriction $x = (x_1, \ldots, x_i, 0, 0, \ldots, 0)$. Choose k coordinates $h_1, \ldots, h_k$ such that $0 < h_1 < \ldots h_j \le i < h_{j+1} < \cdots < h_k \le n$, and let D be the set of vectors of weight k having nonzero coordinates exactly at the h_j. Then $\sum_{z\in D} \zeta^{x\cdot z} = (v-1)^{k-j}(-1)^j$. Summing over all D, we get our statement.   ▯

If we interpret i as a variable X, the right side of the equation in Lemma 15.24 describes a polynomial, the Kravchouk polynomial $K_k(X)$. We find it best to work with two-variable K-polynomials as follows:

15.25 Definition. *The two-variable degree k Kravchouk polynomial is defined by*

$$K_k(X,Y) = \sum_{j=0}^{k} (-1)^j (v-1)^{k-j} \binom{X}{j} \binom{Y}{k-j}.$$

One advantage is that it does not involve the length explicitly. The one-variable K-polynomial is obtained by substituting $Y = n - X$.

What is the transform of the transform? In order to answer that, let us generalize the K transform. The fact that $M = \sum A_i$ suggests how this is done in the most natural way:

15.26 Definition. *Let*

$$S = \{x = (x_i) \mid x_i \in \mathbb{R}\}$$

and $S_0 \subset S$ the subspace defined by $x_0 = 1, \sum_j x_j \neq 0$. Define the K transform on S_0 by $x' = (x'_k)$, where

$$x'_k = \sum_i x_i K_k(i) / (\sum_j x_j).$$

15.27 Lemma.

$$x \in S_0 \longrightarrow x' \in S_0.$$

PROOF   Observe $K_0(i) = 1$. It follows that $x'_0 = 1$. Definition 15.22 shows that $\sum_k K_k(i) = 0$ if $i > 0$, whereas $\sum_k K_k(0) = g$. This shows that $\sum_k x'_k = gx_0(\sum x_i)$. $\square$

Here are some trivially computed special values which will often be used:

15.28 Lemma.

$$K_0(X) = 1, \ K_1(X) = n(v-1) - vX,$$

$$K_k(0) = \binom{n}{k}(v-1)^k, \ K_k(n) = (-1)^k \binom{n}{k}.$$

The leading term of $K_k(X)$ is $\dfrac{(-v)^k}{k!} X^k$.

In order to see what happens if the K tranform is applied twice, the first of the following orthogonality relations is needed:

15.29 Theorem (orthonormality relations).   1. $\sum_k K_k(i) K_j(k) = g\delta_{i,j}$.

2. $K_i(0) K_k(i) = K_k(0) K_i(k)$.

3. $\sum_i K_i(0) K_k(i) K_l(i) = g K_k(0) \delta_{k,l}$.

PROOF For the first statement we have to fix a vector x of weight i and obtain $\sum_k K_k(i) K_j(k) = \sum_{z:wt(z)=k} \zeta^{x \cdot z} \sum_{u:wt(u)=j} \zeta^{z \cdot u} = \sum_{z,u} \zeta^{z \cdot (x+u)}$. If $i \neq j$, then $x + u \neq 0$ and the sum vanishes. If $i = j$, we can choose $u = -x$ and obtain the value g.

Both sides of the middle equation equal the sum of $\zeta^{x \cdot z}$ over all x of weight i and all z of weight k.

The third equation is obtained by applying the second to the left side, then using the first. □

A routine calculation shows that the K transform is involutorial on $\mathcal{S}_0$. The K transform of the weight distribution of code C is a special case. The duality of LP bounds will be used in the form of Theorem 15.12.

Now let

$$\alpha = (\alpha_0 = 1, \alpha_1 = 0, \ldots, \alpha_{r-1} = 0, \alpha_r, \ldots, \alpha_n) \in \mathcal{S}$$

for some r which also satisfies $\alpha \geq 0$ and $\alpha' \geq 0$. Consider the dual problem of Theorem 15.12 with unknowns $(x_r, \ldots, x_n)$, matrix $A = (-K_k(i))$ for $k = 1, \ldots, n$ and $i = r, \ldots, n$, where $p_k = K_k(0)$ and b is constant 1. Then $(\alpha_r, \ldots, \alpha_n)$ is an instance of this problem and $\sum_{j=r}^n \alpha_j$ is bounded by its maximum. Instances of the primal problem are described by $(c_1, \ldots, c_n) \geq 0$ such that $\sum_{k \geq 1} c_k K_k(i) \geq 1$, the corresponding value being $\sum_{k \geq 1} c_k K_k(0)$. This can be conveniently expressed in terms of the polynomial $p(X) = \sum_{k=0}^n c_k K_k(X)$, where $c_0 = 1$. The conditions are $c_k \geq 0$ and $p(i) \leq 0$ for $i = r, r+1, \ldots, n$. The value is $p(0) - 1$. We obtain the following form of the LP bound:

15.30 Theorem. *Let $p(X)$ be a real polynomial of degree $\leq n$ which satisfies the following conditions:*

- *If $p(X)$ is expressed in terms of the Kravchouk basis: $p(X) = \sum_{k=0}^n c_k K_k(X)$, then $c_0 = 1$, $c_k \geq 0$ for all k.*

- *$p(i) \leq 0$, $i = r, r+1, \ldots n$.*

Then for every $(n+1)$-tuple $(\alpha_0, \alpha_1, \ldots, \alpha_n)$ of nonnegative real numbers with nonnegative Kravchouk transform, satisfying $\alpha_0 = 1, \alpha_1 = \alpha_2 = \ldots \alpha_{r-1} = 0$, we have

$$\sum_{k=0}^n \alpha_k \leq p(0).$$

Call the c_k above the **K coefficients** of $p(X)$. Now let $p(X)$ be a polynomial that satisfies the conditions of Theorem 15.30. Two bounds are obtained: first, application to the distance distribution of an $(n, M, d)_v$ code shows $M = |C| \leq p(0)$, where $d = r$. Second, application to the K transform of the distance distribution shows that a v-ary strength $t = r - 1$ orthogonal array has $|C| \geq v^n / p(0)$.

In order to facilitate calculations with concrete values of the K polynomials, consider the product $K_1(X)K_k(X)$. This is a polynomial of degree $k+1$ and therefore has to be a linear combination of the $K_i(X)$ for $i \le k+1$. In fact, go back to the definition of the K coefficients and fix a vector x of weight i :

$$K_1(i)K_k(i) = \left(\sum_{wt(u)=1} \zeta^{u \cdot x} \right)\left(\sum_{wt(v)=1} \zeta^{v \cdot x} \right) = \sum_{u,v} \zeta^{(u+v) \cdot x}.$$

Clearly $wt(u+v)$ is between $k-1$ and $k+1$. Let $wt(z) = k-1$. The number of pairs u, v as above such that $u + v = z$ does not depend on the choice of z. Denote this number by $p_{1,k}^{k-1}$. Analogously the numbers $p_{1,k}^k$ and $p_{1,k}^{k+1}$ are defined. The following is obtained:

$$K_1(i)K_k(i) = p_{1,k}^{k+1}K_{k+1}(i) + p_{1,k}^k K_k(i) + p_{1,k}^{k-1}K_{k-1}(i).$$

Also, it is an easy combinatorial exercise to compute those coefficients:

15.31 Theorem.

$$K_1(i)K_k(i) = (k+1)K_{k+1}(i) + (v-2)kK_k(i) + (v-1)(n-k+1)K_{k-1}(i).$$

In general, the existence of constants $p_{i,j}^k$ as above expresses the fact that Hamming space is an **association scheme**.

15.32 Definition. *Let z be a vector of weight k in n-dimensional v-ary Hamming space. Let $p_{i,j}^k$ be the number of vectors x such that $wt(x) = i$ and $d(x, z) = j$.*

The values used in Theorem 15.31 are

$$p_{1,k}^{k+1} = k+1, \quad p_{1,k}^k = (v-2)k, \quad p_{1,k}^{k-1} = (v-1)(n-k+1).$$

Here is an obvious generalization of this argument:

15.33 Theorem. *For every natural number r between 0 and n and for $0 \le i, j \le n$, we have*

$$K_i(r)K_j(r) = \sum_{k=0}^{n} p_{ij}^{(k)} K_k(r). \tag{15.7}$$

If $i+j \le n$, then it follows that the polynomials themselves satisfy this relation:

$$K_i(X)K_j(X) = \sum_{k=0}^{n} p_{ij}^{(k)} K_k(X) \text{ if } i + j \le n. \tag{15.8}$$

The following obvious corollary will be used in Section 15.4.

15.34 Corollary. *If $p_1(X)$ and $p_2(X)$ have nonnegative K coefficients, then the same is true of $p_1(X)p_2(X)$.*

Optimality of the Best code

In the final chapter we will encounter the **Best code,** a $(10, 40, 4)_2$ code. In the present section we wish to prove its optimality, as an application of the **LP** method. Here we follow the lines of [14].

15.35 Theorem. *There is no $(10, 41, 4)_2$ code and no $(9, 21, 4)_2$ code.*

PROOF   By shortening it suffices to prove the second statement of the theorem. Assume a $(9, 21, 4)_2$ code exists. Then there exists such a code all of whose weights are even. The unknowns are the rational numbers

$$x = A_4, \ y = A_6, \ z = A_8$$

and $M = 1 + x + y + z$. An obvious inequality is $z = A_8 \leq 1$. Consider the transforms A'_k for $k = 1, 2, 3, 4$. The K coefficients can be calculated recursively from Theorem 15.31:

$$(k + 1)K_{k+1}(i) = (9 - 2i)K_k(i) - (10 - k)K_{k-1}(i)$$

using $K_0(i) = 1, \ K_1(i) = 9 - 2i$. For example,

$$A'_1 = (1/21)(9 + x - 3y - 7z) \geq 0.$$

Likewise, as $K_2(i) = 2i^2 - 18i + 36$, we have

$$A'_2 = (1/21)(36 - 4x + 20z) \geq 0.$$

The inequality $A'_k \geq 0$ takes the form

$$-K_k(4)x - K_k(6)y - K_k(8)z \leq \binom{9}{k}.$$

Use these inequalities for $k = 1, 2, 3, 4$ together with the obvious inequality $z \leq 1$. This leads to the following linear program:

$-x$	$+3y$	$+7z$	$\leq$	9
x		$-5z$	$\leq$	9
x	$-2y$	$+7z$	$\leq$	21
$-3x$	$+3y$	$-7z$	$\leq$	63
		z	$\leq$	1

In order to obtain a basic LP problem, introduce five additional variables y_1, y_2, y_3, y_4, y_5 as described after Definition 15.1. The starting vertex is then $\mathbf{x}_0 = (0, 0, 0, 9, 9, 21, 63, 1)$, the basis is $Z = \{4, 5, 6, 7, 8\}$ and clearly we have the objective function $Q = -x - y - z$. The starting array is

	1	2	3	x_0
4	-1	3	7	9
5	**1**	0	-5	9
6	1	-2	7	21
7	-3	3	-7	63
8	0	0	1	1
	1	1	1	$Q_0 = 0$

As the bottom entry in the first column is positive, we choose this column as the pivot column. The pivot element is printed in **bold.** Proceeding in this fashion the following arrays are obtained:

	5	2	3	x_1
4	1	**3**	2	18
1	1	0	-5	9
6	-1	-2	12	12
7	3	3	-22	90
8	0	0	1	1
	-1	1	6	$Q_1 = -9$

	5	4	3	x_2
2	1/3	1/3	2/3	6
1	1	0	-5	9
6	-1/3	2/3	40/3	24/3
7	2	-1	-24	72
8	0	0	**1**	1
	-4/3	-1/3	16/3	$Q_2 = -15$

	5	4	8	x_3
2			-2/3	16/3
1			5	14
6			-40/3	
7			24	
3			1	1
	-4/3	-1/3	-16/3	$Q_3 = -20\frac{1}{3}$

As the entries d_i in the last row are negative, we have reached the minimum. Observe that there is no need to calculate all the entries in this last array. We conclude that $M = 1 + x + y + z \le 1 - Q_3 = 21\frac{1}{3}$, hence $M \le 21$.

Now assume a code $(9, 21, 4)$ exists. As M is odd, the basic relation between A'_k and the Fourier transform (see the calculation after Definition 15.22) shows that in fact $A'_k \ge \binom{9}{k}/M^2$. This leads to a strengthened inequality

$$-K_k(4)x - K_k(6)y - K_k(8)z \le \binom{9}{k}(M-1)/M,$$

with a factor of $(M-1)/M = 20/21$ on the right side. If you think about it, also the last inequality $z \leq 1$ can be strengthened. As M is odd, it cannot be that all codewords have a partner at distance 8, so in fact $z \leq 20/21$. The right sides of the inequalities pick up a factor of $20/21$. The same is true of the objective value: the contradiction $M \leq 1 + 20 \cdot 61/63 < 21$ is obtained. $\square$

15.3   Some explicit LP bounds

The easiest way to derive an explicit bound from the general LP bound of Theorem 15.30 is to use polynomials of small degree. Degree 1 leads to the Plotkin bound: recall $K_1(X) = (v-1)n - vX$. We must have $p(X) = 1 + cX$ where $c \geq 0$. The second condition to satisfy is then $p(r) \leq 0$. This is possible only if $K_1(r) < 0$. The optimal choice is then $c = -1/K_1(r)$ and the bound is $p(0) = 1 + c(v-1)n = vr/(vr - (v-1)n)$. In the case of codes, this is the Plotkin bound. As for OA, the following is obtained:

15.36 Theorem (dual Plotkin bound). *Let M be the number of rows of a v-ary OA of strength t. Then*

$$\frac{v^n}{M} \leq \frac{v(t+1)}{v(t+1) - n(v-1)}$$

provided the expression on the right is > 0.

This is a strong bound on OA of large strength $((t+1)/n > (v-1)/v)$ just as the Plotkin bound on codes is useful for codes of large minimum distance. Reconsider the family of additive codes and OA from Theorem 16.29. They have length $n = q^a$, are q^b-ary (where $a \geq b > 0$) and have $M = q^{a+b}$ codewords. The minimum distance is $d = q^a - q^{a-b}$, and the strength is 2. We encountered them again in Section 18.1 (see Theorem 18.3), where they were called $\mathcal{C}(a,b)$ and used in the Bose-Bush construction. The minimum distance is rather large but not large enough to apply the Plotkin bound (the Plotkin bound would require $\dfrac{q^a - q^{a-b}}{q^a} > \dfrac{q^b - 1}{q^b}$, which is not satisfied: we have equality instead of strict inequality). This motivates us to generalize the Plotkin bound, applying Theorem 15.30 not to linear but to quadratic polynomials.

Choose $p(X) = a(X - r)(X - n)$ for $a > 0$. The nonpositivity condition of Theorem 15.30 is then satisfied. As $p(0) = arn$, we want to minimize a. Let $p(X) = c_0 + c_1 K_1(X) + c_2 K_2(X)$. We have to choose a such that $c_0 = 1$, c_1, $c_2 \geq 0$. Recall the value of $K_1(X)$ given above and observe that

$$K_2(X) = \frac{v^2}{2}X - v[(v-1)(n-1) + v/2]X + \binom{n}{2}(v-1)^2.$$

Comparing coefficients, we obtain $a = c_2 v^2/2$,

$$c_1 = c_2\{\frac{v}{2}(r+1-n)+n-1\}, \quad c_2 = \frac{2}{n[vr - (n-1)(v-1)]}$$

provided the denominator of the last expression is positive. It follows that $p(0) = \frac{v^2 r}{vr - (n-1)(v-1)}$. This bound is never better than the Plotkin bound as long as the Plotkin bound is applicable. The quadratic bound is interesting when $(n-1)\frac{v-1}{v} < r \le n\frac{v-1}{v}$.

15.37 Theorem. *Let* $(n-1)\frac{v-1}{v} < d \le n\frac{v-1}{v}$. *If an* $(n, M, d)_v$ *code exists, then*

$$M \le \frac{v^2 d}{vd - (n-1)(v-1)}.$$

Let $(n-1)\frac{v-1}{v} < t+1 \le n\frac{v-1}{v}$. *If there is a* v-*ary OA of length* n *and strength* t *with* M *rows, then*

$$\frac{v^n}{M} \le \frac{v^2(t+1)}{v(t+1) - (n-1)(v-1)}.$$

The quadratic bound applies to our $(q^a, q^{a+b}, q^a - q^{a-b})_{q^b}$ codes $\mathcal{C}(a, b)$. The bound on M is in fact reached with equality.

The Singleton bound on codes is also an example of this mechanism. Surprisingly its derivation is rather involved. Our direct proof in Chapter 4 was clear and short in comparison. The corresponding polynomial is

$$p(X) = v^{n-r+1} \cdot \binom{n-X}{n-r+1} / \binom{n}{r-1} = \sum_{k=0}^{n} \frac{\binom{n-k}{r-1}}{\binom{n}{r-1}} \cdot K_k(X). \qquad (15.9)$$

The dual Singleton bound states that v^t is a lower bound on the number of rows of a v-ary OA of strength t. This is a triviality.

For the sphere-packing bound, the situation is even more extreme. Our direct proof from Section 1.6 is all but trivial whereas the derivation from the LP bound is rather involved. The underlying polynomial is $p(X) = \sum_{k=0}^{n} c_k K_k(X)$, where $c_k = (\Psi_s(k)/V)^2$ and $\Psi_s(X) = \sum_{l=0}^{s} K_l(X)$, $V = V_v(s, n)$ (see Definition 1.11) and s is maximal such that $2s < r$. However, the dual sphere-packing bound is interesting. It is known as the Rao bound:

15.38 Theorem (Rao bound). *Let* $M = \lambda v^t$ *be the number of rows of a* v-*ary OA of strength* t *and length* n. *Let* $e' = \lfloor t/2 \rfloor$. *Then*

$$M \ge \sum_{i=0}^{e'} \binom{n}{i}(v-1)^i.$$

For odd t the Rao bound can be strengthened. In this text we only needed the strength 2 case of the Rao bound in Theorem 15.38. This is the Bose-Bush bound Theorem 9.6, which we derived from the Plotkin bound for codes.

15.4   The bound of four

In 1977 McEliece, Rodemich, Rumsey and Welch [140] proved a powerful bound on binary codes. It is well known that their method extends to codes over arbitrary alphabets. As the bound is a consequence of the LP bound, our basic observation applies and we obtain asymptotic bounds on orthogonal arrays. The main ingredient of the proof is a classical formula in the theory of orthogonal polynomials, the Christoffel-Darboux formula. As a starting point use the recursion formula:

15.39 Lemma (Recursion formula).

$$(k + 1)K_{k+1}(X, Y) = \{-X + (v - 1)Y - (v - 2)k\}K_k(X, Y)$$
$$-(v - 1)(X + Y + 1 - k)K_{k-1}(X, Y).$$

This is the two-variable version of Theorem 15.31, obtained by reading i as a variable X and substituting $Y = n - X$. See $Y = a$ as a constant again. The Christoffel-Darboux formula answers the following natural question: consider $f(X) = K_{k+1}(X)K_k(a) - K_k(X)K_{k+1}(a)$. It is expressed in the K-basis and has $X = a$ as a zero. It follows that $f(X)/(X - a)$ is a polynomial again. What is the expression of this polynomial in terms of the K-basis? The answer is as follows:

15.40 Theorem (Christoffel-Darboux formula).

$$(k+1)(K_{k+1}(X)K_k(Y) - K_k(X)K_{k+1}(Y)) = K_k(0)v(Y-X)\sum_{j=0}^{k} K_j(X)K_j(Y)/K_j(0)$$

PROOF   Consider

$$m_k = K_{k+1}(X)K_k(Y) - K_k(X)K_{k+1}(Y).$$

Observe $m_0 = K_1(X) - K_1(Y) = v(Y - X)$. Apply the recursion formula of Lemma 15.39 to the terms of degree $k + 1$:

$$(k + 1)m_k = (K_1(X) - K_1(Y))K_k(X)K_k(Y) + (v - 1)(n + 1 - k)m_{k-1} =$$

$$= m_0 K_k(X)K_k(Y) + (v - 1)(n + 1 - k)m_{k-1}.$$

Continue this process. Collect the coefficients of the terms $K_j(X)K_j(Y)$. As j decreases, the power of $v - 1$ increases. The remaining terms combine to $\binom{n}{k}/\binom{n}{j}$. □

Another fundamental ingredient in the proof of the bound of four is the interlacing property of the zeroes of the K-polynomials. This makes use of the recursion formula.

15.41 Lemma. *Let* $f_i(X), i = 0, 1, \ldots$ *be real polynomials of degree* i, *where* $f_0(X) = C > 0, f_1(X) = Ax + B(A > 0)$ *and for* $i \geq 2$

$$f_i(X) = (A_i X + B_i)f_{i-1}(X) - C_i f_{i-2}(X), A_i > 0, C_i > 0.$$

Then $f_i(X)$ *has* i *different real zeroes, and the zeroes of* $f_i(X)$ **interlace** *those of* $f_{i-1}(X)$. *This means that, if* $\alpha_1 > \cdots > \alpha_i$ *are the zeroes of* $f_i(x)$ *and* $\beta_1 > \ldots, \beta_{i-1}$ *are those of* $f_{i-1}(X)$, *then*

$$\alpha_1 > \beta_1 > \alpha_2 > \beta_2 > \cdots > \beta_{i-1} > \alpha_i.$$

PROOF   Use induction from f_i to $f = f_{i+1}$. Observe that f has leading coefficient > 0. We have $f(\alpha_j) = -C_{i+1}f_{i-1}(\alpha_j), f(\alpha_{j+1}) = -C_{i+1}f_{i-1}(\alpha_{j+1})$. These numbers have different signs, as f_{i-1} has exactly one (simple) zero between α_j and α_{j+1}. The same argument applies to α_1 and α_i.   □

Using $K_0(X) = 1, K_1(X) = -v \cdot X + n(v - 1)$ (see Lemma 15.28) and the recursion formula, we see that $(-1)^k K_k(X), k = 0, 1, 2, \ldots, n$ satisfy the conditions of the preceding lemma. Recall $K_k(0) = \binom{n}{k}(v - 1)^k > 0, K_k(n) = \binom{n}{k}$. We obtain the following:

15.42 Theorem (Interlacing of zeroes). $K_k(X), k = 0, 1, \ldots, n$ *has* k *different real zeroes between* 0 *and* n. *If these zeroes are* $v_1 < v_2 < \ldots, < v_k$, *and the zeroes of* K_{k-1} *are* $u_1 < u_2 < \ldots, u_{k-1}$, *then*

$$0 < v_1 < u_1 < v_2 < u_2 < \cdots < u_{k-1} < v_k < n.$$

We are ready to prove the bound of four. Start from

$$p_*(X) = K_{t+1}(X)K_t(a) - K_t(X)K_{t+1}(a).$$

The natural number t and the real number a will be determined later. It follows from the Christoffel-Darboux formula that $p_*(X)$ may be written as

$$p_*(X) = \frac{v(a - X)}{t + 1}K_t(0) \sum_{k=0}^{t} \frac{K_k(X)K_k(a)}{K_k(0)}.$$

Let $p(X) = \dfrac{p_*(X)^2}{a - X}$. Using the two expressions above for $p_*(X)$, we obtain

$$p(X) = \frac{v}{t + 1}K_t(0)[K_{t+1}(X)K_t(a) - K_t(X)K_{t+1}(a)] \sum_{k=0}^{t} \frac{K_k(X)K_k(a)}{K_k(0)}.$$
$$(15.10)$$

We will use $p(X)$ in Theorem 15.30. The second condition, the non-positivity condition, of that theorem, is automatically satisfied by choosing $a < r$. This is the reason for choosing the $p(X)$ of this form, with a square in the numerator. It remains to make sure that the first condition of Theorem 15.30 is satisfied: the K coefficients have to be positive. The K-coefficients of $p_*(X)$ are $K_t(a)$ and $-K_{t+1}(a)$. The interlacing property shows that the smallest root $x_1^{(t)}$ of $K_t(X)$ decreases when t increases. Choose $x_1^{(t+1)} < a < x_1^{(t)}$. Then the K-coefficients of $p_*(X)$ are positive. The Christoffel-Darboux formula shows that the same is true of $p_*(X)/(x - X)$. It follows from Corollary 15.34 that $p(X)$ has nonnegative K coefficients. Let c_0 be its coefficient at $K_0(X) = 1$. Then $p(X)/c_0$ satisfies the conditions of Theorem 15.30.

In order to determine c_0, use Equation 15.10 again. By definition,

$$p_{ij}^{(0)} = 0 \text{ if } i \neq j, \ p_{ii}^{(0)} = K_i(0).$$

It follows that $c_0 = -\dfrac{v}{t+1}K_t(0)K_t(a)^2 \cdot Q$. In order to simplify further, choose a such that $Q = -1$. Then our LP bound is

$$\frac{p(0)}{c_0} = K_t(0)\frac{[(v-1)(n-t)+t+1]^2}{av(t+1)}.$$

Definition 15.22 shows the following:

15.43 Lemma. *Let $k \geq 1$. Then*

$$K_k(1) = \binom{n-1}{k}(v-1)^k - \binom{n-1}{k-1}(v-1)^{k-1}$$

and $K_k(1) \geq 0$ if and only if $k \leq \frac{v-1}{v} \cdot n$.

Choose t such that $t+1 \leq \frac{v-1}{v} \cdot n$. Then $K_{t+1}(1) \geq 0$. Using the Christoffel-Darboux formula again it can be shown that roots of $K_k(X)$ differ by more than 1. It follows $x_1^{(t+1)} \geq 1$. As $a > x_1^{(t+1)}$ it follows $a \geq 1$. We use this to simplify.

15.44 Theorem (MRRW bounds). *Let $t < \frac{v-1}{v}n$.*
If $d \geq x_1^{(t)}$, where $x_1^{(t)}$ is the smallest root of $K_t(X)$, then

$$|\mathcal{C}| \leq \binom{n}{t}(v-1)^t\frac{[(v-1)(n-t)+t+1]^2}{v(t+1)}.$$

If $d' \geq x_1^{(t)}$, then

$$\frac{v^n}{|\mathcal{C}|} \leq \binom{n}{t}(v-1)^t\frac{[(v-1)(n-t)+t+1]^2}{v(t+1)}.$$

Further study of the position of the smallest roots of the Kravchouk polynomials leads to asymptotic bounds. In the binary case the following is proved in [140]:

15.45 Theorem. *Consider a family of binary codes $\mathcal{C}_n$ such that the lengths $n \longrightarrow \infty$ and the relative minimum distances $\frac{d}{n} \longrightarrow \delta$. Then the rate $\frac{1}{n}log_2|\mathcal{C}_n|$ is asymptotically bounded by $h(\frac{1}{2} - \sqrt{\delta(1 - \delta)})$. Here*

$$h(x) = -xlog_2(x) - (1 - x)log_2(1 - x)$$

is the binary entropy function. In particular, asymptotic rates $R > 0$ are possible only when $\delta < 1/2$.

Our basic observation shows that a corresponding statement holds for binary orthogonal arrays. Instead of the relative minimum distance, we have to consider the relative dual distance $\delta' = d'/n$ (asymptotically equivalent to the relative strength of the array). This yields the following:

15.46 Theorem. *Consider a family of binary orthogonal arrays $\mathcal{C}_n$ such that the lengths $n \longrightarrow \infty$ and the relative dual distances $\frac{d'}{n} \longrightarrow \delta'$. Then the rate $\frac{1}{n}log_2|\mathcal{C}_n|$ is asymptotically lower bounded by $1 - h(\frac{1}{2} - \sqrt{\delta'(1 - \delta')})$. In particular, asymptotic rates $R < 1$ are possible only when $\delta' < 1/2$.*

The statement shows which asymptotic statement is needed: If $x_1^{(t_n)}/n \longrightarrow \delta$, then $t_n/n \longrightarrow \tau = 1/2 - \sqrt{\delta(1 - \delta)}$. The transformation $\delta \longrightarrow \tau$ is in fact involutorial: $\delta < 1/2$ is expressed as $\delta = 1/2 - \sqrt{\tau(1 - \tau)}$ in terms of τ. It suffices therefore to prove the following lemma (in Appendix A of [140]):

15.47 Lemma. *Let $v = 2$. If $t_n/n \longrightarrow \tau$, then $limsup(x_1^{(t_n)}/n) \leq 1/2 - \sqrt{\tau(1 - \tau)}$.*

1. All LP problems can be brought into standard form:

2. $\mathbf{x} = (x_1, \ldots, x_n) \geq 0$, $A\mathbf{x} = \mathbf{b} = (b_1, \ldots, b_m) \geq 0$,
   $Q(x) = \mathbf{p} \cdot \mathbf{x} = p_1x_1 + \cdots + p_nx_n \longrightarrow Min.$

3. Let $\mathcal{M} = \{\mathbf{x} \geq 0 | A\mathbf{x} = \mathbf{b}\}$.

4. A **vertex** is a point $\mathbf{x} \in \mathcal{M}$ such that the columns of A where $x_k > 0$ are independent.

5. The minimum (if it exists) is reached in a vertex.

6. The **Simplex algorithm** walks from vertex to vertex and finds the minimum.

7. **Duality theorems:** the min of the original problem is the max of the dual problem (if both exist).

1. The **Fourier transform** is based on characters of abelian groups.

2. The strength of an OA is expressed by the Fourier coefficients of its characteristic function.

3. Let (A_i) be the distance distribution of a (length n, v-ary) code C with M codewords.

4. The **Kravchouk transform** is $A'_k = (1/M) \sum_i A_i K_k(i)$.

5. Here $K_k(i)$ are values of the **Kravchouk polynomials** $K_k(X)$.

6. C has strength $\geq t$ if and only if $A'_1 = \cdots = A'_t = 0$.

7. The fact that all $K'_k \geq 0$ leads to the LP bound on codes.

8. Transforming again shows that this also is a bound on OA.

9. Examples of pairs of explicit bounds on codes/OA following from the LP bound include:

10. The Plotkin/dual Plotkin bound.

11. The quadratic bound.

12. The sphere-packing/Rao bound.

13. The asymptotic bound of four.

Exercises 15

15.1. *Explain why in the standard form of the LP problem it is usually assumed that $m \leq n$.*

15.2. *Consider the LP-problem in standard form where*
$$m = 2, n = 3, A = \begin{pmatrix} 2 & 1 & 1/2 \\ -5 & -1 & 0 \end{pmatrix} \text{ and } \mathbf{b} = (2,1), \mathbf{p} = (1,0,5).$$
Determine the region $\mathcal{M}$ of admissible points and solve the problem.

15.3. *Draw the convex hull of points*

$$P_1 = (2,0), \ P_2 = (2,1), \ P_3 = (2,2), \ P_4 = (3,1), \ P_5 = (0,-2).$$

Write $(1.5,0)$ as a convex combination of $P_1, \ldots, P_5$.

15.4. *Determine the vertices of the first toy example.*

15.5. *Let*
$$A = \begin{pmatrix} 0 & 1 & 2 & 2 & 0 & 0 & 3 \\ 1 & 1 & 5 & 1 & 1 & 1 & -5 \\ 0 & 1 & 7 & 1 & 0 & 1 & -2 \\ 1 & 0 & -2 & 1 & 0 & -3 & -1 \end{pmatrix}.$$

Then $I = \{4,5,6,7\}$ describes a basis. Express columns c_1, c_2, c_3 in terms of c_4, c_5, c_6, c_7.

15.6. *Use Gauß elimination to determine the rank of*

$$A = \begin{pmatrix} 0 & 2 & 1 & 4 & 0 & -3 \\ -2 & 3 & 5 & 6 & -7 & 2 \\ 3 & -1 & -4 & 0 & 8 & -2 \\ 1 & 0 & 0 & 2 & 1 & 3 \end{pmatrix}.$$

15.7. *Consider the linear program*

$$\boxed{\begin{aligned} x_1 + x_2 - 3x_3 &\leq 5 \\ x_1 - 2x_2 + 2x_3 &\geq 6 \\ x_1, x_2, x_3 &\geq 0 \end{aligned}}$$

with objective function $Q(x) = x_1 + x_2 + x_3 \to max$.
Write it in standard form and determine the dual LP.

15.8. *Show how n unconstrained variables can be replaced by $n+1$ constrained variables.*

15.9. *Consider the LP with unknowns* x_1, x_2, x_3 *and conditions*

$$x_1, x_2 \geq 0, \ 2x_1 - 3x_3 \geq 3, \ x_2 + 5x_3 \leq 6, \ x_1 + x_2 + x_3 = 1$$

where $x_1 - x_3$ *has to be minimized.*
Determine the dual problem.

15.10. *Find a vertex of the polyhedron* $Ax = b, x \geq 0$ *where*

$$A = \begin{pmatrix} 1 & 0 & 1 & 2 \\ 0 & 1 & 1 & -1 \\ 0 & 0 & 1 & 3 \end{pmatrix} \quad and \quad b = \begin{pmatrix} 2 \\ 1 \\ 1 \end{pmatrix}.$$

15.11. *Let*

$$A = \begin{pmatrix} 1 & 0 & 0 & 0 & 2 & -2 \\ 0 & 1 & 0 & 0 & 4 & 2 \\ 0 & 0 & 1 & 0 & 3 & 0 \\ 0 & 0 & 0 & 1 & 1 & 7 \end{pmatrix} \quad b = \begin{pmatrix} 1 \\ 2 \\ 3 \\ 4 \end{pmatrix}.$$

Find the vertex X *with basis* $\{1, 2, 3, 4\}$ *and use it to find a vertex with basis* $I' \subset \{1, 2, 3, 4, 6\}, I' \neq I.$

15.12. *Find the minimum of* $x_1 + 2x_2$ *subject to*

$$\begin{array}{rcl} 2x_1 - 5x_2 & \geq & -40 \\ 5x_1 + 8x_2 & \leq & 720 \\ -5x_1 + 3x_2 & \geq & -830 \\ 8x_1 + 10x_2 & \geq & 440 \end{array}$$

15.13. *Apply the Simplex algorithm to* $Q(x) = -x_1 - x_2 - x_3 - x_4 \rightarrow min$ *subject to*

$$\begin{array}{rcl} x_1, x_2, x_3, x_4 & \geq & 0 \\ -x_1 + 3x_2 + 7x_3 + 11x_4 & \leq & 13 \\ 6x_1 + 2x_2 - 18x_3 - 54x_4 & \leq & 78 \\ x_3 + 4x_4 & \leq & 4 \end{array}$$

15.14. x *five-story and* y *two-story buildings are to be constructed on a swampy plot, where the need for adequate foundations greatly increases the cost of a taller building. The work produced by one person in one month will be denoted a "person-month." The remaining information is in the following table:*

number of stories	costs in $	person-months	area in yd²	number of occupants per building	number of buildings
5	600,000	120	800	30	x
2	200,000	60	600	12	y
available:	18,000,000	4,500	42,000		

How should x *and* y *be chosen if we want to maximize the number of people who can live on this plot of land?*

Chapter 16

Orthogonal arrays in statistics and computer science

16.1 OA and independent random variables

Basic concepts: The strength as a measure of uniformity. Pseudorandomness, OA and perfect local randomizers. The concept of universality.

Most prominent among the applications of error-correcting codes is the transmission of messages via noisy channels. From this point of view the **minimum distance** is the most important parameter. In the linear case the duality theorem leads to a different approach to codes and to a different type of application. The dual of a linear code of minimum distance d is a space (code) of strength $t = d-1$. The **strength** is a uniformity parameter. We view the general notion of an orthogonal array as a combinatorial generalization of "dual codes" (traditionally one speaks of codes if the minimum distance is the important parameter, of orthogonal arrays if the strength is the center of interest). We start from a statistical interpretation of orthogonal arrays. In fact the very name **orthogonal array** was coined in a statistical context. Consider the Reed-Solomon code $\mathcal{RS}(2,4)$ again. It is an $OA_1(2,4,4)$.

0 0 0 0	0 ω $\overline{\omega}$ 1
1 1 1 1	1 $\overline{\omega}$ ω 0
ω ω ω ω	ω 0 1 $\overline{\omega}$
$\overline{\omega}$ $\overline{\omega}$ $\overline{\omega}$ $\overline{\omega}$	$\overline{\omega}$ 1 0 ω
0 1 ω $\overline{\omega}$	0 $\overline{\omega}$ 1 ω
1 0 $\overline{\omega}$ ω	1 ω 0 $\overline{\omega}$
ω $\overline{\omega}$ 0 1	ω 1 $\overline{\omega}$ 0
$\overline{\omega}$ ω 1 0	$\overline{\omega}$ 0 ω 1

View the rows as elements of a uniform sample space. In the example we have a sample space Ω of 16 elements, each with a probability of $1/16$. Each column of the array can be seen as a function defined on Ω, with values in the alphabet. In other words, each column describes a random variable (see Definition 8.10). In our example we have four random variables defined on Ω, with values in a set of four points, $\mathbb{F}_4$. As in each column each entry occurs the same number of times (our array has strength 1), each of these random variables takes on each value with the same probability. Such random variables are called **balanced**. A basic notion of statistics is **statistical independence**. We generalize it from the case of two random variables (Definition 8.11) to an arbitrary number of random variables:

16.1 Definition. *A family $f_1, \ldots, f_t$ of random variables defined on Ω with values in finite sets is* **statistically independent** *if, for every choice of images a_i, the following holds:*
the probability that $f_i(x) = a_i$, $i = 1, 2, \ldots, t$ equals the product

$$\prod_{i=1}^{t} Prob(f_i(x) = a_i) \text{ over the individual probabilities.}$$

In our situation of a uniform sample space with balanced random variables with values in the same set (the alphabet), we have that $f_1, \ldots, f_t$ are statistically independent if, for every choice $a_1, \ldots a_t$ of values in the alphabet, the number of rows (elements $x \in \Omega$) satisfying $f_1(x) = a_1, \ldots, f_t(x) = a_t$ is always the same, independent of the choice of the images. This leads precisely to the definition of an OA of strength t.

16.2 Theorem. *Consider the rows of an $OA_\lambda(t, n, q)$ as elements of a uniform sample space, each column as a random variable. This family of n balanced random variables has the property that any subset of t of our random variables is statistically independent.*

It should be clear that the converse of Theorem 16.2 is true as well. Given a uniform probability space and n random variables with values in a q-set defined on it, we can write out the corresponding array. It will have strength 1 if all the random variables are balanced, strength $t > 1$ if in addition any set of t of our random variables is statistically independent.

We note that the notion of statistical independence is in harmony with the notion of linear independence from linear algebra. In fact, assume the alphabet of the array is a finite field $\mathbb{F}_q$ and the array is linear (a vector subspace of $\mathbb{F}_q^n$). We can represent the array by a generator matrix (a basis). Any subset of random variables will then be **statistically** independent if and only if the corresponding columns of the generator matrix are **linearly** independent. In our example we have strength 2, meaning that any two of our four random variables described by the columns are statistically independent. A generator matrix is

$$\begin{pmatrix} 1\ 1\ 1\ 1 \\ 0\ 1\ \omega\ \overline{\omega} \end{pmatrix}$$

and any two columns are linearly independent.

The most prominent statistical application of orthogonal arrays is in the design of experiments. The book *Orthogonal Arrays: Theory and Applications* by Hedayat, Sloane and Stufken [109] was written from that perspective. We are not going to dwell on these purely statistical applications.

Pseudorandomness is a context where OA come in naturally. Here is a first example.

Perfect local randomizers

Perfect local randomizers were defined by Maurer and Massey [148]. The situation is the following: assume we can generate k **random bits** (a randomly generated element of $\mathbb{F}_2^k$). This corresponds to tossing a fair coin k times. As random bits are a resource just as time and space, we should use them in an economic way. One idea is to use **pseudorandom sequences.** These are sequences that have not been generated randomly but share certain properties of random sequences. The situation considered by Maurer and Massey is the following:

We are searching an injective mapping $G : \mathbb{F}_2^k \longrightarrow \mathbb{F}_2^n$, where $k < n$. It is assumed that the bitstring of length k is chosen at random. The image under G of this random seed is used as a pseudorandom sequence. Which properties should G have? Consider the corresponding array with 2^k rows and n columns where the entries are given by the corresponding images under G. This is the array considered in Theorem 16.2, where each column describes a random variable. Maurer and Massey's condition is that each t of these random variables is statistically independent (in other words completely random). In this case they speak of a (k, n, t) **perfect local randomizer.** Comparing this with our probabilistic interpretation of OA (Theorem 16.2), we see that a (k, n, t) perfect local randomizer is equivalent to an $OA_{2^{k-t}}(t, n, 2)$.

This gives us another angle under which to view OA. A binary OA with 2^k rows, n columns and strength t can be viewed as a function which stretches a random seed of k bits (randomly chosen elements from $\mathbb{F}_2^k$) to n pseudorandom bits having the property that any subfamily of t bits is perfectly random.

Linear shift register sequences, the most popular and frequently used family of pseudorandom sequences, are the object of the next section. Section 16.3 discusses the use of distance and strength in cryptography. An application of OA of strength 2 in the testing of chips is described in Section 16.4. This is an example of a general paradigm of applications of OA in computer science, which in the pertinent literature is denoted as **universality** (an unfortunate expression). The OA in question (with n columns) is viewed as a microcosmos which shares important features with the macrocosmos formed by all n-tuples. For example, the macrocosmos $\mathbb{F}_q^n$ certainly is an OA of strength n. If the application in question requires only strength t for some $t < n$, we can replace $\mathbb{F}_q^n$ by a smaller structure, a q-ary OA with n columns, of strength t. The term **universal hashing** is based on this concept. Another application of the same idea is derandomization of algorithms.

16.2   Linear shift register sequences

and Simplex codes, pseudorandomness, correlation properties.

16.3 Definition. *An infinite sequence* $\mathcal{S} = (a_n)_{n \geq 0}$ *is a **linear shift register sequence** (LSRS) if it satisfies the linear recurrence relation*

$$a_n = \sum_{i=1}^{r} c_i \cdot a_{n-i},$$

where $a_n,\ c_i \in \mathbb{F}_q$.

We assume without restriction $c_r \neq 0$ and call r the **depth** of the recurrence relation, c_i the **feedback coefficients.** Obviously $\mathcal{S}$ is uniquely determined by two ingredients of very unequal importance, the all-important feedback coefficients and the **initial conditions,** which may conveniently be written as $a_{-1}, \ldots a_{-r}$. If the recurrence relation is fixed, then the future of $\mathcal{S}$ is determined once a set of r consecutive terms of $\mathcal{S}$ is known. The basic theory is rather an easy application of some elementary number theory. All this is explained very well in Golomb's book [96], which, however, suffers from the vice that too much weight is put on the binary case. It is natural to use the generating function.

16.4 Definition. *The **generating function** of* (a_n) *is*

$$G(X) = \sum_{n=0}^{\infty} a_n X^n.$$

The recurrence relation implies that the generating function of an LSRS is a rational function:

16.5 Lemma. *Let* $S = (a_n)$ *be an LSRS, $G(X)$ its generating function. Then*

$$G(X) = g(X)/f(X),$$

where the denominator, the **characteristic polynomial** *of S, is*

$$f(X) = 1 - \sum_{i=1}^{r} c_i X^i$$

of degree r depending only on the feedback coefficients, whereas

$$g(X) = \sum_{k=0}^{r-1} \left(\sum_{i=k+1}^{r} c_i \cdot a_{k-i} \right) \cdot X^k,$$

a polynomial of degree $< r$, depends on the feedback coefficients and on the initial conditions.

PROOF $G(X) = \sum_n a_n X^n = \sum_n \sum_{i=0}^{r} c_i a_{n-i} X^n$ $= \sum_{i=1}^{r} c_i X^i \cdot \sum_n a_{n-i} X^{n-i}$. The last term in this expression is $G(X) + \sum_{j=1}^{i} a_{-j} X^{-j}$. Collect all the terms containing $G(X)$ on the left side. Then the left side has the form $G(X) \cdot f(X)$. We have the desired expression as a rational function. The numerator $g(X)$ is our right-hand side: $g(X) = \sum_{i=1}^{r} c_i X^i \cdot \sum_{j=1}^{i} a_{-j} X^{-j}$. Put $k = i - j$ and simplify. $\square$

Observe the special case when $a_{-1} = \cdots = a_{-(r-1)} = 0, a_{-r} = 1$. In this case the numerator is a constant $g(X) = c_r$.

16.6 Lemma. *Let $G(X) = g(X)/f(X) \neq 0$ be the generating function of an LSRS S, where $f(X)$ is the characteristic polynomial. Then S is periodic and the (smallest)* **period** *$\rho = \rho(S)$ of S is the smallest natural number such that*

$$f(X) \text{ divides } g(X) \cdot (1 - X^\rho).$$

Moreover, $\rho(S) \leq q^r - 1$.

PROOF First of all, there are only $q^r - 1$ possible different segments $\neq 0$ of length r. As soon as one occurs twice, we have periodicity. This shows that the period is $\leq q^r - 1$. The fact that S is periodic with period ρ is equivalent to $G(X) = e(X) \cdot (1 + X^\rho + X^{2\rho} + \ldots) = e(X)/(1 - X^\rho)$, where $e(X)$ is a polynomial of degree $< \rho$. This is equivalent to

$$e(X) \cdot f(X) = g(X) \cdot (1 - X^\rho),$$

where all the terms are polynomials. If such an equation holds, then the degree of $e(X)$ is necessarily $< \rho$.   □

The criterion of Lemma 16.6 becomes particularly handy when $f(X)$ and $g(X)$ are coprime. This will be the case when the initial conditions are chosen as above such that $g(X)$ is a constant. It will also be the case if $f(X)$ is irreducible, as $g(X)$ has a smaller degree than $f(X)$. In this coprime case (which clearly is the generic case) the period of $\mathcal{S}$ is the smallest natural number ρ such that $f(X)$ divides $X^\rho - 1$.

16.7 Definition. *The smallest natural number i such that $f(X)$ divides $X^i - 1$ is the* **exponent** *of $f(X)$. An LSRS $\mathcal{S}$ has* **maximum length** *if its period is maximal: $\rho(\mathcal{S}) = q^r - 1$.*

16.8 Theorem. *If an LSRS $\mathcal{S}$ has maximum length, then its characteristic polynomial $f(X)$ is* **irreducible.**

PROOF   Because of maximum length, all the nonzero segments of length r must occur. We may choose a segment of $r - 1$ zeroes preceded by 1 as initial conditions. This has the effect of reducing the numerator $g(X)$ to a nonzero constant. Now assume $f(X)$ factors: $f(X) = h_1(X) \cdot h_2(X)$ of degrees r_1 and r_2, respectively, where $r_1 + r_2 = r$. Consider first the case where $h_1(X)$ and $h_2(X)$ are coprime. Then $G(X)$ can be written as a sum of two rational functions with denominators $h_1(X)$ and $h_2(X)$, respectively. We know that the periods of the summands cannot exceed $q^{r_1} - 1$ and $q^{r_2} - 1$, respectively. The period of $G(X)$ is at most the product of these two numbers, but this is too small, a contradiction. So assume $f(X) = h(X)^m$, where $h(X)$ is irreducible of degree s (hence $r = ms$) and $m \geq 2$. We have $h(X)|X^{q^s-1} - 1$. Choose j such that $q^{j-1} < m \leq q^j$. Then $h(X)^m|h(X)^{q^j}|X^{(q^s-1)q^j} - 1$. The length of the period is therefore $\leq (q^s - 1)q^j$. We claim that $(q^s - 1)q^j < q^r - 1$. This is then the desired contradiction. In fact, we have $r = sm$, so the claim is equivalent to $q^j < 1 + q^s + \cdots + q^{(m-1)s}$. It suffices therefore to show $q^j \leq q^{(m-1)s}$. As $j = \lceil log_q(m) \rceil$ and $m > 1$, we even have $j \leq m - 1$.   □

We are interested in sequences whose characteristic polynomial $f(X)$ is irreducible and in the subset of sequences of maximum length. How large is the step from irreducible $f(X)$ to maximum length? Assume $f(X)$ is irreducible. We know that the period of the sequence is then the exponent of $f(X)$. Which information is contained in this exponent? Well, the irreducible polynomial $f(X)$ describes the field $F = \mathbb{F}_{q^r}$ and X is mapped onto a certain element of F. The exponent of $f(X)$ is the order of X in the multiplicative group of this field (see Section 3.2).

16.9 Lemma. *Let $f(X)$ be an* **irreducible** *polynomial. Then the period of a nonzero LSRS with characteristic polynomial $f(X)$ is the order of X in the multiplicative group of the field $F = \mathbb{F}_{q^r} = \mathbb{F}_q[X]/(f(X))$. In particular, the period divides $q^r - 1$.*

The trace representation

An LSRS can be described equivalently using the trace. As we made heavy use of field extensions and the trace in the theory of cyclic codes, this description probably is the most natural for us.

Consider an LSRS $\mathcal{S} = (a_n)$ with entries in $\mathbb{F}_q$ whose characteristic polynomial $f(X)$ is irreducible. We use the corresponding extension field $F = \mathbb{F}_{q^r}$ and the trace $tr : F \longrightarrow \mathbb{F}_q$. Let $\epsilon \in F$ be the element corresponding to X, hence a root of $f(X)$. The powers $1, \epsilon, \ldots, \epsilon^{r-1}$ form a basis of F over $\mathbb{F}_q$. There is therefore a uniquely determined element $u \in F$ such that $tr(\epsilon^i u) = a_i$ for $i = 0, \ldots, r - 1$ (see Exercise 16.2.2). In fact, this formula describes not just the initial values of the LSRS but the LSRS as a whole:

16.10 Theorem. *Let $\mathcal{S} = (a_n)$ be an LSRS over $\mathbb{F}_q$ and $f(X)$ its characteristic polynomial, irreducible of degree r. Let $F = \mathbb{F}_{q^r}$ be the extension field described by $f(X)$, $tr : F \longrightarrow \mathbb{F}_q$ the trace and $\epsilon \in F$ a root of $f(X)$. There is an element $u \in F$ such that*

$$a_n = tr(\epsilon^n u) \text{ for all } n.$$

PROOF The initial segment $a_0, \ldots, a_{r-1}$ is right. It suffices to show that the sequence $(tr(\epsilon^n u))$ satisfies the same recurrence relation as (a_n). Let $n \geq r$. Then

$$tr(\epsilon^n u) - \sum_{i=1}^{r} c_i tr(\epsilon^{n-i} u) = tr((\epsilon^r - \sum_i c_i \epsilon^{r-i})\epsilon^{n-r} u) = 0$$

as $\epsilon^r - \sum_i c_i \epsilon^{r-i} = f(\epsilon) = 0$. □

Theorem 16.10 is Theorem 6.24 of Lidl and Niederreiter's *Introduction to Finite Fields and Their Applications* [134], a standard textbook in this area.

The link to Simplex codes

Let $f(X)$ be irreducible of degree r over $\mathbb{F}_q$ and M an $(q^r, q^r - 1)$ matrix whose rows are all the LSRS with $f(X)$ as characteristic polynomial. The rows of M are the codewords of a cyclic code $\mathcal{C}$. Theorem 16.10 shows that $\mathcal{C} = tr(\mathcal{B}(\{1\}))$ in the terminology of Chapter 13 (a q-ary cyclic code of length $q^r - 1$). In the binary case $\mathcal{C}$ is an $[2^r - 1, r]_2$ code of strength 2, the binary Simplex code. We studied the binary Simplex codes in Chapter 2 and rediscovered them as cyclic codes in Chapter 13.2; see also Exercise 13.2.5. For general q, the situation is similar. By Theorem 13.28, $\mathcal{C}$ can be contracted ($u = q - 1$). The contracted code $\mathcal{K}$ of length $(q^r - 1)/(q - 1)$ has as columns of a generator matrix representatives of the points in $PG(r - 1, q)$. It follows that $\mathcal{K}$ is equivalent to the q-ary r-dimensional Simplex code. As $\mathcal{K}$ has constant weight q^{r-1} (the most natural proof of this fact will be given in Section 17.1, after Proposition 17.2), it follows that all nonzero codewords of the cyclic code $\mathcal{C}$ have weight $(q - 1)q^{r-1}$.

This also serves to demystify the LSRS. All we do is use the initial conditions to individuate a codeword. Then we read the rest of the codeword from left to right. This is the LSRS.

Randomness properties of LSRS

Golomb formulated three properties which maximum length binary LSRS have in common with random sequences. We are a little more general in that we consider entries from an arbitrary finite field $\mathbb{F}_q$. Observe that the period (best visualized as a circle) has length $q^r - 1$ and that the segments of length r run through every nonzero r-tuple exactly once. The first obvious fact is that every symbol except 0 occurs precisely q^{r-1} times in a period, whereas 0 occurs $q^{r-1} - 1$ times:

16.11 Proposition (R1). *Let $f(X)$ (irreducible, degree r) be the characteristic polynomial of a maximum length LSRS. Then each nonzero element occurs q^{r-1} times in a period, whereas 0 occurs $q^{r-1} - 1$ times. In particular, the frequencies of elements in a period differ by at most one.*

The second property is concerned with the frequencies of **runs** in a period. Here a run of length k is a segment of length k, with identical entry u, which is preceded and followed by an entry $\neq u$. Observe that all this happens in a fixed period. In particular, the lengths of all the runs add up to $q^r - 1$. It is easy to determine the number $r(u^k)$ of runs with entry u of length k, for all

u and k. The second randomness property (R2) states that the distribution of runs is as expected in a random sequence.

The third randomness property is of a different nature.

16.12 Definition. *Let* $S_1 = (a_n)$ *and* $S_2 = (b_n)$ *be sequences, defined over the same alphabet, and periodic with the same period* ρ*. Define* ϵ*, where*

$$\epsilon(x, y) = \begin{cases} +1 \in \mathbb{R} & \text{if } x = y \\ -1 \in \mathbb{R} & \text{if } x \neq y. \end{cases}$$

The correlation function $C(\tau), \tau = 0, 1, \ldots \rho - 1$ *is defined by*

$$C(\tau) = \frac{1}{\rho} \cdot \sum \epsilon(a_n, b_{n+\tau}) = \frac{1}{\rho} \cdot \sum_n (-1)^{a_n + b_{n+\tau}}$$

where the sum is over a full period. If $S_1 = S_2$*, then* $C(\tau)$ *is the* **autocorrelation function** *of our periodic sequence.*

The autocorrelation $C(\tau)$ certainly satisfies $-1 \leq C(\tau) \leq 1$. For $\tau = 0$ (the **in-phase autocorrelation**) it is maximal: $C(0) = 1$. In general, $C(\tau)$ is small if the sequence and its shift disagree in many coordinates. It may be considered as typical of random sequences that the **out-of-phase autocorrelation** $C(\tau), \tau > 0$ is close to being constant. The difference of S and a cyclic shift is a nonzero element of the code C. It therefore has weight $(q-1)q^{r-1}$. As S and its shift agree in precisely $q^{r-1} - 1$ positions, it follows that $C(\tau) = (q^{r-1} - 1) - (q-1)q^{r-1}/(q^r - 1) = (-q^r + 2 \cdot q^{r-1} - 1)/(q^r - 1)$, when $\tau \neq 0$.

16.13 Proposition. *Let* S *be a maximum length LSRS. Then the out-of-phase autocorrelation of* S *is*

$$C(\tau) = \frac{-q^r + 2 \cdot q^{r-1} - 1}{q^r - 1}, \quad 0 < \tau < q^r - 1.$$

In particular the following is satisfied:
(R3) The autocorrelation of S *is two valued.*

Extremal correlation properties

Families of sequences with low correlation have applications in communications and cryptography; see the survey article by Helleseth and Kumar [110]. Let $(a_n) = (tr(\epsilon^n u))$ be an LSRS of period $N = 2^r - 1$. Consider a **decimation** of this sequence: (b_n), where $b_n = a_{tn}$ for some t coprime to N. The correlation $C(\tau)$ of these sequences can be expressed in terms of the weight of the binary word $a + b \in \mathbb{F}_2^N$; see Exercise 16.2.5. This word has entry $a_n + b_{n+\tau} = tr(ue^n + ue^{t(n+\tau)})$ in coordinate n. It is a codeword of $tr(\mathcal{B}(\{1,t\}))$. It follows that, when this cyclic code has only a small number of different weights, then the cross-correlation function of (a_n) and (b_n) has only a small number of different values. This is a desirable property for the applications. Another related link to cyclic codes was discovered by Carlet, Charpin and Zinoviev [42]: the property known as APN (*almost perfect non-linear*) is equivalent to the binary cyclic code $tr(\mathcal{B}(\{1,t\}))$ having strength 4, which, as we know, is equivalent to $\mathcal{C}(\{1,t\})$ having minimum distance 5. This link triggered increased interest in the question of when binary cyclic codes have $d = 5$. Some popular cases when $\mathcal{C}(\{1,t\})$ of length $2^r - 1$ has distance 5 are the following:

1. $t = -1$ when r is odd (see Exercise 16.2.6) (the **Mélas** case).

2. $t = 4^h - 2^h + 1$ when h is coprime to r (the **Kasami** case).

3. $t = 2^h + 1$ when h and r are coprime (the **Gold** case).

4. $t = 2^m + 3$ if r, m both are odd (the **Welch** case).

Applications of LSRS

Linear shift register sequences are also known as *linear feedback register sequences*. These terms indicate how these sequences can be efficiently implemented. One needs r registers, where at any point in time a segment of length r of the sequence is stored, say from a_{n-r} in the register on the left, up to a_{n-1} on the right. Next a_n is calculated as a linear combination of the entries in these registers. A fixed wiring representing the coefficients c_i is used. Next comes the shift procedure: the entries of the registers are shifted one to the left, with a_{n-r} dropping out, and a_n is entered in the rightmost register. This concludes one application of the algorithm. This is computationally very efficient, as only a small number of elementary calculations are needed, and the wiring is fixed.

Most applications of LSRS are inspired by the concept of pseudorandomness. However, in spite of the randomness properties, an LSRS of depth r is uniquely determined by a segment of length $2r$: the first half of the segment serves as initial values, and each of the entries of the second half yields a linear equation for the feedback coefficients. In the presence of an opponent (see the following section), it is therefore necessary to use LSRS of sufficiently large depth. The depth of an LSRS is also known as its **linear complexity.** In fact, any infinite sequence can be seen as an LSRS. An opponent will be able to detect the construction of the sequence if the linear complexity is not high enough. One method to increase the linear complexity is to use the output of several LSRS and combine them in a suitable way. It is easy to show that the linear complexity of the termwise sum is at most the sum of the complexities (see Exercise 16.2.3). It is much harder to prove that the linear complexity of the product is at most the product of the complexities; see Zierler and Mills [226]. Under mild conditions, equality holds in both cases. It is therefore possible to use LSRS of linear complexities $r_1, r_2, \ldots$ and construct LSRS of complexity $p(r_1, r_2, \ldots)$, where $p(X_1, X_2, \ldots)$ is a polynomial. For a more complete description of what is known about the linear complexity of the sum and the product, we refer to Camion and Canteaut [41] and the literature cited therein.

Exercises 16.2

16.2.1. *Let $f(X)$ be irreducible of degree r over $\mathbb{F}_q$.*
When is it true that every LSRS with $f(X)$ as characteristic polynomial has maximum length?

16.2.2. *Let $F = \mathbb{F}_{q^r}$ and $tr : F \longrightarrow \mathbb{F}_q$ the trace. Show that, for every $\mathbb{F}_q$ linear mapping $f : F \longrightarrow \mathbb{F}_q$, there is exactly one $u \in F$ such that $f(x) = tr(xu)$ for all u.*

16.2.3. *Show that the linear complexity of the termwise sum of two q-ary LSRS is at most the sum of the complexities.*

16.2.4. *Find a maximal length binary LSRS of depth (linear complexity) 4.*

16.2.5. *Let $x = (x_i)$, $y = (y_i) \in \mathbb{F}_2^n$. Prove that the correlation $\frac{1}{n}\sum_i (-1)^{x_i + y_i}$ is given by $\frac{1}{n}(n - 2wt(x + y))$.*

16.2.6. *Use the Roos bound to prove that the binary cyclic code $\mathcal{C}(\{1, -1\})$ of length $n = 2^r - 1$ has minimum distance 5 when r is odd.*
These are the Mélas codes; their duals (the corresponding trace codes) are known as **Kloosterman codes.**

16.2.7. *What are the smallest code parameters of the family of binary codes constructed in the previous exercise?*

16.2.8. *The parameters $[128, 113, 6]_2$ could be obtained from a $[129, 114, 6]_2$ code, by shortening. Does such a code exist?*

16.3 Cryptography and S boxes

Basic concepts: The nature of cryptography. Block ciphers. Strength and minimum distance providing confusion and diffusion. The McEliece system.

In the original historically first application of error-correcting codes, a transmitter wishes to send a message via a noisy channel to a remote receiver. In cryptography the situation is similar insofar as again we wish to guarantee the integrity of a message sent via a channel to a distant receiver. The important difference is that the obstacle is not the **technical quality** of the channel but an **opponent,** who may use his access to the channel to obtain information about the message or to alter the message. A rough subdivision of cryptography according to its aims may be described as follows:

secrecy (also called **privacy** or **confidentiality**): the message should be encrypted in such a way that only the legitimate receiver is able to decipher.

authentication (in a broad sense:) we wish to enable the
legitimate receiver to convince himself that the message
received is identical to what was sent (**data integrity**) and to identify
the correct source of the message.

Cryptography is an ancient art. D. Kahn's *The Codebreakers* [122] is centered around the use of ciphers in the two world wars. In the modern world dominated by electronic communication, the subject has been revitalized and is probably more important than ever before in history. Stinson's book *Cryptography-Theory and Practice* [200] is a self-contained introduction to modern cryptography. The *Handbook of Applied Cryptography* [107] provides technical descriptions and background material.

Gustavus J. Simmons describes in [191] coding theory and authentication theory as dual theories in the sense that they are diametrically opposed. This is based on the following observation: when we use an error-correcting code for the transmission of messages via a noisy channel, then the code is chosen in such a way that the most probable alterations (caused by noise) form clusters around the codewords. In contrast to this, if we wish to enable the receiver to detect alterations of the message caused by interference by an opponent, we want to guarantee that such alterations will spread uniformly throughout space. Despite this "duality," we will see in some of the following sections that our basic parameters, the minimum distance and the strength, play a fundamental role throughout cryptography. Other basic mathematical ingredients which keep being used in cryptography are information theory (see Chapter 8) and elementary number theory. Naturally we will concentrate on the coding-theoretic aspects.

Just as much of modern coding theory originated in Shannon's paper *A mathematical theory of communication* [183], another long paper by Shannon, *Communication theory of secrecy systems* [184] (though far from being flawless) has had a deep influence on modern cryptography, in particular the secrecy aspect. It is also valuable for its descriptions of many classical ciphers. We emphasize the following classification, given by Shannon, of cryptographic tools according to their aims:

confusion: destroy any structure (regularity) that the opponent could make use of in an attack, and

diffusion: make sure that every local change in the input spreads through the whole system, preventing the opponent from conducting a local attack that uses only parts of the input or output.

The paradigmatic example of a classical cryptosystem is the **Data Encryption Standard** (DES). The heart of the system and its only nonlinear ingredients are the **substitution boxes** (S boxes). Each S box is a mapping $: \mathbb{F}_2^6 \longrightarrow \mathbb{F}_2^4$. The S-boxes provide the confusion and some diffusion. The point of view of boxes (mappings) providing diffusion gives another interpretation of our basic parameters, the minimum distance and the strength of a linear code.

Let $\mathcal{C}$ be an $[n, k]_q$ code described by a generator matrix G. Choose notation such that the first k columns of G form the unit matrix I. Define the linear mapping

$$\pi : \mathbb{F}_q^k \longrightarrow \mathbb{F}_q^{n-k}$$

such that $\pi(x)$ is the projection of the codeword starting with x to the last $n - k$ coordinates. In particular, the $(x, \pi(x))$, where $x \in \mathbb{F}_q^k$, are precisely the codewords of $\mathcal{C}$. We describe x as the **input** and $y = \pi(x)$ as the **output** of our cryptographic function. In our cryptographic perspective, the minimum distance d and the strength t have the following meaning:

minimum distance: For every $i < d$, whenever i values among inputs and outputs are changed, then it is guaranteed that at least $d - i$ more such values will change. This is a diffusion property.

strength: Whenever the values of at most $t-1$ among inputs and outputs are known, this gives no information on any single further input or output.

The construction in Schnorr and Vaudenay [181] is based on a (nonlinear) $OA_1(2, 4, q)$. We know from Theorem 3.16 of Section 3.4 that an $OA_1(2, 4, q)$ is equivalent to a pair of orthogonal Latin squares of order q. The corresponding function π is also called a **multipermutation**. The variant when $\mathcal{C}$ is a Reed-Solomon code in characteristic 2 is described in Camion and Canteaut [41]. This has been used in the design of the ciphers SHARK and SQUARE ([185, 188]).

We illustrate with the Reed-Solomon code $\mathcal{RS}(2,4)$:

$$
\begin{array}{ccc}
0\ 0 & \mapsto & 0\ 0 \\
1\ 1 & \mapsto & 1\ 1 \\
\omega\ \omega & \mapsto & \omega\ \omega \\
\overline{\omega}\ \overline{\omega} & \mapsto & \overline{\omega}\ \overline{\omega} \\
0\ 1 & \mapsto & \omega\ \overline{\omega} \\
1\ 0 & \mapsto & \overline{\omega}\ \omega \\
\omega\ \overline{\omega} & \mapsto & 0\ 1 \\
\overline{\omega}\ \omega & \mapsto & 1\ 0 \\
0\ \omega & \mapsto & \overline{\omega}\ 1 \\
1\ \overline{\omega} & \mapsto & \omega\ 0 \\
\omega\ 0 & \mapsto & 1\ \overline{\omega} \\
\overline{\omega}\ 1 & \mapsto & 0\ \omega \\
0\ \overline{\omega} & \mapsto & 1\ \omega \\
1\ \omega & \mapsto & 0\ \overline{\omega} \\
\omega\ 1 & \mapsto & \overline{\omega}\ 0 \\
\overline{\omega}\ 0 & \mapsto & \omega\ 1 \\
\end{array}
$$

The corresponding function $\pi : \mathbb{F}_4^2 \longrightarrow \mathbb{F}_4^2$ maps the left pair of entries to the right pair of entries. As $d = 3$, we have, for example, that a change in an input or an output of π changes at least two more (inputs or outputs).

Cryptosystems such as DES are nowadays known as **private key** systems. Transmitter and receiver need to agree upon a secret key before transmission starts. This key is used for encryption by the transmitter, and for decryption by the receiver. Recently DES has been replaced by a new standard, the **Advanced Encryption Standard** (AES), a private key system which is also known as **Rijndael**. For a nice introduction see Trappe and Washington [209]. The fruitful idea of **public key** systems was first described by Diffie and Hellman [70]. It relies upon **one-way functions**. For example, given two primes p and q, it is easy to multiply them and obtain $n = pq$.

However, given the number n, it may be computationally very hard to find those factors. Assume e is the encryption function and d its inverse, the decryption function. Assume further that e has the one-way property, that it is computationally infeasible to compute d on the basis of e. If this is the case, then e can be published (e is a public key). The receiver's private key consists of the knowledge of the inverse d. Anyone wishing to send a message to the receiver can use e as the key. Only the receiver will be able to decrypt it. The most famous public key system is RSA, first described in [175]. It is based on the computational infeasability of factoring.

The **McEliece** public key cryptosystem [139] is based on decoding algorithms of codes. The best known decoding algorithms for general linear codes are very slow when compared to Euclidean algorithm based decoding for BCH codes as described in Section 13.3. In fact, there is a larger class of codes, the **Goppa codes,** which possess the same type of fast decoding algorithm as the BCH codes. We will not discuss them in this book. The idea behind the McEliece system may be described as follows: use the private key to disguise a Goppa code. Decryption involves error correction. A non-legitimate user does not have a fast decoding algorithm, whereas the receiver can use the Euclidean algorithm based decoding algorithm of the Goppa code. For more details see Stinson [200].

In general, an S box simply is a mapping $f : \mathbb{F}_2^r \longrightarrow \mathbb{F}_2^s$. The most important case is when $r = s$ (although we saw that in DES the S boxes have $r = 6, s = 4$). The cryptographic theory of S boxes is the study of those functions when used as ingredients in a private key cryptosystem under the point of view of protection against various types of attacks; see [201]. In order to protect against a type of attack which is known as the **differential attack,** the following parameters are essential:

$$N_D(a, b) = |\{x \mid f(x) + f(x + a) = b\}|.$$

Here $0 \neq a, b \in \mathbb{F}_2^r, x \in \mathbb{F}_2^r$. A differential attack is possible when at least one of those parameters is large (the average is $= 1$). As all of those numbers are even, we see that the ideal situation is when all those numbers are 0 or 2. This is exactly the definition of APN functions:

16.14 Definition. *Let $F = \mathbb{F}_{2^r}$. An APN function (APN=almost perfectly nonlinear) is a mapping $f : F \longrightarrow F$ such that $N_D(a, b)$ is either 0 or 2 for each $0 \neq a \in F, b \in F$ (in other words, the function $x \mapsto f(x) + f(x + a)$ is a two-to-one function for each $0 \neq a \in F$).*

16.15 Definition. *Let $F = \mathbb{F}_{2^r}$. A **Boolean function** is a mapping $f : F \longrightarrow \mathbb{F}_2$.*

Here a corresponds to the input-XOR, b is the output-XOR. APN functions are S boxes yielding optimal protection against the differential attack. When, in an earlier section on sequences with extremal correlation properties, we talked about the APN property of cyclic codes $\mathcal{C}(\{1, t\})$, the corresponding APN function was the monomial function X^t.

Turning to the **linear attack,** the basic parameters are

$$N_L(a, b) = |\{x \mid a \cdot x + b \cdot f(x) = 0\}|.$$

The aim is to have all those numbers about $2^r - 1$, equivalently $N_L(a, b)/2^r \sim 0.5$ Any pair a, b such that $N_L(a, b)/2^n$ is bounded away from 0.5 may lead to a linear attack. Here $a \cdot x$ is a bit extracted from the input and $b \cdot f(x)$ is a bit extracted from the output. Let $g(x) = b \cdot f(x) : F \longrightarrow \mathbb{F}_2$. We want the random variable $a \cdot x + g(x)$ to have a small bias. Stinson defines the bias of a binary random variable as the probability of 0 minus 0.5. An obvious comparison with the Fourier coefficients (see Definition 15.18) shows that $\hat{g}(a)$ is the double of the bias of $a \cdot x + g(x)$. This shows that optimal protection against the linear attack is equivalent to the following property: for each $b \neq 0$, the Boolean function $b \cdot f(x) = tr(bf(x))$ has all its Fourier coefficients at $a \neq 0$ of small absolute value (close to 0). For a single such Boolean function, the ideal is a **bent function**. What the correct definition in general should be seems to be open. **Almost bent** is a related notion.

16.4   Two-point-based sampling

and chips testing.

Our motivational example will be **chips testing.** Let a chip with n inputs be given. We want to devise methods to test if all the circuits of the chip are faultless. The safest way would of course be to test all 2^n possible input vectors individually. This would be all too costly. To fix notation, let $U = \mathbb{F}_2^n$, the "universe" of all possible input vectors. Assume a circuit is faulty and let $W \subset U$ be the set of all input vectors which detect that fault (W is the set of **witnesses** of the fault). Let $\epsilon = 1 - \dfrac{|W|}{2^n}$. If one input vector is chosen at random (this requires n random bits, n tosses of a fair coin), then ϵ is the probability that the fault will not be detected. The objective is to devise an algorithm which generates a sequence of k test vectors (elements of U) such that the probability that the fault escapes attention is low. The bound on this **escape probability** should not depend on the set W of witnesses itself, but only on k and the size of W. We want a worst case bound on the escape probability in terms of ϵ.

An obvious method based on pure randomness is to generate k elements of U, independently. This requires kn random bits (an element of $\mathbb{F}_2^{kn}$ has to be chosen at random). The escape probability is then ϵ^k. In practice one would use a pseudorandom sequence, but this would be dangerous, as the analysis leading to the bound on the escape probability would not be valid.

A practitioner would be tempted to use LSRS. So consider an LSRS of length $2^n - 1$. This will eventually generate all nonzero input vectors on a circle. The random seed of n bits will determine the starting point. However, we see that the worst case is really bad. If the witnesses should be on a segment of length $|W| + n - 1$ on this circle, then the escape probability will be $(2^n - 1 - |W| - k + 1)/2^n = \epsilon - k/2^n$.

Spenser [197] suggests using field arithmetic in $F = \mathbb{F}_{2^n}$. The structure underlying this method is $\mathcal{RS}(2, q)$, the two-dimensional Reed-Solomon code, where $q = 2^n$. If k test vectors are to be used, we restrict to the first k columns of the Reed-Solomon code. The random seed ($2n$ bits) is used to determine a row of the array at random. The k entries in this row are then used as test vectors. The analysis leading to the desired bound on the escape probability does not make use of the structure of the Reed-Solomon code. All it uses is the fact that we have an orthogonal array of strength 2 and index 1.

16.16 Theorem. *Let A be an $OA_1(2, k, q)$ with a set of entries Q. Let $W \subset Q$, $|W| = m$ and $\epsilon = 1 - m/q$. Choose a row of A uniformly at random. Then the probability that the row does not contain any element of W is at most*

$$\frac{\epsilon}{1 + (k - 1)(1 - \epsilon)}.$$

PROOF We will use quadratic counting. Let x_i be the number of rows precisely i of whose entries are from W. The escape probability is x_0/q^2. Using the defining properties of an OA of strength 2, we obtain three equations for the x_i :

$$\begin{array}{rcl} \sum_{i=0}^{k} x_i &=& q^2 \\ \sum_{i=0}^{k} i x_i &=& kmq \\ \sum_{i=0}^{k} i(i - 1)x_i &=& k(k - 1)m^2. \end{array}$$

For any real number z, we have $\sum_{i=1}^{k}(i - z)^2 x_i \geq 0$. Using the above equations, we can rewrite this as

$$z^2 \sum_{i=1}^{k} x_i + k(k - 1)m^2 + kmq - 2zkmq \geq 0.$$

Solve this for $\sum_{i=1}^{k} x_i$:

$$\sum_{i=1}^{k} x_i \geq \frac{2kmqz - kmq - k(k - 1)m^2}{z^2}.$$

The right hand side is maximal when $z = \{q + (k - 1)m\}/q$. This yields a lower bound on $\sum_{i=1}^{k} x_i$. The corresponding upper bound on x_0/q^2 is

$$x_0/q^2 \leq \frac{\epsilon}{1 + (k - 1)(1 - \epsilon)}.$$

⧠

This analysis is in Gopalakrishnan and Stinson [98]. In practice one would probably not make use of the constant rows of the Reed-Solomon code. A brief description is also given in the article **Derandomization** by the same authors, in the *Handbook of Combinatorial Designs* [106].

16.5   Resilient functions

as large sets of orthogonal arrays. The Nordstrom-Robinson code. The wire-tap channel. Generation of random bits.

There are many ways to motivate this topic. Start from the following cryptographic scenario. Several users of a network agreed on a common randomly generated bitstring of length n. Assume that, after a certain time, up to t of these bits may have leaked to an opponent (we do not know which bits have leaked, but there is this threshold of t). The random string should therefore not be used any more. The idea is to map the n bits to some $m < n$ bits such that the resulting shorter bitstring is still "random" and can be used without compromising the security of the system.

By now we have enough experience to translate from the language of "randomness" to "structure." The original bitstring lives in $\mathbb{F}_2^n$. The fact that it is chosen in a perfectly random way means that we consider $\mathbb{F}_2^n$ as a uniform sample space. What we are looking for is a mapping $f : X = \mathbb{F}_2^n \to Y = \mathbb{F}_2^m$ with suitable properties. Our first requirement is that each $y \in Y$ occurs as an image equally often; in other words, $|f^{-1}(y)| = 2^{n-m}$ for every $y \in Y$. We consider f and all functions $g : \mathbb{F}_2^{n-t} \to \mathbb{F}_2^m$ obtained from f by fixing the values of some t input bits. For example, if $n = 6$, then $f(x_1, x_2, x_3, x_4, x_5, x_6)$ depends on six input bits. If $t \geq 2$, we could prescribe, for example, $x_1 = 0, x_3 = 1$. Consider $f(0, x_2, 1, x_4, x_5, x_6)$ as a function defined on $\mathbb{F}_2^4$. This is a typical case of a restriction function g.

The central randomness property of f translates as follows: every $y \in Y$ occurs equally often as image under any mapping g as above obtained from prescribing some t input bits of f, formally, $|g^{-1}(y)| = 2^{n-m-t}$ for all such g.

What does this have to do with codes and orthogonal arrays? We are almost there: consider $f^{-1}(y)$, and imagine it written as an array with 2^{n-m} rows and n columns. The central randomness property says that it is an orthogonal array of strength t. Its parameters are therefore $OA_{2^{n-m-t}}(t, n, 2)$. As this has to be true for every $y \in Y$, it follows that $\mathbb{F}_2^n$ must be partitioned into OA, all with the above parameters. Here is the formal definition:

16.17 Definition. *A binary resilient function $RF(n, m, t)_2$ is a balanced mapping $f : X = \mathbb{F}_2^n \to Y = \mathbb{F}_2^m$ with the property that, whenever $y \in Y$ and some t input bits of f are prescribed, there are exactly 2^{n-m-t} elements $x \in X$ which have those prescribed input values and satisfy $f(x) = y$.*

Our discussion showed that $\mathbb{F}_2^n$ must be partitioned into 2^m orthogonal arrays, each of strength t, with the above parameters. It is clear that this is an equivalent description of binary resilient functions. Given the partition, we simply define f in such a way that $f^{-1}(y)$ is a part of the partition for each $y \in Y$. We have seen that we can describe binary resilient functions in terms of orthogonal arrays, as follows.

16.18 Theorem. *The following are equivalent:*

- *A binary resilient function $RF(n, m, t)_2$,*

- *a partition of $\mathbb{F}_2^n$ in 2^m orthogonal arrays, each with parameters $OA_{2^{n-m-t}}(t, n, 2)$.*

The definition of resilient functions can be generalized in an obvious way from the binary to an arbitrary finite alphabet size. In the literature most often case $m = 1$ is considered. The case $m > 1$ is then referred to as **vector resilient functions.** There is another slight generalization: it is not strictly necessary to demand that f is balanced. In the language of Theorem 16.18 one demands that $\mathbb{F}_2^n$ be partitioned into 2^m orthogonal arrays of strength t, but not necessarily with the same index. These functions are known as **correlation-immune functions.**

The characterization of resilient and correlation-immune functions in terms of orthogonal arrays is due to D. R. Stinson [202]. It generalizes to arbitrary alphabets, of course.

Here is another possible application of resilient functions in the same style as our motivational application also taken from Chor et al. [54]; n processors wish to generate a common random bitstring. Unfortunately, there is a possibility that up to t of those may generate dangerous nonsense. In that case, let each processor generate a string of length c, write those strings side by side to obtain a string of length cn, finally apply an $RF(cn, m, ct)_2$ to obtain a perfectly random string of length m.

The definition of correlation-immune and resilient functions goes back to Siegenthaler [189]. The next application explains why those functions have been termed correlation-immune.

Construction of stream ciphers

For an introduction to stream ciphers consult Rueppel's book *Analysis and Design of Stream Ciphers* [178]. The plaintext arrives in the form of a bit-string. We think of the plaintext as a potentially (countably) infinite sequence of binary symbols

$$\mathcal{T} = t(1), t(2), t(3), \ldots$$

At each unit of time one of the binary symbols arrives and has to be transmitted via a channel to a remote receiver. The channel used for transmission is public. Our concern is with **secrecy**. This means that the text has to be encrypted in such a way that an opponent will be unable to reconstitute the plaintext from his knowledge of the enciphered text. As the text arrives in the form of a bit stream, each symbol has to be enciphered seperately. We can therefore think of enciphering as the componentwise addition of a bit string $\mathcal{C} = (c(1), c(2), c(3), \ldots)$ to the plaintext. At time m the symbol $t(m) + c(m)$ will appear in the channel. The quality of the process will be determined by the properties of the string $\mathcal{C}$. The ideal solution would be to choose each $c(m) \in \{0, 1\}$ uniformly at random. This method is known as the **Vernam Chiffre** or **Vernam one-time pad** in cryptography (**one-time** because such a random bit string should be used only once). However, this is an impractical method as the information on $c(m)$ has to be transmitted to the receiver via a secure channel, and the amount of information having to be transmitted in a secure way is not less than the information contained in the plaintext itself. Another idea would be to use a linear shift register sequence for $\mathcal{C}$. Here the difficulty arises that we would need a high linear complexity in order to be protected against a brute-force attack of the opponent. In order to get around this difficulty while still making use of pseudo random sequences with a short random seed, the following procedure has been proposed:

Use n processors. Processor i generates $Z_i(m)$, $m = 1, 2, \ldots$, an LSRS which is determined by a key $K_i \in \mathbb{F}_2^{n_i}$. At time m we have the binary numbers $Z_i(m)$, $i = 1, 2, \ldots, n$ at our disposal. Then use a **combining function** $f : \mathbb{F}_2^n \longrightarrow \mathbb{F}_2$:

$$c(m) = f(Z_1(m), Z_2(m), \ldots, Z_n(m)).$$

We wish to discuss which properties the function f should have in order to safeguard against an opponent's attempt to break the cipher. This method brings a new aspect into play: instead of using a pseudorandom sequence based on a short random seed, we use several of these and combine them with the help of a suitable (deterministic) function f, which may be publicly known.

The brute-force attack forces us (brutally) to make sure the resulting sequence $f(Z_1(m), Z_2(m), \ldots, Z_n(m))$ has large linear complexity. Write f as a polynomial in n variables. In order to obtain a large linear complexity, this polynomial needs to have a large degree (see Section 16.2).

We turn to another type of attack, the **correlation attack**. Instead of searching through the entire key space of size $2^{\sum_i n_i}$, the opponent attempts to determine at first K_1, then K_2 and so on. This is conceivable if there is a correlation between one (or a small number) of the Z_i and the resulting sequence $f(Z_1, Z_2, \ldots, Z_n)$.

16.19 Example. *Consider $f(x_1, x_2, x_3) = x_1 x_2 + x_1 x_3 + x_2 x_3$ and the expression $x_1 \cdot f(x_1, x_2, x_3)$. For three of the $8 = 2^3$ inputs, the result is $= 1$, namely, for $(x_1, x_2, x_3) = (1, 1, 0), (1, 0, 1)$, or $(1, 1, 1)$. This is a marked difference from the case of the product of two independent uniformly distributed binary sequences, as in that situation the result would be $= 1$ in one quarter of all cases. If this function f was used as the combining function, then the opponent could proceed as follows: search through the entire key space of sequence Z_1 and compute $Z_1(m) \cdot f(Z_1(m), Z_2(m), Z_3(m))$ in a not too small number of cases. As long as the partial key is not correct, we expect the behavior of independent uniformly distributed sequences. The product will be $= 1$ in about one quarter of all cases. If the partial key (for Z_1) is correct, then the product will be $= 1$ in about three eighths of all cases. In that case the opponent can be almost sure to have found the correct partial key.*

More generally, the opponent can divide the n input sequences into handy parts and use a correlation attack to determine each of the partial keys at a time. Let us speak of a correlation attack of **order** t if a partial key consists of the keys for t sequences taken together. In order to be protected against such a correlation attack, we have to choose f such that, for t as large as possible, correlation attacks of order t are impossible. This led Siegenthaler [189] to introduce the notion of a **correlation-immune function**.

A correlation attack of order t will be possible if there is a value y of f (in the example 0 or 1) and a set of t columns such that the projection of $f^{-1}(y)$ on this set of columns is not uniformly distributed. Such an attack will be impossible precisely if $f^{-1}(y)$ is an orthogonal array of strength t. Comparison with Theorem 16.18 shows that we arrive at the notion of a correlation-immune (resilient) function. Correlation immunity can also be viewed as a version of Shannon's requirement of confusion.

We saw that the combining function f needs to be a polynomial of high degree in order to obtain high linear complexity, and it needs to be a resilient function of large strength t in order to safeguard against correlation attacks. It can be shown that these two requirements are in conflict. Also, additional cryptographic requirements may come into play. This leads to subtle tradeoff problems.

The wire-tap channel

Here is another natural application of resilient functions. We describe the setting following the lines of Ozarow and Wyner [162]. The source states (plaintext) are binary m-tuples. The encoder associates with the m-bit binary data sequence $y \in \mathbb{F}_2^m$ an n-bit binary transmitted sequence $x \in \mathbb{F}_2^n$, where $n > m$. It is required that a decoder can correctly obtain y with high probability by examining x. The intruder can examine a subset of his choice of size t of the n positions in x and the system designer's task is to make the intruder's uncertainty about the data as large as possible. The encoder is allowed to introduce **randomness** into the transformation : $\mathbb{F}_2^m \longrightarrow \mathbb{F}_2^n$.

More specifically, the scenario is as follows. We partition $\mathbb{F}_2^n$ into 2^m parts, each of size 2^{n-m}, and use the elements of $\mathbb{F}_2^m$ to label the parts of this partition:

$$\mathbb{F}_2^n = \cup_y X_y, \text{ where } y \in \mathbb{F}_2^m, |X_y| = 2^{n-m}.$$

Let the plaintext be y. The encoder rolls dice and chooses $x \in X_x$ uniformly at random. This is the use of randomness mentioned above. Message x is sent via the channel. We assume that no transmission errors occur. The receiver decodes x as the label of the part of the partition it belongs to. The wire-tapper has access to some t of the n positions of x, the set of positions being his choice. We want to limit the amount of information the intruder can receive about the plaintext y. Let us be ambitious and aim at giving the intruder no information at all. Let $(x_1, x_2, \ldots, x_t)$ be the segment of x known to the intruder. What we need is that the 2^{n-m} possible messages x with that same segment are uniformly distributed on the X_y. This means that each X_y must contain precisely 2^{n-m-t} such continuations. As the set of coordinates is arbitrary, we see that we require each X_y to be an orthogonal array of strength t, an $OA_{2^{n-m-t}}(t, n, 2)$. Comparison with Theorem 16.18 shows that this is equivalent to an $RF(n, m, t)_2$.

Construction of resilient functions

Resilient functions are equivalent not to orthogonal arrays but to partitions of the ambient space into orthogonal arrays. In the case of linear OA, the partition exists automatically: it is given by the cosets of the linear OA. Generalizing from the binary to the q-ary in an obvious way, we see that a **linear** $RF(n, m, t)_q$ is an $\mathbb{F}_q$ linear surjective mapping $f : \mathbb{F}_q^n \to \mathbb{F}_q^m$ whose kernel is an $[n, n-m]_q$ of strength t. The dual is an $[n, m, t+1]_q$-code. In particular we see that a linear $RF(n, m, t)_q$ exists if and only if an $[n, m, t+1]_q$ code exists.

As the optimal parameters of binary linear codes of small dimension are known, the best parameters of linear $RF(n, m, t)_2$ are known for small m (in fact, for $m \leq 8$). It can happen that these parameters are superseded by those of nonlinear RF.

16.20 Definition. *An* $OA_{2^{k-t}}(t, n, 2)$ *is* **systematic** *if there are some* k *coordinates such that, in the projection of the rows from the OA to these* k *coordinates, each* k-*tuple occurs (exactly once).*

Each linear OA is systematic: just use k coordinates such that the corresponding (k, k) submatrix of a generator matrix has a nonzero determinant. The reader is asked (see Exercise 16.5.2) to prove that each systematic OA can be embedded in a partition of the ambient space $\mathbb{F}_2^n$ into orthogonal arrays, all of the above parameters. Theorem 16.18 shows that a systematic $OA_{2^{k-t}}(t, n, 2)$ can be used to obtain $RF(n, n - k, t)_2$.

A Z_4 linear binary code: the Nordstrom-Robinson code

The single most famous nonlinear binary code is the Nordstrom-Robinson code $\mathcal{NR}$. As it is a very good systematic OA, it also yields an interesting resilient function (see Definition 16.20 and Exercise 16.5.2).

In order to understand the Nordstrom-Robinson code, it is best to start from the extended binary Hamming code $[8, 4, 4]_2$. We know this self-dual code quite well. Its most symmetric generator matrix (and check matrix) is

$$\begin{pmatrix} 1000 & 0111 \\ 0100 & 1011 \\ 0010 & 1101 \\ 0001 & 1110 \end{pmatrix}.$$

The same matrix, when interpreted quaternary, generates a self-dual $[8, 4, 4]_4$ (see Exercise 14.1.4).

We construct a code whose alphabet is the ring $Z_4 = \mathbb{Z}/4\mathbb{Z} = \{0, 1, 2, 3\}$. There is a close relationship between linear codes defined over Z_4 and binary linear codes. One reason is that the factor ring of Z_4 modulo its zero divisors $\{0, 2\}$ is $\mathbb{F}_2$. Denote the image of $x \in Z_4$ under this projection $: Z_4 \longrightarrow \mathbb{F}_2$ by $\bar{x}$. We have

$$\bar{0} = \bar{2} = 0 \in \mathbb{F}_2, \quad \bar{1} = \bar{3} = 1 \in \mathbb{F}_2.$$

It follows that each quaternary code which is linear over Z_4 has a binary linear code as the image under projection. Equivalently we can consider quaternary Z_4 linear codes as liftings of binary linear codes.

Let us try to lift the binary Hamming code, using the generator matrix above. We are looking for a Z_4 generator matrix G such that

$$\overline{G} = \begin{pmatrix} 1000 & 0111 \\ 0100 & 1011 \\ 0010 & 1101 \\ 0001 & 1110 \end{pmatrix}.$$ This means that each entry 0 of $\overline{G}$ has to be replaced

by 0 or 2, and, for each entry 1 of $\overline{G}$, the corresponding entry of G is 1 or $-1 = 3$. Let us copy the unit matrix and choose the lifting P on the right half of $\overline{G}$ such that the quaternary code is self-dual with respect to the dot product over Z_4. In order to do this, it suffices to lift the 0 entries on the right to 2 and to choose the lifting of the entries 1 such that, after lifting for any two rows in the two coordinates in the right half where no 2 appears, the entries are equal in one, nonequal in the other. The following is an antisymmetric solution for P ($P^t = -P$):

$$G = (I|P) = \begin{pmatrix} 1000 & 2333 \\ 0100 & 1231 \\ 0010 & 1123 \\ 0001 & 1312 \end{pmatrix}.$$

The quaternary Z_4 linear code $\mathcal{N}$ generated by G is known as the **octacode.** Clearly its codewords $x(a, b, c, d)$ are

$$(a, b, c, d | 2a+b+c+d, -a+2b+c-d, -a-b+2c+d, -a+b-c+2d), \quad (16.1)$$

where a, b, c, d vary in Z_4. We have $\mathcal{N} = \mathcal{N}^\perp$. It contains the repetition code: $x(1, 1, 1, 1) = 1^8$.

Our aim is to construct a binary code. This is done by applying the Gray map $\gamma : Z_4 \longrightarrow \mathbb{F}_2 \times \mathbb{F}_2$, which maps $0 \in Z_4$ to the 0 pair, the units of Z_4 to pairs of weight 1 and the zero divisor 2 to the pair of weight 2.

16.21 Definition. *The Nordstrom-Robinson code $\mathcal{NR}$ is the image of the octacode $\mathcal{N}$ under the Gray map. Here the Gray map γ is defined by*

$$\gamma : \ 0 \mapsto 00, \ 1 \mapsto 10, \ 3 \mapsto 01, \ 2 \mapsto 11.$$

Observe that $\mathcal{NR}$ is a nonlinear code. The nonlinear element in its construction is the Gray map. As the nonzero codewords of the extended Hamming code have weights 4 and 8, it follows that all nonzero words of $\mathcal{NR}$ have even weight ≥ 4. It is not hard to see that weight 4 does not occur. In the combinatorial Exercise 16.5.6 the reader is asked to determine the weight distribution of $\mathcal{NR}$. In particular, $\mathcal{NR}$ is a $(16, 2^8, 6)_2$-code. It is easy to see that a linear $[16, 8, 6]_2$ code cannot exist; see Exercise 16.5.4.

Application of the Gray map is a variant of concatenation. The underlying alphabet shrinks, the length doubles, the number of codewords remains the same. In our context we are more interested in orthogonal arrays than in

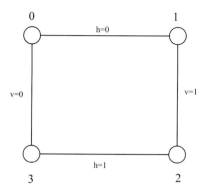

FIGURE 16.1: Z_4 as a graph

codes. In fact, the Nordstrom-Robinson code behaves like a self-dual binary code. It has strength 5. Let us see why.

Denote by $v(i)$ the first digit of $\gamma(i)$, by $h(i)$ its second digit:

$$v(0,3) = 0, \ v(1,2) = 1, \ h(0,1) = 0, \ h(2,3) = 1.$$

This means $\gamma(i) = (v(i), h(i))$ for all $i \in Z_4$. Observe also that $h(i) = v(-i)$. Introduce the structure of a graph on Z_4, where two elements form an edge if and only if their difference is ± 1. The reason is that two different elements of Z_4 have the same prescribed value under v or under h if and only if they form a certain edge of this graph; see Figure 16.1.

Two elements form a nonedge of this graph if they differ by 2. For every $i \in Z_4$, we have that $v(i) \neq v(2 + i)$, analogously for h.

Now consider $\mathcal{NR}$ as an orthogonal array. We want to show it has strength 5. Denote the coordinates of the octacode by s_i, $i = 1, \ldots, 8$ and by v_i and h_i for $i = 1, \ldots, 8$ the coordinates of $\mathcal{NR}$. Here the entry in v_i is the image of the entry in s_i under v, analogously for h_i. By definition, the eight columns $\{v_1, h_1, \ldots, v_4, h_4\}$ are independent.

Also use the result of Exercise 16.5.5: the automorphism group of the octacode is 3-transitive on the eight coordinates, and it has only two orbits on unordered 4-sets of coordinates, with representatives $\{1, 2, 3, 4\}$ and $\{1, 2, 3, 8\}$.

Let M be a set of five columns of the Nordstrom-Robinson code. If they are derived from only three coordinates s_i it can be assumed these are s_1, s_2, s_3 because of the triple transitivity of the automorphism group. It follows that M is independent. The same is true if M is derived from four coordinates s_i if those are in the orbit of $\{s_1, s_2, s_3, s_4\}$. Assume they are in the orbit of s_1, s_2, s_3, s_8. We can assume $M = \{v_1, h_1, v_2, v_3, v_8\}$ in this case. The projection of $x(a, b, c, d)$ (see Equation 16.1) on these coordinates is $(v(a), h(a), v(b), v(c), v(a - b + c + 2d))$. Fix a, b, c. When d varies in Z_4, the entry in the last coordinate is twice 0 and twice 1. It follows that M is independent.

The last case to consider is that M is derived from five different columns of the octacode. As the automorphism group of the octacode is 3-transitive, it is also transitive on unordered sets of five coordinates. We can assume $s_1, \ldots, s_5$ are the columns involved. As $h(i) = v(-i)$, it can be assumed $M = \{v_1, v_2, v_3, v_4, v_5\}$. The projection of $x(a, b, c, d)$ is $(v(a), v(b), v(c), v(d), v(2a + b + c + d))$. Once again the entry 2 is decisive. In fact, fix b, c, d. The entries in the first and last coordinate of M are $(v(a), v(2a + const))$. As a varies in Z_4, this expression varies through all of $\mathbb{F}_2^2$, no matter what the constant is. We have $v(a) = 0$ if and only if $a \in \{0, 3\} = e$, an edge of the graph. Then $2a$ varies through $0, 2$ and $v(2a + const)$ through $0, 1$.

16.22 Theorem. *The Nordstrom-Robinson code $\mathcal{NR}$ is an OA of strength 5. Its parameters are therefore $OA_8(5, 16, 2)$.*

By definition, $\mathcal{NR}$ is systematic. Therefore it describes a resilient function $RF(16, 8, 5)_2$.

Exercises 16.5

16.5.1. *Let C be an $[n, m, t + 1]_q$ code. Show how to obtain an $RF(n, m, t)_q$.*

16.5.2. *Show that a systematic $OA_{2^{k-t}}(t, n, 2)$ can be embedded in a partition of $\mathbb{F}_2^n$ into $OA_{2^{k-t}}(t, n, 2)$.*

16.5.3. *Give a definition of $RF(n, m, t)_q$ for arbitrary q and prove $m + t \leq n$. Why do we call this the Singleton bound?*

16.5.4. *Show that there is no linear $[16, 8, 6]_2$ code.*

16.5.5. *Determine the automorphism group of the octacode.*

16.5.6. *Determine the weight distribution of the Nordstrom-Robinson code.*

16.5.7. *Let $\mathcal{G}$ be the binary Golay code $[24, 12, 8]_2$, and choose notation such that the first eight coordinates form an octad. Let $\mathcal{D} \subset \mathcal{G}$ be the subcode consisting of the codewords with entries 0 in the first eight coordinates. Show that $dim(\mathcal{D}) = 5$ and the image of the projection from $\mathcal{G}$ to the first eight coordinates is the sum-0 code $[8, 7, 2]_2$.*

16.5.8. *In the situation of the preceding exercise, define $\mathcal{D}_1 = \mathcal{D}$ and for $i = 2, \ldots 8$ the subcode $\mathcal{D}_i \subset \mathcal{G}$ as the set of codewords having entry 1 in coordinates 1 and i, entry 0 in the remaining among the first eight coordinates. Let $\mathcal{E}$ be the projection of $\mathcal{D}_1 \cup \cdots \cup \mathcal{D}_8$ on the last 16 coordinates. Determine the code parameters of $\mathcal{E}$.*

16.6   Derandomization of algorithms

Basic concept: The maximal independent set problem.

The general procedure may best be understood by studying a particular problem in some detail. We choose the problem of finding a maximal independent set of vertices in a given graph.

The maximal independent set problem

Consider algorithms which accept as input an arbitrary graph G (with set V of vertices and set E of edges) and produce as output a maximal set I of independent vertices. So no two vertices in I are connected by an edge, and I is maximal with this property. The method we describe here stems from M. Luby [137]. Our algorithms will all follow the same general scheme: in step k an independent set I_k is given. Let $N(I_k)$ be the set of vertices of G which are not in I_k but are neighbors of at least one vertex from I_k. Let $V_k = V \setminus (I_k \cup N(I_k))$ and let $G_k = (V_k, E_k)$ be the graph induced by G on the vertices of V_k. The decisive step is the construction of an independent set I'_k in G_k. We then put $I_{k+1} = I_k \cup I'_k$. This is an independent set again. The algorithm terminates when the set G_k is empty. The output is the maximal independent set $I = I_k$.

A parallel Monte Carlo algorithm

All we need is an algorithm which quickly produces a large independent set in a given graph. Our algorithm for the construction of a **maximal** independent set as described above reduces to a repeated application of this algorithm. Let us call this step **CHOICE**. The algorithm we describe in this subsection makes use of parallelization and random bits. We call it **MC-CHOICE**. Here the letters MC stand for **Monte Carlo**.

So let a graph $G = (V, E)$ be given, and denote by $d_i, i \in V$ the valency of vertex $i \in G$. Recall that the **CHOICE** step will be applied to graph G_k, for every k. The algorithm **MC-CHOICE** is defined as follows:

- Compute in parallel the valencies d_i for the vertices $i \in V$.

- Define $|V|$ statistically independent random variables
  $f_i : V \longrightarrow \{0,1\}$, as follows: If $d_i = 0$, then $f_i = 1$. If $d_i \neq 0$, then
  $f_i = 1$ with probability $1/(2d_i)$. In this way V and its power set become
  probability spaces. The higher the valency of vertex i the smaller the
  probability that we include it in a large independent set. Compute in
  parallel for each $i \in V$ a value $f_i \in \{0,1\}$, according to these probabilities. This defines the set $X = \{i \mid i \in V, f_i = 1\}$.

- Compute in parallel for each edge $\{i,j\} \in E$:
  if $\{i,j\} \in X$, then remove i from X such that $d_i \leq d_j$.

- The resulting set X of vertices is the output of the algorithm, unless X
  is empty. In that case, choose as X a set with just one vertex.

Now consider the algorithm which uses **MC-CHOICE** in each step. We
want to bound the number of steps needed. This is the value of k for which
G_k becomes empty. Because of the random element in the algorithm, this
number is not a function of the input data but a random variable. We are
interested in its expected value. In order to do this, one has to study how
the number e_k of edges in E_k decreases with k. The details are in [137]. The
result is as follows: the expected value of the e_k satisfies $E(e_{k+1}) \leq \frac{7}{8} E(e_k)$.
The expected number of steps k needed to obtain an empty graph G_k, and
hence to terminate the algorithm, is therefore $O(log|E|) = O(log|V|)$.

Derandomization of the Monte Carlo algorithm

The aim is to reduce or ideally to eliminate the stochastic element from
the **CHOICE** step without changing the expected behavior of the resulting
algorithm. Let us look back at what we did in the preceding subsection: let
k be fixed and $H = (V, E) = (V_k, E_k)$ the given graph. We gave $\mathbb{F}_2^V$ the
structure of a probability space as a direct product of the spaces $\Omega_i = \mathbb{F}_2$,
with probabilities $p_i(1) = p_i = \frac{1}{2d_i}, p_i(0) = 1 - p_i$. The algorithm is based on
a random choice of an n-tuple from this space $\mathbb{F}_2^n$, where $n = |V|$. The cor-
responding random variables f_i where $f_i(x_1, x_2, \ldots, x_n) = x_i$ are statistically
independent.

We want to change the f_i just a little bit and represent the new random
variables on a much smaller probability space. The new algorithm will be
based on a random choice from this small sample space. This will yield a
drastic reduction of the stochastic element. Alternatively we can test each
element of the probability space in parallel. This then yields a deterministic
algorithm. One speaks of **complete derandomization** in that case. The

crucial point is to prove that the new algorithm has essentially the same behavior as the Monte Carlo algorithm considered earlier.

Here is the construction: choose a prime q satisfying $n \leq q \leq 2n$. Replace the probability $p_i = 1/(2d_i)$ by the rational number $p'_i = \lfloor p'_i \cdot q \rfloor / q = n_i/q \leq p_i$. Use the Reed-Solomon code $\mathcal{RS}(2, q)$, which is an $OA_1(2, q, q)$ with q^2 rows. Eventually we omit some columns and obtain an $OA_1(2, n, q)$, which we call Ω. The rows of Ω are the elements of our probability space, with uniform distribution. The random variables f'_i are constructed as follows: choose $N_i \subseteq \mathbb{F}_q, |N_i| = n_i$, put $f'_i(x_1, x_2, \ldots, x_n) = 1$ if $x_i \in N_i$, $f'_i = 0$ otherwise. Here $(x_1, x_2, \ldots, x_n) \in \Omega$. Then f'_i has almost the same distribution as f_i. While the f_i are statistically independent, the f'_i are still **pairwise independent** (see Theorem 16.2). Just as in the preceding subsection, one has to study the behavior of the algorithm where the construction of I'_k is determined by random choice of a row of Ω.

Here is the algorithm as proposed by Luby, which we want to call **RS-CHOICE,** as the stochastic element has been reduced to the choice of a word of the Reed-Solomon code. Let $n = |V|$, choose a prime $q, n \leq q \leq 2n$, and put p_i, p'_i as described above.

- If there is a vertex i such that $d_i \geq n/16$, put $I' = \{i\}$ (observe that this case can occur for at most 16 values of k).

- Assume $d_i < n/16$ for all i. Then $p'_i \geq 8/q$. Moreover, the p'_i can be bounded by p_i in both directions; more precisely, $\frac{7}{8} \cdot p_i \leq p'_i \leq p_i$. Carry out the Monte Carlo algorithm, which differs from **MC-CHOICE** only in that the binary n-tuples are chosen from the rows of the Reed-Solomon code, with uniform distribution.

If the same analysis as before is carried out, qualitatively similar results to the case of **MC-CHOICE** are obtained. More precisely, one has

$$E(e_k - e_{k+1}) \geq \frac{1}{18} \cdot e_k.$$

In conclusion:

We considered algorithms for the maximal independent set problem in graphs. All the algorithms follow the general scheme described in the beginning. The decisive point is then the **CHOICE** algorithm, which is applied for each step to determine an independent set in the graph at hand. Two variants, **MC-CHOICE** and **RS-CHOICE**, have been considered. Both are parallel algorithms and use a stochastic element. The behavior of the global algorithm is very similar in both cases. The most important point here is the expected number of steps in the iteration, as a function of the number n of vertices of the graph. In both cases the expected number of steps in the iteration is $O(n)$. The decisive difference is in the size of the underlying probability space. The original Monte Carlo algorithm **MC-CHOICE** uses

$O(n)$ random bits (the size of the probability space is 2^n), whereas the algorithm **RS-CHOICE**, which is based on a two-dimensional Reed-Solomon code, uses only $O(log(n))$ random bits. It is even possible to construct a deterministic algorithm by testing all the rows of the Reed-Solomon code. The method of reducing the number of random bits by representing the random variables describing a Monte Carlo algorithm on a smaller probability space, while at the same time maintaining the essential features of the algorithm, is known as **derandomization**. **Complete derandomization** is obtained if the stochastic element is completely eliminated. In the case studied here we found that it suffices when the random variables are pairwise independent. This means that we can use an orthogonal array of strength 2 for the derandomization. This result had to be expected. The reason is that we were considering a graph problem. As a graph is a structure which is defined by a pairwise relation, it may be expected that pairwise independence is sufficient.

The formalism of derandomization

The algorithms for the maximal indepedent set problem as discussed above indicate how and when derandomization may be expected to work in general. The point of departure is a randomized algorithm using a large number of statistically independent random variables.

- Assume the analysis of the algorithm shows that t-wise independence of the random variables suffices to guarantee the essential features, above all, the order of magnitude of the expected running time.

As the maximal independent set problem is a problem involving only edges of graphs, it is not surprising that pairwise independence is sufficient. However, the analysis leading to this conclusion is not short or trivial. The crucial step is to replace the two-valued random variables f_i (with probabilities p_i and $1 - p_i$) by random variables f'_i (with probabilities p'_i and $1 - p'_i$), which do not differ very much from f_i, are pairwise independent and represented on a small probability space. This is made more precise by the following definitions:

16.23 Definition. *Let f, f' be random variables with the same finite set Z of values. Then f, f' are ϵ close if, for every $z \in Z$, we have*

$$|Pr(f = z) - Pr(f' = z)| \leq \epsilon. \tag{16.2}$$

In our example we replaced probabilities p_i by p'_i and made sure in the derandomized algorithm **RS-CHOICE** that we could assume $\frac{7}{8}p_i \leq p'_i \leq p_i$. This shows that the corresponding random variables are $\frac{1}{8}$ close.

16.24 Definition. *Let random variables $f_i, i = 1, 2 \ldots, n$ be defined on finite probability spaces Ω_i. A* **representation** *of $(f_1, f_2, \ldots, f_n)$ is a probability space Ω with random variables $g_1, \ldots, g_n$ defined on it such that the distribution of g_i is the same as the distribution of f_i, for every i. The representation is* **uniform** *if Ω has the uniform distribution, and it is t independent if any t of the g_i are statistically independent.*

In the case of the maximal independent set problem, we used an orthogonal array of strength 2, more precisely, the two-dimensional Reed-Solomon code, to construct a two-independent uniform representation of $(f_1', f_2', \ldots, f_n')$ on the rows of the OA. In the light of our interpretation of an orthogonal array of strength t as a family of t-wise independent random variables (see Section 16.1), the following is trivial:

16.25 Proposition. *Let $f_i, i = 1, 2, \ldots, n$ be random variables defined on finite sample spaces and let Ω be a finite sample space. Further, let v be a natural number such that $p \cdot v$ is an integer for every probability p occurring in the distribution of the f_i. If an $OA_\lambda(t, n, v)$ exists, then there exists a t-independent uniform representation of $(f_1, \ldots, f_n)$ on the rows of Ω.*

We can now formulate the remaining conditions for the derandomization of a Monte Carlo algorithm. This still is vague because it remains open what the analysis which shows that t-wise independence suffices, looks like.

- Let $f_1, f_2, \ldots, f_n$ be the independent random variables describing the Monte Carlo algorithm. Assume there are random variables f_i' such that f_i and f_i' are ϵ close for some $0 \leq \epsilon < 1$ and such that $(f_1', f_2', \ldots, f_n')$ have a t-independent representation on some probability space Ω.

If these conditions are satisfied, then we can replace the stochastic choice of the original Monte Carlo algorithm by the choice of an element from Ω. As f_i' and f_i are ϵ close, it can probably be proved that the expected number of steps of the new algorithm will have the same order of magnitude as for the Monte Carlo algorithm. The value of ϵ should only affect the constants. Naturally we wish ϵ to be as small as possible. In our example we had $\epsilon = 1/8$, and Ω consisted of the rows of the Reed-Solomon code, with uniform distribution. The effect of derandomization is then measured by the cardinality of Ω as compared with the cardinality of the direct product of the probability spaces on which the f_i were defined. If Ω is small enough, then a deterministic algorithm can be obtained by searching through all the elements of Ω and continuing with the element which gives the best result.

In Section 16.7 we will exploit another degree of liberty. This arises from weakening the notion of t-wise independence. The case $t = 2$ will be discussed in some detail. This idea will turn out to be very fruitful. It is possible to construct such arrays whose properties differ only slightly from 2-independence, but of a much smaller cardinality than orthogonal arrays of strength 2.

M columns

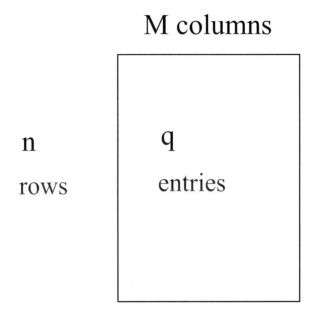

n

rows

q

entries

FIGURE 16.2:   The language of arrays

16.7   Authentication and universal hashing

In Section 16.1 orthogonal arrays were interpreted from a probabilistic point of view. Its rows are elements of a probability space, and each column describes a random variable with values in the set of entries. The basic parameter t, the strength of the OA, guarantees a certain statistical independence condition for this family of random variables.

In this section we will look at an array from the point of view introduced in Chapter 6, as a family of mappings. We index the columns with the elements of a set (the ground set) and interpret each row as a mapping from the ground set to the set of entries. Recall from Definition 6.1 that an $(n, M)_q$ array (or $(n, M)_q$ **hash family**) is an array (a matrix) with n rows and M columns, where each entry is taken from an alphabet Y of size q. Now we interpret each row of the array as a mapping $f : X \longrightarrow Y$, where $|X| = M$ (the elements of X parametrize the columns).

In Chapter 6 an $(n, M)_q$ hash family was defined to be ϵ universal if, for any choice of two different elements in the ground set, the probability of collision is bounded by ϵ (see Definition 6.3). It turned out that ϵ universal hash families are equivalent to error-correcting codes. An application to error detection was given.

Our next area of application is authentication. Before getting into this, let

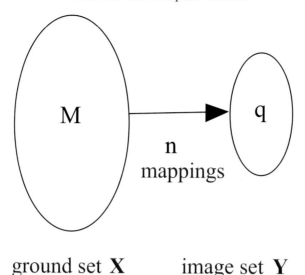

ground set **X**   image set **Y**

FIGURE 16.3: Equivalently, families of mappings

us define the type of hash families that we want to use now.

16.26 Definition. *An* $(n, M)_q$ *hash family is* ϵ **strongly universal** *for some* $0 < \epsilon < 1$ *if the following are satisfied:*

- *For any* $x \in X$ *and* $y \in Y$, *the number of mappings* $f \in \mathcal{F}$ *such that* $f(x) = y$ *is precisely* n/q.

- *For any two distinct elements* $x_1, x_2 \in X$ *and for any two (not necessarily distinct) elements* $y_1, y_2 \in Y$, *the number* ν *of functions* $f \in \mathcal{F}$ *such that* $f(x_1) = y_1$, $f(x_2) = y_2$ *satisfies* $\dfrac{\nu}{n/q} \le \epsilon$.

The first condition says that in the corresponding array each entry occurs with the same frequency in each column (equivalently, the array is an orthogonal array of strength 1). This facilitates the probabilistic interpretation of this structure: we know that there are precisely n/q mappings f satisfying $f(x_1) = y_1$. Restrict attention to these functions only. Choose one of them at random. The second condition says that the chance of finding a function which satisfies $f(x_2) = y_2$, is $\le \epsilon$. This means that ϵ is an upper bound on a conditional probability: $Pr(f(x_2) = y_2 \mid f(x_1) = y_1) \le \epsilon$ (the probability that $f(x_2) = y_2$ provided that $f(x_1) = y_1$ is bounded by ϵ).

Authentication

Let $\mathcal{A}$ be an ϵ strongly universal (ϵ SU) hash family. We want to show how to use it as an authentication system. Identify $\mathcal{A}$ with the corresponding array. A source produces messages to be sent from the originator (the **transmitter**) to a remote **receiver**, via a public channel. In order to be able to distinguish two types of messages, those produced by the source and those (possibly encrypted or otherwise changed) sent via the channel, we denote the messages produced by the source as **source states**, and those sent over the channel as **messages**. Identify the source states with the columns of the array (the elements of the ground set X). Our concern here is not with secrecy (protecting the knowledge as to which source state is being transmitted) nor with error correction (the channel may be noisy and we may wish to use an error-correcting code), but with **authentication**. This means that we want to devise schemes which enable the receiver to verify if the message really originated from the proper transmitter and has not been altered on the way. This is where the defining properties of an ϵ SU family come in. Use the mappings $f \in \mathcal{A}$ as **keys**. The elements $y \in Y$ are **authenticators**. Assume $x \in X$ is the source state to be transmitted. Transmitter and receiver will agree, via a secure private channel, on a key $f \in \mathcal{A}$ to be used. Then the message (consisting of x, possibly in encrypted form, and encoded for error correction purposes) will contain the **authentication tag** $f(x) = y$. Upon receiving a message which decodes to $x' \in X$, with authentication tag $y' \in Y$, the receiver will accept it as authentic if and only if $f(x') = y'$. Observe that, if no problems occurred, then $x' = x, y' = y = f(x)$, and the message will be accepted as authentic.

Why would we want to use an ϵ SU hash family, and what is the meaning of ϵ? Imagine an opponent has access to the communication channel. His or her aim is to disturb the system, more precisely, to have nonauthentic messages accepted as authentic. Everything hinges upon the assumptions concerning the power of this opponent. Consider two types of attack:

The impersonation attack

It is always assumed that the opponent knows the system. His only problem is that he has no access to the secure channel, and so has no a priori information on the key used. Assume the opponent forges a message and sends it to the receiver, claiming it originated from the transmitter. He tries to maximize his chances of success. However, the first property of Definition 16.26 shows that his probability of success is precisely $1/q$, where $q = |Y|$ is the number of authenticators.

The substitution attack

Assume the opponent is able not only to read messages but also to alter them and send the altered version to the receiver. Knowledge of the message gives him some information on the key used. Assume he knows the source state x (we do not want to assume anything concerning secrecy). Then the secret key is one of the mappings f satisfying $f(x) = y$. How can the opponent use this knowledge to improve his chances of cheating? He wishes to have a different message, corresponding to a source state $x' \neq x$, accepted as authentic. He has to choose an authentication tag y' such that, with high probability, $y' = f(x')$. However, the second property of Definition 16.26 shows that his probability of success will always be $\leq \epsilon$. Observe that, when $x' \neq x$ has been chosen, then the opponent's probability of success can be written as a conditional probability: $Pr(f(x') = y' \mid f(x) = y)$. Here x, x', y, y' are given $(x' \neq x)$, and the probability refers to the choice of $f \in \mathcal{A}$.

The problem of efficiency

We have seen how an ϵ SU hash family can be interpreted as an authentication system (often called **authentication code**) with the property that an opponent's probability of success with a substitution attack is provably limited by ϵ. It should be observed that, for each transmission of a source state, a new key has to be generated at random and this information transmitted to the receiver beforehand, via a secure channel. In order to obtain a system which can be used in practice, we have to limit the amount of time and space required by authentication.

If, for example, there are $n = 2^{100}$ keys and $q = 2^{40}$, then the authenticator tag will have 40 bits and the information on the key, to be sent via the secure channel, will be a bit string of length 100. It would not make much sense to use such a system when the number of messages is $M = |X| = 2^{100}$. We cannot afford to spend half the transmission time for authentication purposes. This leads to the problem of efficiency.

The parameters M (the number of source states) and ϵ (the level of security) can be thought of as given. We want to construct authentication systems, equivalently ϵ-SU hash families, such that the number n of keys and the number q of authenticators are as small as possible. The most reasonable measure seems to be $log_2(n) + log_2(q) = log_2(nq)$. This is the parameter we wish to minimize, where M and ϵ are given.

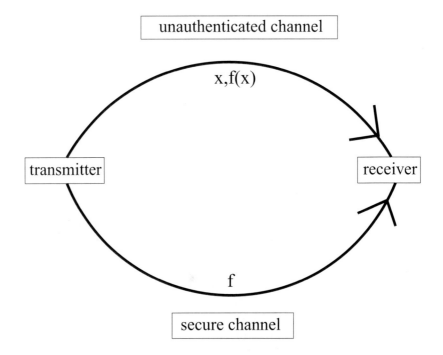

x source state, f key, f(x) authentication tag

FIGURE 16.4:   Authentication with ϵ SU

Basic theory

Recall that ϵ U hash families are nothing but error-correcting codes in disguise. In particular each bound on codes implies a bound on ϵ-U hash families. As we are interested in small values of ϵ, the codes corresponding to ϵ-U hash families need to have extremely large minimum distances in order to be interesting for our applications. The typical bound in these situations is the Plotkin bound, Theorem 9.3. In the language of ϵ U hash families, it says the following:

16.27 Theorem (Plotkin bound). *Let $\mathcal{A}$ be an $(n, M)_q$ hash family, which is ϵ U. If $\epsilon < 1/q$, then*

$$M \le \frac{q(1 - \epsilon)}{1 - q\epsilon}.$$

We turn to ϵ-SU hash families. It is left to the reader to verify that an ϵ-SU hash family is also ϵ-U (see Exercise 16.7.1). It follows that ϵ-SU families satisfy stronger conditions than ϵ-U families (recall that the latter are equivalent to error-correcting codes). As the defining property is in terms of pairs of columns of the corresponding array, it is natural to compare with orthogonal arrays of strength 2.

16.28 Theorem. *Let $\mathcal{A}$ be an $(n, M)_q$ hash family, which is ϵ SU. Then $\epsilon \ge 1/q$. Equality holds if and only if $\mathcal{A}$ is an orthogonal array of strength 2.*

PROOF   Let x, y, x' be given as before. As y' varies over the q-set Y, we see that the maximum of the conditional probabilities $Pr(f(x') = y' \mid f(x) = y)$ must be $\ge 1/q$. It follows that $\epsilon \ge 1/q$. In the case of equality it must be the case that, for all x, y, x', y' as before ($x, x' \in X, x' \ne x; y, y' \in Y$), the number of mappings f satisfying $f(x) = y, f(x') = y'$ must be a constant, say λ. We compare with the definition of an orthogonal array (Definition 2.20) and see that we have an OA of strength 2.   ☐

Theorem 16.28 shows that OA of strength 2 are examples of strongly universal hash families. However, the problem of efficiency arises. The Bose-Bush bound (Theorem 9.6) shows that the number n of rows (keys) of a strength 2 orthogonal array must exceed the number M of columns (source states).

It follows that OA of strength 2 are very inefficient when used as authentication schemes. This raises the problem if, for values of ϵ slightly larger than $1/q$, we can construct ϵ SU $(n, M)_q$ hash families such that n is much smaller than M.

Direct constructions

Next we construct a family of nonlinear codes which have a large minimum distance and also yield interesting OA of strength 2. One member of the family was constructed in Proposition 9.7 as an illustration of the Bose-Bush bound.

Let $\Phi : \mathbb{F}_{q^a} \longrightarrow \mathbb{F}_q^b$ be a mapping, which is linear over $\mathbb{F}_q$ and onto (this simply is a mapping from an a-dimensional to a b-dimensional vector space. It is described by an (a, b)-matrix of rank b). Clearly the assumption $a \geq b$ is needed. We construct a q^b-ary code of length $n = q^a$ with q^{a+b} codewords, as follows:

- The coordinates are parametrized by $u \in \mathbb{F}_{q^a}$,

- The codewords are parametrized by pairs (x, y), where $x \in \mathbb{F}_{q^a}, y \in \mathbb{F}_q^b$.

- The entry of word (x, y) in position u is $\Phi(ux) + y$. Here ux is the product in the field $\mathbb{F}_{q^a}$.

Let us determine the minimum distance of this code. Let $(x_1, y_1) \neq (x_2, y_2)$. These codewords agree in coordinate u if and only if $\Phi(u(x_1 - x_2)) = y_2 - y_1$. Here we have used the $\mathbb{F}_q$ linearity of Φ. If $x_1 = x_2$, then $y_1 \neq y_2$. The codewords are at distance $n = q^a$ in this case. Assume $x_1 \neq x_2$. As the kernel of Φ has dimension $a - b$ over $\mathbb{F}_q$, the number of u satisfying this equation is precisely q^{a-b}. It follows that our code has minimum distance $d = q^a - q^{a-b}$.

We leave it to the reader to check that this code also is an orthogonal array of strength 2; see Exercise 16.7.2.

16.29 Theorem. *For every prime-power q and natural numbers $a \geq b$, there is an $OA_{q^{a-b}}(2, q^a, q^b)$ whose rows form a code with minimum distance $d = q^a - q^{a-b}$.*

In particular, we obtain an $(q^{a+b}, q^a)_{q^b}$ hash family, which is q^{-b} U and even q^{-b} SU.

In terms of mappings, we have q^{a+b} mappings from a q^a set to a q^b set, and this family is q^{-b} strongly universal. As a toy example we can use the $OA_2(2, 8, 4)$ from Proposition 9.7 (case $q = 2, b = 2, a = 3$). Its rows describe a family of 32 mappings from an 8 set to a 4 set. It is $\frac{1}{4}$ SU.

A generalization of Theorem 16.29 yields ϵ SU hash families which are not strength 2 orthogonal arrays. Later on we will see that the OA from Theorem 16.29 can be used as ingredients in recursive constructions.

16.30 Theorem. *Let q be a prime-power and $k, a \geq b$ natural numbers. There is a $(q^{a+b}, q^{ka})_{q^b}$ hash family, which is $\dfrac{k}{q^b}$ SU.*

PROOF As before, fix a surjective $\mathbb{F}_q$ linear mapping $\Phi : \mathbb{F}_{q^a} \longrightarrow \mathbb{F}_q^b$. Form an array $\mathcal{A}$ with q^{ka} columns, q^{a+b} rows, with entries in $\mathbb{F}_q^b$ as follows:

- The columns are indexed by the polynomials $p(X)$ with coefficients in $\mathbb{F}_{q^a}$, of degree $\leq k$ and with constant term $= 0$.

- The rows are indexed by the pairs (u, v), where $u \in \mathbb{F}_{q^a}, v \in \mathbb{F}_q^b$.

- The entry in row (u, v) and column $p(X)$ is

$$\Phi(p(u)) + v.$$

It is clear that each column contains each tuple from $\mathbb{F}_q^b$ with frequency q^a. Fix two different columns $p_1(X), p_2(X)$ and two not necessarily different elements $y_1, y_2 \in \mathbb{F}_q^b$. We have to count the rows (u, v) which satisfy the following:

$$\Phi(p_1(u)) + v = y_1$$
$$\Phi(p_2(u)) + v = y_2.$$

Subtract and use the linearity of Φ. This leads to the equivalent equation

$$\Phi((p_1 - p_2)(u)) = y_1 - y_2.$$

Each solution u of this equation gives us a uniquely determined row (u, v). As each tuple from $\mathbb{F}_q^b$ has precisely q^{a-b} preimages under Φ, and $p_1(X) - p_2(X)$ is a nonconstant polynomial of degree $\leq k$, the number of solutions is $\leq k \times q^{a-b}$. $\qquad\Box$

Case $k = 1$ of Theorem 16.30 is the family from Theorem 16.29. Theorem 16.30 produces an authentication scheme for q^{ka} source states, using q^b authenticators and only q^{a+b} keys, which limits the opponent's probability of success in a substitution attack by $\epsilon = k/q^b$. Theorem 16.28 and the Bose-Bush bound show that more than q^{ka} keys are needed if we insist on the minimum possible value $\epsilon = q^{-b}$.

Several recursive constructions are known. They all have in common that error-correcting codes are essential ingredients.

The q-twisted construction

This construction is due to Kabatianskii, Smeets and Johansson [121]. Start from a linear $[n, k + 1, d]_q$ code, which contains the all-1 word $\mathbf{1}$. Let $\mathcal{C}$ be a k-dimensional subcode which does not contain $\mathbf{1}$. The presence of $\mathbf{1}$ shows that $\mathcal{C}$ has the following property:
for each nonzero codeword $0 \neq x \in \mathcal{C}$ and each element $\alpha \in \mathbb{F}_q$, the frequency $\nu(\alpha)$ of α as an entry of x is $\leq n - d$.

The construction idea may be described as randomization or uniformization, similar to the construction of the Justesen codes (the idea is also used in the family of Theorem 16.30). Each codeword $x \in C$ is blown up to a qn-tuple $\tilde{x}$ whose coordinates are indexed by pairs (i, α), where i is a coordinate of x and $\alpha \in \mathbb{F}_q$, the entry being $x_i + \alpha$. Our hash family is the $(qn, q^k)_q$ array whose columns are the $\tilde{x}$. It is clear that each field element occurs n times as entry in each column. This verifies the first axiom of Definition 16.26.

As for the second axiom of Definition 16.26, fix two different columns (codewords) x, y, two not necessarily different entries a, b and count rows (i, α) such that $\tilde{x}$ has entry a and $\tilde{y}$ has entry b in this row. This means $x_i + \alpha = a$, $y_i + \alpha = b$, equivalently, $(x - y)_i = a - b$. The number of such rows is therefore bounded by $n - d$. It follows that we have an ϵ-SU array, where $\epsilon = 1 - \dfrac{d}{n}$. This proves the following:

16.31 Theorem. *Assume there exists a linear $[n, k + 1, d]_q$ code containing the all-1 word. We can construct a $(qn, q^k)_q$ hash family which is $(1 - \frac{d}{n})$ SU.*

Apply Theorem 16.31 to Reed-Solomon codes. As these are $[q, k+1, q-k]_q$ codes containing **1**, we can construct $(q^2, q^k)_q$ hash families, which are $\frac{k}{q}$-SU, for every $k \leq q - 1$. However, these parameters are not new to us. They correspond to the special case $a = b = 1$ of Theorem 16.30.

Stinson composition

Here is another construction principle for ϵ-SU hash families, Stinson composition. It is best understood using the language of mappings.

16.32 Theorem. *Let $\mathcal{A}_1$ be an $(N_1, n)_m$ hash family which is $\epsilon_1 U$ and $\mathcal{A}_2$ an $(N_2, m)_q$ hash family which is $\epsilon_2 SU$.*

*The **composition** $\mathcal{A} = \mathcal{A}_2 \circ \mathcal{A}_1$ is defined as the multiset of compositions $f = f_2 \circ f_1$, where $f_i \in \mathcal{A}_i$. Then $\mathcal{A}$ is an $(N_1 N_2, n)_q$ hash family which is $\epsilon - SU$, where $\epsilon = \epsilon_1 + \epsilon_2 - \epsilon_1 \epsilon_2 < \epsilon_1 + \epsilon_2$.*

PROOF As $\mathcal{A}_i$ is a multiset of N_i mappings $(i = 1, 2)$, it is clear that the multiset $\mathcal{A}$ of all compositions consists of $N_1 N_2$ mappings. We have that $\mathcal{A}$ is an $(N_1 N_2, n)_q$ hash family. The first condition of Definition 16.26 is satisfied. In order to prove the main condition, we use the language of probabilities. So let x_1, x_2 in the ground set $(x_1 \neq x_2)$ and z_1, z_2 in the q-set be given. Write $f \in \mathcal{A}$ as $f = f_2 \circ f_1$, let $y_1 = f_1(x_1)$, $y_2 = f_1(x_2)$ and $P = Prob(f(x_1) = z_1 \mid f(x_2) = z_2)$. We want to show that $P \leq \epsilon_1 + \epsilon_2 - \epsilon_1 \epsilon_2$. Consider at first the case $z_1 \neq z_2$. We have $P \leq Prob(y_1 \neq y_2)\epsilon_2 < \epsilon_2$. The case $z_1 = z_2$ is harder:

$$P \leq Prob(y_1 \neq y_2)\epsilon_2 + Prob(y_1 = y_2).$$

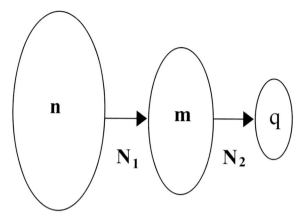

FIGURE 16.5: Stinson composition

Let $\alpha = Prob(y_1 = y_2)$. We have $P \leq \alpha + (1 - \alpha)\epsilon_2$, which we denote by $h(\alpha)$. As $h'(\alpha) = 1 - \epsilon_2 \geq 0$ and $\alpha \leq \epsilon_1$ by the defining property of $\mathcal{A}_1$, we obtain $P \leq h(\epsilon_1)$, which is our claim. □

This construction technique can be used to obtain rather efficient authentication codes. It turns out that, by increasing the security parameter ϵ only slightly above its theoretical minimum, ϵ-SU hash families can be obtained with a dramatically smaller number of mappings (keys). We call this effect as the **Wegman-Carter effect,** as it was first observed in their joint papers [43, 218].

Let us apply Stinson composition Theorem 16.32 with the OA of Theorem 16.29 in the role of $\mathcal{A}_2$. We know that $\mathcal{A}_1$, an ϵ_1-U hash family, is equivalent to an error-correcting code. In particular, the Reed-Solomon code $\mathcal{RS}(k, q)$ yields a $(q, q^k)_q$ hash family which is $(k - 1)/q$ universal.

Stinson composition, using as ingredients Reed-Solomon codes and the OA from Theorem 16.29 (which are themselves very closely related to Reed-Solomon codes), yields interesting strongly universal hash families:

16.33 Theorem. *Let q be a prime-power and $a \geq b$.*
There is a $(q^{2a+b}; q^{aq^{a-b}})_{q^b}$ hash family which is $2q^{-b}$ SU.

PROOF $\mathcal{RS}(q^{a-b}, q^a)$ yields a $(q^a, q^{aq^{a-b}})_{q^a}$ hash family which is ϵ_1-U. Here $\epsilon_1 = (q^{a-b} - 1)/q^a < q^{-b}$. An OA from Theorem 16.29 yields a $(q^{a+b}, q^a)_{q^b}$ hash family which is q^{-b}-U. The composition has q^{2a+b} members. Its ϵ-value as a strongly universal family is $\epsilon \leq \epsilon_1 + q^{-b} < 2q^{-b}$. □

Here is an example to illustrate:

16.34 Example. *Let* $q = 2, b = 20, a = 32$. *Theorem 16.33 yields a* $(2^{84}; 2^{2^{17}}, 2^{20})$ *hash family which is* 2^{-19}-*SU. In the language of authentication (Section 16.7), we see that we can authenticate a* 2^{17}-*bit source with 20 authenticator bits and 84 key bits such that the deception probability is limited to* 2^{-19}. *This illustrates how dramatic the Wegman-Carter effect is. Had we insisted on a deception probability* 2^{-20}, *while still using only 20 authenticator bits, then we would have been forced to use an OA of strength 2 by Theorem 16.28, leading to more than* 2^{17} *key bits because of the Bose-Bush bound, as mentioned earlier.*

Associative Memories

In all applications the domain X is chosen to be considerably larger than the range Y. In computer science functions $f : X \longrightarrow Y$ are often called **hash functions** if the intention is to partition (to hash) X into small pieces (those with the same value under f). Suitably chosen hash functions may be used in search algorithms in the following way: whenever some $x \in X$ is stored, then we also store the value $f(x)$ (the hash value). If one wants to check if some $x \in X$ has been stored already, then one computes the hash value at first and searches only through the shorter list of those elements of X with this hash value. This is a very innocent application. For example, imagine a compiler which in a table of symbols lists all identifiers that have been used in a certain program. The number of all possible identifiers is very large, about $1.62 \cdot 10^9$ in FORTRAN. One can use an ϵ SU family $\mathcal{A}$ with the set of possible identifiers as the domain. Choose some $f \in \mathcal{A}$ at random and use it as a hash function. Assume x is an identifier. We want to check if it is new. Instead of searching through our huge set, compute $f(x)$ at first and search through the list of all x' satisfying $f(x') = f(x)$. Although only one hash function f is really used, this may be seen as an application of universal hashing, as the properties of the hash family determine the expected behavior of the algorithm.

Multiple authentication with counters

A drawback of our method as given in Section 16.7 is that each key may be used only once. Suppose we want to economize on the number of key bits by using each key a certain number of times for the authentication of several messages. It would certainly be possible to use OA of higher strength. However, the problem of efficiency (see Subsection 16.7) imposes severe restrictions.

The Rao bound on orthogonal arrays, a generalization of the Bose-Bush bound (see Theorem 15.38), shows that the number of keys would be much too large. Here is a method to get around this problem:

Let $\mathcal{A}$ be an $(N, n)_m$ hash family which is ϵ-SU, and impose the structure of an additive group on the set of entries: $Y = \mathbb{Z}/m\mathbb{Z}$. We wish to be able to authenticate up to w source states with the same key. The keys are tuples $(f, y_1, \ldots, y_{w-1})$, where f is chosen randomly from $\mathcal{A}$ and the y_i are chosen independently and uniformly at random from Y. Let $(x_1, x_2, \ldots, x_w)$ be a tuple of source states ($x_i \in X$, pairwise different) to be authenticated with the same key. The messages have the form (i, x_i). Here i is the **counter**.

The authentication tag is then $f(x_1)$ if $i = 1$; it is $f(x_i) + y_{i-1}$ if $i > 1$. Assume the opponent observes up to w messages which have been authenticated with the same key. He uses this information to alter the last of these messages in such a way that the source state is replaced by some $x \notin \{x_1, x_2, \ldots, x_w\}$, his aim being to have x accepted as authentic. It is clear that his probability of success is still bounded by ϵ.

Privacy amplification by public discussion

Bennett, Brassard and Roberts [10] consider the following scenario: Assume that transmitter and receiver have two channels at their disposal: a **private** channel which is subject to the attacks of an opponent, and a well-authenticated **public** channel. The opponent gathers information from both channels and also has the power to alter the messages sent via the private channel. We want to devise methods for using communication via the public channel to improve **authentication** or **privacy** of the messages sent via the private channel. This explains the apparent paradox of the title. In the introduction of Bennett et al. [9] the following description of privacy amplification is offered:

Privacy amplification is the art of distilling highly secret shared information, perhaps for use as a cryptographic key, from a larger body of shared information that is only partially secret.

Let us consider the situation where transmission errors via the private channel do not occur, so that no error-detection protocol is carried out. Assume transmitter and receiver wish to agree on a secret bitstring of length a, possibly for use as a cryptographic key. We are working under the simplifying assumption that the bitstring $x \in \mathbb{F}_2^a$, which is generated by the transmitter and sent via the public channel, is received in unaltered form. We wish to limit the amount of information which may leak to an opponent by eavesdropping on the private channel. Everything hinges on the assumptions made on the power of the eavesdropper. Consider the situation where the opponent has free choice of an **eavesdropping function** $e : \mathbb{F}_2^a \longrightarrow \mathbb{F}_2^s$. The designers of the system know s, but not the function e, which is completely under the eavesdropper's control. In Section 16.5 we considered the situation when the eavesdropper's information is limited to the knowledge of a predetermined number of physical bits. In that case it was possible to completely wipe out the eavesdropper's information by using the shorter bitstring $f(x)$, where f is a suitable publicly known **resilient function.** In our more difficult situation we will use the same idea of shortening the string in order to limit the opponent's information. We will make use of an ϵ U hash family $\mathcal{A}$ with domain $X = \mathbb{F}_2^a$. A mapping $f \in \mathcal{A}$ is chosen at random by the transmitter and sent via the public channel. The common bitstring to be used by transmitter and receiver is then $y = f(x)$. In our situation it will no longer be possible to eliminate the opponent's information on y altogether, but the properties of $\mathcal{A}$ will give a substantial reduction.

16.35 Definition. *Let X be a finite probability space. The* **collision probability** *is defined as*

$$p_c(X) = \sum_{x \in X} p(x)^2.$$

The (second-order) **Rényi entropy** *is $R(X) = -log(p_c(X))$.*

Recall that *log* is the binary logarithm. $p_c(X)$ is the probability that, upon two independent draws from X, the same element is drawn. This is obviously related to the defining property of an ϵ-U hash family (see Definition 6.3 and the interpretation following it). The Jensen inequality (Theorem 8.7) shows that the Rényi entropy $R(X)$ is upper bounded by the Shannon entropy $H(X)$.

16.36 Lemma. *Let X be a finite probability space, $H(X)$ the Shannon entropy and $R(X)$ the Rényi entropy. Then*

$$H(X) \geq R(X).$$

16.37 Lemma. *Let X be a probability space with n elements and $\mathcal{A}$ an $(N, n)_m$ hash family which is ϵ-U. Each $f \in \mathcal{A}$ is a mapping $f : X \longrightarrow Y$. Choose $f \in \mathcal{A}$ uniformly at random (probability $1/N$). Let $H(Y|f)$ and $R(Y|f)$ be the entropies of the probability distribution on Y as defined by X and f ($p(y) = \sum_x p(x)$, where the sum is over all x such that $f(x) = y$). Then*

$$H(Y|\mathcal{A}) \geq R(Y|\mathcal{A}) > -log(p_c(X) + \epsilon).$$

Here $H(Y|\mathcal{A})$ is the expected value of the $H(Y|f)$, $f \in \mathcal{A}$, likewise for $R(Y|\mathcal{A})$.

PROOF The inequality $H(Y|\mathcal{A}) \geq R(Y|\mathcal{A})$ follows from Lemma 16.36. Jensen's inequality shows that $R = R(Y|\mathcal{A})$ is

$$R = -\frac{1}{N} \sum_{f \in \mathcal{A}} log(\sum_y (\sum_{f(x)=y} p(x))^2) \geq -log(\frac{1}{N} \sum_{f,y} (\sum_{f(x)=y} p(x))^2).$$

The expression in parentheses is the probability that $f(x_1) = f(x_2)$, when $f \in \mathcal{A}$ is chosen uniformly at random and x_1, x_2 are chosen independently from X. The probability that $x_1 = x_2$ is $p_c(X)$. If $x_1 \neq x_2$ are given, then the probability that $f(x_1) = f(x_2)$ is $\leq \epsilon$, by definition of ϵ-U. It follows that the expression in parentheses is $< p_c(X) + \epsilon$. $\qquad\square$

Return to our scenario: let $X = \mathbb{F}_2^a$, with uniform distribution. The designers of the system know that the eavesdropper Eve has a function $e : X \longrightarrow \mathbb{F}_2^s$ at her disposal. Whenever $x \in X$ is sent, she learns $z = e(x)$. Transmitter and receiver choose a security parameter $\sigma < a - s$. Let $r = a - s - \sigma$ and $\mathcal{A}$ the 2^{-r}-U hash family $(2^{a+r}, 2^a)_{2^r}$ as constructed in Theorem 16.29. The transmitter chooses $f \in \mathcal{A}$ uniformly at random and sends f via the public channel. The common key to be used by transmitter and receiver is $y = f(x)$. We have to show that the eavesdropper has very little information on y, equivalently, that her entropy of Y is close to r, despite her knowledge of f and z.

Eve knows the set $X(z) = \{x \mid e(x) = z\}$, of size $n(z)$. Restrict $\mathcal{A}$ to the columns from $X(z)$. By Lemma 16.37, $H(Y|z) > -log(\frac{1}{n(z)} + 2^{-r})$. Here we use that $\epsilon = 2^{-r}$ and that the collision probability of $X(z)$ (with uniform distribution) is $1/n(z)$. We cannot guarantee that $H(Y|z)$ is close to r, but we can guarantee that this is true for its expected value. Observe that the probability of z is $n(z)/2^a$. Using Jensen's inequality again:

$$E(H(Y|z)) \geq -log(2^{-r} + 2^s/2^a) = r - log(1 + 2^{-\sigma}).$$

If we increase σ, then Eve's expected information on $y = f(x)$ becomes arbitrarily small.

At this point recall the basic interpretation of Shannon entropy. Y is a uniform sample space with 2^r elements, so $H(Y) = r$. The conditional entropy $H(Y|z)$ is the residual entropy of Y when $z = e(x)$ is known. It follows that the difference $H(Y) - H(Y|z)$ represents the amount of information which z has revealed about Y. In the present situation we can make the expected value of this quantity arbitrarily small. The main ingredient in the construction is an ϵ-U hash family and once again, this is equivalent to an error-correcting code.

1. **Hash families** may be viewed either as arrays
   or as families of mappings from a large set to a small set.

2. **Universal hash families** are formally equivalent to codes. The
   relative minimum distance is the central parameter.

3. **Strongly universal hash families** are a
   generalization of strength 2 orthogonal arrays.

4. Stinson composition uses as ingredients universal families
   (hence codes) and strength 2 orthogonal arrays.

5. An important application is **unconditional authentication.**

Exercises 16.7

16.7.1. *Let $\mathcal{A}$ be an $(N, n)_m$ hash family, which is ϵ SU.
Show that $\mathcal{A}$ is ϵ U.*

16.7.2. *Prove that the codes from Theorem 16.29
are $OA_{q^{a-b}}(2, q^{a+b}, q^b)$, (where $a \geq b$).*

Chapter 17

The geometric description of linear codes

17.1   Linear codes as sets of points

Basic concepts: Projective geometry. The main theorem. Hexacode and Simplex codes. Extended Reed-Solomon codes and normal rational curves. Arcs, ovals, hyperovals and their codes. Low-dimensional codes and the Griesmer bound. Ternary Golay code. Arcs of species s, Barlotti arcs. The Denniston arcs.

We describe linear codes from a geometric point of view.

Projective geometry

Let $V = V(k,q) = \mathbb{F}_q^k$ the space of k-tuples, a k-dimensional vector space over $\mathbb{F}_q$. It makes sense, intuitively, to view the one-dimensional subspaces of V as **points** and the two-dimensional subspaces as **lines.** The main reason is the following: any two points are on precisely one common line. This is clear (two different one-dimensional subspaces generate a two-dimensional subspace) and it is a familiar axiom in geometry. In fact, this is the whole point: we give geometric names to the subspaces of the vector space V and use geometric intuition.

Observe the shift in dimension: we view one-dimensional subspaces as points (zero-dimensional geometric objects), two-dimensional subspaces as lines (one-dimensional geometric objects) and so forth. Consequently, the geometry derived from $V(k,q)$ is considered to be $(k-1)$-dimensional: $PG(k-1,q)$, the $(k-1)$-dimensional **projective geometry.** It has $k-1$ types of objects, from points (one-dimensional subspaces) to **hyperplanes** ($(k-1)$-

dimensional vector subspaces). The most frequently used objects are points, lines, planes (three-dimensional subspaces) and hyperplanes.

In Chapter 10 we considered a special case, the **projective planes** $PG(2, q)$ (of **order** q). We found that $PG(2, q)$ has some very convenient geometric properties: any two points are on a unique line, and any two lines intersect in a unique point. Also, we were able to enumerate the geometric objects: there are $q^2 + q + 1$ points and equally many lines, each line has $q + 1$ points, each point is on $q + 1$ lines.

A "geometric" reason why the number of points must be $q^2 + q + 1$ is the following: fix a point P. It is on $q+1$ lines. Each line contains q points different from P. In the resulting picture each point of $PG(2, q)$ shows up exactly once. We count $1 + (q + 1)q$ points.

The lowest-dimensional case of interest is the projective line $PG(1, q)$. It has $q + 1$ points (why? The number of one-dimensional subspaces of a two-dimensional space is $(q^2 - 1)/(q - 1) = q + 1$).

We wish to get some familiarity with $PG(k - 1, q)$ in general. The enumeration of objects is just as easy as in the case of the projective plane. For example, the number of points is

$$(q^k - 1)/(q - 1) = q^{k-1} + q^{k-2} + \cdots + q + 1,$$

equal to the number of hyperplanes (there are $q^k - 1$ nonzero vectors in $V(k, q)$, each such vector determines a point, but each point contains $q - 1$ nonzero vectors).

As an example, consider $PG(3, 2)$. It has $2^4 - 1 = 15$ points and therefore also 15 planes. It is somewhat more interesting to count lines. As any two points determine a line, we count 15×14 (the number of ordered pairs of different points), which counts each line 3×2 times (as a line has 3 points). The number of lines in $PG(3, 2)$ is therefore $15 \times 14/6 = 35$.

In the same manner, the number of lines of $PG(3, q)$ can be determined. As there are $(q^4 - 1)/(q - 1) = q^3 + q^2 + q + 1$ points, the number of lines is

$$\frac{(q^3 + q^2 + q + 1) \times (q^3 + q^2 + q)}{(q + 1)q} = (q^2 + 1)(q^2 + q + 1).$$

Another way to count the lines of $PG(3, q)$ is the following: there are $q^3 + q^2 + q + 1$ planes, each containing $q^2 + q + 1$ lines. The product $(q^3 + q^2 + q + 1)(q^2 + q + 1)$ counts each line x times, where x is the number of planes through a given line. We have that x is the number of three-dimensional subspaces of a four-dimensional vector space, which contain a fixed two-dimensional subspace. This is the same as the number of points on a line: $x = q + 1$. The number of lines of $PG(3, q)$ is therefore $(q^3 + q^2 + q + 1)(q^2 + q + 1)/(q + 1) = (q^2 + 1)(q^2 + q + 1)$.

Fix a hyperplane H in $PG(k-1, q)$. The points of $PG(k-1, q)$ outside H are called **affine points** (with respect to H). They form the $(k - 1)$-dimensional **affine geometry** $AG(k - 1, q)$. The number of points in $AG(k - 1, q)$ is q^{k-1}.

We are now confident it should be possible to solve each counting problem of this type whenever this is required.

The main theorem

Let $\mathcal{C}$ be a linear $[n, k, d]_q$ code, described by a generator matrix G with entry a_{ij} in row i and column j. The idea is to consider the columns of G as generators of points in $PG(k-1, q)$. So let P_j be the point generated by $(a_{1j}, a_{2j}, \ldots, a_{kj})$, where $j = 1, 2, \ldots, n$. Here we assume that there is no 0 column in G (if there were, the corresponding coordinate would be superfluous as we could puncture with respect to this coordinate and obtain an $[n-1, k, d]_q$ code). Consider the set $\mathcal{P} = \{P_1, P_2, \ldots, P_n\}$ of points in $PG(k-1, q)$. In fact, $\mathcal{P}$ is not necessarily a set but rather a multiset of points, as it can happen that more than one column of G yields the same point (different columns of G can be identical or scalar multiples of each other). As a multiset, each point P counted with its multiplicity $w(P)$, the number of columns of G that generate point P, $\mathcal{P}$ has cardinality n. These are two ways of expressing the same concept. Either we speak of a multiset of n points in $PG(k-1, q)$ or of a mapping w assigning a nonnegative integer weight $w(P)$ to each point $P \in PG(k-1, q)$, such that $\sum_P w(P) = n$.

It is the idea of the geometric description to read off the code parameters from the multiset $\mathcal{P}$ of points in $PG(k-1, q)$. The main theorem shows how to determine the weight distribution.

The codewords of $\mathcal{C}$ are the linear combinations of rows of the generator matrix G. Let the rows of G be $v_i = (a_{i1}, a_{i2}, \ldots, a_{in})$. Each codeword has the form $x = c_1 v_1 + c_2 v_2 + \cdots + c_k v_k$ and is uniquely determined by the tuple $c = (c_1, c_2, \ldots, c_k)$ of coefficients. Fix a coordinate j, indexed by point P_j. What does it mean that x has entry 0 in coordinate j? It means that

$$c_1 a_{1j} + c_2 a_{2j} + \cdots + c_j a_{kj} = 0.$$

However, this linear relation with the c_i as coefficients defines a hyperplane. In other words, we have $x_j = 0$ if and only if point P_j is contained in that hyperplane. It follows that $wt(x)$ equals the number of points from $\mathcal{P}$ which are **not** in this hyperplane. This is already the main theorem.

17.1 Theorem. *Let $\mathcal{C}$ be a linear $[n, k]_q$ code and $G = (a_{ij})$ a generator matrix, with notation as in the paragraph above and such that G has no 0 column. For each point $P \in PG(k-1, q)$, let $w(P)$ be the number of columns of G which define P ($\sum_P w(P) = n$).*

For each nonzero tuple $c = (c_1, c_2, \ldots, c_k) \in \mathbb{F}_q^k$, let $c^\perp$ be the hyperplane of $PG(k-1, q)$, which consists of all $y = (y_1, y_2, \ldots, y_k)$ such that $c \cdot y = 0$.

Then the weight of $x = c_1v_1 + c_2v_2 + \cdots + c_kv_k$ is

$$wt(x) = \sum_{P \notin c^\perp} w(P).$$

The v_i are the rows of G. In particular, the minimum weight d is the minimal number of points from $\mathcal{P}$ outside a hyperplane (in the multiset sense).

It is harder to write down this theorem than to prove it. Even case $d = 0$ makes sense. This will be the case when all points from $\mathcal{P}$ are contained in a hyperplane, which is equivalent to the dimension of $\mathcal{C}$ being $< k$.

Observe that each linear code $\mathcal{C}$ can be described by point sets in many different ways. Each step of Gauß elimination (row operations) will produce a different point set without changing the code. When describing codes geometrically, the appropriate notion of equivalence of codes is monomial equivalence (see Section 12.4). In fact, the multiset $\mathcal{P}$ of points is unchanged not only under permutations of coordinates but also when columns are multiplied by nonzero scalars. One drawback of this approach was noted in Section 12.4: the parameters of trace codes and subfield codes are not invariant under these operations. In other words, the geometric description is too coarse if we want to control the parameters of trace codes and of subfield codes.

It is desirable to distinguish the cases when a code can be described by a set of points instead of a multiset ($w(P) = 0$ or $w(P) = 1$ for each point $P \in PG(k-1,q)$). This can easily be read off from the dual code. In fact, if two columns of G describe the same projective point, the dual code $\mathcal{C}^\perp$ contains a word of weight 2, and vice versa. This shows the following:

17.2 Proposition. *A linear $[n, k]_q$ code $\mathcal{C}$ is called **projective** if there is a generator matrix whose columns generate different projective points. The following are equivalent:*

- *$\mathcal{C}$ is a projective code.*

- *The columns of a generator matrix describe different projective points.*

- *$\mathcal{C}^\perp$ does not contain words of weight 2.*

The assumption that no column of G is 0 can be expressed in terms of dual weights as well. It is equivalent to the absence of weight 1 words in $\mathcal{C}^\perp$. A code is projective if and only if its dual has minimum weight ≥ 3.

The **Simplex codes** are particularly easy to understand geometrically. The description of $\mathcal{S}_r(q)$ given in Section 3.4 simply says that $\mathcal{S}_r(q)$ is described by the set of **all** points in $PG(r-1,q)$. Theorem 17.1 says that the weight of each nonzero word of $\mathcal{S}_r(q)$ equals the number of points in $AG(r-1,q)$, which is q^{r-1}. We conclude that $\mathcal{S}_r(q)$ is a $[\frac{q^r-1}{q-1}, r, q^{r-1}]_q$ code, where, in addition, all nonzero codewords have the same weight q^{r-1}. Such codes are known as (linear) **constant weight codes.**

As an example, the columns of a generator matrix of $\mathcal{S}_3(3)$ are representatives of the 13 points of $PG(2,3)$, the projective plane of order 3. All nonzero weights are 9. In particular, we have a $[13,3,9]_3$ code. A generator matrix was given in Section 3.4.

It is the charm of the geometric description of linear codes that it gives a completely different angle under which to view the construction problem. What we need are sets (or multisets) of points in projective geometry which are in some sense uniformly distributed: no hyperplane contains very many of our points, in other words: outside every hyperplane we always need to find sufficiently many points.

The hexacode

Consider an example in the projective plane $PG(2,4)$ of order 4. It is customary to denote by $(a:b:c)$ the point generated by (a,b,c). One speaks of **homogeneous coordinates** for points. Consider the following set $\mathcal{O}$ of six points in $PG(2,4)$:

$$P_1 = (1:0:0), \ P_2 = (1:1:1), \ P_3 = (1:\omega:\overline{\omega}),$$

$$P_4 = (1:\overline{\omega}:\omega), \ P_5 = (0:0:1), \ P_6 = (0:1:0)$$

It is easy to see that no more than two of these points are on a common line, in other words, the line connecting two of our points does not gobble up any of the remaining 4 points. As an example, consider the line $P_1 P_2$. It consists of all points $(a:b:b)$. No third point of $\mathcal{O}$ has this property.

Let $\mathcal{C}$ be a code which has representatives of the P_i as columns (a code which is geometrically described by the point set $\mathcal{O}$). It is a quaternary code of length 6. Observe that, in projective planes, lines are hyperplanes. We see that outside any line there are at least four points of $\mathcal{O}$. In particular, $\mathcal{C}$ has dimension 3 (as the six points of $\mathcal{O}$ are not on a line), and minimum distance 4. We conclude that $\mathcal{C}$ is a $[6,3,4]_4$ code. This is the hexacode which we saw on several occasions, earliest in Exercises 3.4.5 and 3.4.6. In fact, the geometric description makes it easy to determine the weight distribution of the hexacode. There are $\binom{6}{2} = 15$ lines meeting $\mathcal{O}$ in two points. This gives us $15 \times 3 = 45$ codewords of weight 4. There is no line meeting $\mathcal{O}$ in precisely one point (fix $P \in \mathcal{O}$, there are five lines connecting P to the remaining five points; as P is on five lines in $PG(2,4)$, there is no space left for tangents). It follows that the remaining $21 - 15 = 6$ lines of $PG(2,4)$ avoid $\mathcal{O}$ altogether. This yields $6 \times 3 = 18$ words of weight 6. The hexacode's weight distribution is therefore

$$A_0 = 1, \ A_4 = 45, \ A_6 = 18.$$

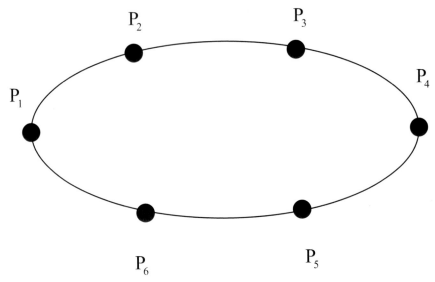

FIGURE 17.1: The hexacode as point set

Check that $1 + 45 + 18 = 64 = 4^3$ is the number of codewords, as it has to be.

What can we say about $\mathcal{C}^\perp$? Clearly it is a $[6,3]_4$ code. The reader was asked in Exercise 4.3 to prove that the dual of a linear MDS code is MDS again. It follows that $\mathcal{C}^\perp$ has parameters $[6,3,4]_4$, the same as $\mathcal{C}$.

Here is a geometric argument: As $\mathcal{C}$ is projective, we have $d' = d(\mathcal{C}^\perp) \geq 3$. A codeword of weight 3 in $\mathcal{C}^\perp$ would mean that some three of the P_i are linearly dependent, hence on a line. This is not the case. We have $d' \geq 4$. Because of the Singleton bound (or because any four columns of the generator matrix are linearly dependent), d' cannot be larger yet, so $d' = 4$.

Because of duality, the hexacode has strength 3. It is an $OA_1(3,6,4)$.

Extended RS-codes and normal rational curves

Reconsider Reed-Solomon codes (Chapter 4) from a geometric point of view. As generator matrices we used Vandermonde matrices. The columns of a generator matrix of $\mathcal{RS}(k,q)$ generate the points $P_u = (1 : u : u^2 : \cdots : u^{k-1})$, where $u \in \mathbb{F}_q$. As $\mathcal{RS}(k,q)$ is a $[q,k,q-k+1]_q$ code, it follows from Theorem 17.1 that we find at least $q - (k-1)$ of the points P_u outside each hyperplane. This means that each hyperplane of $PG(k-1,q)$ contains at most $k-1$ of these points. On the other hand, any $k-1$ points in $PG(k-1,q)$

are contained in a hyperplane. We conclude that any $k - 1$ of our points must generate a hyperplane (not a space of lower dimension): any two points generate a line (this is true anyway), any three of the points generate a plane (equivalently, no three are on a line), and so forth.

Define $P_\infty = (0 : 0 : \cdots : 0 : 1)$. The code of length $q + 1$ whose generator matrix has as columns representatives of the P_u and of P_∞ is known as the extended k-dimensional Reed-Solomon code. It is clear that any k columns of this matrix are linearly independent. We conclude that the k-dimensional extended Reed-Solomon code has strength k, and its dual is a $[q + 1, q + 1 - k, k + 1]_q$ code. As the dual is MDS, the original code is MDS as well. Our extended Reed-Solomon code is a $[q + 1, k, q + 2 - k]_q$ code.

In geometric terminology, the point set $\{P_u | u \in \mathbb{F}_q\} \cup \{P_\infty\}$ is known as the **normal rational curve** in $PG(k - 1, q)$. We saw that this set of points has the extremal property that any subset of $k - 1$ points generates a hyperplane (not a lower-dimensional space).

Ovals and hyperovals in $PG(2, q)$

Specialize to the projective plane $PG(2, q)$. The normal rational curve consists of the points $P_u = (1 : u : u^2)$ for $u \in \mathbb{F}_q$ and $P_\infty = (0 : 0 : 1)$. The matrix with representatives for these points as columns generates the extended Reed-Solomon code $[q + 1, 3, q - 1]_q$. Its dual is a $[q + 1, q - 2, 4]_q$-code. The extended Reed-Solomon code has strength 3, in particular, no three points of the normal rational curve are on a line.

This seems to be a recurrent theme. The normal rational curve in $PG(2, q)$ consists of $q + 1$ points, no three of which are on a line, and the hexacode is a set of six points in $PG(2, 4)$, no three of which are on a line. What is the maximum number of points with this property?

17.3 Definition. *We call a set of points* **collinear** *if the points are on a common line. A set of points in $PG(2, q)$ is an* **arc** *if no three are collinear. An n-arc is an arc of n points.*

What is the maximum number of points of an arc in $PG(2, q)$? Fix a point P of a hypothetical arc. As P is on $q + 1$ lines and each line through P can contain at most one further point of the arc, the size of the arc is at most $q + 2$. Is it possible that we have equality? Yes, as we know. We just saw that the hexacode in $PG(2, 4)$ is equivalent to a set of six points forming an arc. We also have the normal rational curve, which shows that we can always find an arc of $q + 1$ points. In order to solve this problem, it remains to decide when arcs of $q + 2$ points do exist.

An elementary geometric counting argument shows that arcs of size $q + 2$ can never exist if q is odd. Assume it does exist. Any line through one of the points must contain a second point. In other words, there are no tangents to our arc. Now fix a point Q which does not belong to the arc. Each line through Q contains either zero or two points of the arc. In particular, the size $q + 2$ must be even. We have proved most of the claims of the following theorem:

17.4 Theorem. *A $(q + 1)$-arc in $PG(2, q)$ is called an* **oval**; *a $(q + 2)$-arc is a* **hyperoval**. *The maximum size of an arc in $PG(2, q)$ is $q + 2$ if q is even; it is $q + 1$ if q is odd.*

In order to complete the proof of Theorem 17.4, we still have to prove that hyperovals exist in characteristic 2. More precisely, we will give a geometric proof for the fact that any oval in $PG(2, q)$ can be embedded in a hyperoval, provided q is a power of 2.

Let $\mathcal{O} \subset PG(2, q)$ be an oval (ovals always exist, the normal rational curve being an example). Every point $P \in \mathcal{O}$ is on exactly one **tangent** (a line meeting $\mathcal{O}$ only in P). Call this tangent t_P. We are looking for a point $N \notin \mathcal{O}$ such that $\mathcal{O} \cup \{N\}$ is a hyperoval. This means that NP must be a tangent to $\mathcal{O}$ for every $P \in \mathcal{O}$, hence $NP = t_P$. Here NP denotes the line determined by these two points N and P. It follows that we have to prove the following: the tangents to the points of $\mathcal{O}$ all intersect in one common point. This intersection point N of all tangents complements $\mathcal{O}$ to a hyperoval. In particular, each oval is contained in at most one hyperoval.

Now use the assumption that q is even. Start considering lines which do not contain a point of $\mathcal{O}$ (**exterior lines**). Let g be an exterior line and $Q \in g$. As $q + 1$ is odd, each point $Q \in g$ is on at least one tangent to $\mathcal{O}$. As there are $q + 1$ such tangents and also $q + 1$ points on a line, we must have equality: each point on an exterior line is on precisely one tangent t_P to $\mathcal{O}$.

Now let N be the intersection of two tangents to $\mathcal{O}$. By what we have just seen, there is no exterior line through N. How can that be possible? There are $q + 1$ lines through N and also $q + 1$ points on $\mathcal{O}$. It must be that all lines NP where $P \in \mathcal{O}$ are tangents. This is what we wanted to see.

The proof of Theorem 17.4 is complete. Moreover, we saw that, in characteristic 2, each oval in $PG(2, q)$ is contained in precisely one hyperoval, where the point complementing the oval to a hyperoval (some call it the **nucleus**) is the intersection of the tangents to the oval.

The hexacode is an example of this process. The points $P_1, \ldots, P_5$ form an oval; in fact, they form the normal rational curve in $PG(2, 4)$. Point P_6 is the nucleus of this oval.

Let us sum up: we have seen that the Reed-Solomon codes from Chapter 4 can be extended (really lengthened). These extended Reed-Solomon codes are $[q + 1, k, q + 2 - k]_q$ codes (for arbitrary $k \leq q$). The geometric description is given by normal rational curves. In the special case $k = 3$ this yields an oval in $PG(2, q)$. When q is even, each oval can be embedded in a hyperoval. Use the points of the hyperoval as columns of a generator matrix. This yields a $[q + 2, 3, q]_q$ code.

17.5 Proposition. *Let q be a power of 2. Using representatives of a hyperoval in $PG(2,q)$ as columns of a generator matrix defines a $[q+2,3,q]_q$ code.*

The first example beyond the familiar hexacode is a $[10,3,8]_8$ code. Here is its generator matrix:

$$\begin{pmatrix} 1 & 1 & 1 & 1 & 1 & 1 & 1 & 1 & 0 & 0 \\ 0 & 1 & \epsilon & \epsilon^2 & \epsilon^3 & \epsilon^4 & \epsilon^5 & \epsilon^6 & 0 & 1 \\ 0 & 1 & \epsilon^2 & \epsilon^4 & \epsilon^6 & \epsilon & \epsilon^3 & \epsilon^5 & 1 & 0 \end{pmatrix}$$

Codes of dimension 2

Our newly gained geometric insight makes it easy to decide the existence of two-dimensional linear codes. When will a code $[n,2,d]_q$ exist? Apply Theorem 17.1. The projective line $PG(1,q)$ has $q+1$ points. Each point is a hyperplane. Let w_i be the weight of point P_i, $i = 1,\ldots,q+1$, where notation is chosen such that $0 \le w_1 \le w_2 \cdots \le w_{q+1}$. We must have $w_1+\cdots+w_{q+1} = n$ and $w_1 + \cdots + w_q = d$. In particular, the largest weight of a point is $w_{q+1} = n - d$. As $w_1 + \cdots + w_q$ is bounded by $qw_{q+1} = q(n-d)$, the code will exist if and only if $q(n-d) \ge d$, equivalently, if $n \ge \dfrac{q+1}{q}d$. Compare this with the Griesmer bound (Theorem 5.7). It says that a necessary condition for an $[n,2,d]_q$ code to exist is $n \ge d + \lceil d/q \rceil$. This is equivalent to $n \ge d + d/q = \frac{q+1}{q}d$. It follows that the Griesmer bound is always met with equality in the case of two-dimensional codes.

17.6 Theorem. *An $[n,2,d]_q$ code exists if and only if the Griesmer bound does not exclude it. The precise condition is $n \ge d + d/q = (q+1)d/q$.*

As an example, what is the smallest n such that $[n,2,8]_7$ exists? The Griesmer bound shows $n \ge 8 + 8/7$, so $n \ge 10$. It follows from Theorem 17.6 that a $[10,2,8]_7$ does indeed exist. The largest weight of a point is $n - d = 2$. We can use as weights of points $2,2,1,1,1,1,1,1$. This is not the only choice. Another possibility is to choose weights $2,2,2,2,2,0,0,0$. One possible generator matrix corresponding to the former choice of weights is

$$G = \begin{pmatrix} 0 & 0 & 1 & 1 & 1 & 1 & 1 & 1 & 1 & 1 \\ 1 & 1 & 0 & 0 & 1 & 2 & 3 & 4 & 5 & 6 \end{pmatrix}.$$

Codes of dimension 3

This is the lowest dimension where open problems remain. We want to prove that, for d large enough, there are always $[n,3,d]_q$ codes meeting the

Griesmer bound, Theorem 5.7, with equality. More precisely, we claim that this is the case when $d \geq (q-1)^2$. Naturally, codes for large values of d cannot be projective. After all, $PG(2,q)$ has only q^2+q+1 points. According to Theorem 17.1, we think of the code as described by nonnegative integer weights assigned to the points of $PG(2,q)$. What is the effect of adding 1 to all the weights? The length will increase by q^2+q+1, the minimum distance by q^2. It follows that this translation operation transforms an $[n,3,d]_q$ code into an $[n+q^2+q+1,3,d+q^2]_q$ code. If the former code meets the Griesmer bound with equality, the latter code does likewise. It will therefore suffice to prove our claim for values of d between $(q-1)^2$ and $(q-1)^2+q^2$.

At first concentrate on values of d which are multiples of q, so $d = qr$, where $q-1 \leq r \leq 2q-1$. If $r = q-1$, we obtain parameters $[q^2,3,q^2-q]_q$. Such a code is easily constructed; see Exercise 17.1.2. Case $r = q$ leads to the Simplex code. We can assume $q+1 \leq r \leq 2q-1$ and need to construct $[(q+1)r+2,3,qr]_q$ codes.

This is done geometrically, using the following trick: let $s = 2q-r$. Observe $1 \leq s \leq q-1$. Choose lines $g_1,\ldots,g_s$, which form a **dual arc.** It is clear what this means: no more than two of our lines intersect in a common point; in other words, no point of $PG(2,q)$ is on more than two of the lines g_i. Why does such a dual arc exist? Simply start from an s-arc (of points) in $PG(2,q)$ (we know it exists). If we read the homogeneous parameters $(a:b:c)$ of these points as homogeneous parameters $[a:b:c] = (a:b:c)^\perp$ of lines, we obtain a dual arc.

The weight function is $w = 2PG(2,q)-(g_1\cup\cdots\cup g_s)$. This means that a point P is assigned weight 2 if it is not contained in any of the lines g_i, weight 1 if it is contained in precisely one of those lines and weight 0 if it is the intersection of two such lines. The sum of the weights is $n = 2(q^2+q+1)-s(q+1) = (q+1)r+2$, as required. Let h be a line which does not belong to the dual arc. Then $\sum_{P\in h} w(P) = 2(q+1)-s = r+2$. If $g = g_i$, we obtain $\sum_{P\in g} w(P) = (q+1)-(s-1) = r+2-q$. It follows that the weight distribution w defines a $[(q+1)r+2,3,qr]_q$ code, as required. This code has in fact only two nonzero weights; it is a **two-weight code.** Now assume we are given an $[n_0,3,qr]_q$ code meeting the Griesmer bound with equality. Let $d = qr-\delta$, where $\delta < q$. A corresponding Griesmer code has parameters $[n_0-\delta,3,d]_q$. Such a code certainly exists, as it can be constructed by repeated puncturing. We have proved the following:

17.7 Theorem. *For every $d \geq (q-1)^2$, there is an $[n,3,d]_q$-code meeting the Griesmer bound, Theorem 5.7, with equality; equivalently $n = d + \lceil d/q \rceil + \lceil d/q^2 \rceil$.*

Thanks to the dual arc construction and the normal rational curves (Reed-Solomon codes), we also have a recipe for how to construct these codes. Theorem 17.7 was first proved by Hamada and Tamari [103].

Let us have a look at case $d = (q-1)^2 - 1 = q^2 - 2q$. A code meeting the Griesmer bound with equality would have parameters $[q^2-q-1,3,q^2-2q]_q$.

Exercise 17.1.3 shows that such a code cannot exist.

For most smaller values of d, a three-dimensional code meeting the Griesmer bound with equality will have to be projective. Such codes are therefore described by sets of points (not multisets) in $PG(2, q)$ such that there is an upper bound on line intersections. This leads to the following natural generalization of the notion of an arc.

17.8 Definition. *An n-arc of species s in $PG(2, q)$ is a set of n points, no more than s of which are collinear.*

In terms of codes this corresponds to

17.9 Theorem. *An n-arc of species s in $PG(2, q)$ exists if and only if there is a projective $[n, 3, n - s]_q$ code.*

We expect $[n, 3, d]_q$ codes for small d and minimal n to be projective in most cases. Determining the minimal length n will then be equivalent to the determination of the largest cardinality n of arcs of species s (here $s = n - d$). This is a popular but deep problem of finite geometry, which is nowhere near solution. We conclude that there is little hope of ever obtaining the complete list of all parameters of three-dimensional codes over all finite fields.

An anticode construction

The geometric method describes a k-dimensional projective linear code as a set of points $\mathcal{P} \subset PG(k - 1, q)$. Sometimes it is easier to work instead with the complement $\mathcal{Q} = PG(k - 1, q) \setminus \mathcal{P}$. In order to control the minimum distance of the code, we need an upper bound on hyperplane intersections of $\mathcal{P}$. This is equivalent to a lower bound on the hyperplane intersections of $\mathcal{Q}$. This method of describing a code is known as the **anticode construction** in the coding community. Apparently the term was coined by Farrell [83]. The code geometrically described by $\mathcal{Q}$ is seen as an anticode, as we control not its minimum distance but its maximum distance. From a geometrical perspective such point sets with a lower bound on hyperplane intersections are also known as **min-hypers.**

Clearly the method generalizes to nonprojective codes. If a code is described geometrically by a weight function w on $PG(k - 1, q)$ and l is the maximum weight, then its anticode is described by the weight function $l - w$.

17.10 Example. *Choose $PG(u - 1, q) \subset PG(k - 1, q)$. Let the anticode be described by the points of the subspace. This means that the code itself is described by the points in $PG(k - 1, q) \setminus PG(u - 1, q)$. The length is therefore*

$$n = (q^k - 1)/(q - 1) - (q^u - 1)/(q - 1) = (q^k - q^u)/(q - 1).$$

The dimension is of course k, as the complement of a proper subspace is not contained in a hyperplane. Each hyperplane either contains our $PG(u-1,q)$ or intersects it in a $PG(u-2,q)$. This shows that each word of our code has weight either q^{k-1} or $q^{k-1} - q^{u-1}$. We have constructed a code

$$[(q^k - q^u)/(q-1), k, q^{k-1} - q^{u-1}]_q$$

for every q, k and $u < k$. These codes meet the Griesmer bound with equality.

A thorough study of the anticode construction in the binary case is in Belov, Logachev and Sandimirov [8]. We give one of the two major constructions from [8]. Let $1, 2, \ldots, k$ be a basis of the k-dimensional binary vector space underlying $PG(k-1,q)$. Let $u_1, u_2, \ldots u_a$ be natural numbers such that $1 \le u_i < k$ for all i and $\sum_i u_i \le l$ for some l. Each of the numbers u_i determines a u_i-dimensional vector subspace U_i and therefore a $PG(u_i - 1, 2)$ $\langle U_i \rangle \subset PG(k-1,2)$. Here U_1 is generated by the first u_1 basis vectors, U_2 by the next u_2 basis vectors and so forth, where the basis is ordered cyclically. This explains why we need $u_i < k$.

Now fix a number $x \ge l$. The multiplicity of $P \in PG(k-1,2)$ is defined as x minus the number of i such that $P \in \langle U_i \rangle$. This means that we are taking away the points from the union of the subspaces (in the multiset sense: take the sum of the multiplicities) from x copies of $PG(k-1,2)$. As $\sum_i u_i \le l \le x$, all points have nonnegative multiplicity, as it has to be. The length of the resulting code is $n = x(2^k - 1) - \sum_i(2^{u_i} - 1)$; the minimum distance is $d \ge x2^{k-1} - \sum_i 2^{u_i - 1}$.

17.11 Theorem. *Let $u_1, u_2, \ldots, u_a$ be natural numbers such that $1 \le u_i < k$, $\sum_i u_i \le l$. Pick $x \ge l$. If either $x > l$ or $k > 2$, then the binary code $\mathcal{C}(B1)$ constructed above has parameters*

$$[x(2^k - 1) - \sum_i(2^{u_i} - 1), \ k \ , \ge x2^{k-1} - \sum_i 2^{u_i - 1}]_2.$$

To complete the proof of Theorem 17.11, it remains to show that the dimension of $\mathcal{C}(B1)$ is not $< k$. This is left to the reader; Exercise 17.1.15. Theorem 17.11 has been used primarily to find binary linear codes meeting the Griesmer bound with equality; see Exercise 17.1.16.

The ternary Golay code

It has been mentioned repeatedly that a perfect code $[11, 6, 5]_3$ exists and has a self-dual lengthening $[12, 6, 6]_3$. We illustrate the geometric method by starting from parameters $[12, 6, 6]_3$ and arriving at a construction for the code. Along the way we will find useful information about this code.

Let C be a $[12,6,6]_3$ code and Q a corresponding multiset of twelve points in $PG(5,3)$ which has the property that it intersects each hyperplane $PG(4,3)$ in at most six points. Denote subspaces $PG(3,3)$ as **solids.** An i-hyperplane is a hyperplane intersecting Q precisely in cardinality i.

17.12 Lemma. *Each solid intersects Q in at most four points. In the case of equality each hyperplane containing the solid is a 6-hyperplane.*

PROOF Let S be a solid. If it intersects Q in cardinality 6, we can find a hyperplane with a higher intersection cardinality (always in the multiset sense). Assume $|S \cap Q| = 5$. As S is contained in four hyperplanes, the cardinality of Q is bounded by $5 + 4 = 9$, a contradiction. The same counting argument shows that, in the case of intersection cardinality four, each hyperplane containing S must have six points of Q. □

Assume Q is not projective. A point Q of weight ≥ 2 is contained in a line of intersection cardinality ≥ 3 and so forth until we reach a solid with intersection cardinality ≥ 5, which is impossible by Lemma 17.12. So Q is a **set** of 12 points.

The same elementary counting argument shows that any three points of Q generate a plane, any four points generate a solid and any five points generate a hyperplane. In other words, any five points of Q are in general position, meaning that vectors representing those points are linearly independent. In coding terms this means that the dual distance (the distance of the dual code $C^\perp$) is $d' \geq 6$. Clearly $d' = 6$. It follows that $C^\perp$ has the same parameters $[12,6,6]_3$ as C itself.

By Lemma 17.12 there are no 5-hyperplanes. Each hyperplane containing five points of Q meets Q in precisely six points. There are therefore precisely $\binom{12}{5}/6 = 132$ such 6-hyperplanes. The number of minimum weight words in C is $A_6 = 2 \cdot 132 = 264$. The situation can be expressed in terms of designs. If we use the intersections with 6-hyperplanes as blocks, a Steiner system $S(5,6,12)$ is obtained. This design is known as the **small Witt design** (maybe we should wait until we know that the underlying code exists).

As there are no 5-hyperplanes and no 4-hyperplanes, we have $A_7 = A_8 = 0$. Consider a set T of three points in Q, and E the plane generated by T. Each hyperplane through E is either one of the 6-hyperplanes or a 3-hyperplane. Let x be the number of 6-hyperplanes containing E. Count pairs (Q,H) where $Q \in Q \setminus T$, H is a 6-hyperplane containing E and $Q \in H$. Counting two ways, the result is $9 \times 4 = x \times 3$, hence $x = 12$. As there are $1+3+9 = 13$ hyperplanes containing E, we conclude that any three points of Q are on precisely one 3-hyperplane. The number of 3-hyperplanes is therefore $\binom{12}{3} = 220$, and we have $A_9 = 440$. As 2-hyperplanes and 1-hyperplanes cannot exist ($A_{10} = A_{11} = 0$), the number of 0-hyperplanes follows from the number $(3^6 - 1)/2 = 364$ of all hyperplanes. We have determined the weight distribution of $\mathcal{C}$:

$$A_0 = 1, \ A_6 = 264, \ A_9 = 440, \ A_{12} = 24.$$

Write a generator matrix in the form $G = (I|P)$, where I is the $(6, 6)$ unit matrix and P is a $(6, 6)$ matrix. The points in $\mathcal{Q}$ are those generated by the columns of G. As all weights are multiples of 3, each codeword is orthogonal to itself. Lemma 17.12 shows that no column of P has more than one 0. Consider five points from the left, corresponding to elementary vectors (columns of the unit matrix). The hyperplane generated by these five points must contain a point on the right, generated by a column of P. It follows that we can arrange the columns of P such that P has zeroes in the main diagonal, and all entries off the main diagonal are nonzero.

Let $v_1, \ldots, v_6$ be the rows of G, and write $v_i = (e_i|w_i)$, where w_i is the i-th row of P. We have $wt(v_i - v_j) = 6$ for $i \neq j$, so $wt(w_i - w_j) = 4$. In other words, w_i and w_j agree in precisely two coordinates. Of the four coordinates where both w_i and w_j are nonzero, in two coordinates the entries are equal, and in two they are different. This shows $w_i \cdot w_j = 0 = v_i \cdot v_j$. As we also have $v_i \cdot v_i = 0$, it follows that all words of $\mathcal{C}$ are orthogonal to one another: $\mathcal{C} \subseteq \mathcal{C}^\perp$. For dimensional reasons $\mathcal{C} = \mathcal{C}^\perp$: our code is self-dual. Different rows of P are orthogonal to each other, whereas $w_i \cdot w_i = 2$. This shows in particular that P is a nonsingular matrix: $det(P) \neq 0$; equivalently, the rows of P are linearly independent.

We show now that P can be brought in a special form when we use the following operations:

- Row operations on G (change of basis of $\mathcal{C}$), and

- Monomial operations on the columns of $\mathcal{C}$ (permutations of the columns and multiplication of columns by nonzero constants).
  Compare Section 12.4. The monomial group has order $2^{12}12!$

As columns of P are determined only up to scalar multiples, we can choose $w_1 = (0, 1, 1, 1, 1, 1)$. By changing the sign of one or several rows of P, we can make sure the first column of P is $(0, 1, 1, 1, 1, 1)^t$. This will change some of the entries 1 on the diagonal of I into 2, but we can correct that by multiplying the corresponding columns by 2. So far we have $P = \begin{pmatrix} 0 & 1 & 1 & 1 & 1 & 1 \\ 1 & 0 & \pm 1 & \pm 1 & \pm 1 & \pm 1 \\ 1 & \pm 1 & 0 & \pm 1 & \pm 1 & \pm 1 \\ 1 & \pm 1 & \pm 1 & 0 & \pm 1 & \pm 1 \\ 1 & \pm 1 & \pm 1 & \pm 1 & 0 & \pm 1 \\ 1 & \pm 1 & \pm 1 & \pm 1 & \pm 1 & 0 \end{pmatrix}$.

Because of orthogonality, w_2 has two entries 1 and two entries 2 in the last four coordinates. We can choose $w_2 = (1, 0, 1, 1, 2, 2)$. Eventually the last four columns have to be suitably permuted. This moves the 0 entries of those columns into undesired positions, which can be corrected by a suitable permutation of rows. The resulting permutation of the entries 1 in the first six columns can be corrected by another column permutation. Observe that each $w_i, i > 1$ has two entries 1 and two entries 2 in the last five coordinates. This suffices to show that the entries in column 2 of P are uniquely

determined. We have $P = \begin{pmatrix} 0 & 1 & 1 & 1 & 1 & 1 \\ 1 & 0 & 1 & 1 & 2 & 2 \\ 1 & 1 & 0 & \pm1 & \pm1 & \pm1 \\ 1 & 1 & \pm1 & 0 & \pm1 & \pm1 \\ 1 & 2 & \pm1 & \pm1 & 0 & \pm1 \\ 1 & 2 & \pm1 & \pm1 & \pm1 & 0 \end{pmatrix}$. The entries in the $(2,2)$

squares in the center and in the southeast are uniquely determined as well:

$$P = \begin{pmatrix} 0 & 1 & 1 & 1 & 1 & 1 \\ 1 & 0 & 1 & 1 & 2 & 2 \\ 1 & 1 & 0 & 2 & \pm1 & \pm1 \\ 1 & 1 & 2 & 0 & \pm1 & \pm1 \\ 1 & 2 & \pm1 & \pm1 & 0 & 1 \\ 1 & 2 & \pm1 & \pm1 & 1 & 0 \end{pmatrix}.$$

Up to making the two last columns change places (and following the above procedure) we can choose $p_{35} = 1, p_{36} = 2$. The remaining entries of P are uniquely determined. We have

$$P = \begin{pmatrix} 0 & 1 & 1 & 1 & 1 & 1 \\ 1 & 0 & 1 & 1 & 2 & 2 \\ 1 & 1 & 0 & 2 & 1 & 2 \\ 1 & 1 & 2 & 0 & 2 & 1 \\ 1 & 2 & 1 & 2 & 0 & 1 \\ 1 & 2 & 2 & 1 & 1 & 0 \end{pmatrix}.$$

As P is symmetric, it follows from strength 5 that any five of its rows are independent. The generator matrix is

$$G = \begin{pmatrix} 1 & 0 & 0 & 0 & 0 & 0 & 0 & 1 & 1 & 1 & 1 & 1 \\ 0 & 1 & 0 & 0 & 0 & 0 & 1 & 0 & 1 & 1 & 2 & 2 \\ 0 & 0 & 1 & 0 & 0 & 0 & 1 & 1 & 0 & 2 & 1 & 2 \\ 0 & 0 & 0 & 1 & 0 & 0 & 1 & 1 & 2 & 0 & 2 & 1 \\ 0 & 0 & 0 & 0 & 1 & 0 & 1 & 2 & 1 & 2 & 0 & 1 \\ 0 & 0 & 0 & 0 & 0 & 1 & 1 & 2 & 2 & 1 & 1 & 0 \end{pmatrix}.$$

Starting from matrix G, it is easy to check that it generates a self-dual $[12, 6, 6]_3$ code. In fact, because of self-duality, all weights are divisible by 3, so it suffices to exclude weight 3. Linear combinations involving more than three rows of G present no danger. As the rows of P are linearly independent, linear combinations with exactly three nonzero coefficients have weight > 3 as well. As the rows of G have weight 6, it suffices to show that $w_i + w_j$ and $w_i - w_j$ have weight > 1 for $i \neq j$, which of course is true. It follows that G is a generator matrix of a self-dual $[12, 6, 6]_3$ code, the ternary Golay code $\mathcal{G}_{12}$.

The automorphism group of $\mathcal{G}_{12}$.

The discussion in the preceding subsection shows that there is up to monomial equivalence exactly one $[12, 6, 6]_3$ code. We can in fact determine its automorphism group. Let $Aut(\mathcal{G}_{12})$ be the automorphism group of the ternary Golay code $\mathcal{G}_{12}$. More precisely, it is the stabilizer of $\mathcal{G}_{12}$ under the monomial group of order $2^{12}12!$

We need some more practice with monomial operations. Each element of the monomial group will be written $g = \pi m$, where π is a permutation of $\{1, 2, \ldots, 12\}$ and $m = m(1, 2, \ldots, 12) = (m_1, m_2, \ldots, m_{12}) \in \{1, 2\}^{12}$. For the permutation π we use the cycle notation introduced in Section 12.4. Element g acts by at first permuting the columns of G according to π and then multiplying column i by m_i, for all i. An element g belongs to $Aut(\mathcal{G}_{12})$ if it stabilizes the ternary Golay code $\mathcal{G}_{12} = \langle v_1, v_2, v_3, v_4, v_5, v_6 \rangle$, in other words, if the image of v_i under g is contained in $\mathcal{G}_{12}$, for all $i = 1, \ldots, 6$.

What happens if π is the identity permutation? We leave it to the reader to verify that $m = (m_1, \ldots, m_{12}) \in Aut(\mathcal{G}_{12})$ if and only if either all $m_i = 1$ or all $m_i = 2$. Denote the latter element by -1.

The columns of the generator matrix G generate the point set $\mathcal{Q} = \{Q_1, Q_2, \ldots, Q_{12}\} \subset PG(5, 3)$. Each element $g \in Aut(\mathcal{G}_{12})$ induces a permutation of $\mathcal{Q}$. This permutation is of course π, where $g = \pi m$.

17.13 Definition. *Let $Aut(\mathcal{G}_{12})$ be the stabilizer of the ternary Golay code $\mathcal{G}_{12}$ in the monomial group. Write $g = \pi m$ as above. Let*

$$M_{12} = \{\pi | g = \pi m \in Aut(\mathcal{G}_{12}) \text{ for some } m\}$$

*the permutation group on $\{1, \ldots, 12\}$ (which we identify with the set $\mathcal{Q}$ of points in $PG(5, 3)$ describing the columns of the generator matrix G) determined by $Aut(\mathcal{G}_{12})$. This subgroup $M_{12} \subset S_{12}$ of the symmetric group on 12 objects is known as the **Mathieu group** on 12 symbols.*

As mentioned above, whenever $g = \pi m \in Aut(\mathcal{G}_{12})$, then $-g = \pi(-m)$ is the only other element of $Aut(\mathcal{G}_{12})$ with permutation π. The **orders,** the number of elements of these groups, are therefore related by $|Aut(\mathcal{G}_{12})| = 2 \times |M_{12}|$. In practice we can concentrate on the permutation part π and adapt m by setting $m_1 = 1$.

We already have a lot of information about these groups. Consider subsets of six coordinates. The 132 blocks of the Witt design $S(5, 6, 12)$ correspond to the 6-hyperplanes and to $(6, 6)$ submatrices of G which have rank 5 (hence determinant 0). The remaining $\binom{12}{6} - 132 = 12 \times 11 \times 6$ sets of six columns correspond to regular $(6, 6)$ submatrices of G, **information sets** of coordinates. Examples of such information sets are the first six and the last six coordinates. We have shown above that the generator matrix can be given a special form starting from any information set. This shows that M_{12} is transitive on information sets. By a basic principle of permutation groups, we have that the order of M_{12} is $|M_{12}| = 12 \times 11 \times 6 \times |H|$, where H is the subgroup of M_{12} consisting of those elements that map the set $\{1, 2, 3, 4, 5, 6\}$ of coordinates into itself. In order to determine the order of M_{12}, it suffices to determine the order of H.

We have $h \in H$ if H permutes the points corresponding to the first six columns and also the points corresponding to the last six columns. The zeroes along the main diagonal of P show that the action of π on the last six coordinates is the same as on the first six: if $\pi : i \mapsto j$ where $i, j \leq 6$, then $\pi : 6 + i \mapsto 6 + j$. The elements $h_1 = (2, 3, 5, 6, 4)(8, 9, 11, 12, 10)$, and $h_2 = (2, 3)(4, 5)(8, 9)(10, 11)$ are both in H. The group $\langle h_1, h_2 \rangle$ generated by h_1, h_2 has order 10. Another element of H fixing 1 is

$$h_3 = (3, 5, 4, 6)(9, 11, 10, 12)m(1, 2, 2, 2, 2, 2, 2, 1, 1, 1, 1, 1).$$

The group $\langle h_1, h_2, h_3 \rangle$ has order 20. It is most important to find an element in H mapping 1 to a column $\neq 1$. An example for such an element is

$$h_4 = (1, 2)(5, 6)(7, 8)(11, 12)m(1, 1, 1, 1, 2, 2, 1, 1, 1, 1, 2, 2).$$

It maps $v_1 \leftrightarrow v_2, v_3 \mapsto v_3, v_4 \mapsto v_4, v_5 \leftrightarrow -v_6$.

Let us collect the information on M_{12} we have so far. As H is transitive on $\{1, 2, 3, 4, 5, 6\}$ and the stabilizer of 1 in H has order at least 20, it follows $|H| \geq 120$, and consequently $|M_{12}| \geq 12 \times 11 \times 6 \times 120 = 12 \times 11 \times 10 \times 9 \times 8$.

On the other hand, consider the action of the permutation group M_{12} on the 5-tuples from $\{1, 2, \ldots, 12\}$. What is the stabilizer of the 5-tuple $(1, 2, 3, 4, 5)$? As Q_{12} is the sixth point in the hyperplane generated by $Q_1, \ldots, Q_5$, this point must be fixed as well. Inspection shows that the permutation must be the identity. In fact, the assumption that 6 is not fixed leads to a matrix which, after permutation, has $\pm Q_6$ in the second half of the coordinates, which is not possible. In particular, the stabilizer of a 5-subset acts faithfully on the 5-subset and therefore cannot be larger than 5! This shows $|M_{12}| \leq \binom{12}{5} \times 5! = 12 \times 11 \times 10 \times 9 \times 8$. It follows that we must have equality everywhere. In particular, the stabilizer of any 5-subset is S_5.

17.14 Theorem. *The permutation group M_{12} on 12 objects is 5-transitive. Its order is*
$$|M_{12}| = 12 \times 11 \times 10 \times 9 \times 8.$$

The Mathieu group is much older than the Witt design and the ternary Golay code. We have here a prime example for the interplay of group theory, coding theory and design theory. The Mathieu group M_{12}, the ternary Golay code and the small Witt design $S(5, 6, 12)$ are equivalent objects, but this statement has to be taken with a grain of salt. We saw how to construct the code in a canonical way from its parameters and how to obtain the design and the group. However, as we cannot operate with permutations alone, it is not all that easy to determine the group. Starting from the group, the design is immediate but the code poses a problem. One has to guess that one should use the field of three elements and to find the right ternary vectors. Starting from the design, the group is immediate (it is the group of automorphisms of the design) but once again the code does not seem to be canonically determined from that point of view.

Let us collect the most important information obtained:

17.15 Theorem. *There is a* $[12, 6, 6]_3$ *code. It is self-dual and uniquely determined under the action of the monomial group. Its automorphism group (the stabilizer in the monomial group) has order* $2 \times 12 \times 11 \times 10 \times 9 \times 8$. *It induces a permutation group* M_{12} *on 12 objects, the* **small Mathieu group,** *which is 5-transitive of order* $12 \times 11 \times 10 \times 9 \times 8$.

Squares and nonsquares

Before we can describe a nice family of arcs of high species (and therefore of good three-dimensional codes), it is necessary to get a picture of the distribution of squares and non-squares in finite fields of odd characteristic.

Consider the quadratic equation

$$x^2 + ax + b = 0$$

with coefficients $a, b \in \mathbb{F}_q$. There is a sharp dichotomy between odd characteristic and characteristic 2.

Consider the case when q is odd. By completing the square, we obtain the equivalent equation $(x + \frac{a}{2})^2 = \frac{1}{4}(a^2 - 4b)$. It follows that the equation has a solution if and only if $a^2 - 4b$ is a square in $\mathbb{F}_q$. It all comes down to the distinction between squares and nonsquares in $\mathbb{F}_q$.

Let ϵ be a primitive element of $\mathbb{F}_q$. The powers of ϵ with even exponent are squares. It follows that at least $(q - 1)/2$ nonzero elements of $\mathbb{F}_q$ are squares. On the other hand, each square is the square of two different elements in $\mathbb{F}_q$. An obvious counting argument shows that we have equality: the nonzero squares in $\mathbb{F}_q$ are precisely the powers of ϵ^2. There are $(q - 1)/2$ squares and also $(q - 1)/2$ nonsquares in $\mathbb{F}_q^*$. The squares form a subgroup (clearly the product of two squares is a square). All this sounds familiar. The situation is analogous to the real case.

When is -1 a square? As the multiplicative group of $\mathbb{F}_q$ is cyclic, -1 is the uniquely determined element of order 2. If $q \equiv 3 \pmod 4$, the group order is not divisible by 4, so -1 cannot be a square. Because of cyclicity, -1 is a square if $q \equiv 1 \pmod 4$.

17.16 Proposition. *Let q be odd. Then -1 is a square in the multiplicative group of $\mathbb{F}_q$ if and only if $q \equiv 1 \pmod 4$.*

The Barlotti arcs

A nice construction of arcs of high species is related to normal rational curves in the plane; see Barlotti [6]. Denote by

$$\mathcal{Q} = \{P_\infty\} \cup \{P_y \mid y \in \mathbb{F}_q\}$$

the normal rational curve. This is the set of all points $(x : y : z)$ satisfying $y^2 = xz$. Here $P_\infty = (0 : 0 : 1)$ and $P_y = (1 : y : y^2)$. The oval $\mathcal{Q}$ is also known as a **conic section** in $PG(2, q)$.

Let q be odd. Theorem 17.4 shows that $\mathcal{Q}$ cannot be embedded in a hyperoval. There are three types of lines, relative to their intersection with $\mathcal{Q}$: **exterior lines, secants** (two points in common with $\mathcal{Q}$) and **tangents** (containing one point from $\mathcal{Q}$). Clearly the number of secants is $\binom{q+1}{2}$. As every point of $PG(2, q)$ is on precisely $q + 1$ lines and every point of $\mathcal{Q}$ is on q secants, we see that every point of $\mathcal{Q}$ is on precisely one tangent. The number of tangents is therefore $q + 1$. It follows that the number of exterior lines is $\binom{q}{2}$. Let t_P be the tangent through $P \in \mathcal{Q}$.

Each line (hyperplane) has the form $(a, b, c)^\perp$ for some nonzero tuple (a, b, c). As $(a, b, c)^\perp = (\lambda a, \lambda b, \lambda c)^\perp$ for $\lambda \neq 0$, the lines are parametrized in just the same way as points are. Let $[a : b : c] = (a, b, c)^\perp$. In particular, point $(x : y : z)$ is on line $[a : b : c]$ if and only if the dot product vanishes: $ax + by + cz = 0$.

It is easy to calculate everything we wish in coordinates now. We have

$$t_\infty = [1 : 0 : 0] \quad \text{and} \quad t_y = [y^2 : -2y : 1].$$

It is no problem to determine the intersections of the tangents:

$$t_\infty \cap t_y = (0 : 1 : 2y), \quad t_y \cap t_{y'} = (1 : (y + y')/2 : yy').$$

Assume three tangents meet in a common point.
This would mean $(1 : (y+y')/2 : yy') \in t_{y''}$. Writing this out, we would obtain

$$0 = y''^2 - y''(y + y') + yy' = (y'' - y)(y'' - y'),$$

which is impossible as y, y', y'' were chosen as different. There are three types of points in $PG(2, q)$, with respect to their position relative to $\mathcal{Q}$: points of $\mathcal{Q}$, **interior points** (on no tangent at all) and **exterior points** (on precisely two tangents). In fact, we have seen that no point off $\mathcal{Q}$ is on more than 2 tangents. That such a point cannot be on precisely one tangent follows from a trivial parity argument. The terminology is motivated by the real case: points inside an ellipse are on no tangent to the ellipse, while each point outside is on exactly two tangents.

As we have $q + 1$ tangents and each contains q exterior points, we see that there are precisely $(q + 1)q/2$ exterior points. It follows that the number of interior points must be $\binom{q}{2}$. Let us determine the exterior points explicitly: these are the points $(0 : 1 : 2y)$ with arbitrary y, and the points of the form $(1 : (y + y')/2 : yy')$, where $y \neq y'$. The point $(1 : v : w)$ is exterior if and only if we can find $y \neq y'$ such that $y + y' = 2v, yy' = w$. Eliminating y' from the first equation, we arrive at a quadratic equation for y, which, after completing the square, becomes $(y - v)^2 = v^2 - w$. It follows that $v^2 - w$ must be a square. $v^2 - w = 0$ is not possible, as this would lead to $y = y' = v$, violating the condition $y \neq y'$. Each square $v^2 - w$ gives two solutions. This gives us a complete description.

17.17 Lemma. *Point $(x : y : z)$ is exterior if and only if $y^2 - xz$ is a nonzero square; it is interior if and only if $y^2 - xz$ is a nonsquare.*

A trivial counting argument yields the distribution of types of points on types of lines:

17.18 Lemma. *Every secant has $(q - 1)/2$ exterior points and $(q - 1)/2$ interior points.*
Each tangent has q exterior points and no interior points, each exterior line has $(q + 1)/2$ exterior points and equally many interior points.

This allows the construction of some good three-dimensional codes. Take all interior points and one point of $\mathcal{Q}$. This defines a $(\binom{q}{2} + 1)$-arc of species $\leq (q + 1)/2$ (see Lemma 17.18). Theorem 17.9 produces a projective code with minimum distance $d = \binom{q}{2} + 1 - (q + 1)/2 = (q - 1)^2/2$.

17.19 Theorem. *Let $\mathcal{Q}$ be a quadric (normal rational curve, conic section) in $PG(2, q)$, q odd. The set of interior points together with one point of $\mathcal{Q}$ defines a code*

$$[\binom{q}{2} + 1, 3, (q - 1)^2/2]_q.$$

The union of all interior points and points of $\mathcal{Q}$ forms an arc of species $(q + 3)/2$.

17.20 Theorem. *Let $\mathcal{Q}$ be a quadric in $PG(2, q)$, q odd. The set of interior points together with the points of $\mathcal{Q}$ define a code*

$$[\binom{q + 1}{2} + 1, 3, (q^2 - 1)/2]_q.$$

Before we can give a final family of examples, quadratic equations in characteristic 2 need to be considered.

Quadratic equations in characteristic 2

Consider the quadratic equation

$$x^2 + ax + b = 0$$

with coefficients $a, b \in \mathbb{F}_q$. Let q be a power of 2. The distinction between squares and nonsquares is no longer available. In fact, as squaring is a field automorphism (see Section 12.1, the Frobenius automorphism), every element of $\mathbb{F}_q$ is a square and possesses a unique square root. In particular, if $a = 0$, the equation $x^2 + b = 0$ has a unique solution. If $a \neq 0$, we divide by a^2. With $y = x/a$ the quadratic equation becomes $y^2 + y + (b/a^2) = 0$. The absolute trace tr yields a condition: as $tr(y^2) = tr(y)$, we must have $tr(b/a^2) = 0$ if there is a solution. As each such quadratic equation cannot have more than two solutions, an obvious counting argument shows now that this condition is also sufficient: whenever $tr(b/a^2) = 0$, our quadratic equation has two solutions.

17.21 Proposition. *Let q be a power of 2. The quadratic equation $x^2 + ax + b = 0$, where $a, b \in \mathbb{F}_q, a \neq 0$, has no solution in $\mathbb{F}_q$ if $tr(b/a^2) = 1$. It has two solutions in $\mathbb{F}_q$ if $tr(b/a^2) = 0$.*

Proposition 17.21 is also known as the additive version of **Hilbert's theorem 90.**

Maximal arcs, Denniston arcs and Denniston codes

How many points can an arc of species s in $PG(2, q)$ have? It was noted that this is a very basic and difficult problem in general. An upper bound is obtained by fixing a point P of the arc and considering the distribution of the points on the $q + 1$ lines through P. Each of these lines contains at most $s - 1$ points $\neq P$ of the arc. It follows that the arc has at most $1 + (q + 1)(s - 1) = qs + s - q$ points. In the case of equality, each line of the plane either avoids the arc or intersects it in precisely s points.

17.22 Definition. *An arc $K \subseteq PG(2, q)$ of species s with $qs + s - q$ points is a **maximal s arc**.*

Once again the terminology is unfortunate (it can happen that maximal s arcs do not exist and that s arcs of maximal cardinality are not maximal s arcs). When do maximal s arcs exist? There are some trivial cases: a point is a maximal 1 arc, all of $PG(2, q)$ is a maximal $(q + 1)$ arc, and each affine subplane (the points off a line) is a maximal q arc.

Let $2 \leq s \leq q - 1$. We know already that maximal s arcs will not always exist. In fact, a maximal 2 arc is nothing but a hyperoval and we know that those exist only in characteristic 2. Assume K is a maximal s arc. Fix a point $Q \notin K$ (such a point exists as $s < q + 1$). As each line through Q meets K in 0 or in s points, the cardinality of K must be a multiple of s. It follows that q is a multiple of s.

17.23 Proposition. *A maximal s arc in $PG(2, q)$, $s \leq q$ can exist only if s divides q.*

We are led to the following question: let $q = p^a$, p prime, $a > 1$. For which i, where $0 < i < a$, does there exist a maximal p^i arc in $PG(2, q)$? An elegant construction due to Denniston shows that maximal arcs always exist when $p = 2$. We find it handy to work in the affine plane.

The points $P(x, y) = (x : y : 1)$ are the points of the affine plane corresponding to the line g consisting of points with vanishing last coordinates. Call g the line at infinity, and its points the points at infinity. The $q + 1$ points at infinity are $(1 : m : 0)$, $m \in \mathbb{F}_q$ and $(0 : 1 : 0)$. The line through $(0 : 1 : 0)$ and $(c : y : 1)$ has affine points $(c : y : 1)$, $y \in \mathbb{F}_q$. This line has equation $x = c$. The line through $(1 : 0 : 0)$ and $(x : c : 1)$ has equation $y = c$. For $m \neq 0$ the line through $(1 : m : 0)$ and $(a : b : 1)$ consists of the points $(\lambda + a : \lambda m + b : 1)$. This line has equation $y = mx + c$, where $c = b - am$. Naturally we call m the **slope** of the line $y = mx + c$. The lines $x = c$ have slope ∞.

We can define the Denniston arcs:

17.24 Definition (Denniston arcs). *Let $q = 2^a$, $0 < i < a$. Denote by $tr = tr : \mathbb{F}_q \longrightarrow \mathbb{F}_2$ the trace, choose an element $b \in \mathbb{F}_q$ such that $tr(b) = 1$, put $Q(X, Y) = X^2 + XY + bY^2$. Let $E \subset \mathbb{F}_q$ be an i-dimensional $\mathbb{F}_2$ vector space. The corresponding Denniston arc $\mathcal{D}$ is a subset of the affine plane. It is defined as*

$$\mathcal{D} = \{P(x, y) \mid Q(x, y) \in E\}$$

In order to prove that $\mathcal{D}$ is a maximal 2^i arc we need to prepare the ground.

17.25 Lemma. *Let $q = 2^a$, $F = \mathbb{F}_q$. The image $f(F)$ of a quadratic function $f(X) = aX^2 + bX + c$, where $a \neq 0$, is either a hyperplane ($\mathbb{F}_2$ subspace H of dimension $a - 1$) or the complement $F \setminus H$ of such a subspace. Each image under f has precisely two preimages.*

PROOF   Consider $g(X) = aX^2 + bX$. As $x \longrightarrow x^2$ is a field automorphism, we see that $g(X)$ yields an $\mathbb{F}_2$ linear mapping. Its kernel consists of 0 and b/a, so it has dimension 1. Its image must be an $(a-1)$-dimensional vector subspace H of F (over $\mathbb{F}_2$). If $c \in H$, then $f(F) = H$. If $c \notin H$, then $f(F) = F \setminus H$. $\qquad\square$

17.26 Theorem. *The point set $\mathcal{D}$ of Definition 17.24 is a maximal 2^i arc in $PG(2, 2^a)$.*

PROOF We have defined $\mathcal{D}$ as a point set in the affine plane $AG(2, q)$. As $AG(2, q) \subset PG(2, q)$, we may consider it as a subset of the projective plane. It suffices to show that every line intersects $\mathcal{D}$ either in 0 or in 2^i points (fix a point $P \in \mathcal{D}$, count the points of $\mathcal{D}$ on the lines through P). $Q(X, Y)$ is a homogeneous polynomial, in particular $Q(\lambda X, \lambda Y) = \lambda^2 Q(X, Y)$. For which pairs $(x, y) \in F \times F$ is $Q(x, y) = 0$? If $y = 0$, then $Q(x, 0) = x^2 = 0$ implies $x = 0$. If $y \neq 0$, then without restriction $y = 1$, hence $Q(x, 1) = x^2 + x + b$. By the choice of b and Proposition 17.21, we have $Q(x, 1) \neq 0$ for all x. We conclude that $Q(x, y) = 0$ if and only if $x = y = 0$. Such quadratic forms are called **anisotropic.** They will be studied in more detail in the following section.

Consider at first lines through the origin: In the case of the line $x = 0$, we get $Q(0, y) = by^2$, so this line intersects $\mathcal{D}$ in 2^i points. In the case of the lines $y = mx$, we get $Q(x, mx) = x^2 + mx^2 + bm^2x^2 = Q(bm, 1)x^2/b$. As $Q(bm, 1) \neq 0$, we see that with $x \in F$ this varies over all elements in F. The intersection size is 2^i again.

So consider lines not through the origin. If (x, y) varies through the q affine points of such a line l, then $Q(x, y)$ is a quadratic expression. As $P(0, 0)$ is not on l, we conclude from Lemma 17.25 that $Q(l)$ is the complement of a vector subspace H, where every image occurs exactly twice. To finish the proof we have to consider two cases:

If $E \subseteq H$, then clearly l contains no point of $\mathcal{D}$. In the contrary case we have $|E \cap H| = 2^{i-1}$. As every image occurs twice, we conclude that l intersects $\mathcal{D}$ in cardinality 2^i. ∎

The construction makes heavy use of the fact that we are in characteristic 2. Every hope of generalizing this construction to an odd characteristic has been shattered by a result of Ball, Blokhuis and Mazzocca [4], where it is shown that nontrivial maximal arcs never exist in $PG(2, q)$ when q is odd.

The Denniston arcs yield three-dimensional codes over $\mathbb{F}_q$, where $q = 2^a$, of length $n = 2^{a+i} + 2^i - 2^a$, with minimum distance $d = 2^{a+i} - 2^a$, for every $i < a$. Moreover these are 2-weight codes, the only nonzero weights being d and n. Naturally we will call them **Denniston codes.** These $[2^{a+i} + 2^i - 2^a, 3, 2^{a+i} - 2^a]_{2^a}$ codes meet the Griesmer bound with equality.

The Denniston arcs are also interesting from the design-theoretic point of view. We see the nonempty intersections with lines as blocks. Then every block has 2^i points, and any two points of the Denniston arc are on exactly one block. The Denniston arcs are Steiner systems $S(2, 2^i, 2^{a+i} + 2^i - 2^a)$, which are embedded in a Steiner system $S(2, 2^a, 2^{2a})$, the affine plane (see Definition 7.3).

In the special case $i = 1$, the Denniston arc is a hyperoval. In particular, we recover a version of the hexacode as a Denniston code over $\mathbb{F}_4$.

Consider the case $a = 2, i = 1$ of the Denniston arcs. As usual, write $\mathbb{F}_4 = \{0, 1, \omega, \overline{\omega}\}$. As $tr(\omega) = 1$, we can take the quadratic form Q to be $Q(X, Y) = X^2 + XY + \omega Y^2$. The one-dimensional subspace of $\mathbb{F}_4$ will be chosen to be $\mathbb{F}_2$. The hyperoval is then

$$\mathcal{D} = \mathcal{O} = \{P(0,0), P(0,\omega), P(1,0), P(1,\overline{\omega}), P(\omega,\omega), P(\omega,\overline{\omega})\}.$$

In the case $i = 1, a = 3$, we can choose $b = 1$ as $tr(1) = 1$. The quadratic form is therefore $Q(X, Y) = X^2 + XY + Y^2$. As before, choose $E = \mathbb{F}_2$. Observe that Q is symmetric in X and Y, so the same is true of the points $P(x, y) \in \mathcal{O}$. We give the generator matrix whose columns are the points of $\mathcal{O}$:

$$\begin{pmatrix} 0 & 0 & 1 & 1 & \epsilon^3 & \epsilon^5 & \epsilon^3 & \epsilon^6 & \epsilon^5 & \epsilon^6 \\ 0 & 1 & 0 & 1 & \epsilon^5 & \epsilon^3 & \epsilon^6 & \epsilon^3 & \epsilon^6 & \epsilon^5 \\ 1 & 1 & 1 & 1 & 1 & 1 & 1 & 1 & 1 & 1 \end{pmatrix}.$$

1. The **points** of the $(k-1)$-dimensional projective geometry are the 1-dim subspaces of the vector space $V(k,q)$ over $\mathbb{F}_q$, the **lines** are the 2-dim subspaces and so forth. The **hyperplanes** are the $(k-1)$-dimensional vector subspaces.

2. The number of points (and of hyperplanes) in $PG(k-1,q)$ is $(q^k - 1)/(q-1)$.

3. Any two points are on exactly one line.

4. Geometric description of linear codes: $[n,k,d]_q$ is equivalent to a multiset of n points in $PG(k-1,q)$ such that outside each hyperplane there are at least d points of the multiset. A generator matrix of the code has as columns representatives of the points.

5. The Simplex codes $[\dfrac{q^k - 1}{q - 1}, k, q^{k-1}]_q$ are geometrically described by the set of all points of $PG(k-1,q)$.

6. The **extended Reed-Solomon codes** $[q+1, k, q+2-k]_q$ are geometrically described by the **normal rational curves.**

7. The maximum size of an arc (no three points collinear) in $PG(2,q)$ is $q+1$ (an **oval**) for odd q, and it is $q+2$ (a **hyperoval**) for even q.

8. The corresponding codes: $[q+1, 3, q-1]_q$ (oval codes, odd q) and $[q+2, 3, q]_q$ (hyperoval codes, even q).
   Case $q = 4$ is the **hexacode** $[6, 3, 4]_4$.

9. Two-dimensional code parameters exist if and only if the Griesmer bound is satisfied.

10. The same is true of three-dimensional codes $[n, 3, d]_q$ provided $d \geq (q-1)^2$.

1. Half of the nonzero elements of a finite field in odd characteristic are squares, half are nonsquares.

2. Arcs of **species** s are point sets in $PG(2, q)$ no more than s of which are collinear (on a line).

3. The Barlotti arcs are large arcs of species $(q+1)/2$ and $(q+3)/2$ in $PG(2, q)$ for odd q. They are related to conic sections (normal rational curves).

4. The **Denniston arcs** are Steiner systems $S(2, 2^i, 2^{a+i} + 2^i - 2^a)$ embedded in the affine plane $AG(2, q)$, where $q = 2^a$, $i < a$. Case $i = 1$ are hyperovals. The corresponding Denniston codes meet the Griesmer bound with equality.

Exercises 17.1

17.1.1. *Express the assumption that a generator matrix of the linear code $\mathcal{C}$ has no zero column in terms of the dual distance d'.*

17.1.2. *Give a geometric description of $[q^2, 3, q^2 - q]_q$ codes.*

17.1.3. *Show that a $[q^2 - q - 1, 3, q^2 - 2q]_q$ code cannot exist.*
Hint: use the geometric method. Show at first that such a code would have to be projective. Then show that each line of $PG(2, q)$ would have to intersect the corresponding point set in either 0 or $q - 1$ points.

17.1.4. *Prove the following: a $[n, 3, d]_q$ code meeting the Griesmer bound with equality such that $d \leq q^2$ is a multiple of q is necessarily projective.*

17.1.5. *Construct $[\dfrac{q^{2m} - 1}{q - 1}, 3m, q^{2m-1} - q^{m-1}]_q$ for all q and m.*
Hint: use concatenation with an oval code (an extended three-dimensional Reed-Solomon code) as outer code.

17.1.6. *Show that the code constructed in the previous exercise has only three nonzero weights. Determine the weight distribution.*

17.1.7. *Construct the Barlotti arcs (and with them the generator matrices of the corresponding three-dimensional codes) over $\mathbb{F}_5$ and $\mathbb{F}_7$.*

17.1.8. *The Griesmer bound is a consequence of Proposition 5.6, which states that from an $[n, k, d]_q$ code with minimum distance precisely d we can construct an $[n - d, k - 1, \lceil d/q \rceil]_q$ code, the residual code.*

Give a geometric proof of Proposition 5.6.

Hint: imagine the code described by a multiset $\mathcal{P}$ of cardinality n in $PG(k - 1, q)$. By assumption there is a hyperplane H such that $\mathcal{P} \cap H$ has cardinality $n - d$ (in the multiset sense, including weights). It has to be proved that $\mathcal{P} \cap H$ generates H and that, for every hyperplane L of H, the intersection $\mathcal{P} \cap L$ has cardinality at most ... what? This is a simple counting exercise. Observe that L is contained in precisely $q + 1$ hyperplanes of $PG(k - 1, q)$, one of them being H.

17.1.9. *Give a geometric proof for the nonexistence of $[13, 7, 6]_4$ codes and of $[13, 6, 7]_4$ codes.*

17.1.10. *Give a geometric proof for the fact that the dual of an $[11, 6, 5]_4$ code is an $[11, 5, 6]_4$ code. Use this and the cyclic $[11, 6, 5]_4$ from Section 13.4 to construct a $[12, 6, 6]_4$ code.*

17.1.11. *Find the weight distribution of the Denniston codes.*

17.1.12. *Give a geometric proof for the fact that the dual of an $[11, 5, 6]_3$ code is an $[11, 6, 5]_3$ code.*

17.1.13. *Construct the ternary Golay codes $[11, 6, 5]_3$ and $[12, 6, 6]_3$ using cyclic codes of length 11.*

17.1.14. *Fix disjoint subspaces $PG(u - 1, q)$ and $PG(k - u - 1, q)$ in $PG(k - 1, q)$. Find the parameters of the q-ary k-dimensional linear codes described by the points of $PG(k - 1, q)$, which are not contained in the union of those subspaces. Compare with the Griesmer bound.*

17.1.15. *Complete the proof of Theorem 17.11.*

17.1.16. *Show that the codes constructed in Theorem 17.11 meet the Griesmer bound with equality if and only if the u_i are pairwise different.*

17.1.17. *Use Theorem 17.11 to construct a $[12, 4, 6]_2$ code.*

17.1.18. *The Reed-Muller code $\mathcal{RM}(1, r)$ from Section 14.1 is a $[2^r, r + 1, 2^{r-1}]_2$ code. Give a geometric description for such a code.*

17.1.19. *Show that the subgroup H of the Mathieu group M_{12}, the stabilizer of $\{1, 2, 3, 4, 5, 6\}$, is sharply 3-transitive on $\{1, 2, 3, 4, 5, 6\}$.*

17.1.20. *Find a set of generators of M_{12}.*

17.1.21. *Show that the automorphism group of the hexacode $[6, 3, 4]_4$ is isomorphic to $Z_3 \times A_6$.*

17.1.22. *Do we know the quaternary Denniston code?*

17.1.23. *Determine the parameters of the $\mathbb{F}_8$ linear Denniston codes.*

17.1.24. *Determine the parameters of the $\mathbb{F}_{16}$ linear Denniston codes.*

17.1.25. *Construct a $[196, 9, 96]_2$ code.*

17.2 Quadratic forms, bilinear forms and caps

Basic concepts: Caps and distance 4 codes. Introduction to geometric algebra: symmetric bilinear forms, quadratic forms, symplectic forms.

17.27 Definition. *A set of points in $PG(k-1,q)$ such that no more than two points are collinear is a* **cap.**

Caps in $PG(2,q)$ are called arcs (see Definition 17.3). This is a terminological absurdity (different names in dimension 2 and in dimension > 2), but it may be too late to change that. Let $\mathcal{K}$ be a point set in $PG(k-1,q)$ and $\mathcal{C}$ a code whose generator matrix has representatives of the points from $\mathcal{K}$ as columns. Can we read off the fact that $\mathcal{K}$ is a cap from the code parameters of $\mathcal{C}$? The fact that no three points of $\mathcal{K}$ are on a line is equivalent to the fact that no three columns of the generator matrix are linearly dependent. This means that $\mathcal{C}$ has strength 3. By duality it is equivalent to $\mathcal{C}^{\perp}$ having minimum distance at least 4.

17.28 Proposition. *The following are equivalent:*

- *A cap of n points in $PG(k-1,q)$.*

- *A code $[n, n-k, 4]_q$.*

It is natural to ask what is the maximum number of points of a cap in a given space $PG(k-1,q)$. By Proposition 17.28 this is equivalent to the existence of certain codes with $d = 4$. For $PG(2,q)$ the answer is in Theorem 17.4. In coding terms, $[q+2, q-1, 4]_q$ exists in characteristic 2 (a hyperoval), whereas $[q+3, q, 4]_q$ does not exist. In odd characteristic, $[q+1, q-2, 4]_q$ exists, and $[q+2, q-1, 4]_q$ does not exist.

How about caps in larger dimensions? There is a feature of normal rational curves (Reed-Solomon codes) in projective planes $PG(2,q)$ which gives us some hope that a suitable generalization may produce large caps.

What do the points $P_u = (1 : u : u^2)$ for $u \in \mathbb{F}_q$ and $P_\infty = (0 : 0 : 1)$ of the normal rational curve have in common? Write an arbitrary point of $PG(2,q)$ with homogeneous coordinates as $P = (x : y : z)$. Then P belongs to the normal rational curve if and only if $xz = y^2$. In other words, there is a homogeneous quadratic polynomial $f(X,Y,Z) = XZ - Y^2$ in three variables such that the points P of the normal rational curve are precisely those points satisfying $f(P) = 0$.

The sets of zeroes of homogeneous quadratic polynomials are also known as **quadrics.** We have seen that the normal rational curve in $PG(2, q)$ is a quadric. Can we find other quadrics which are caps?

First of all, why did we choose a homogeneous polynomial? The reason is that we want the notion $f(P) = 0$ to be independent of the choice of the representative of point P.

How general is this approach? The set of zeroes of homogeneous polynomials $f(X_1, X_2, \ldots, X_k)$ of some degree in k variables, with coefficients in $\mathbb{F}_q$, is a point set in $PG(k - 1, q)$, an **algebraic variety.** Varieties are studied in **algebraic geometry.** We do not want to proceed in such generality, but it is interesting to note that, because of Theorem 17.1, the study of algebraic varieties is expected to have an impact on coding theory.

The case of quadrics is special. They are closely related to quadratic forms and bilinear forms. The algebraic discipline where these objects are studied is known as **geometric algebra.** We want to follow the basic theory until a point is reached where we can construct interesting codes and caps.

Symmetric bilinear forms

Symmetric bilinear forms are conceptually the easiest to understand. In fact, the idea is not new to us. Our concept of duality relied on the dot product in $\mathbb{F}_q^k$. This is a somewhat arbitrary choice. All we need are the magical properties of the dot product given in the beginning of Section 2.4. In general, these properties define symmetric bilinear forms. We have reached a point where it is no longer immaterial which bilinear form we work with.

Let $V = V(k, q)$ be a k-dimensional vector space over $\mathbb{F}_q$. We prefer this neutral notation now, as $\mathbb{F}_q^k$ suggests that a certain basis has been chosen and kept fixed. Consider mappings

$$(,) : V \times V \longrightarrow \mathbb{F}_q.$$

We speak of a **bilinear form** (or scalar product) if it satisfies the following conditions:

- Biadditivity: $(x_1 + x_2, y) = (x_1, y) + (x_2, y)$ and
  $(x, y_1 + y_2) = (x, y_1) + (x, y_2)$
  (everything in V, in particular $(0, y) = (x, 0) = 0$),

- $(\lambda x, y) = (x, \lambda y) = \lambda(x, y)$ for all $x, y \in V$, $\lambda \in \mathbb{F}_q$.

A bilinear form is **nondegenerate** if the only element x_0 satisfying $(x_0, y) = 0$ for all $y \in V$ is $x_0 = 0$ and analogously the only $y_0 \in V$ satisfying $(x, y_0) = 0$

for all $x \in V$ is $y_0 = 0$. A vector $v \in V$ is **isotropic** if $(v, v) = 0$. A subspace $U \subset V$ is **totally isotropic** if $(u, u') = 0$ for all $u, u' \in U$.

We think of the relation $(v, w) = 0$ as v being orthogonal to w and write $v \perp w$ in this case. In particular, $(,)$ is nondegenerate if only the 0 vector is orthogonal to the whole space. In contrast to the situation in Euclidean spaces (real vector spaces with the dot product), it can happen that a nonzero vector is orthogonal to itself. These are the isotropic vectors.

Given the inner product $(,)$ and a basis $\{v_1, v_2, \ldots, v_k\}$ of V, we can form the **Gram matrix** $A = ((v_i, v_j))_{i,j}$. Clearly the bilinear form is nondegenerate if and only if the Gram matrix is invertible.

Bilinear forms can be represented in matrix notation. Let $V = \mathbb{F}_q^k$ and $e_1, \ldots, e_k$ be the standard basis. Choose constants a_{ij} such that $(e_i, e_j) = a_{ij}$. Let $x = (x_1, x_2, \ldots, x_k)$ and $y = (y_1, y_2, \ldots, y_k)$. Then

$$(x, y) = \sum_{i,j} x_i y_j a_{ij}.$$

This means that a k-dimensional bilinear form is described by choosing k^2 coefficients a_{ij}. The bilinear form can then also be written in matrix notation as follows: define the matrix $A = (a_{ij})$ whose entries are the chosen coefficients. Then

$$(x, y) = \sum_{i,j=0}^{k} a_{ij} x_i y_j = x A y^t.$$

Then A is the Gram matrix with respect to the standard basis.

As an example, consider the following three-dimensional bilinear form over $\mathbb{F}_q$ for odd q :

$$(x, y) = \frac{1}{2} x_1 y_3 + \frac{1}{2} x_3 y_1 - x_2 y_2.$$

Its Gram matrix is $A = \begin{pmatrix} 0 & 0 & 1/2 \\ 0 & -1 & 0 \\ 1/2 & 0 & 0 \end{pmatrix}$ of determinant $1/4$. In particular, it is nondegenerate.

As indicated earlier, we are particularly interested in the scalar products of vectors with themselves. These are the values (x, x), in our example $(x, x) = x_1 x_3 - x_2^2$. This leads to a homogeneous quadratic polynomial, the polynomial of Section 17.1. The isotropic points in $PG(2, q)$ with respect to this bilinear form are precisely the points of the normal rational curve (here only in odd characteristic). The Gram matrix A of the example has another special property: it is symmetric. This has the following effect on the bilinear form: $(x, y) = (y, x)$ for all x, y. Such bilinear forms are called (guess what?) symmetric.

17.29 Definition. *A bilinear form on V is **symmetric** if $(x, y) = (y, x)$ for all $x, y \in V$. This is equivalent to the Gram matrix being a symmetric matrix.*

17.30 Definition. *Let* $(,)$ *be a symmetric bilinear form on* V. *The corresponding* **quadratic form** *is* $Q(x) = (x, x)$. *Here* $Q : V \to \mathbb{F}_q$ *has the property* $Q(\lambda x) = \lambda^2 Q(x)$ *for all* $x \in V$.

It may seem that the quadratic form Q carries less information than the symmetric bilinear form it was derived from. However, this is not so. We can recover the symmetric bilinear form from the quadratic form. Let us see why:

$$Q(x + y) = (x + y, x + y) = Q(x) + Q(y) + 2(x, y).$$

Here we used that the bilinear form is symmetric. This shows that

$$(x, y) = \frac{1}{2}(Q(x + y) - Q(x) - Q(y)) \text{ for all } x, y \in V.$$

However, this works only in odd characteristic (we divided by 2). Let us collect all this, starting from the definition of a quadratic form in odd characteristic.

17.31 Definition. *Let* q *be odd and* $V = V(k, q)$. *A* **quadratic form** *on* V *is a mapping* $Q : V \to \mathbb{F}_q$ *satisfying* $Q(\lambda x) = \lambda^2 Q(x)$ *for all* $\lambda \in \mathbb{F}_q, x \in V$ *and such that*

$$(x, y) = \frac{1}{2}(Q(x + y) - Q(x) - Q(y))$$

is a bilinear form (by force symmetric). Call Q *nondegenerate if the corresponding bilinear form is.*

Let us check this for our example. Start from $Q(x) = x_1 x_3 - x_2^2$:

$$2(x, y) = (x_1 + y_1)(x_3 + y_3) - (x_2 + y_2)^2 - x_1 x_3 + x_2^2 - y_1 y_3 + y_2^2$$

$$= x_1 y_3 + x_3 y_1 - 2x_2 y_2 \text{ (as predicted)}.$$

The quadratic form corresponding to a symmetric bilinear form is always described by a homogeneous quadratic polynomial. We have seen that, in odd characteristic, quadratic forms are equivalent to symmetric bilinear forms.

Concentrate on symmetric bilinear forms in odd characteristic. So $V = V(k, q)$ for odd q and $(,)$ a nondegenerate symmetric bilinear form defined on V. The structure we have in mind is the quadric consisting of the isotropic points, that is, the points generated by vectors $x \neq 0$ such that $(x, x) = 0$.

17.32 Definition. *Let* $(,)$ *be a symmetric bilinear form on* V. *For every subset* $W \subseteq V$, *define* $W^\perp = \{v | v \in V, (v, W) = 0\}$. *Here* $(v, W) = 0$ *is short for* $(v, w) = 0$ *for all* $w \in W$.

17.33 Proposition. *Let* $(,)$ *be a symmetric bilinear form on* V. *For every subset* $W \subseteq V$, *we have that* $W^\perp$ *is a subspace. If* $(,)$ *is nondegenerate and* W *a subspace, then*

$$dim(W) + dim(W^\perp) = dim(V).$$

PROOF   The first statement is obvious. Now consider the nondegenerate case. By definition we have $v^\perp \neq V$ for all $v \neq 0$. Let $dim(V) = k$. We claim $dim(v^\perp) = k - 1$. It suffices to show that $v^\perp$ intersects every two-dimensional subspace of V in dimension ≥ 1. Let v_1, v_2 be a basis of a two-dimensional subspace, and $(v, v_1) = \lambda, (v, v_2) = \mu$. Then $(v, \mu v_1 - \lambda v_2) = 0$, hence $\mu v_1 - \lambda v_2 \in v^\perp$.

We have shown that the orthogonal of a one-dimensional subspace has dimension $k - 1$. Now let $v_1, v_2, \ldots, v_k$ be a basis of V. We have

$$\langle v_1 \rangle^\perp \supseteq \langle v_1, v_2 \rangle^\perp \supseteq \ldots \langle v_1, v_2, \ldots, v_k \rangle^\perp = V^\perp = 0.$$

As at each step the dimension decreases by at most one (intersection with a hyperplane, a $(k-1)$-dimensional subspace), and in the last step we reach the 0 space, the dimension must decrease by precisely one at each step. In particular, $dim(\langle v_1, \ldots, v_m \rangle^\perp) = k - m$ for each m. As this is true for an arbitrary basis, the statement is proved.   □

Lemma 2.9 is the special case of Proposition 17.33 when the symmetric bilinear form is the ordinary dot product on vector spaces over $\mathbb{F}_2$.

The space $V^\perp$ of vectors in V which are orthogonal to the whole space is also called the **radical** of V. The symmetric bilinear form is nondegenerate if and only if its radical is $\{0\}$.

The one-dimensional case

It may sound silly, but we start with case $k = 1$. Let $x \neq 0$. Then $(x, x) \neq 0$ because of nondegeneracy. Recall $(\lambda x, \lambda x) = \lambda^2 (x, x)$. This means that, by changing the "basis", we can introduce an arbitrary quadratic factor in the "Gram matrix". This shows that there are two nonequivalent bilinear forms in the case of dimension 1. Either we can find a "basis" v such that $(v, v) = 1$ or we can find a basis such that $(v, v) = \nu$ is our favorite nonsquare in $\mathbb{F}_q$.

The two-dimensional case

Let $V = V(2, q)$ with a nondegenerate form $(,)$. Assume there is a vector $v \neq 0$ such that $(v, v) = 0$. Let w be linearly independent of v. We have $(v, w) \neq 0$. Multiplying w by a suitable scalar, we can assume $(v, w) = 1$. Upon replacing w by $w + sv$ for suitable s, we can assume $(w, w) = 0$. This shows that we can find a basis such that the Gram matrix is $\begin{pmatrix} 0 & 1 \\ 1 & 0 \end{pmatrix}$ in this case. These two-dimensional spaces are known as **hyperbolic planes.**

Can it happen that there is no nonzero isotropic vector? Such spaces are called **anisotropic.** Examples of anisotropic spaces are Euclidean (real) spaces. Assume at first (v, v) is a square for all $v \neq 0$. Choose $w \neq 0$ such that $(v, w) = 0$. As $(sv + tw, sv + tw) = s^2(v, v) + t^2(w, w)$ must be a square unless $s = t = 0$, we see in particular that the sum of two squares must be a square. This leads to the contradiction that the squares together with 0 form a subfield of $\mathbb{F}_q$. Now assume (v, v) is a nonsquare for all $v \neq 0$. Replace the bilinear form $(,)$ by $\nu(,)$, where ν is a fixed nonsquare. This new form is still symmetric bilinear anisotropic, and each value $\nu(v, v)$ for $v \neq 0$ is a square, by assumption. This case has just been excluded.

We can choose $(v, v) = 1$ and $(v, w) = 0$. As $(sv + tw, sv + tw) = s^2 + t^2(w, w) \neq 0$ unless $s = t = 0$, it must be that $-(w, w)$ is a nonsquare, without restriction $(w, w) = -\nu$, where ν is our favorite nonsquare. The Gram matrix is $\begin{pmatrix} 1 & 0 \\ 0 & -\nu \end{pmatrix}$. As $Q(sv + tw) = s^2 - t^2\nu$ we see that indeed this describes an anisotropic space. Observe also that every field element is represented. In fact, we saw that it is the uniquely determined anisotropic two-dimensional space. Multiplying the Gram matrix by a constant c yields another anisotropic space, where c is represented as often as 1 is represented in the original space. As these spaces are equivalent, every nonzero element is represented the same number of times.

It is also interesting to consider the degenerate two-dimensional case. If the radical is two-dimensional, the bilinear form is identically 0 and all points of the projective line are isotropic.

The remaining case is when $V^\perp = \langle v \rangle$ is one-dimensional. Then V is the orthogonal sum of the radical and a one-dimensional nondegenerate space. In particular the line contains precisely one isotropic point in this case.

17.34 Theorem. *Let $(,)$ be a symmetric bilinear form on $V = V(2, q)$, for odd q. The number of isotropic points on the projective line $PG(1, q)$ corresponding to V is $0, 1, 2$ or $q + 1$. It is $q + 1$ if the form is identically zero; it is 1 if the radical has dimension 1.*

There are up to choice of basis only two nondegenerate bilinear forms, the hyperbolic plane with Gram matrix $\begin{pmatrix} 0 & 1 \\ 1 & 0 \end{pmatrix}$ and the anisotropic space with Gram matrix $\begin{pmatrix} 1 & 0 \\ 0 & -\nu \end{pmatrix}$.

If the Gram matrix is not given in one of the standard forms of Theorem 17.34, how can we decide on the fly what the type is? For that purpose it is good to know what happens to the Gram matrix under change of basis.

17.35 Theorem. *Let $v_1, \ldots, v_k$ be a basis of V and $A = (a_{ij})$ the Gram matrix of a bilinear form ($a_{ij} = (v_i, v_j)$). Let $w_1, \ldots, w_k$ be another basis of*

V with Gram matrix $B = (b_{ij})$. Let $w_i = \sum_{l=1}^{k} t_{il}v_l$ and $T = (t_{ij})$ be the matrix describing this change of basis. Then $B = TAT^t$.

PROOF   Simply compute

$$b_{ij} = (w_i, w_j) = \left(\sum_l t_{il}v_l, \sum_r t_{jr}v_r\right) = \sum_{l,r} t_{il}a_{lr}t_{jr}.$$

This is the (i, j) entry of TAT^t.   □

The most important point about the base change is what remains invariant. Consider the determinants: $det(B) = det(A)det(T)^2$. It changes but squares are mapped to squares and nonsquares are mapped to nonsquares.

17.36 Definition. *Let A be a (k, k) matrix with entries from $\mathbb{F}_q$, where q is odd. The* **discriminant** *of A is*

$$disc(A) = \begin{cases} +1 & if \quad det(A) \text{ is a square} \\ -1 & if \quad det(A) \text{ is nonsquare} \\ 0 & if \qquad det(A) = 0. \end{cases}$$

The discriminant of a bilinear form is the discriminant of its Gram matrix.

Theorem 17.35 shows that the discriminant of the Gram matrix is invariant under change of basis. It is therefore an invariant of the bilinear form. This shows how to distinguish hyperbolic planes from anisotropic two-dimensional spaces. As the determinants of the Gram matrices in standard form are -1 and $-\nu$, respectively, they have different discriminants.

Our analysis of the anisotropic two-dimensional case suggests that anisotropic spaces of higher dimensions cannot exist.

17.37 Proposition. *There is no anisotropic symmetric bilinear form in odd characteristic on a vector space of dimension $k > 2$.*

PROOF   We can find v_1 such that $(v_1, v_1) = 1$. The space V is the orthogonal sum of $\langle v_1 \rangle$ and $\langle v_1 \rangle^{\perp}$, and the latter space still has dimension ≥ 2. We can find v_2 such that $(v_1, v_2) = 0$ and $(v_2, v_2) = -1$. However $Q(v_1 - v_2) = 0$, a contradiction.   □

We can now complete the classification in arbitrary dimension. Let $k \geq 3$. By Proposition 17.37 there is an isotropic vector $v_1 \neq 0$. Because of non-degeneracy, there is $w_1' \neq 0$ such that $(v_1, w_1') \neq 0$. After multiplying w_1' by a suitable constant, we have $(v_1, w_1') = 1$. Let $w_1 = w_1' + sv_1$. We have $(v_1, w_1) = 1$ independent of the value of s. As $Q(w_1) = Q(w_1') + 2s$, we can choose s such that $Q(w_1) = 0$. It follows that $H_1 = \langle v_1, w_1 \rangle$ is a hyperbolic plane. As H_1 is itself nondegenerate, we obtain $V = H_1 \perp H_1^{\perp}$, and the $(k-2)$-dimensional space $H_1^\perp$ is nondegenerate. Apply induction. This procedure can be repeated until we are left with a space of dimension 1 (if k is odd) or 2 (if k is even). As we have already done the classification in dimensions 1 and 2, the picture is complete.

17.38 Theorem. *Let $V = V(k, q)$ be a k-dimensional vector space for odd q and $\nu \in \mathbb{F}_q$ a nonsquare. There are up to choice of basis precisely two nondegenerate symmetric bilinear forms $(,)$ on V.*
These have different discriminants.
   Let $k = 2m$ be even. Then either

- $V = H_1 \perp \cdots \perp H_m$ *is the orthogonal sum of m hyperbolic planes (a Gram matrix has determinant $(-1)^m$), or*

- $V = H_1 \perp \cdots \perp H_{m-1} \perp A$ *is the orthogonal sum of $m-1$ hyperbolic planes and a two-dimensional anisotropic space (a Gram matrix has determinant $(-1)^m \nu$).*

   Let $k = 2m+1$ be odd. Then $V = H_1 \perp \cdots \perp H_m \perp \langle v \rangle$, where either $(v, v) = 1$ or $(v, v) = \nu$.

The isotropic points of the nondegenerate symmetric bilinear forms are interesting sets of points in the corresponding projective spaces. They are quadrics, sets of roots of degree 2 homogeneous polynomials. When $k = 2m+1$, we have two different bilinear (quadratic) forms, but the sets of isotropic points are identical. This is clear, as multiplication of the bilinear form by a nonsquare transforms one type into the other.

17.39 Definition. *Let q be odd. A **parabolic quadric** $Q(2m, q) \subset PG(2m, q)$ is the set of isotropic points of a nondegenerate symmetric bilinear form in $V(2m+1, q)$.*
   *$Q^+(2m-1, q) \subset PG(2m-1, q)$ (a **hyperbolic quadric**) is the set of isotropic points of a nondegenerate symmetric bilinear (quadratic) form in $V(2m, q)$ such that $V(2m, q)$ is the orthogonal sum of m hyperbolic planes.*
   *When $V(2m, q)$ is the orthogonal sum of an anisotropic space and $m-1$ hyperbolic planes, the **elliptic quadric** $Q^-(2m-1, q) \subset PG(2m-1, q)$ is obtained.*

It is not hard to count the points on these quadrics. Recall the normal rational curve. It is a quadric in $PG(2, q)$ and has $q+1$ points. This shows $|Q(2, q)| = q + 1$. When dealing with the normal rational curve, there was no need to distinguish between the characteristic 2 and the odd characteristic case. This indicates that the mechanism of the present section should generalize to cover the characteristic 2 case as well.

17.40 Definition. *Let q be odd. Denote by $h_m(c)$ the number of vectors $v \in V(2m, q)$ such that $(v, v) = c$ for the hyperbolic quadratic form. Analogously, $e_m(c)$ denotes the representation number of c for the elliptic quadratic form.*

Let $p_m(c)$ be the number of vectors $v \in V(2m + 1, q)$ such that $(v, v) = c$ when $V(2m + 1, q)$ is the orthogonal sum of m hyperbolic planes and $\langle v_0 \rangle$, where $Q(v_0) = 1$. The corresponding representation numbers for the case when $Q(v_0) = \nu$ is a nonsquare are denoted by $p'_m(c)$.

Let us compute the representation numbers in Definition 17.40. Observe at first that a scalar multiple of a symmetric bilinear form is a symmetric bilinear form again. If the original form is elliptic, then all scalar multiples are elliptic. The same holds for hyperbolic forms. This shows that $h_m(c) = h_m(1)$ and $e_m(c) = e_m(1)$ for all $c \neq 0$. Also $h_1(0) = 2q - 1$ (there are two isotropic points in a hyperbolic plane) and $h_1(1) = q - 1$, analogously $e_1(0) = 1$ and $e_1(1) = q + 1$. We have the following obvious recurrence relation for the hyperbolic case:

$$h_m(0) = h_{m-1}(0)h_1(0) + (q^{2(m-1)} - h_{m-1}(0))h_1(1)$$

$$= qh_{m-1}(0) + (q - 1)q^{2(m-1)}.$$

We obtain $h_2(0) = q(q^2 + q - 1)$, $h_3(0) = q^2(q^3 + q - 1)$, by induction $h_m(0) = q^{m-1}(q^m + q - 1)$. The corresponding number of isotropic points is $(h_m(0) - 1)/(q - 1)$. Fortunately $h_m(0) - 1 = (q^m - 1)(q^{m-1} + 1)$. For $m = 1$, we obtain two isotropic points, as it should be. The next case is $h_2(0) = q(q^2 + q - 1)$, leading to $(q + 1)^2$ isotropic points.

17.41 Theorem. *Let q be odd. We have $h_m(0) = q^{m-1}(q^m + q - 1)$, $h_m(1) = q^{m-1}(q^m - 1)$ and*

$$|Q^+(2m - 1, q)| = \frac{(q^m - 1)(q^{m-1} + 1)}{q - 1}.$$

For the elliptic type, we can use this result:

$$e_m(0) = h_{m-1}(0) \cdot 1 + (q^{2(m-1)} - h_{m-1}(0))(q + 1) = q^{2m-1} - q^m + q^{m-1}.$$

As $e_m(0) - 1 = (q^m + 1)(q^{m-1} - 1)$, we obtain the following result:

17.42 Theorem. *Let q be odd. We have $e_m(0) = q^{m-1}(q^m - q + 1)$, $e_m(1) = q^{m-1}(q^m + 1)$ and*

$$|Q^-(2m - 1, q)| = \frac{(q^m + 1)(q^{m-1} - 1)}{q - 1}.$$

By the same argument, $p_m(0) = h_m(0) \cdot 1 + \dfrac{q^{2m} - h_m(0)}{2} \cdot 2 = q^{2m}$ and $p_m(1) = h_m(0) \cdot 2 + h_m(1)(q - 2) = q^m(q^m + 1)$.

17.43 Theorem. *Let q be odd. We have $p_m(0) = p'_m(0) = q^{2m}$, $p_m(1) = p'_m(\nu) = q^m(q^m + 1)$ and $p_m(\nu) = p'_m(1) = q^m(q^m - 1)$. Further*

$$|Q(2m, q)| = \frac{q^{2m} - 1}{q - 1}.$$

It is a natural question to determine the maximum dimension of totally isotropic subspaces.

17.44 Definition. *The* **Witt index** *of a symmetric bilinear form in odd characteristic is the largest vector space dimension of a totally isotropic subspace.*

17.45 Proposition. *Let q odd, $(,)$ be a nondegenerate symmetric bilinear form. If W is an i-dimensional totally isotropic subspace, we can find i hyperbolic planes such that*

$$V = H_1 \perp H_2 \perp \cdots \perp H_i \perp R,$$

where R is nondegenerate. The Witt index is the largest such i.

PROOF Let $0 \neq v_1 \in W$. Find a hyperbolic plane $H_1 = \langle v_1, w_1 \rangle$. We have $V = H_1 \perp H_1^\perp$ and $H_1^\perp \cap W = w_1^\perp \cap W$ of dimension at least $m - 1$. The first claim follows by induction. The second claim is immediate, as we see i-dimensional totally isotropic subspaces if we have a subspace of the form $H_1 \perp H_2 \perp \cdots \perp H_i$. □

Proposition 17.45 shows that the hyperbolic and the elliptic quadric in $PG(2m - 1, q)$ have different Witt indices. The hyperbolic quadric $Q^+(2m - 1, q)$ has Witt index m (the space is the orthogonal sum of m hyperbolic planes), whereas the elliptic quadric $Q^-(2m - 1, q)$ has Witt index $m - 1$ (we can split off $m - 1$ hyperbolic planes).

1. A **bilinear form** on $V = V(k, q)$ is a mapping $: V \times V \to \mathbb{F}_q$, which is $\mathbb{F}_q$ linear in both arguments.

2. Choose a basis $e_1, \ldots, e_n$ of V. The **Gram matrix** A with entries $a_{ij} = (e_i, e_j)$ determines the bilinear form.

3. In matrix notation, $(x, y) = xAy^t$.
   Here $x = (x_1, \ldots, x_n)$, analogously for y.

4. The bilinear form is **symmetric** $((x, y) = (y, x)$ for all $x, y)$ if and only if its Gram matrix is symmetric.

5. If $(x, y) = (y, x) = 0$, we call x, y **orthogonal.** x is **isotropic** if $(x, x) = 0$ (x is orthogonal to itself).

6. The **radical** of the bilinear form consists of the vectors, which are orthogonal to all of V. The bilinear form is **nondegenerate** if the radical is $\{0\}$.

7. The bilinear form is nondegenerate if and only if the Gram matrix is regular $(det(A) \neq 0)$.

8. A symmetric bilinear form $(,)$ determines the **quadratic form** $Q(x) = (x, x)$.

9. A quadratic form is described (with respect to a fixed basis) by a homogeneous polynomial of degree 2 in n variables: $Q(x_1, \ldots, x_n) = \sum_{1 \leq i \leq j \leq n} q_{ij} x_i x_j$.

10. Let q be odd, $dim(V) = k$ over $\mathbb{F}_q$ and $(,)$ a symmetric bilinear form with Gram matrix A.

11. The quadratic form determines the equivalent symmetric bilinear form via $(x, y) = \frac{1}{2}(Q(x + y) - Q(x) - Q(y))$ (this is also called the **polarization** of Q).

12. The polarization in coordinates: each q_{ii} yields entry $a_{ii} = q_{ii}$, each $q_{ij}, i < j$, yields $a_{ij} = a_{ji} = q_{ij}/2$ in A.

13. If $(,)$ is nondegenerate and $W \subset V$ is a subspace of dimension m, then the orthogonal $W^\perp$ has complementary dimension $k - m$.

14. A two-dimensional nondegenerate space either is a hyperbolic plane (2 isotropic points) or is anisotropic (no isotropic point). In each case the space is uniquely determined.

15. The **discriminant** is 1 if $det(A)$ is a square, it is -1 otherwise. It is an invariant of the space.

16. The **Witt index** d is the largest dimension of a totally isotropic subspace (meaning that Q vanishes on it).

17. Let $k = 2m$. Either V (nondegenerate) is the orthogonal sum of m hyperbolic planes (the **hyperbolic case** or $+$ case, Witt index m) or it is the orthogonal sum of $m - 1$ hyperbolic planes and an anisotropic two-dimensional space (the **elliptic** case or $(-)$ case, Witt index $m - 1$). The discriminants are different.

18. Let $k = 2m + 1$. Then V is the orthogonal sum of m hyperbolic planes and a one-dimensional space (the **parabolic** case, Witt index m).

19. The corresponding sets of isotropic points (quadrics) in $PG(k - 1, q)$ are denoted $Q^+(2m - 1, q)$ (hyperbolic), $Q^-(2m - 1, q)$ (elliptic), $Q(2m, q)$ (parabolic).

20. We determine the representation numbers of the quadratic forms (given $c \in \mathbb{F}_q$, how many vectors have $Q(x) = c$?) and the number of points of the quadrics.

Symplectic bilinear forms

Before covering the characteristic 2 case of quadratic forms, it is natural to consider symplectic bilinear forms first. A bilinear form is **symplectic** if all vectors are isotropic:

$$(x, x) = 0 \text{ for all } x \in V.$$

As $0 = (x + y, x + y) = (x, x) + (y, y) + (x, y) + (y, x) = (x, y) + (y, x)$, a symplectic form is also skew symmetric:

$$(y, x) = -(x, y) \text{ for all } x, y.$$

This shows that the **radical** of V is

$$V^{\perp} = \{x | (x, V) = 0\} = \{x | (V, x) = 0\}.$$

A symplectic form is nondegenerate if $V^{\perp} = \{0\}$. Symplectic forms are bilinear forms which are described by skew-symmetric Gram matrices A. The form is nondegenerate if and only if $det(A) \neq 0$. Also, the same proof as in the symmetric case shows that the dual

$$W^{\perp} = \{x | (x, W) = 0\} = \{x | (W, x) = 0\}$$

of a subspace W has complementary dimension, provided $(,)$ is nondegenerate.

Clearly there is no nondegenerate symplectic form in dimension $k = 1$. Let $k \geq 2$ and $v_1 \neq 0$. Find w_1 such that $(v_1, w_1) = 1$. The two-dimensional subspace $\langle v_1, w_1 \rangle$ is nondegenerate. It follows $V = \langle v_1, w_1 \rangle \perp \langle v_1, w_1 \rangle^{\perp}$. By induction we see that $k = 2m$ must be even and V is the orthogonal sum of $\langle v_i, w_i \rangle, i = 1, 2, \ldots, m$ such that the Gram matrix with respect to this basis has $(2, 2)$ submatrices $\begin{pmatrix} 0 & 1 \\ -1 & 0 \end{pmatrix}$ along the main diagonal. All other entries are 0. Such a basis $\{v_1, v_2, \ldots, v_m\} \cup \{w_1, w_2, \ldots, w_m\}$ is a **symplectic basis**.

---

1. A bilinear form is **symplectic** if all vectors are isotropic $((x, x) = 0$ for all $x)$.

2. It is then **skew symmetric**: $(x, y) = -(y, x)$ for all x, y.

3. A nondegenerate symplectic form on $V = V(k, q)$ exists only if $k = 2m$ is even.

4. V is then the orthogonal sum of m two-dimensional spaces, each with Gram matrix $\begin{pmatrix} 0 & 1 \\ -1 & 0 \end{pmatrix}$.

5. A corresponding basis is a **symplectic basis**.

Quadratic forms in characteristic 2

A quadratic form on $V = V(k, q)$ is a homogeneous polynomial of degree 2 in k unknowns:

$$Q(x_1, \ldots, x_k) = \sum_{i=1}^{k} a_{ii} x_i^2 + \sum_{i<j} a_{ij} x_i x_j.$$

It can be described by a symmetric matrix $A = (a_{ij})$. As we know, quadratic forms are equivalent to symmetric bilinear forms when the underlying field has odd characteristic. In characteristic 2 this is not true. In fact, it is to be expected that quadratic polynomials should behave in a special way in characteristic 2. The reason is that squaring is a field automorphism.

Let $A = \begin{pmatrix} a & b \\ b & d \end{pmatrix}$ be a symmetric $(2, 2)$ matrix in characteristic 2 and $(,)$ the corresponding bilinear form. Let $x = (x_1, x_2)$. Then $Q(x) = ax_1^2 + dx_2^2$. The nondiagonal entry b does not show up at all. This shows that (x, x) does not give us the general case of a quadratic form. We should start from quadratic forms.

In dimension 2 we take matrix A to describe the quadratic form $Q(x) = ax_1^2 + dx_2^2 + bx_1x_2$. We have $Q(x + y) + Q(x) + Q(y) = bx_1y_2 + bx_2y_1$, a symmetric bilinear form, which is also symplectic. This suggests a formal definition of quadratic forms in characteristic 2.

17.46 Definition. *Let $V = V(k, q)$, where q is a power of 2. A* **quadratic form** *on V is a mapping $Q : V \longrightarrow \mathbb{F}_q$ such that*

- $Q(\lambda x) = \lambda^2 Q(x)$ *for all $\lambda \in \mathbb{F}_q, x \in V$, and*

- $(x, y) = Q(x + y) + Q(x) + Q(y)$ *is a bilinear form.*

Observe that the bilinear form is by force symplectic. The quadratic form carries more information. The underlying symplectic form is uniquely determined by the quadratic form, but not the other way around. In fact, if the quadratic form is described by the symmetric matrix A, then the Gram matrix of the corresponding symplectic bilinear form is obtained by putting zeroes in the main diagonal of A.

17.47 Definition. *A vector v is* **singular** *with respect to the quadratic form Q in characteristic 2, if $Q(v) = 0$. A subspace is* **totally singular** *if Q vanishes on it. The dimension of the radical V_0 of the underlying symplectic bilinear form is the* **index** *$i = i(Q)$ of Q. The quadratic form Q is* **nondegenerate** *if Q is asingular on the radical of the bilinear form.*

The one-dimensional case

In case $k = 1$ we have without restriction $Q(v_1) = 1$. Recall that, in contrast to the odd characteristic case, every field element is a square.

The two-dimensional case

We know that $(,)$ is either identically 0 or nondegenerate. Assume it is identically 0. Then $Q(x+y) = Q(x) + Q(y)$ and $Q(\lambda x) = \lambda^2 Q(x)$ (Q is **semi-linear**). It follows that Q must be degenerate. Either Q is identically 0 on V or it has a radical of dimension 1 (and the projective line corresponding to $V = V(2, q)$ has one singular point).

So let $(,)$ be nondegenerate. Assume there is v such that $Q(v) = 0$. Choose w' such that $(v, w') = 1$. Consider $w = w' + tv$. We have $(v, w) = (v, w') = 1$ and (see Definition 17.46) $Q(w) = Q(w') + 0 + (w', tv) = Q(w') + t$. Choosing $t = Q(w')$ yields $Q(w) = 0$. This shows that we can find a symplectic basis v, w such that $Q(v) = Q(w) = 0$. Call $\langle v, w \rangle$ a **quadratic hyperbolic plane.**

Now assume there is no singular nonzero vector. Choose v such that $Q(v) = 1$ and w such that $(v, w) = 1$. Let $Q(w) = a$. An arbitrary vector $sv + tw$ has $Q(sv + tw) = s^2 + t^2 a + st$. This expression has to be $\neq 0$ whenever $(s, t) \neq 0$. This is satisfied when $t = 0$. Let $t \neq 0$. Division by t shows that we can assume without restriction $t = 1$. We must have $s^2 + s \neq a$ for all $s \neq 0$. By our introductory considerations concerning quadratic equations (Hilbert's theorem 90), this is equivalent to $tr(a) = 1$. Replacing w by $w' = w + sv$, we obtain $Q(w') = a + s^2 + s$. When s varies, $Q(w')$ varies over all elements of trace 1. This shows that Q is uniquely determined.

To sum up, there are precisely two nondegenerate quadratic forms in dimension 2, the quadratic hyperbolic space and the **asingular** space. The general procedure is rather similar to the odd characteristic case, with the role played by the distinction between squares and nonsquares in the odd case replaced by the distinction between field elements of traces 0 or 1.

Next we show that asingular spaces have dimension ≤ 2 and use induction to describe all nondegenerate quadratic forms, just as in odd characteristic.

17.48 Proposition. *There is no asingular quadratic form on a vector space of dimension $k > 2$ in characteristic 2.*

PROOF   Choose $0 \neq v \in V$ and $w \notin \langle v \rangle$ such that $(v, w) = 0$. Then $Q(v + tw) = Q(v) + t^2 Q(w)$. Either $Q(w) = 0$ or we can choose t such that $Q(v + tw) = 0$.   ☐

The Witt index is defined and determined just as in the odd characteristic case.

17.49 Definition. *The* **Witt index** *of a quadratic form is the largest dimension of a totally singular subspace.*

17.50 Proposition. *Let Q be a nondegenerate quadratic form in characteristic 2. If there is an i-dimensional totally singular subspace, we can find i quadratic hyperbolic planes such that*

$$V = H_1 \perp H_2 \perp \cdots \perp H_i \perp R,$$

where R is nondegenerate.

The proof is just as the proof of Proposition 17.45. We are in the same position as in the odd characteristic case.

When $k \geq 3$, we can split off a quadratic hyperbolic plane. Let $k = 2m$. After splitting off $m - 1$ quadratic hyperbolic planes, the remaining two-dimensional space is either another hyperbolic plane (this is the hyperbolic case) or it is asingular (we have an elliptic quadratic form in this case). If $k = 2m + 1$, we can split off m quadratic hyperbolic planes and are left with a one-dimensional space.

17.51 Definition. *Let q be even. Denote $Q(2m, q) \subset PG(2m, q)$ the set of singular points of a nondegenerate quadratic form in $V(2m + 1, q)$. This is a* **parabolic quadric**.

$Q^+(2m - 1, q) \subset PG(2m - 1, q)$ (a **hyperbolic quadric***) is the set of singular points of a quadratic form in $V(2m, q)$ such that $V(2m, q)$ is the orthogonal sum of m quadratic hyperbolic planes.*

In the **elliptic** *case, the quadric $Q^-(2m - 1, q) \subset PG(2m - 1, q)$ is the set of singular points of a nondegenerate quadratic form in $V(2m, q)$ such that $V(2m, q)$ is the orthogonal sum of $m - 1$ hyperbolic planes and an asingular space.*

The representation numbers $h_m(c)$, $e_m(c)$, $p_m(c)$ are defined as the number of vectors x such that $Q(x) = c$ in the respective cases.

Observe that in characteristic 2 there is only one one-dimensional quadratic form and therefore only one nondegenerate quadratic form in odd dimension. It follows from Proposition 17.50 that the two nondegenerate quadratic forms in even dimension have different Witt indices.

With respect to the standard basis for the quadratic hyperbolic plane, we have $Q(sv + tw) = st$ and it follows

$$h_1(0) = 2q - 1, \; h_1(c) = q - 1 \text{ for } c \neq 0.$$

By the usual argument, $e_1(c) = q + 1$ for all $c \neq 0$. This shows that the representation numbers $h_m(c)$, $e_m(c)$ are the same as in the odd characteristic case. For odd dimension, things are easier yet. As the one-dimensional nondegenerate quadratic form represents each field element precisely once, we obtain $p_m(c) = q^{2m}$ for all c in characteristic 2. In particular, we conclude that the formulas for the number of points on our quadrics are the same as in odd characteristic.

1. Let $q = 2^f$ and $V = V(k, q)$.

2. In characteristic 2 quadratic forms (described by homogeneous quadratic polynomials) and symmetric bilinear forms are not equivalent. The quadratic form carries more information.

3. The bilinear form underlying the quadratic form Q is $(x, y) = Q(x + y) + Q(x) + Q(y)$. It is symplectic (and symmetric).

4. If $Q(x_1, \ldots, x_n) = \sum_{1 \le i \le j \le n} q_{ij} x_i x_j$, then each q_{ij} for $i < j$ contributes $a_{ij} = a_{ji} = q_{ij}$ in the Gram matrix of $(,)$. Terms q_{ii} do not contribute to A. If A is given, then the coefficients $q_{ij}, i < j$ of the quadratic form are determined. The diagonal terms q_{ii} are arbitrary.

5. A vector x is **singular** if $Q(x) = 0$, a subspace is **totally singular** if all its vectors are singular. A space is **asingular** if $Q(x) \ne 0$ for all $x \ne 0$.

6. Q is **nondegenerate** if it is asingular on the radical of $(,)$.

7. A two-dimensional nondegenerate space V is either a **quadratic hyperbolic plane** (two singular points) or it is asingular (no singular point).

8. Let $tr : \mathbb{F}_q \to \mathbb{F}_2$ be the trace. In the analysis the distinction between elements of trace 0 and of trace 1 replaces the distinction between squares and nonsquares in odd characteristic.

9. The results are as in odd characteristic, the words singular, asingular replacing isotropic, anisotropic.

10. The formulas for the number of singular points are the same as in odd characteristic.

Quadrics in $PG(2, q)$ and in $PG(3, q)$

Let us go back to the quadratic form $Q(x) = x_1 x_3 - x_2^2$ defining the normal rational curve in $PG(2, q)$. In odd characteristic the corresponding symmetric bilinear form is

$$(x, y) = \frac{1}{2} x_1 y_3 + \frac{1}{2} x_3 y_1 - x_2 y_2.$$

Its Gram matrix is $A = \begin{pmatrix} 0 & 0 & 1/2 \\ 0 & -1 & 0 \\ 1/2 & 0 & 0 \end{pmatrix}$. We saw this example in the opening stages of the present section. As $det(A) = 1/4$, we have that Q is non-degenerate. Its isotropic points form $Q(2, q)$, a set of $q + 1$ points (this is the case $m = 1$ of Theorem 17.43). As Q has Witt index 1, there is no totally isotropic line. It follows that no more than two points of $Q(2, q)$ are collinear. Lines l containing just one point P of $Q(2, q)$ (tangents) correspond to degenerate two-dimensional spaces. As P is in the radical of l, we must have $l = P^\perp$. This shows that every point of $Q(2, q)$ is on precisely one tangent. All this confirms results obtained earlier by calculations with coordinates and by combinatorial counting.

This example also illustrates the bijection between symmetric bilinear forms and quadratic forms in odd characteristic. Terms $c_{ii} X_i^2$ lead to entry $a_{ii} = c_{ii}$ in the Gram matrix, whereas each mixed term $c_{ij} X_i X_j$ yields entries $a_{ij} = a_{ji} = c_{ij}/2$ in the Gram matrix.

Now let q be a power of 2. The Gram matrix is $A = \begin{pmatrix} 0 & 0 & 1 \\ 0 & 0 & 0 \\ 1 & 0 & 0 \end{pmatrix}$. Recall that A describes the underlying symplectic bilinear form, not the quadratic form. We have $det(A) = 0$, so $(,)$ is degenerate. Recall that this is always so in odd dimension. Nondegenerate symplectic forms exist only in even dimension. The radical of $(,)$ is $V_0 = \langle e_2 \rangle$. We have that Q is nondegenerate as $Q(e_2) = Q(0, 1, 0) = 1$. As $(e_1, e_3) = 1$ and $Q(e_1) = Q(e_3) = 0$, the space $H_1 = \langle v_1, v_3 \rangle$ is a quadratic hyperbolic plane.

Recall that each oval in $PG(2, q)$ is embedded in a uniquely determined hyperoval when q is a power of 2. The tangents of the oval meet in a unique point, the nucleus N, which complements the oval to a hyperoval. Can we confirm that from our quadratic point of view? Well, the nucleus corresponds to the radical of the underlying symplectic form. This radical is one-dimensional, so we talk about a point in $PG(2, q)$, and Q is nondegenerate on it, so $N \notin Q(2, q)$. More importantly, all tangent lines meet in N, as the tangent lines are the duals of the points of the quadric, and vectors from N are orthogonal to everything.

We have recovered all the information that we gathered earlier concerning $Q(2, q)$. Let us go one dimension higher.

Consider $Q(x_1, x_2, x_3, x_4) = x_1^2 + x_2^2 + x_3^2 + x_4^2$. In characteristic 2 we have $Q(x) = (\sum_i x_i)^2$. As Q vanishes on a three-dimensional subspace, it clearly is degenerate. Let q be odd.

In dimension $k = 4$ and odd characteristic, the hyperbolic quadric has discriminant 1, and the elliptic quadric has discriminant -1. This means that the determinant of the Gram matrix is a square in the hyperbolic case, a non-square in the elliptic case. The Gram matrix of our form is the unit matrix. It follows that the quadric is hyperbolic. We know $|Q(3, q)| = (q + 1)^2$. It is not all that easy to confirm this by concrete calculations with coordinates.

Let us use the standard form corresponding to a decomposition in hyperbolic planes, in odd characteristic. The Gram matrix is $A = \begin{pmatrix} 0 & 1 & 0 & 0 \\ 1 & 0 & 0 & 0 \\ 0 & 0 & 0 & 1 \\ 0 & 0 & 1 & 0 \end{pmatrix}$.

The quadratic form is $Q(x) = 2x_1x_2 + 2x_3x_4$. We can use just as well $Q(x) = x_1x_2 + x_3x_4$. There should be $(q + 1)^2$ isotropic points.

There are $4q$ points such that $x_1x_2 = x_3x_4 = 0$. All remaining points must have all coordinates $\neq 0$. We can choose $x_1 = 1$. For arbitrary nonzero values of x_2, x_3, we can find precisely one value of x_4 such that $x_4x_3 = x_2$. This gives us $(q - 1)^2$ points all of whose coordinates are $\neq 0$. We count $|Q^+(3, q)| = (q - 1)^2 + 4q = (q + 1)^2$, as predicted.

An elliptic four-dimensional quadratic form in odd characteristic is described by the Gram matrix $A = \begin{pmatrix} 0 & 1 & 0 & 0 \\ 1 & 0 & 0 & 0 \\ 0 & 0 & 1 & 0 \\ 0 & 0 & 0 & -\nu \end{pmatrix}$. The quadratic form is $Q(x) = 2x_1x_2 + x_3^2 - \nu x_4^2$. We expect $q^2 + 1$ isotropic points. This is in fact easy to count directly. Assume at first $x_3 = x_4 = 0$. As either $x_1 = 0$ or $x_2 = 0$, this gives us the two points $\langle e_1 \rangle, \langle e_2 \rangle$. Now let $(x_3, x_4) \neq (0, 0)$. Then $x_3^2 - \nu x_4^2 \neq 0$. We can choose $x_1 = 1$ and obtain a unique solution x_2. This gives us $q^2 - 1$ points. We have confirmed that $|Q^-(3, q)| = q^2 + 1$.

As the Witt index is 1, there are no totally isotropic lines. It follows that each line intersects $Q^-(3, q)$ in at most two points. In other words, the elliptic quadric in $PG(3, q)$ is a cap.

Consider planes and how they intersect the elliptic quadric. Let $P \in Q^-(3, q)$ and $E = P^\perp$. Then E intersects $Q^-(3, q)$ only in P. We see that each point $P \in Q^-(3, q)$ is contained in precisely one such **tangent plane.** If E is not one of the $q^2 + 1$ tangent planes, then the restriction of the quadratic form to E is nondegenerate. It follows that $|E \cap Q^-(3, q)| = q + 1$ in this case, and the $q + 1$ points of intersection form an oval. Observe that all this is true in any characteristic.

17.52 Proposition. *The elliptic quadric $Q^-(3, q) \subset PG(3, q)$ is a cap. Each point $P \in Q^-(3, q)$ is on precisely one plane $P^\perp$ meeting $Q^-(3, q)$ only in P.*

Any plane which is not one of the $q^2 + 1$ tangent planes meets $Q^-(3, q)$ in $q + 1$ points, which form an oval.

Now consider the hyperbolic quadric in $PG(3, q)$. As the Witt index is 2, totally singular lines exist. Let us count them.

Fix $P \in Q^+(3, q)$. Then $P^\perp$ is the orthogonal sum of P and a hyperbolic plane. It follows that P is on precisely two totally singular lines (in odd characteristic we can speak equivalently of isotropy instead of singularity). By double counting we see that the total number of totally singular lines is $2(q + 1)^2/(q + 1) = 2(q + 1)$.

Can we determine the structure of this family of $2(q+1)$ lines? Fix one such line l, let $P_1, \ldots, P_{q+1}$ be the points of l and g_i, $i = 1, \ldots, q + 1$ the second totally singular line through P_i (aside from l). Assume lines g_i, g_j intersect in a point R. Then the plane E through the points P_i, P_j, R contains the lines l, g_i, g_j. It is clear that E must be totally singular, which is impossible, as the Witt index is 2. We conclude that the lines g_i are mutually disjoint (one says, skew). It follows that the lines $g_i, i = 1, 2, \ldots, q + 1$ partition the points of $Q^+(3, q)$. Starting from one of the g_i, we see that l also is part of such a parallel class.

17.53 Proposition. *Let $Q^+(3, q) \subset PG(3, q)$ be the hyperbolic quadric. There are $2(q + 1)$ totally singular lines. These come in two parallel classes of $q + 1$ each. Each parallel class partitions the points of the quadric, whereas two lines from different parallel classes intersect in a point.*

Exercises 17.2

17.2.1. *Let $(,)$ be the bilinear form with $A = \begin{pmatrix} 1 & 0 & 2 \\ \mu & 1 & 0 \\ \lambda & 0 & 1 \end{pmatrix}$ as the Gram matrix in odd characteristic. For which λ, μ is it degenerate?*

17.2.2. *Let $Q(x_1, x_2, x_3) = x_1^2 - x_3^2 + 2x_1x_3 + 3x_2x_3$ over $\mathbb{F}_p, p \neq 2$. Determine the corresponding bilinear form. In which characteristic is this nondegenerate?*

17.2.3. *Consider the preceding example in characteristic 5. Find a standard basis.*

17.2.4. *Let $(,)$ be the bilinear form with $A = \begin{pmatrix} 0 & 1 & 1 & 1 \\ 1 & 0 & 1 & 1 \\ 1 & 1 & 0 & a \\ 1 & 1 & a & 1 \end{pmatrix}$ as the Gram matrix, in odd characteristic. When is it degenerate?*

17.2.5. *Consider the preceding problem with $a = 1$. Over which fields is the bilinear form hyperbolic (respectively, elliptic)?*

17.2.6. *Consider the preceding problem over $\mathbb{F}_5$. Find a totally isotropic line.*

17.2.7. *Let $Q(x_1, x_2, x_3) = x_1^2 + x_1 x_2 + \omega x_2 x_3$, a quadratic form on $V(3, 4)$. Determine the Gram matrix of the underlying symplectic form and its radical. Is Q degenerate?*

17.2.8. *Find the number of totally isotropic lines in $PG(3, q)$ of a nondegenerate symplectic bilinear form on $V(4, q)$.*

17.2.9. *Consider $V(4, 2)$ with a nondegenerate symplectic form. Count the symplectic bases.*

17.2.10. *Define a design as follows: the points are the points of $Q^-(3, q) \subset PG(3, q)$, the blocks are the intersections with nontangent planes. Determine the parameters of this design.*

17.2.11. *Show that the trace $\mathrm{tr} : F = \mathbb{F}_{q^m} \longrightarrow K = \mathbb{F}_q$ defines a nondegenerate symmetric bilinear form, the **trace form** $\langle\rangle$ on F (seen as a vector space over K) by $\langle x, y \rangle = \mathrm{tr}(xy)$.*

17.2.12. *Show that a quadric meets a line in either 0 or 1 or 2 or in $q + 1$ points.*

17.2.13. *Show that a quadric in $PG(2, q)$ which is not identically zero has at most $2q + 1$ points.*

17.2.14. *Describe the parameters of the codes geometrically described by the hyperbolic quadrics on $PG(3, q), q$ odd.*

17.2.15. *Describe the parameters of the codes geometrically described by the elliptic quadrics on $PG(3, q), q$ odd.*

17.3   Caps: Constructions and bounds

Basic concepts: Ovoids and their codes. Caps in $PG(4, q)$. Bounds and recursive constructions. The Hill cap. An asymptotic problem.

The problem we want to study is the largest size of a cap in projective space $PG(k, q)$ and in affine space $AG(k, q)$. For an introduction to $PG(k, q)$, see the beginning of Section 17.1. The points of **affine** k-space $AG(k, q)$ are the points outside a fixed hyperplane of $PG(k, q)$. Represent the points of $PG(k, q)$ by homogeneous coordinates $(x_1 : x_2 : \cdots : x_{k+1})$ and choose the hyperplane H defined by $x_{k+1} = 0$. The affine points, the points outside H, are the points $(x_1 : x_2 : \cdots : x_k : 1)$. Each such point can be uniquely represented by a k-tuple: $P = (x_1, x_2, \ldots, x_k)$. When considering linear dependencies, the hidden coordinate should not be forgotten. For example, three points $P_1 = (x_1, x_2, \ldots, x_k)$, $P_2 = (y_1, y_2, \ldots, y_k)$, $P_3 = (z_1, z_2, \ldots, z_k)$ of $AG(k, q)$ are collinear if and only if there are coefficients $\lambda_1, \lambda_2, \lambda_3$, not all $= 0$, such that

$$\lambda_1 P_1 + \lambda_2 P_2 + \lambda_3 P_3 = 0 \text{ and } \lambda_1 + \lambda_2 + \lambda_3 = 0.$$

The hyperplanes of $AG(k, q)$ are the intersections with $AG(k, q)$ of the hyperplanes $\neq H$ of $PG(k, q)$.

In terms of codes, the largest number n of points of a cap in $PG(k - 1, q)$ is the largest n such that an $[n, n - k, 4]_q$ code exists (see Proposition 17.28). Recall that case $q = 2$ is not interesting, as $AG(k, 2)$ is a cap. In projective dimension 2 we know the answer to our problem. The ovals and hyperovals of $PG(2, q)$ each avoid lines and are therefore contained in an affine plane $AG(2, q)$. It follows that the largest size of a cap in $PG(2, q)$ and in $AG(2, q)$ is $q + 1$ for odd q and $q + 2$ for even q (see Theorem 17.4).

In $PG(3, q)$ we have the elliptic quadrics, which form $(q^2 + 1)$ caps (see Proposition 17.52).

17.54 Theorem. *The maximum size of a cap in* $PG(3, q)$, $q > 2$ *is* $q^2 + 1$.

PROOF Let $\mathcal{O} \subset PG(3, q)$ be an n-cap, where n is maximal. Choose a secant g (a line meeting $\mathcal{O}$ in two points). Assume each of the $q + 1$ planes E through g meets $\mathcal{O}$ in at most $q+1$ points. Then $|\mathcal{O}| \leq 2+(q+1)(q-1) \leq q^2+1$. This is the case in particular when q is odd. So assume $q > 2$ is a power of 2. We show first that every $P \in \mathcal{O}$ is on a tangent. Assume this is not the case. As every point is on $q^2 + q + 1$ lines, we obtain $n = q^2 + q + 2$. Also, every plane must intersect $\mathcal{O}$ in either 0 or $q + 2$ points. It is clear that there must be a line h which does not contain a point of $\mathcal{O}$. The usual counting argument shows that $q + 2$ divides $q^2 + q + 2$. This leads to the contradiction $q + 2 \mid 4$. We have seen that tangents exist and that $n < q^2 + q + 2$. Let t be a tangent to $\mathcal{O}$ and $t \cap \mathcal{O} = \{P\}$. Consider the $q + 1$ planes through t. The existence of t shows that no such plane can meet $\mathcal{O}$ in $q + 2$ points. If they all contain at most q points of $\mathcal{O}$ we obtain $n \leq 1 + (q + 1)(q - 1) = q^2$. So there is a plane E through t meeting $\mathcal{O}$ in a $(q+1)$ cap. We saw in the proof of Theorem 17.4

that $\mathcal{O} \cap E$ is contained in a hyperoval. Let N (the **nucleus** of $\mathcal{O} \cap E$) be the point outside $\mathcal{O}$ on this hyperoval. N must be on some secant g, as otherwise $\mathcal{O} \cup \{N\}$ would be a cap, contradicting the maximality of n. Every plane E' through g intersects E in a line through N, and hence in a tangent to $\mathcal{O}$. It follows that none of these planes can have more than $q+1$ points in common with $\mathcal{O}$. The usual counting argument shows $n \leq 2 + (q+1)(q-1) \leq q^2 + 1$. $\square$

This proof has been taken from Hirschfeld [114]. The elliptic quadric $Q^-(3, q)$ intersects planes (hyperplanes) either in one or in $q+1$ points. This shows that the code described by $Q^-(3, q)$ is an $[q^2 + 1, 4, q^2 - q]_q$ code, an **ovoid code.** Its dual is a $[q^2 + 1, q^2 - 3, 4]_q$ code (because $Q^-(3, q)$ is a cap). We saw ovoid codes earlier in this text. In Section 14.1 we constructed them using cyclic codes and construction XX. In Section 13.2 the Roos bound was used to give a cyclic construction. In the ternary case we saw an ovoid code in Exercise 3.6.2.

In projective dimensions > 3, no canonical models for caps are known. What we know is that quadrics can never be caps in those dimensions. The reason is that the Witt index is > 1, implying that there are lines all of whose points are on the quadric.

Caps in dimension 4

It is trivial to construct caps of $2q^2$ points in $PG(4, q)$. Choose two hyperplanes H_1, H_2 and let $\mathcal{O}_i \subset H_i$ be an elliptic quadric. The union $\mathcal{O}_1 \cup \mathcal{O}_2$ is a cap in $PG(4, q)$ of about $2q^2$ points. The next construction shows that we can do better.

Let $q \equiv 3 \pmod 4$. Use homogeneous coordinates such that $x = (x_1 : x_2 : x_3 : x_4 : x_5)$ in $PG(4, q)$. Consider the hyperplanes $H_1 = (x_3 = 0)$, $H_2 = (x_4 = 0)$ and $H_3 = (x_5 = 0)$. The quadrics Q_i, $i = 1, 2, 3$ are given by the following:

$$\begin{aligned} Q_1(x) &= x_1^2 + x_2^2 - x_4^2 + x_5^2 \\ Q_2(x) &= x_1^2 + x_2^2 + x_3^2 - x_5^2 \\ Q_3(x) &= x_1^2 + x_2^2 + 2x_3^2 - 2x_4^2. \end{aligned}$$

The symmetric bilinear form corresponding to Q_1 is

$$(x, y)_1 = x_1 y_1 + x_2 y_2 - x_4 y_4 + x_5 y_5,$$

analogously for Q_2 and Q_3. The radicals of Q_i are $Rad(Q_i) = \langle e_{i+2} \rangle, i = 1, 2, 3$. In particular, the restriction of Q_i to H_i is nondegenerate. We know that we can tell the elliptic from the hyperbolic case using the discriminant. As -1 is a nonsquare (see Proposition 17.16), $Var(Q_i) \cap H_i$ has discriminant

-1 (see Definition 17.36) and is therefore elliptic, in particular, a cap in H_i, $i = 1, 2, 3$. Start from the point set

$$(Var(Q_1) \cap H_1) \cup (Var(Q_2) \cap H_2) \cup (Var(Q_3) \cap H_3)$$

of $\approx 3q^2$ points. Removing the points in the intersections $H_i \cap H_j$ does not change the asymptotic size. In the resulting set any three collinear points must be on a line l, which is not contained in any of the hyperplanes H_i (a **generic** line). The plan is to find a subset $U \subset Var(Q_3)$ of size $\approx 0.5q^2$ such that no generic line l meets $Var(Q_1) \cup Var(Q_2) \cup U$ in three points.

Let l be a generic line and $P_i = l \cap H_i, i = 1, 2, 3$. Assume $P_i \in Var(Q_i)$, and write $P_i = \langle v_i \rangle$. Choose representatives v_i such that $v_1 + v_2 + v_3 = 0$. Then

$$v_1 = x = (x_1, x_2, 0, x_4, x_5)$$
$$v_2 = y = (y_1, y_2, y_3, 0, y_5)$$
$$v_3 = z = (z_1, z_2, z_3, z_4, 0).$$

Recall $x + y + z = 0$ and $Q_1(x) = Q_2(y) = Q_3(z) = 0$. The equation $2Q_1(x) + 2Q_2(y) - Q_3(z) = 0$ shows $(x_1 - y_1)^2 = -(x_2 - y_2)^2$. As -1 is a nonsquare, we must have $x_1 = y_1, x_2 = y_2$. The relation $Q_1(x) - Q_2(y) = 0$ yields

$$z_3^2 + z_4^2 = 2x_5^2.$$

This is impossible provided $2(z_3^2 + z_4^2)$ is a nonsquare. Let U consist of all $P = (z_1 : z_2 : z_3 : z_4 : 0)$ such that $2(z_3^2 + z_4^2)$ is nonsquare and $Q_3(P) = z_1^2 + z_2^2 + 2z_3^2 - 2z_4^2 = 0$. We have to show $|U| \approx 0.5q^2$ (as a function of q). We can impose the conditions $z_3 z_4 \neq 0$ and $z_3^2 - z_4^2 \neq 0$ without changing the asymptotics. For each of the $(q-1)/2$ nonsquares u there are $e_1(1) = q + 1$ pairs (z_3, z_4) such that $z_3^2 + z_4^2 = 2u$ ($x^2 + y^2$ is anisotropic). For each such choice of u, z_3, z_4 there are $q + 1$ pairs (z_1, z_2) such that $z_1^2 + z_2^2 = 2z_4^2 - 2z_3^2$. The number of (projective) points in U is therefore $\approx 0.5q^2$.

In the case $q \equiv 1 \pmod 4$ the construction is similar. Precise values and an extension construction can be found in [23].

17.55 Theorem. *There are caps in $PG(4, q), q$ odd, whose number of points is a polynomial in q with leading term $2.5q^2$.*

In characteristic 2 we get larger caps sometimes. The next construction is in $AG(4, q)$, where $q = 2^f$ and f is odd.
Consider $F = \mathbb{F}_{q^4}$, and identify the points in $AG(4, q)$ with the points $(1, x)$, $x \in F$. We claim that the points $(1, x)$ where $x = 0$ or x is a $3(q^2 + 1)$-th root of unity form a $(3q^2 + 4)$ cap $\mathcal{K}_q \subset AG(4, q)$.
Consider also the subfield $K = \mathbb{F}_{q^2} \subset F$ and the traces

$$T : F \to K, \; Tr : K \to \mathbb{F}_4, \; tr : K \to \mathbb{F}_2, \; tr_0 : \mathbb{F}_q \to \mathbb{F}_2$$

(see Figure 17.2).

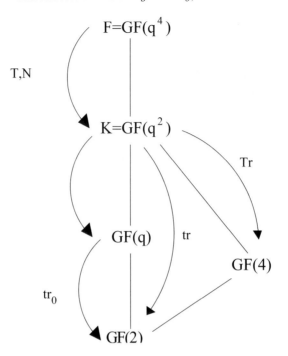

FIGURE 17.2: Fields and traces

Recall in this context the **transitivity of the trace,** proved in Exercise 12.1.16: if $K \subseteq L \subseteq F$ is a tower of three fields, then $tr_{F|K}$ is the concatenation $tr_{L|K} \circ tr_{F|L}$. Let $W \subset F$ be the group of $(q^2 + 1)$-st roots of unity. Denote the points of $\mathcal{K}_q$ as $P(0) = (1,0)$ and $P(a,w) = (1,aw)$, where $0 \neq a \in \mathbb{F}_4$ and $w \in W$.

The numbers 3, $q - 1$ and $q^2 + 1$ are pairwise coprime. This implies that the points $P(0)$ and $P(a,w)$ are pairwise different. It is left as Exercise 17.3.3 to prove that $P(0)$ is not collinear with two of the remaining points of $\mathcal{K}_q$. We show that the $P(a,w)$ form a cap. At first we need a criterion to decide which elements of K are images under the trace T of elements of W. This follows from the additive version of Hilbert's theorem 90, Proposition 17.21.

17.56 Lemma. *Let $0 \neq \alpha \in \mathbb{F}_{q^2}$. The following are equivalent:*

- *There exists $w \in W$ such that $T(w) = \alpha$.*

- $Tr(1/\alpha) \in \mathbb{F}_4 \setminus \mathbb{F}_2$.

- $tr(1/\alpha) = 1$.

PROOF Because of the transitivity of the trace, the second and third condition are equivalent. We have $w^{q^3} = 1/w$ for $w \in W$, hence $T(w) =$

$w + 1/w$. It follows that $T(w) = \alpha$ is equivalent to $(w/\alpha)^2 + (w/\alpha) + 1/\alpha^2 = 0$. Clearly, $w = 1$ is not a solution. Let $tr(1/\alpha) = 0$. By Proposition 17.21 the solutions of $x^2 + x + 1/\alpha^2 = 0$ are in K. It follows that $w \in K$, a contradiction. This shows that there is no solution in this case. The statement now follows from counting: $w = 1$ is the only element $w \in W$ such that $T(w) = 0$, and each element $\alpha \neq 0$ can have at most two preimages under T in W. It follows that each $\alpha \in K$ satisfying $tr(1/\alpha) = 1$ must have two such preimages.   □

Let $P(a_1, w_1)$, $P(a_2, w_2)$, $P(a_3, w_3)$ be different points of $\mathcal{K}_q$ which are collinear. Let $1, \lambda, \lambda + 1$ be the coefficients of an affine linear combination, where $\lambda \in \mathbb{F}_q \setminus \mathbb{F}_2$. This yields the equation

$$(\lambda + 1)a_1 w_1 = a_2 w_2 + \lambda a_3 w_3.$$

Assume at first $a_i = a_j$ for some $i \neq j$. We can assume $a_2 = a_3 = 1$ and $w_3 = 1$. The equation is

$$(\lambda + 1)a_1 w_1 = w_2 + \lambda.$$

As the second and third points are different, it follows that $w_2 \neq 1$. Application of the norm $(N(x) = x^{q^2+1})$ to both sides yields $(\lambda^2 + 1)a_1^2 = \lambda^2 + 1 + \lambda\alpha$, where $\alpha = T(w_2)$. We have $\alpha \neq 0$ as $w_2 \neq 1$. It follows that $a_1 \neq 1$. The equation

$$\frac{1}{\alpha}(a_1^2 + 1) = \frac{\lambda}{\lambda^2 + 1} = \frac{1}{\lambda + 1} + \frac{1}{(\lambda + 1)^2}$$

shows, after application of Tr,

$$(a_1^2 + 1)Tr(1/\alpha) = Tr(\frac{1}{\lambda + 1} + \frac{1}{(\lambda + 1)^2}) = tr(\frac{1}{\lambda + 1})$$

(the last equation follows from the definition of the trace). The transitivity of the trace shows that tr factorizes over $tr_{K|\mathbb{F}_q}$. As $tr_{K|\mathbb{F}_q}$ vanishes on $\mathbb{F}_q$, it follows that $(a_1^2 + 1)Tr(1/\alpha) = 0$, hence $Tr(1/\alpha) = 0$, contradicting Lemma 17.56.

Assume now the a_i are pairwise different (and nonzero) elements of $\mathbb{F}_4$. We can choose notation such that

$$(\lambda + 1)w_1 = ww_2 + \lambda\overline{w}.$$

Application of N yields $\lambda^2 + 1 = \overline{\omega} + \lambda^2\omega + \lambda\alpha$, where $\alpha = T(w_2)$. We have $\alpha \neq 0$, as otherwise $\lambda \in \mathbb{F}_4$, which is impossible, as $\mathbb{F}_q \cap \mathbb{F}_4 = \mathbb{F}_2$. An equivalent form of the equation is $\lambda^2\overline{\omega} + \lambda\alpha + \omega = 0$. Multiplication by $\overline{\omega}/\alpha^2$ yields $x^2 + x + 1/\alpha^2 = 0$, where $x = \lambda\overline{\omega}/\alpha$. This yields $tr(1/\alpha) = 0$, contradicting Lemma 17.56 again.

The constructions presented in this subsection show that we can find caps of $\approx 2.5q^2$ points in $PG(4, q)$ for odd q and caps of size $\approx 3q^2$ in $AG(4, q)$ for

$q = 2^{odd}$. An expression like $\approx 3q^2$ for the number of points means that the number of points is a polynomial in q with $3q^2$ as the leading term. These seem to be the largest caps known in $PG(4, q)$ aside from constructions which are valid only for individual small q. Observe that our cap in $AG(4, q)$ can be seen as a dual BCH code (see Exercise 17.3.4). No family of caps of $\approx cq^2$ for $c > 2$ seems to be known in $PG(4, q)$ for $q = 2^{even}$. Moreover, all these families of caps are far from the known upper bounds.

Consider the code C_q geometrically described by the $(3q^3 + 4)$ cap $\mathcal{K}_q \subset AG(4, q)$, where $q = 2^f, f$ odd. Due to a close link with the Kloosterman codes, the weight distribution of C_q can be determined. In the special case $q = 8$, this yields an $[196, 5, 164]_8$ code. As it contains a subcode of dimension 4 and minimum distance 168, we can apply lengthening and obtain a $[200, 5, 168]_8$ code (see [81]).

General upper bounds on caps

We start from a general upper bound on affine caps.

17.57 Definition. *Denote by $C_k(q)$ the maximum size of a cap in $AG(k, q)$, and by $c_k(q) = C_k(q)/q^k$ its relative size.*

As $AG(k, q)$ is the disjoint union of q copies of $AG(k - 1, q)$ (see Exercise 17.3.5), it follows that $C_k(q) \leq qC_{k-1}(q)$, hence $c_k(q) \leq c_{k-1}(q)$. We have $c_3(q) = 1/q$ (see Exercise 17.3.7). We will derive a lower bound on $c_{k-1}(q) - c_k(q)$. This shows in particular that a maximal cap in $AG(k, q)$ cannot intersect each hyperplane in a maximal cap.

As usual, assume $q = p^f > 2$. Let $A \subset AG(k, q)$ be a cap. As $q > 2$, we can find nonzero elements $\lambda_i \in \mathbb{F}_q$ such that $\lambda_1 + \lambda_2 + \lambda_3 = 0$. The λ_i are fixed throughout the proof. Let $V = \mathbb{F}_q^k = AG(k, q)$ and $T : \mathbb{F}_q \longrightarrow \mathbb{F}_p$ be the trace function. Put $Q = |V| = q^k$. Finally, the complex number $\zeta = exp(2\pi i/p)$ is a p-th root of unity. The aim is an upper bound on $|A|$.

To this end, study the complex number

$$S = \sum_{y \in V \setminus \{0\}} \sum_{a_1, a_2, a_3 \in A} \zeta^{T((\sum_i \lambda_i a_i) \cdot y)}.$$

The number is easily determined: reverse the order of summation and extend the inner sum over all $y \in V$. The additional term corresponding to $y = 0$ is $|A|^3$. As the sum over all powers of ζ vanishes, the inner sum will vanish unless $a_1 = a_2 = a_3$. We obtain

$$S = |A|(Q - |A|^2).$$

Now we obtain an upper bound on $|S|$. Let $0 \neq \lambda \in \mathbb{F}_q$ and $0 \neq y \in V$. Consider the complex number $U(\lambda)_y = \sum_{a \in A} \zeta^{T((\lambda a) \cdot y)}$. Let $u(\lambda)_y = |U(\lambda)_y|$. Define a real vector $u(\lambda)$ of length $Q - 1$ whose coordinates are parametrized by the $0 \neq y \in V$, the corresponding entry being $u(\lambda)_y$. We have $S = \sum_{y \neq 0} U(\lambda_1)_y U(\lambda_2)_y U(\lambda_3)_y$, in particular,

$$|S| \leq \sum_{y \neq 0} u(\lambda_1)_y u(\lambda_2)_y u(\lambda_3)_y.$$

What do we know about the real vectors $u(\lambda)$? A calculation similar to the determination of S shows that we know the length of these vectors:

$$\|u(\lambda)\|^2 = |A|(Q - |A|).$$

The combinatorial information is in the following lemma, an upper bound on the entries $u(\lambda)_y$:

17.58 Lemma. *Let $0 \neq \lambda \in \mathbb{F}_q$ and $0 \neq y \in V$. Then*

$$u(\lambda)_y \leq qC_{k-1}(q) - |A| = c_{k-1}(q)Q - |A|.$$

PROOF As λA is a cap, we can assume $\lambda = 1$. Denote by ν_c the number of elements $a \in A$ such that $a \cdot y = c$. As the $v \in V$ satisfying $v \cdot y = c$ form a subspace $AG(k - 1, q)$ (see Exercise 17.3.6), we have $\nu_c \leq C_{k-1}(q)$. It follows that

$$u(\lambda)_y = |\sum_{c \in \mathbb{F}_q} \nu_c \zeta^{T(c)}| = |\sum_{c \in \mathbb{F}_q} (C_{k-1}(q) - \nu_c) \zeta^{T(c)}|$$

(here we use again that the sum of all powers of ζ vanishes)

$$\leq \sum_c (C_{k-1}(q) - \nu_c) = qC_{k-1}(q) - |A|.$$

$\square$

Now we can obtain an upper bound on $|S|$, using the bound from the preceding lemma and the classical Cauchy-Schwartz inequality. The best known version of this inequality states that for real vectors $x = (x_1, \ldots, x_n)$, $y = (y_1, \ldots, y_n)$, we have

$$|x \cdot y| \leq \|x\| \|y\|.$$

Recall that the proof of the Plotkin bound made use of a consequence of this inequality; see Chapter 9. We have

$$|S| \leq (c_{k-1}(q)Q - |A|)(|A|(Q - |A|)).$$

Comparison of this upper bound and the precise value of $|S|$ yields the following bound on $C_k(q)$, which is the main result in this subsection.

17.59 Theorem. *Let $q > 2$ be a prime-power. If $k \geq 3$, then*

$$c_k(q) \leq \frac{q^{-k} + c_{k-1}(q)}{1 + c_{k-1}(q)},$$

equivalently

$$(1 - c_k(q))(c_{k-1}(q) - c_k(q)) \geq c_k^2 - q^{-k}.$$

In particular we have the desired lower bound on the decrease $c_{k-1}(q) - c_k(q)$. Theorem 17.59 was first obtained by R. Meshulam [152] in odd characteristic, using the mechanism of the Fourier transform. The direct approach sketched in this section is from [24]. It also covers the characteristic 2 case.

An asymptotic bound can be derived from Theorem 17.59:

17.60 Theorem. *Let $q > 2$ and $k \geq 3$. Then*

$$c_k(q) \leq \frac{k+1}{k^2}.$$

In order to derive bounds on projective caps from Theorem 17.59, the following recursive construction is needed:

17.61 Theorem. *Let $q > 2$. If there is an n-cap in $PG(k-1, q)$, then $AG(k, q)$ contains a $(2n)$ cap.*

PROOF   For each point $(x_1 : \cdots : x_k)$ of the cap in $PG(k - 1, q)$, choose a representative $x = (x_1, \ldots, x_k) \in \mathbb{F}_q^k$. Let $c \in \mathbb{F}_q, c \neq 0, 1$. We claim that the $2n$ points of the form $(x, 1)$, $(cx, 1)$ form a cap in $AG(k, q)$. It is clear that these are $2n$ different points. They are in affine space, as they avoid the hyperplane $x_{k+1} = 0$. The elementary proof that they form a cap is left to the reader.   □

Theorems 17.61 and 17.60 together show the following:

17.62 Theorem. *Let $q > 2$ and $k \geq 3$. The size of a cap in $PG(k, q)$ is bounded by $q^{k+1} \dfrac{k + 2}{2(k + 1)^2}$.*

The doubling construction Theorem 17.61 is an elementary example of a recursive cap construction. More recursive constructions will be given in the next subsection.

Recursive constructions for caps

A general product construction for caps is due to Mukhopadhyay [154]. We start from a slight generalization.

17.63 Theorem. *If there is an n-cap $\mathcal{A} \subset AG(k, q)$ and an m-cap $\mathcal{B} \subset PG(l, q)$, then we can construct an nm-cap in $PG(k + l, q)$. Moreover, if $\mathcal{A}$ is avoided by $i \geq 1$ hyperplanes in general position and $\mathcal{B}$ is avoided by $j \geq 0$ hyperplanes in general position, then the product cap is avoided by $i + j - 1$ hyperplanes in general position.*

PROOF   Let $(a|1)$ be the typical representative of the affine cap, and b the typical representative of the cap in $PG(l, q)$. Here $a \in \mathbb{F}_q^k$, $b \in \mathbb{F}_q^{l+1}$. The typical representative of a point of the product cap is $(a|b)$. Assume $\sum_{i=1}^{3} \lambda_i(a_i|b_i) = 0$. The second coordinate section shows $b_1 = b_2 = b_3$ and $\lambda_1 + \lambda_2 + \lambda_3 = 0$. This shows $\sum_{i=1}^{3} \lambda_i(a_i|1) = 0$, a contradiction.

An affine cap is a cap which is avoided by a hyperplane. We can represent the cap by a matrix with a row all of whose entries are nonzero. The fact that $\mathcal{A}$ is avoided by i hyperplanes **in general position** simply means that we can represent it by a matrix which possesses i rows all of whose entries are nonzero, likewise for $\mathcal{B}$. This shows the second assertion.   ☐

Two points on the projective line form a 2-cap. For $q > 2$ these two points form an affine cap (they avoid a point $=$ hyperplane). It follows that the doubling construction Theorem 17.61 is a special case of the product construction Theorem 17.63.

Ovals or hyperovals in $PG(2, q)$ clearly are affine. They are in fact avoided by $i = 3$ lines (hyperplanes) in general position. Application of the product construction yields, among others, $(q + 1)(q^2 + 1)$ caps in $AG(5, q)$ when q is odd and $(q + 2)(q^2 + 1)$ caps in $AG(5, q)$ in characteristic 2.

Here is a generalization of the product construction from [77]:

17.64 Theorem. *Assume there is an n-cap $\mathcal{A} \subset PG(k, q)$ intersecting a hyperplane in $n - w$ points, and an m-cap $\mathcal{B} \subset PG(l, q)$. We can construct an $\{wm + (n - w)\}$ cap in $PG(k + l, q)$.*

PROOF   With notation as in Theorem 17.63, the $(a|b)$ where $(a|1)$ varies over the affine points of $\mathcal{A}$ form the product cap, a wm-cap. The $(\alpha|0)$, where $(\alpha|0)$ varies over the points of $\mathcal{A}$ from the hyperplane, extend it to a $\{wm + (n - w)\}$ cap.   ☐

If in Theorem 17.64 we choose $\mathcal{A}$ to be an ovoid (a (q^2+1) cap in $PG(3, q)$ intersecting a hyperplane in one point), then a classical construction by B. Segre [182] is obtained: if there is an n-cap in $PG(k, q)$, then there is an $\{q^2n + 1\}$ cap in $PG(k + 3, q)$.

17.65 Theorem. *Assume there is an n-cap $\mathcal{A} \subset PG(k, q)$ and an m-cap $\mathcal{B} \subset PG(l, q)$, both possessing tangent hyperplanes. We can construct an $\{nm - 1\}$ cap in $PG(k + l, q)$.*

PROOF   A **tangent hyperplane** is a hyperplane containing precisely one point of the cap. With notation as before, let $(a|1)$ and $(b|1)$ be the affine points of $\mathcal{A}$ and $\mathcal{B}$, respectively. Denote by $(\alpha|0)$ the point of $\mathcal{A}$ on the tangent hyperplane, $(\beta|0)$ the point of $\mathcal{B}$ on the tangent hyperplane. The points of the cap in $PG(k+l,q)$ are those of the form

$$(a|b|1), \ (a|\beta|0) \text{ and } (\alpha|b|0).$$

▯

Application to ovoids yields a $\{q^4 + 2q^2\}$ cap in $PG(6,q)$. The construction in Theorem 17.65 shows that this cap has a hyperplane intersection of size $q^2 + 1$. Application of Theorem 17.64 with an ovoid as the second ingredient yields a $q^2(q^2+1)^2$ cap in $PG(9,q)$.

The ternary case: the Hill cap

We know from Theorem 17.42 that the elliptic quadric $Q_5^-(q)$ in $PG(5,q)$ has $(q+1)(q^3+1)$ points. In the ternary case we obtain $|Q_5^-(3)| = 112$. The corresponding quadratic form can be represented by $Q(x) = \sum_{i=1}^{6} x_i^2$ (the Gram matrix has determinant 1). It follows that the one-dimensional subspace $\langle x \rangle$ generated by vector $x = (x_1, x_2, \ldots, x_6)$ belongs to $Var(Q) = Q_5^-(3)$ if and only if x has weight 3 or 6. As there are $8\binom{6}{3} = 160$ vectors of weight 3 and $2^6 = 64$ vectors of weight 6 in $V(6,3)$, we obtain $|Q_5^-(3)| = 80+32 = 112$, as predicted. As the Witt index is 2, there are totally isotropic lines but no totally isotropic planes.

We wish to partition $Q_5^-(3)$ into two caps. What do two-dimensional totally isotropic subspaces of Q (lines of $PG(5,3)$ all of whose points belong to $Q_5^-(3)$) look like? It is clear that each such subspace contains at least two points P_1, P_2 generated by vectors of weight 3. Let P_1, P_2 be such points. The line l defined by P_1, P_2 will be totally isotropic (contain only points of weight 3 or 6) if the supports of P_1 and P_2 either are complementary or intersect in cardinality 2. Typical examples are the lines

$$(1^3 : 0^3), \ (0^3 : 1^3), \ (1^6), \ (1^3 : 2^3)$$

for the first case and

$$(1^3 : 0^3), \ (1:2:0:1:0^2), \ (2:0:1:1:0^2), \ (0:2:1:2:0^2)$$

for the second. This indicates how $Q_5^-(3)$ can be partitioned into two caps. It suffices to choose the partition such that each totally isotropic line has two points in each part. This can be done as follows:

Please check that the following 10 blocks form a $2 - (6,3,2)$ design:

123	236
124	245
135	256
146	345
156	346

We denote it by $\mathcal{D}$. It can be described as an orbit of the permutation group $PSL(2,5) \cong A_5$ on the projective line $PG(1,5)$ whose points we identify with coordinates of $V(6,3) = \mathbb{F}_3^6$. As the permutation action of $PSL(2,5)$ is 2-transitive, each orbit defines a 2-design. The complement of a block is not a block. Let $\overline{\mathcal{D}}$ consist of the complements of the blocks of $\mathcal{D}$, equivalently, the blocks of $\overline{\mathcal{D}}$ consist of the 3-subsets of $\{1,2,3,4,5,6\}$, which are not blocks of $\mathcal{D}$. Denote by D the set of vectors of weight 3 whose support is a block of $\mathcal{D}$, analogously for $\overline{D}$. Let $\langle D \rangle$ and $\langle \overline{D} \rangle$ be the corresponding points in $PG(5,3)$.

Let R be the union of vectors of weight 6 whose representatives have an even number of entries 2, and $\overline{R}$ the remaining weight 6 vectors. As before, $\langle R \rangle, \langle \overline{R} \rangle$ denote the sets of corresponding points in $PG(5,3)$. We have partitioned the points of $Q_5^-(3)$ in the form

$$Q_5^-(3) = \langle D \rangle \cup \langle \overline{D} \rangle \cup \langle R \rangle \cup \langle \overline{R} \rangle$$

where each of $\langle D \rangle, \langle \overline{D} \rangle$ has 40 points, and each of $\langle R \rangle, \langle \overline{R} \rangle$ has 16 points. The description of totally isotropic lines shows that each of

$$\langle D \rangle \cup \langle R \rangle, \ \langle D \rangle \cup \langle \overline{R} \rangle, \ \langle \overline{D} \rangle \cup \langle R \rangle, \langle \overline{D} \rangle \cup \langle \overline{R} \rangle$$

is a cap (observe that each set of elements contains precisely two blocks of $\mathcal{D}$ and two blocks of $\overline{\mathcal{D}}$).

17.66 Theorem. *The* 112 *points of* $Q_5^-(3)$ *can be partitioned into two* 56-*caps. In particular, there is a* 56-*cap in* $PG(5,3)$.

Here is a generator matrix of a 56-cap in $PG(5,3)$, which arises from halving $Q_5^-(3)$ in the way just described. The points in $PG(5,3)$ generated by the columns of the matrix are the points of the cap.

The Hill cap
1111112222222222 2211 2211 2211 2211 2211 0000 0000 0000 0000 0000
1122221111222222 2121 2121 0000 0000 0000 2211 2211 2211 0000 0000
1212221222111222 2112 0000 2121 0000 0000 2121 0000 0000 2211 2211
1221222122122112 0000 2112 0000 2121 0000 0000 2121 0000 2121 2121
1222122212212121 0000 0000 2112 0000 2121 0000 2112 2121 2112 0000
1222212221221211 0000 0000 0000 2112 2112 2112 0000 2112 0000 2112

The 56-cap in $PG(5,3)$ is essentially uniquely determined, and there is no larger cap in $PG(5,3)$. It is known as the Hill cap. For the original construction, see R. Hill [111]. The existence of the Hill cap is the main reason why the ternary case displays a particular behavior in most asymptotic questions concerning caps.

Consider the hyperplane S defined by $\sum_{i=1}^6 x_i = 0$. The only point from $\langle R \rangle$ on S is $(1:1:1:1:1:1)$, and for each block of $\mathcal{D}$ there is precisely one point on S having the block as its support. It follows that $\langle D \rangle \cup \langle R \rangle$ intersects the hyperplane S in 11 points. The corresponding 45-cap in $AG(5,3)$ will be called the affine Hill cap. It can be proved that the affine Hill cap is the only 45-cap in $AG(5,3)$, and there is no larger cap in $AG(5,3)$.

So the answer to our question is known in $PG(5,3)$ and in $AG(5,3)$. Clearly the answer was known in dimension 4 before Hill constructed his cap in $PG(5,3)$. It was G. Pellegrino who in [164] conducted an exhaustive enumeration of large caps in $PG(4,3)$ and in $AG(4,3)$. It turns out that there are no 21-caps in $PG(4,3)$. The 20-cap in $AG(4,3)$ is uniquely determined. Such a cap is easily constructed. It arises from doubling the elliptic quadric in

$PG(3,3)$; see Theorem 17.61. We encountered the Pellegrino cap (the 20-cap in $AG(4,3)$) in Section 3.6, as the largest collection of cards in the game of SET which does not contain a SET. It follows from the discussion in Section 3.6 that such a SET-free collection of cards is indeed equivalent to a cap in $AG(4,3)$.

If we accept Pellegrino's result that 21-caps do not exist in $PG(4,3)$, it is in fact easy to prove that $PG(5,3)$ cannot contain a 57-cap. This is an elementary application of the geometric method; see Exercise 17.3.11.

What is known in dimension 6? The doubled Hill cap is a 112-cap in $AG(6,3)$. In the language of Theorem 17.59, we have in particular $c_5(3) = 45/3^5 = 5/27$. An application of Theorem 17.59 yields $C_6(3) \leq 114$. This is a remarkably narrow gap between the lower bound 112 and the upper bound 114 for the size of the largest cap in $AG(6,3)$ (see Definition 17.57). In fact, this gap has been closed by A. Potechin, who showed in [168] that $C_6(3) = 112$ and this affine cap is essentially uniquely determined. Most of the largest known caps over $\mathbb{F}_3$ are related to the Hill cap. This is certainly true in dimension ≤ 7 : the Calderbank-Fishburh 248-cap in $PG(7,3)$ is an extension of the double of the double of the Hill cap. Exceptions to this rule are a 541-cap in $PG(8,3)$ and a 2744-cap in $PG(10,3)$ which resulted from computer searches [26]. A 1216-cap in $PG(9,3)$ and a 6464-cap in $PG(11,3)$ are obtained in [74] from a sophisticated version of the product construction.

We close with a list of the currently best known lower bounds on the largest size of a cap in $PG(k,q)$, for small parameter values.

The values in the quaternary case for dimension ≥ 8 were obtained in [86] and [133] using recursive constructions and computer searches.

$k\backslash q$	3	4	5	7	8	9
2	4	6	6	8	10	10
3	10	17	26	50	65	82
4	20	41	66	132	208	212
5	56	126	195	434	695	840
6	112	288	675	2499	4224	6723
7	248	756	1715	6472	13520	17220
8	541	2136	5069	21555	45174	68070
9	1216	5124	17124	122500	270400	544644
10	2744	15840	43876	323318	878800	1411830
11	6464	36150	130951	1067080	2931457	5580100

There are precise values for $k = 2$ (the ovals and hyperovals), for $k = 3$ (see Theorem 17.54) as well as in $PG(4,3)$, $PG(5,3)$ and $PG(4,4)$. In all other cases the determination of the largest cap size remains an open problem. In $PG(6,3)$ the best known upper bound is 136 [5] and in $PG(4,5)$ it seems to be 88.

An asymptotic problem

17.67 Definition. *Fix q and t. Denote by $n_{t,q}(k)$ the maximal length n of a code $[n, n - k, t + 1]_q$, equivalently the largest length of a linear orthogonal array of dimension k and strength t. Define*

$$\lambda(t, q) = lim_{k \to \infty} \frac{log_q(n_{t,q}(k))}{k}.$$

Consider the columns of a check matrix H for a code $[n, n - k, t + 1]_q$, where n is maximal. If $n = q^{lk}$, then $n \leq q^k$ and $l = log_q(n)/k$. It follows that $\lambda(t, q) \leq 1$. In the case of equality, the columns of H fill the space $\mathbb{F}_q^k$, in an asymptotic sense.

The sphere-packing bound from Section 1.6 yields an upper bound: let $e = \lfloor t/2 \rfloor$. It is clear that in the process (take log, divide by k, let $k \to \infty$) the dominating term on the left side is $\binom{n}{e}$ and yields a contribution $e\lambda(t, q)$. As the right side yields a contribution 1, we obtain the bound $\lambda(t, q) \leq 1/\lfloor t/2 \rfloor$.

In order to obtain a general lower bound, use primitive BCH codes. In the language of Chapter 13 we use codes $tr(\mathcal{B}(A))$, where $n = q^r - 1$ and $A = \{0, 1, \ldots, t - 1\}$. This yields a linear q-ary orthogonal array of strength t and dimension $k \leq 1 + (t - 1)r$. The lower bound is

$$\lambda(t, q) \geq lim_{r \to \infty} \left(\frac{r}{1 + (t - 1)r} \right) = \frac{1}{t - 1}.$$

Once again the binary case is degenerate; see Exercise 17.3.14.

17.68 Theorem. *We have $\lambda(t, 2) = 1/\lfloor t/2 \rfloor$ and, for $q > 2$,*

$$\frac{1}{t - 1} \leq \lambda(t, q) \leq \frac{1}{\lfloor t/2 \rfloor}.$$

Clearly $\lambda(t, q)$ is a nonincreasing function of t, and $\lambda(2, q) = 1$ (see Exercise 17.3.15). It follows from Theorem 17.68 that $\lambda(t, q) < 1$ for $t > 3$. For $t = 3$ it is conceivable that $\lambda(3, q) = 1$. This would, however, be quite surprising, as it would mean that the space can be filled by a cap, in some sense. It has to be conjectured that $\lambda(3, q) < 1$ for all $q > 2$. This is an open problem.

Now let an n-cap in $AG(k, q)$ be given. The product construction Theorem 17.63 shows that there exists an n^l-cap in $AG(kl, q)$. This shows $\lambda(3, q) \geq log_q(n)/k$. We conclude that every affine cap yields a lower bound on the asymptotic value $\lambda(3, q)$. The affine ovoids (q^2-caps in $AG(3, q)$) yield $\lambda(3, q) \geq 2/3$. This is much better than the lower bound of $1/2$ which follows from Theorem 17.68. We can do slightly better.

An application of Theorem 17.65 to ovoids yields a $\{q^4 + 2q^2\}$ cap in $PG(6, q)$ with a hyperplane intersection of size $q^2 + 1$. This yields a $\{q^4 + q^2 - 1\}$ cap in $AG(6, q)$, for every q. The resulting bound

$$\lambda(3, q) \geq \frac{log_q(q^4 + q^2 - 1)}{6}$$

seems to be the best lower bound known for general q.

As mentioned earlier, cases $q = 3$ and $q = 4$ usually are special in questions concerning caps with extremal properties. In the ternary case the reason for this exceptional behavior is the presence of the Hill cap. The doubled Hill cap yields $\lambda(3,3) \geq 0.7158\ldots$ (already considerably larger than $2/3$). A construction due to Calderbank-Fishburn [38] based on the Hill cap and its symmetries shows $\lambda(3,3) \geq 0.7218\ldots$ An ingenious generalization of the product construction is due to Y. Edel [74]. When applied to the Hill cap it yields a lower bound $\lambda(3,3) \geq 0.724851\ldots$

This appears to be the best lower bound currently known.

The reason why the quaternary case is special as well is the presence of a 126-cap in $PG(4,5)$ due to D. Glynn [93]. It intersects a hyperplane in six points. The resulting 120-cap in $AG(5,4)$ yields the lower bound $\lambda(3,4) \geq log_4(120)/5 = 0.6906\ldots$

Finally, we observe that the bound from Theorem 17.60 is far from sufficient to bound $\lambda(3,q)$ away from 1. This bound roughly has the form $c_k(q) \leq 1/k$. What would be needed is a bound of the form $c_k(q) \leq q^{-k\delta}$ for some $\delta > 0$.

As the case of caps (linear OA of strength 3) is wide open, it may come as a surprise that $\lambda(4,3)$ and $\lambda(4,4)$ are known. In the ternary case, use the cyclic codes from Theorem 13.8, in the quaternary case the constacyclic families from Theorem 13.29 or from Theorem 13.30. We obtain $\lambda(4,3) = \lambda(4,4) = 1/2$. More on this question can be found in Dumer and Zinoviev [73].

Exercises 17.3

17.3.1. *Show that lines in $AG(k,q)$ have q points.*

17.3.2. *Determine the weight distribution of an ovoid code.*

17.3.3. *Prove that $P(0) = (1,0)$ extends the cap $\mathcal{K}_q \subset AG(4,q)$ to a $(3q^2 + 4)$ cap (here $q = 2^f$, f odd).*

17.3.4. *Describe the cap $\mathcal{K}_q \subset AG(4,q)$ in terms of cyclic codes.*

17.3.5. *Show that $AG(k,q)$ is the disjoint union of q copies of $AG(k-1,q)$.*

17.3.6. *Show that the hyperplanes of $AG(k,q)$ are represented by pairs (y,c) where $y = (y_1,\ldots,y_k) \in \mathbb{F}_q^k$ varies over representatives of one-dimensional subspaces (in other words points of $PG(k-1,q)$) and $c \in \mathbb{F}_q$.*
Point $x = (x_1,\ldots,x_k) \in AG(k,q)$ is on the hyperplane described by the pair (y,c) if $x \cdot y = c$.

17.3.7. *Prove that the maximum size of a cap in $AG(3,q)$ $(q > 2)$ is q^2.*

17.3.8. *Prove the Cauchy-Schwartz inequality for real vectors:*
Let $x = (x_1, \ldots, x_n)$, $y = (y_1, \ldots, y_n)$. *Then*

$$|x \cdot y| \leq \|x\| \|y\|.$$

17.3.9. *Derive the bound used in the proof of the Plotkin bound in Chapter 9 from the Cauchy-Schwartz inequality.*

17.3.10. *Prove Theorem 17.60 for* $q > 3$.

17.3.11. *Prove that there is no 57-cap in* $PG(5,3)$.
Use the fact that there is no 21-cap in $PG(4,3)$.

17.3.12. *Prove that the following form the points of a 236-cap in* $AG(7,3)$:

$$(1,0,D) \ (1,0,R) \ (1,1,\overline{D} \ (1,1,R) \ (1,2,U).$$

This construction is from Calderbank and Fishburn [38].

17.3.13. *Show that the union of the Calderbank-Fishburn cap from the previous exercise and the points of type* $(0,1,U)$ *form a 248-cap in* $PG(7,3)$. *This construction is from [78].*

17.3.14. *Show* $\lambda(t,2) = 1/\lfloor t/2 \rfloor$.

17.3.15. *Show* $\lambda(2,q) = 1$ *for all* q.

17.3.16. *Accepting the fact that 20 is the largest size of a cap in* $PG(4,3)$, *prove that the largest intersection of the Hill cap with a hyperplane is 20.*

17.3.17. *Show that the Hill cap defines a* $[56,6,36]_3$ *code and compare with the Griesmer bound.*

17.3.18. *Use the generator matrix of the Hill cap to obtain a 20-cap in* $PG(4,3)$.

Chapter 18

Additive codes and network codes

18.1 Basic constructions and applications

Basic concepts: Definition, basic properties, Chen projection, applications in computer memory systems and deep space communication, convolutional codes, direct constructions, Bose-Bush construction of orthogonal arrays.

The main method for constructing good binary codes is to start from good linear codes over an extension field $\mathbb{F}_Q$ and to go down from $\mathbb{F}_Q$ to $\mathbb{F}_2$ using concatenation, subfield codes or the trace. The largest and most useful family of binary linear codes which we encountered in this text, the cyclic codes, are constructed in this fashion. The parent codes are the Reed-Solomon codes and we go down using the trace. More generally, we use parent codes defined over an extension field of $\mathbb{F}_q$ in order to construct linear q-ary codes. In this process the linearity of the parent code over the extension field is not really essential. Imagine a q^m-ary parent code $\mathcal{B}$, which is not necessarily q^m linear but only q linear. In other words, we consider the alphabet of $\mathcal{B}$ not as a field but only as an m-dimensional vector space $V(m, q)$ over $\mathbb{F}_q$. In this situation we can generate q-ary linear codes in many ways, replacing the trace by a linear functional $\Phi : V(m, q) \longrightarrow \mathbb{F}_q$ of our choice (Φ should be surjective). Such codes $\mathcal{B}$ are known as **additive codes.**

18.1 Definition. *A q-linear q^m-ary code $[n, k]_{q^m}$ is a km-dimensional $\mathbb{F}_q$ subspace of $\mathcal{C} \subseteq E^n$, where $E = \mathbb{F}_q^m$. In particular, $\mathcal{C}$ has q^{km} codewords.*

As the alphabet E is a vector space, the minimum distance d equals the minimum weight, where the weight of a codeword of $\mathcal{C}$ is of course the number of coordinates (there are n coordinates) with a nonzero entry (the entries are elements of the vector space $\mathbb{F}_q^m$). Observe that the dimension k need not be an integer. It can have m in the denominator. This is a natural notation, as it facilitates the comparison with q^m-ary linear codes, which form a subfamily.

One way of thinking of a q^m-ary q-linear code of length n is to see it as a linear code over $\mathbb{F}_q$ of length nm, whose coordinates have been grouped into n blocks of m each. One advantage of this point of view is that it shows the use of duality. Define the dual with respect to the dot product on $\mathbb{F}_q^{nm}$. The dual of a k-dimensional q^m-ary q-linear code has complementary dimension $n - k$.

A generator matrix G of a q-linear $[n, k]_{q^m}$ code is a (km, n) matrix with entries from $E = \mathbb{F}_q^m$ whose rows form an $\mathbb{F}_q$ basis of the code (the codewords are the $\mathbb{F}_q$-linear combinations of the rows of G). A check matrix is a generator matrix of the dual code. Observe here that km is of course an integer, but k need not be integer.

So far the main features of the theory of linear codes generalize smoothly to additive codes. This is true of the principle of duality as well. Let $\mathcal{C}$ be a q-linear $[n, k]_{q^m}$ code. The same argument as in the case of linear codes shows that $\mathcal{C}$ is an OA of strength t if and only if $\mathcal{C}^\perp$ has minimum weight $> t$.

In the category of additive codes, it is much easier to go down to smaller alphabets than in the category of linear codes. In the case of linear codes, we are restricted to subfields, and those are rare. For example, a field with 2^p elements, where p is a prime, has only one proper subfield, the prime field $\mathbb{F}_2$. In the case of additive codes, the alphabet forms a vector space over the ground field. We can go down to vector spaces of an arbitrary smaller dimension. Here is a simple construction due to C. L. Chen [49] which makes use of this additional degree of liberty:

Let $\mathcal{C}$ be a q^m-ary q-linear code $[n, k, d]$. Application of an arbitrary regular (m, m) matrix in each coordinate yields a code which is equivalent to $\mathcal{C}$. We can therefore assume that $\mathcal{C}^\perp$ contains a word which in each coordinate has as entry either $(1, 0, \ldots, 0)$ or $(0, \ldots, 0, 0)$. The q-ary dimension of $\mathcal{C}$ is km. Consider the subcode $\mathcal{D} \subset \mathcal{C}$ consisting of those words which vanish in the first position of each coordinate. As there are n conditions to satisfy, we would expect $\mathcal{D}$ to have codimension n in $\mathcal{C}$. However, due to the presence of our special word in $\mathcal{C}^\perp$, the last of these n conditions is automatically satisfied. We have $dim_q(\mathcal{D}) \geq km - (n - 1)$. Clearly we can omit this last position in each coordinate and obtain a q^{m-1}-ary q linear code of q^{m-1}-ary dimension $\geq \frac{km - (n-1)}{m-1} = k - \frac{n-1-k}{m-1}$.

18.2 Theorem (Chen projection). *If there is a q^m-ary q-linear code $[n, k, d]$ with $m > 1$, there is a q^{m-1}-ary q-linear code $[n, k - \dfrac{n - k - 1}{m - 1}, d]$.*

Observe that the proof of Theorem 18.2 uses the same argument as construction Y_1 for linear codes (see Exercise 5.1.8).

Theorem 18.2 allows us to start from a good q^m-linear code and to derive q^{m-1}-ary, q^{m-2}-ary ... q-linear codes by repeated application of Chen projection. As an example, start from the linear $[65, 61, 4]_8$ (the elliptic quadric in $PG(3, 8)$). Theorem 18.2 produces an additive quaternary code $[65, 59.5, 4]_4$. Can we reach these parameters using linear quaternary codes? We would have to construct something better, a linear $[65, 60, 4]_4$ code. This is equivalent to a 65-cap in $PG(4, 4)$, by Proposition 17.28. Such a cap does not exist. The maximum size of a cap in $PG(4, 4)$ is known to be 41 (see [76]).

For another example, start from a hyperoval in $PG(2, 16)$, which is equivalent to a linear $[18, 15, 4]_{16}$ code. Chen projection yields an additive $[18, 14\frac{1}{3}, 4]_8$ code. A linear $[18, 15, 4]_8$ code does not exist, as it would be equivalent to an 18-arc in $PG(2, 8)$, an impossibility. The dual of our $[18, 14\frac{1}{3}, 4]_8$ code has dimension $11/3$ and therefore 2^{11} codewords. Its strength is 3. The parameters as an orthogonal array are therefore $OA_4(3, 18, 8)$.

Caps and additive codes in computer memory systems

Codes with small distance have a long history in computer memory systems. In the literature the binary codes with $d = 4$ are known as SEC-DED codes (*single error-correcting and double error-detecting*). Naturally they are derived by shortening from codes $[2^r, 2^r - (r + 1), 4]_2$ (the extended binary Hamming codes). Hsiao has given a construction of codes with these parameters which is more symmetric than adding a parity check bit to the Hamming codes. These **odd-weight column codes** (the name describes the construction) *have been widely implemented by IBM and the computer industry worldwide* (see Chen and Hsiao [52]).

An interesting situation arises when a multiple-bit-per-chip organization is used. In this case the bits are grouped together in **bytes,** where each byte has m bits. The codes can then be considered as defined over an alphabet with 2^m elements. The class of SBC-DBD codes (*single byte error-correcting and double byte error-detecting*) are by definition 2^m-ary additive codes of minimum distance $d = 4$.

Most constructions in the computer memory literature use $\mathbb{F}_{2^m}$ linearity. In this case we have equivalence with caps in characteristic 2. The idea is that some errors tend to occur in bursts. We want to be able to correct several errors if only they occur within the same byte. That is what 2^m-ary codes with distance ≥ 3 do. Assume $d > 3$. If decoding does not work, then more than one byte must be in error. In this sense these codes (caps) will in the same process correct one byte error and detect two byte errors. Part of the theory of caps and of high-dimensional codes has been developed in parallel in this literature. For example, we find construction Y_1 from coding theory in Chen and Hsiao [52]. C. L. Chen's paper [48] contains a construction of the ovoids as well as the product construction and in particular the doubling construction for caps.

Application of Chen projection to hyperovals and ovoids yields additive codes with $d = 4$ and very high dimension. Here are some examples of $\mathbb{F}_2$-linear code parameters constructed in this way:

$$[65, 59\tfrac{1}{2}, 4]_4, \quad [1025, 1016\tfrac{1}{2}, 4]_4,$$

$$[18, 14\tfrac{1}{3}, 4]_8, \quad [257, 252, 4]_8, \quad [1025, 1019, 4]_8,$$

$$[34, 30\tfrac{1}{2}, 4]_{16}, \quad [1025, 1020\tfrac{1}{4}, 4]_{16}.$$

For example, compare the last two 8-ary additive codes above with corresponding 8-ary linear codes. The largest caps known have 208 points in $PG(4, 8)$, 695 points in $PG(5, 8)$. Equivalently, this yields linear codes $[208, 203, 4]_8, [695, 689, 4]_8$, considerably weaker parameters than those of the additive codes derived from caps via Chen projection. Some of these codes have found multiple applications in computer memory systems; see Chen [49] and Chen and Grosbach [51].

Another interesting feature is the use of symmetry. Kaneda and Fujiwara [123] show how certain symmetries can be used to partition the encoding-decoding circuitry into identical modules and to parallelize the computations. We conclude that practical needs demand the construction of caps and of high-dimensional codes with large groups of automorphisms.

Additive codes in deep-space communication

The old NASA standard for communication with spacecrafts used a concatenation scheme with a rate $1/2$ convolutional code as inner code and a Reed-Solomon code $[255, 223, 33]_{256}$ as outer code. In Hattori, MacEliece and Solomon [108] it is proposed to replace this Reed-Solomon code by an additive 256-ary code. The construction of suitable additive codes is based on the theory of **Reed-Solomon subspace codes.** However, things are simpler than that.

The extended q^m-ary Reed-Solomon codes have parameters $[q^m + 1, q^m + 2 - d, d]_{q^m}$, for all $d \leq q^m + 1$. Theorem 18.2 yields q-linear q^{m-1}-ary codes with parameters

$$[q^m + 1, q^m + 2 - d - \frac{d-2}{m-1}, d]_{q^{m-1}} \text{ for } m > 1.$$

In order to obtain a 2^8-ary code that can rival the Reed-Solomon code, we start from a 2^9-ary extended Reed-Solomon code (case $q = 2, m = 9$). In cases $d = 30$ and $d = 42$, this yields additive 256-ary codes

$$[513, 480\frac{1}{2}, 30]_{256} \text{ and } [513, 467, 42]_{256}.$$

The codes proposed in [108], based on computer simulations, are shortened versions of these codes: $[511, 478\frac{1}{2}, 30]_{256}$, $[511, 465, 42]_{256}$.

Convolutional codes

As we mentioned that deep space communication systems use a concatenation scheme one of whose components is a convolutional code, it should at least be explained what the construction idea of these codes is.

We concentrate on a special case, binary convolutional codes of rate $1/2$. Imagine the incoming bitstream as a (potentially) infinite sequence $(x_0, x_1, x_2, \dots)$. The encoding of each bit depends also on the $m - 1$ preceding bits. That is why we define $x = (0, \dots, 0, x_0, x_1, x_2, \dots)$, where $x_{-1} = \dots = x_{-(m-1)} = 0$.

Fix two elements $g_1, g_2 \in \mathbb{F}_2^m$. When bit x_i comes in, we consider a window W_i of m consecutive bits: $W_i = (x_{i-m+1}, \dots, x_i) \in \mathbb{F}_2^m$. The output generated at that time is the pair $(W_i \cdot g_1, W_i \cdot g_2) \in \mathbb{F}_2^2$. As i increases, the window shifts to the right (or, if you prefer, the window is fixed, the bitstream moves one to the left). The encoded sequence is

$$y = (W_0 \cdot g_1, W_0 \cdot g_2, W_1 \cdot g_1, W_1 \cdot g_2, W_2 \cdot g_1, W_2 \cdot g_2, \dots).$$

As each source bit x_i produces two bits of the encoded message, one speaks of rate $1/2$. Using k instead of 2 tuples g_i would result in a rate $1/k$ code.

Take the example where $m = 4$ and $g_1 = 1101$, $g_2 = 1011$. Let the incoming bitstream, including the prefix of $m - 1 = 3$ zeroes, be

$$x = 000|010011000101\dots$$

The first window is $W_0 = 0000$ and produces 00 of course. The second window is 0001 and produces $(W_1 \cdot g_1, W_1 \cdot g_2) = 11$. The encoded message is

$$y = 00|11|01|10|00|10|11|01|11|11|01|01\dots$$

Observe that each window ends with the $(m - 1)$-tuple that the next window starts with. We can base ourselves on the $(m - 1)$-tuples (the *states*) and set up the state diagram, where an arrow points from state S_1 to S_2 if the $(m - 2)$-tuple ending S_1 is the one starting S_2. It follows that two arrows lead out of each state and two lead into each state. If I know the state S and the next bit x_i, this gives me window $W_i = (S|x_i)$. The next state is

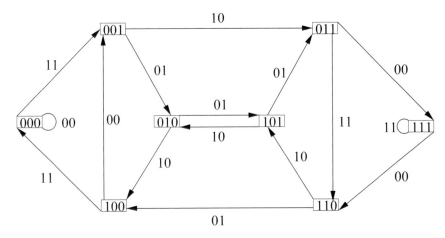

FIGURE 18.1: A state diagram

the $(m-1)$-tuple ending W_i. The state diagram is completed by writing the output $(W \cdot g_1, W \cdot g_2)$ over each arrow; see Figure 18.1. Movement begins in state 000 (courtesy of the prefix). Given a state, the next input bit decides which of the two arrows emanating from the state to follow. This generates two output bits (the ones written alongside the arrow) and leads to the next state.

As applied mathematicians have a predilection for polynomials and power series, it is customary to represent the incoming bitstream as a power series: instead of $0, 0, 0, x_0, x_1, x_2 \ldots$ one writes $m(X) = x_0 X^3 + x_1 X^4 + x_2 X^5 + \ldots$. The convolutional code is then described as a transformation on formal power series, thus giving some superficial dignity to the subject. The m-tupels $g_1, g_2, \ldots$ are represented by polynomials, in our case $g_1(X) = X^3 + X + 1$, $g_2(X) = X^3 + X^2 + 1$. The calculation of the dot products with g_i corresponds to the computation of the power series $g_i(X)m(X)$ and these computations can be done efficiently using shift registers.

Convolutional codes are linear: the sum of two input bitstreams is mapped to the sum of the output bitstreams. The minimum distance therefore equals the minimum weight, which is the smallest weight of a possible output stream different from the all-0 sequence. In our example the minimum distance is $d = 6$. In fact, we have to leave state 000 along an edge generating 11 and to come back via state 100, generating 11 again. There is no path along arrows from state 001 to state 100 generating an output of weight less than 2. Such a path of weight 2 is obtained passing through state 010. As $d = 6$, the usual argument shows that, in general, we can correct up to two errors using this convolutional code. We will not consider the question of decoding algorithms here.

The Bose-Bush construction

Let $a \geq b$ and fix a surjective $\mathbb{F}_q$-linear mapping $\Phi : \mathbb{F}_{q^a} \longrightarrow \mathbb{F}_q^b$, where of course $a \geq b$. We used this setting in Section 16.7 and constructed a q^b-ary code $\mathcal{C}_q(a, b)$ of length q^a, with q^{a+b} codewords, which has strength 2 and minimum distance $q^a - q^b$.

Recall the construction (see Theorem 16.29): the coordinates are parametrized by the elements $x \in \mathbb{F}_{q^a}$, the codewords by pairs (u, v), where $u \in \mathbb{F}_{q^a}$, $v \in \mathbb{F}_q^b$, and the entry of codeword $c(u, v)$ in coordinate x is $c(u, v)_x = \Phi(ux) + v$.

Clearly $\mathcal{C}_q(a, b)$ is q-linear. In the terminology of the present chapter we see $\mathcal{C}_q(a, b)$ as a q-linear q^b-ary code of dimension $(a + b)/b = 1 + a/b$. As it has strength 2, its dual is a q-linear $[q^a, q^a - (1 + a/b), 3]_{q^b}$ code.

Moreover, we can write $\mathcal{C}_q(a, b)$ as a direct sum:

$$\mathcal{C}_q(a, b) = \mathcal{D}_q(a, b) \oplus \mathcal{R}_q(a, b),$$

where $\mathcal{D}_q(a, b)$ has as codewords the $c(u, 0)$ and the codewords of $\mathcal{R}_q(a, b)$ are the constant words $c(0, v)$. Both $\mathcal{R}_q(a, b)$ (which we call the repetition code) and $\mathcal{D}_q(a, b)$ are q linear, the first of dimension 1, the second of dimension $a/b \geq 1$. The repetition code clearly has strength 1.

18.3 Theorem. *The code $\mathcal{C}_q(a, b)$ (recall $a \geq b$) of Theorem 16.29 is a q-linear q^b-ary code of dimension $(a+b)/b \geq 2$, strength 2 and minimum distance $q^a - q^{a-b}$. It is the direct sum of the repetition code $\mathcal{R}_q(a, b)$ (a q-linear $[q^a, 1]_{q^b}$-code of constant weight $q^a - q^{a-b}$ and strength 1) and the q-linear q^b-ary code $\mathcal{D}_q(a, b)$.*

Consider the construction problem of strength 2 orthogonal arrays. If we restrict to linear codes, the situation is familiar. A linear code $[n, r]_q$ of strength 2 exists if $n \leq (q^r - 1)/(q - 1)$. This observation led us to the Hamming and Simplex codes (see Section 3.4). What happens if we generalize from linear codes to additive codes? The question is: given r, a rational number such that mr is integer, determine the maximal length n of a q-linear q^m-ary code of dimension r and strength 2. Using linear q^m-ary codes, we reach length $(q^{mr} - 1)/(q^m - 1)$ if r should be an integer. The following construction shows that our codes $\mathcal{C}_q(a, b)$ and their subcodes $\mathcal{R}_q(a, b)$, the repetition codes, can be used to obtain additive codes of strength 2 with larger length.

18.4 Proposition. *Assume the following q^m-ary q linear codes exist:*

- *An $[n, K]$ code $\mathcal{C}$ of strength 2 containing an $[n, k]$ subcode $\mathcal{R}$ of strength 1, and*

- *a code $[l, K - k]$ of strength 2 (the auxiliary code).*

Then we can construct such a code $[n + l, K]$ of strength 2.

PROOF It may be easiest to work with $\mathbb{F}_q$ generator matrices. Choose such a generator matrix (with Km rows) for $\mathcal{C}$ such that the first km rows form a generator matrix for $\mathcal{R}$. Lengthen this generator matrix by a generator matrix of the auxiliary code in the last $m(K-k)$ rows, and by a 0 matrix in the first rows. The result is a generator matrix of a code with the right dimension, clearly of strength 2. $\qquad\square$

The generator matrix of the code resulting from Proposition 18.4 has the form $G = \begin{pmatrix} A & 0 \\ C & D \end{pmatrix}$, where the left part generates $\mathcal{C}$, A generates $\mathcal{R}$ and D generates the strength 2 auxiliary code. Should the auxiliary code possess a strength 1 subcode again, it will, together with $\mathcal{R}$, generate a strength 1 subcode of the resulting code. It is then possible to apply Proposition 18.4 again.

Start with code $\mathcal{C}_q(a, b)$. As it contains a b dimensional strength 1 subcode ($\mathbb{F}_q$ dimensions), we can apply Proposition 18.4 with $\mathcal{C}_q(a-b, b)$ as auxiliary code (provided $a - b \geq b$). The resulting code has a $2b$ dimensional subcode of strength 1. Apply Proposition 18.4 again, with $\mathcal{C}_q(a-2b, b)$ as auxiliary code. Procede inductively in this fashion. The last inductive step occurs when $a - lb \geq b$, but $a - (l+1)b < b$. Clearly $l = \lfloor a/b \rfloor - 1$. In this last step there is no need to use a strength 2 code with a strength 1 subcode. Instead of $\mathcal{C}_q(a - lb, b)$ we use the code of length $q^{a-lb} + 1$ (see Exercise 18.1.4) and obtain as end product a code of length $1 + \sum_{i=0}^{l} q^{a-ib}$.

18.5 Theorem. *Let $a \geq b$ be integers, $l = \lfloor a/b \rfloor - 1$. Then there is a q-linear q^b-ary code of strength 2, dimension $(a+b)/b$ and length*

$$n = 1 + \sum_{i=0}^{l} q^{a-ib} = 1 + (q^{a+b} - q^{a-lb})/(q^b - 1).$$

The OA parameters of Theorem 18.5 are due to Bose and Bush [36], see also Hedayat, Sloane and Stufken [109]. The approach used in our proof of Theorem 18.5 is due to Y. Edel; see [17].

How much better is the Bose-Bush construction of strength 2 orthogonal arrays than the linear construction? Only when a is a multiple of b is a direct comparison possible. We leave it to the reader to verify that, in these cases, the length in Theorem 18.5 equals the maximum length of linear OA. We gain nothing in these situations. What Theorem 18.5 does: it embeds the linear construction in a vaster family of strength 2 OA where the index λ and the alphabet are powers of the same prime. In OA notation we have

$$OA_\lambda(2, n, Q), \text{ where } Q = q^b, \ \lambda = q^{a-b}$$

and n is the length from Theorem 18.5.

Let $b < a < 2b$. Write $a = b + i$, where $i < b$. Then $l = 0$, $n = q^{b+i} + 1$ and $\lambda = q^i$. We obtain $OA_{q^i}(2, q^{b+i}+1, q^b)$, for example, $OA_2(2, 9, 4), OA_4(2, 17, 8)$.

A direct construction

We want to construct a family of low-dimensional additive codes which represent an additive (generalized) version of an interesting family of linear codes, the generalized Kasami codes. The construction makes use of basic properties of the trace and norm; see Chapter 12.

Consider field extensions

$$\mathbb{F}_q \subset F_0 = \mathbb{F}_{q^{r'}} \subset F = \mathbb{F}_{q^r},$$

where $r = 2r'$. Denote traces and norms by $T : F \longrightarrow F_0$, $N : F \longrightarrow F_0$, $Tr : F_0 \longrightarrow \mathbb{F}_q$. Choose $m \leq r'$ and an $\mathbb{F}_q$-linear surjective mapping $\Phi : F_0 \longrightarrow \mathbb{F}_q^m$.

Φ is a q-linear mapping from the r'-dimensional space F_0 onto the m-dimensional space $\mathbb{F}_q^m$. It is represented simply by an (r', m) matrix of rank m with entries in $\mathbb{F}_q$. A concrete representation is as follows: choose a set $\{\eta_1, \ldots, \eta_m\} \subset F_0$, which is linearly independent over $\mathbb{F}_q$. Then

$$\Phi(y) = (Tr(\eta_1 y), Tr(\eta_2 y), \ldots, Tr(\eta_m y)) \text{ for } y \in F_0.$$

We define an additive code $\mathcal{C}$ of length $q^r - 1$ with entries in $\mathbb{F}_q^m$, which is $\mathbb{F}_q$ linear. The coordinates are parametrized by $x \in F^*$, the codewords by pairs (u, v), where $u \in F, v \in F_0$. The entry of the corresponding codeword $c(u, v)$ in coordinate x is defined as

$$c(u, v)_x = \Phi(T(ux) + vN(x))$$

$$= (Tr(\eta_j \cdot (T(ux) + vN(x)))_j \ , j = 1, 2, \ldots, m.$$

It is clear that the mapping $(u, v) \mapsto c(u, v)$ is $\mathbb{F}_q$ linear. There is no need for an explicit proof that the mapping has zero kernel, as we will determine the weight of $c(u, v)$ in general. As this weight is never 0 unless $u = v = 0$, it will follow that $dim_{\mathbb{F}_q}(\mathcal{C}) = 3r' = r + r'$. All that is needed for the determination of the weight is the following elementary lemma:

18.6 Lemma. *Let $\mathbb{F}_q \subset L = \mathbb{F}_{q^2}$ and denote by T, N the trace and norm, respectively. Let elements $a \in L, b \in \mathbb{F}_q^*, u \in \mathbb{F}_q$ be given. Denote by $\nu(a, b, u)$ the number of elements $x \in L^*$ satisfying*

$$T(ax) + bN(x) = u.$$

Then

$$\nu(a, b, u) = \begin{cases} q & \text{if } u = 0 \\ 1 & \text{if } bu = -N(a) \\ q + 1 & \text{otherwise.} \end{cases}$$

PROOF　　We have a quadratic extension, so $T(x) = x + x^q$, $N(x) = x^{q+1}$. Compute

$$N(x + \frac{a^q}{b}) = (x + \frac{a^q}{b})(x^q + \frac{a}{b}) = N(x) + N(a)/b^2 + T(ax)/b.$$

Our equation is therefore equivalent to

$$N(x + \frac{a^q}{b}) = \frac{u}{b} + \frac{N(a)}{b^2}.$$

If $bu = -N(a)$, there is only one solution $x = -a^q/b$. In all other cases the right side is nonzero. There are therefore $q + 1$ solutions $x \in L$. One of them is $x = 0$ if and only if $u = 0$. 　　　　□

It is easier to count zeroes than nonzeroes. Let $\chi_0(c(u,v))$ be the number of coordinates x such that $c(u,v)_x = 0$ (here 0 represents the m-tuple $(0, 0, \dots, 0)$).

Observe that $Ker(\Phi)$ is an $(r' - m)$-dimensional subspace of F_0. The definition shows that

$$c(u, v)_x = 0 \longleftrightarrow T(ux) + vN(x) \in Ker(\Phi).$$

In the case $u = 0, v \neq 0$, the condition is $vN(x) \in Ker(\Phi)$. As every nonzero element of F_0 has precisely $q^{r'} + 1$ preimages under N, we obtain

$$\chi_0(c(0, v)) = (q^{r'} + 1)(q^{r'-m} - 1).$$

In the case $v = 0, u \neq 0$ the condition is $T(ux) \in Ker(\Phi)$. Each element of F_0 has precisely $q^{r'}$ preimages under T. Observing that we may not count $x = 0$ as a preimage, we obtain

$$\chi_0 c((u, 0)) = q^{r'-m} q^{r'} - 1 = q^{r-m} - 1.$$

Now let $uv \neq 0$. We can use Lemma 18.6. Assume at first $-N(u)/v \in Ker(\Phi)$. We obtain

$$\chi_0(c(u,v)) = q^{r'} + 1 + (q^{r'-m} - 2)(q^{r'} + 1) = q^{r-m} - q^{r'} + q^{r'-m} - 1.$$

The final case is $-N(u)/v \notin Ker(\Phi)$. The result is

$$\chi_0(v(u,v)) = q^{r'} + (q^{r'-m} - 1)(q^{r'} + 1) = q^{r-m} + q^{r'-m} - 1.$$

As $wt(c(u,v)) = q^r - 1 - \chi_0(c(u,v))$, the weights are as given in the following theorem:

18.7 Theorem.　*There is a q^m–ary $\mathbb{F}_q$ linear code of length $q^r - 1$ and dimension $3r'/m$, where $r = 2r'$, whose codewords $c(u,v)$ have the following weights:*

$$\begin{cases} 0 & \text{if } u = v = 0 \\ q^r - q^{r-m} & \text{if } v = 0, u \neq 0 \\ q^r - q^{r-m} + q^{r'} - q^{r'-m} & \text{if } -N(u)/v \in Ker(\Phi) \\ q^r - q^{r-m} - q^{r'-m} & \text{if } -N(u)/v \notin Ker(\Phi). \end{cases}$$

The underlying linear codes are obtained in the case $m = 1$, when we can choose Φ to be the trace $tr_{F_0|F_q}$. These linear codes are generalizations of the Kasami codes. The case of linear $p-$ary codes $(p = q^m)$ makes its appearance in Delsarte and Goethals [67]. The binary linear case $(q^m = 2)$ is the Kasami codes.

The $c(u, 0)$ form an $r-$dimensional subcode with constant weight $q^r - q^{r-m}$, and among the remaining codewords $c(u, v), v \neq 0$ only two different weights occur. Clearly concatenation works for additive codes as well as for linear codes. Apply concatenation with the Simplex code $[(q^m - 1)/(q-1), m, q^{m-1}]_q$. The result is a linear q-ary code. The length picks up a factor of $(q^m - 1)/(q - 1)$, and the weights have to be multiplied by q^{m-1}.

18.8 Theorem. *Concatenation of C with the Simplex code $[(q^m - 1)/(q - 1), m, q^{m-1}]_q$ yields a linear $3r'-$dimensional q-ary code $\tilde{C}$ of length $(q^r - 1)(q^m - 1)/(q - 1)$ whose codewords $c(\tilde{u}, v)$ have the following weights:*

$$\begin{cases} 0 & if\ u=v=0 \\ q^{r+m-1} - q^{r-1} & if\ v = 0, u \neq 0 \\ q^{r+m-1} - q^{r-1} + q^{r'+m-1} - q^{r'-1} & if\ -N(u)/v \in Ker(\Phi) \\ q^{r+m-1} - q^{r-1} - q^{r'-1} & if\ -N(u)/v \notin Ker(\Phi). \end{cases}$$

Finally, we apply construction X with the Simplex code $[(q^{r'} - 1)/(q - 1), r', q^{r'-1}]_q$ as auxiliary code. The length increases by $(q^{r'} - 1)/(q - 1)$. The weights in the subcode consisting of the words with $v = 0$ remain unchanged. The remaining weights increase by $q^{r'-1}$. Only two nonzero weights survive.

18.9 Theorem. *For all $1 \leq m \leq r'$, there is a q-ary linear code of length $q^m(q^{r'} - 1)(q^{r'} - q^{r'-m} + 1)/(q - 1)$ and dimension $3r'$ whose only nonzero weights are $d = q^{r+m-1} - q^{r-1}$ and $w_2 = q^{r+m-1} - q^{r-1} + q^{r'+m-1}$.*

Exercises 18.1

18.1.1. *Let C be a q-linear $[n, k]_{q^m}$ code whose dual has minimum weight $> t$. Determine the parameters of C as an orthogonal array.*

18.1.2. *Formulate a generalization of the method of concatenation (Definition 5.8, Theorem 5.9) valid for additive q-linear codes as outer codes.*

18.1.3. *Prove: if there exists a q-linear $[n, k]_{q^m}$ code of strength t and $m > 1$, then there exists a q-linear $[n, k + \dfrac{k - 1}{m - 1}]_{q^{m-1}}$ code of strength t.*

18.1.4. *Construct q-linear q^b-ary codes of length $q^a + 1$, strength 2 and dimension $(a + b)/b$ for every $b \leq a$.*

18.1.5. *Let $\mathcal{D}_q(a, b)$ (where $a \geq b$) be the q-linear q^b-ary code of dimension a/b and length q^a constructed in this section. Determine its strength and minimum distance.*

18.1.6. *Let $a \geq b$ and $l = \lfloor a/b \rfloor - 1$. Show that*

$$\sum_{i=0}^{l} q^{a-ib} = (q^{a+b} - q^{a-lb})/(q^b - 1).$$

18.1.7. *Determine the parameters of the OA resulting from the Bose-Bush construction when $2b \leq a < 3b$. Give at least two examples.*

18.1.8. *Determine the parameters of the codes in Theorem 18.9 for parameters $m = 1, r' = 2$ and show that they are d-optimal for $q \leq 4$.*

18.1.9. *Construct linear codes $[70, 9, 32]_2$ and $[196, 9, 96]_2$.*

18.2 The cyclic theory of additive codes

We developed the theory of **linear** cyclic codes in earlier chapters. Here we extend the theory to the category of additive codes. Formally, this is self-contained: the earlier linear theory is proved once again in the present chapter. It is instructive that this does not take much space or time. Some of the basic facts from those earlier chapters are used (subfield code, trace code, Galois closure, the Delsarte theorem and of course cyclotomic cosets). In this section we follow [20].

Start from Definition 18.1. We will develop the general theory of additive codes which are cyclic in the monomial sense under the general assumption that the characteristic p of the underlying field is coprime to the length n of the code. The main effect of the assumption is that the action of a (cyclic) group of order n on a vector space over a field of characteristic p is completely reducible, by Maschke's theorem (see Definition 18.14 and Theorem 18.15).

18.2.1 Code equivalence and cyclicity

Recall the notion of permutation equivalence from Section 12.4.

18.10 Definition. *Codes C and $\mathcal{D}$ are **permutation equivalent** if there is a permutation π on n objects such that*

$$(x_1, x_2, \ldots, x_n) \in C \iff (x_{\pi(1)}, x_{\pi(2)}, \ldots, x_{\pi(n)}) \in \mathcal{D}$$

for all $x = (x_1, \ldots, x_n) \in C$.

The notion of permutation equivalence can be used for all codes of fixed block length n. It uses the symmetric group S_n as the group of motions. Two codes are equivalent if they are in the same orbit under S_n. The stabilizer under S_n is the permutation automorphism group of the code.

In the special case of additive codes, the following more general notion of equivalence is more natural.

18.11 Definition. *q-linear q^m-ary codes C and D are* **monomially equivalent** *if there exist a permutation π on n objects and elements $A_i \in GL(m, q)$ such that*

$$(x_0, x_1, \ldots, x_{n-1}) \in C \iff (A_0 x_{\pi(1)}, A_1 x_{\pi(2)}, \ldots, A_{n-1} x_{\pi(n-1)}) \in D$$

for all $x = (x_0, \ldots, x_{n-1}) \in C$.

The group of motions is the wreath product of $GL(m, q)$ and S_n, of order $|GL(m, q)|^n \times n!$ and two additive codes are equivalent in the monomial sense if they are in the same orbit under the action of this larger group. The elements of the wreath product are described by $(n+1)$-tuples $(A_0, \ldots, A_{n-1}, \pi)$ where the coefficients $A_i \in GL(m, q)$ and the permutation $\pi \in S_n$. Write the elements of the monomial group as $g(A_0, \ldots, A_{n-1}, \pi)$.

18.12 Definition. *A code C of length n is* **cyclic** *in the permutation sense if there is an n cycle $\pi \in S_n$ such that $\pi(C) = C$.*
A q-linear q^m-ary code C is **cyclic** *in the monomial sense if it is invariant under an element $g(A_0, \ldots, A_{n-1}, \pi)$ of the monomial group where $\pi \in S_n$ is an n cycle.*

As the n cycles are conjugate in S_n, it is clear that each code which is cyclic in the permutation sense is permutation equivalent to a code which is invariant under the permutation $\pi = (0, 1, \ldots, n - 1)$.

18.13 Proposition. *Let C be q-linear q^m-ary of length n and cyclic in the monomial sense. Then the following hold:*

- *C is monomially equivalent to a code which is invariant under $g(I, \ldots, I, A'_{n-1}, \pi)$ where $\pi = (0, 1, \ldots, n - 1)$.*

- *If C is invariant under $g(A_0, \ldots, A_{n-1}, (0, \ldots, n-1))$ and $A_0 A_1 \ldots A_{n-1} = I$, then C is cyclic in the permutation sense.*

- *If C is invariant under $g(I, \ldots I, A, (0, \ldots, n-1))$ and $\mathrm{ord}(A)$ is coprime to n, then C is cyclic in the permutation sense.*

PROOF We just saw that we can assume C to be invariant under $g(A_0, A_1, \ldots, A_{n-1}, \pi)$ where $\pi = (0, 1, \ldots, n - 1)$. We have that $(x_i) \in C$ implies $(A_i x_{i+1}) \in C$. Now let

$$C' = \{(B_0 x_0, \ldots, B_{n-1} x_{n-1}) | x = (x_0, \ldots, x_{n-1}) \in C\}.$$

Let $y_i = B_i x_i$. Then $(y_i) \in \mathcal{C}'$ implies $(B_i A_i B_{i+1}^{-1} y_{i+1}) \in \mathcal{C}'$. Choose the B_i such that (for $i = 0$) $B_1 = B_0 A_0$, (for $i = 1$) $B_2 = B_1 A_1 = B_0 A_0 A_1$, up to (for $i = n - 2$) $B_{n-1} = B_{n-2} A_{n-2} = B_0 A_0 \ldots A_{n-2}$. It follows that $\mathcal{C}'$ is invariant under $g(I, \ldots, I, A, (0, 1, \ldots, n - 1))$ as claimed. Here $A = B_0(A_0 A_1 \ldots A_{n-1}) B_0^{-1}$, a conjugate of $A_0 A_1 \ldots A_{n-1}$. The first statement concerning cyclicity in the permutation sense follows. Now let $\mathcal{C}$ be invariant under $g = g(I, \ldots, I, A, (0, \ldots, n-1))$ and $ord(A) = u$ coprime to n. Then g^u has u of its coefficients equal to A, and the others $= I$. The permutation of g^u is an n cycle. The first statement implies that $\mathcal{C}$ is cyclic in the permutation sense. $\qquad\qquad\Box$

The cyclic group $G = \langle g(A_0, \ldots, A_{n-1}, (0, \ldots, n-1)) \rangle$ has an order a multiple of n and acts on an n-dimensional vector space V over $\mathbb{F}_q$. The cyclic codes are the G submodules of V. The theory simplifies considerably if the action of G on V is completely reducible.

18.14 Definition. *Let the group G act on a vector space $V = V(n, K)$. An **irreducible submodule** $A \subset V$ is a nonzero G submodule which does not contain a proper G submodule (different from 0 and from A itself). The action of G is **completely reducible** if, for every G submodule $A \subset V$, there is a direct complement B of A which is a G module.*

If the action is completely reducible, it suffices in a way to know the irreducibles, as every G submodule is a direct sum of irreducibles. There is still work to do, as the representation of a module as a direct sum of irreducibles may not be unique. However, the situation is a lot simpler than in the cases where complete reducibility is not satisfied. This is where Maschke's theorem comes in.

18.15 Theorem (Maschke). *In the situation of Definition 18.14, let the order of G be coprime to the characteristic p of the underlying field. Then the action of G is completely reducible.*

See also [130], p. 666. This fundamental theorem is the reason why, in the theory of cyclic codes, it is often assumed that $\gcd(n, p) = 1$.

Additive codes which are cyclic in the permutation sense

Recall the situation: W is the cyclic group generated by the permutation $(0, 1, \ldots, n - 1)$. It acts on the vector space $V(n, q)$ as a cyclic permutation of the coordinates. The W submodules are precisely the cyclic linear codes over $\mathbb{F}_q$. We assume $\gcd(n, p) = 1$. Because of Maschke's theorem, the action of W on $V(n, q)$ is completely reducible.

Let $r = ord_n(q)$ be the order of q when calculating mod n. Then $n | (q^r - 1)$ and r is the smallest natural number with this property. Let $F = \mathbb{F}_{q^r}$. We find then a cyclic group of order n in F^*. It will be profitable to identify W with this subgroup of F^*. Let $W = \langle \alpha \rangle$.

As we are interested in cyclic **additive** codes, our alphabet is $E = \mathbb{F}_q^m$ and we study the action of W on $V_m = E^n = \mathbb{F}_q^{mn}$. Keep in mind, however, that we consider the elements of V_m (the codewords) not as nm-tuples over $\mathbb{F}_q$ but rather as n-tuples over E. Refer to the n entries of a codeword as the outer coordinates (in bijection with the elements $\alpha^i, i = 0, \ldots, n-1$ of W) and to the m coordinates of the alphabet E as the inner coordinates. Let $V_{m,F} = V_m \otimes F = (F^m)^n$ be obtained by constant extension with the action of W on outer coordinates. We will relate the codes $C \subset V_m = E^n = \mathbb{F}_q^{nm}$ to their constant extensions $C \otimes F \subset V_{m,F}$ and use basic facts concerning relations between codes defined over a larger field F and a subfield $\mathbb{F}_q$. In particular, we associate to each subcode $U \subset V_{m,F}$ its trace code $tr(U) \subset V_m$ obtained by applying the trace $tr = tr_{F|\mathbb{F}_q}$ in each inner coordinate.

Project to one of the m inner coordinates. We then obtain spaces $\mathbb{F}_q^n$ and F^n. The elements of F^n can be uniquely described by univariate polynomials $p(X) \in F[X]$ of degree $< n$, where the codeword defined by $p(X)$ is the evaluation $ev(p(X)) = (p(\alpha^i)), i = 0, \ldots, n-1$. Doing this for each inner coordinate, we see that the elements of $V_{m,F}$ can be uniquely described by tuples $(p_1(X), \ldots, p_m(X))$ of polynomials $p_j(X) \in F[X]$ of degree $< n$, where the corresponding codeword is the evaluation $ev(p_1(X), \ldots, p_m(X)) = (p_1(\alpha^i), p_2(\alpha^i), \ldots, p_m(\alpha^i))_i$ (an n-tuple of m-tuples, each of the nm entries in F).

Consider the Galois group $G = Gal(F|\mathbb{F}_q)$ (cyclic of order r) and its orbits on $\mathbb{Z}/n\mathbb{Z}$, the cyclotomic cosets. We interpret the elements of $\mathbb{Z}/n\mathbb{Z}$ as the exponents of α in the description of the elements of the cyclic group W. For each polynomial $p(X) \in F[X]$ of degree $< n$, consider the exponents of the monomials occurring in $p(X)$ and how they distribute on the cyclotomic cosets.

18.16 Definition. *Let $\mathcal{P}$ be the space of polynomials in $F[X]$ of degree $< n$. Then $\mathcal{P}$ is an n-dimensional F vector space. Let $A \subseteq \mathbb{Z}/n\mathbb{Z}$ be a set of exponents. The F vector space $\mathcal{P}(A)$ consists of the polynomials $\in F[X]$ of degree $< n$ all of whose monomials have degrees in A. The **Galois closure** $\tilde{A}$ of A is the union of all cyclotomic cosets that intersect A nontrivially.*

Observe that $\mathcal{P}(A)$ is an F vector space of dimension $|A|$ and it is isomorphic to $ev(\mathcal{P}(A)) \subset V_{1,F}$. The terminology of Definition 18.16 is justified by the obvious fact that the Galois closure of the code $ev(\mathcal{P}(A)) \subset V_{1,F} = F^n$ is $ev(\mathcal{P}(\tilde{A}))$.

Linear cyclic codes (case $m = 1$)

We have $\mathcal{P} = \oplus_Z \mathcal{P}(Z)$ where Z varies over the cyclotomic cosets and correspondingly $V_{1,F} = F^n = \oplus_Z ev(\mathcal{P}(Z))$. As the $ev(\mathcal{P}(Z))$ are Galois closed, we also have $V_1 = \mathbb{F}_q^n = \oplus tr(ev(\mathcal{P}(Z)))$ and $dim_{\mathbb{F}_q}(tr(ev(\mathcal{P}(Z)))) = dim_F(ev(\mathcal{P}(Z)) = |Z|$. Let $V_1(Z) = tr(ev(\mathcal{P}(Z)))$. It is our first task to identify the irreducible W submodules (the irreducible cyclic codes) in V_1.

18.17 Theorem. *In the case $m = 1, \gcd(n, p) = 1$, the irreducible cyclic codes in the permutation sense are precisely the $V_1(Z) = tr(ev(\mathcal{P}(Z)))$ of $\mathbb{F}_q$ dimension $|Z|$ where Z is a cyclotomic coset. Each cyclic code can be written as a direct sum of irreducibles in precisely one way. The total number of cyclic codes is 2^u, where u is the number of different cyclotomic cosets.*

We know Theorem 18.17 from earlier chapters. In order to be self-contained, we will prove it in the remainder of this subsection.

It follows from basic properties of the trace that, in the description of a code $\mathcal{C} \subseteq V_1$ by codewords $tr(ev(p(X)))$, it suffices to use polynomials of the form $p(X) = a_1 X^{z_1} + a_2 X^{z_2} + \dots$ where $z_1, z_2, \dots$ are representatives of different cyclotomic cosets.

18.18 Lemma. *Let Z be a cyclotomic coset, $z \in Z, |Z| = s, L = \mathbb{F}_{q^s}$. The $\mathbb{F}_q$ vector space $\langle W^z \rangle$ generated by the u^z where $u \in W$ is L. The codeword $tr(ev(aX^z))$ where $a \in F$ is identically zero if and only if $a \in L^\perp$ where duality is with respect to the trace form.*

PROOF As $ev(\mathcal{P}(Z))$ is F linear and Galois closed of dimension s, it follows that $tr(ev(\mathcal{P}(Z)))$ is $\mathbb{F}_q$ linear of dimension s. Its generic codeword is $tr(ev(aX^z))$. This codeword is identically zero if and only if $a \in \langle W^z \rangle^\perp$. Comparing dimensions shows $dim(\langle W^z \rangle) = s$. Let $u \in W$. Then $u^{zq^s} = u^z$ by definition of a cyclotomic coset. This shows $W^z \subset L$. It follows $\langle W^z \rangle = L$. $\square$

Lemma 18.18 shows that we can assume $a_i \notin L_i^\perp$ for each i. Let us speak of a polynomial in **standard form** in this case, for the moment. Observe that $tr(ev(p(X))) = tr(a_1 X^{z_1}) + tr(a_2 X^{z_2}) + \dots$ and the summands $tr(a_i X^{z_i})$ are in different parts of the direct sum decomposition $V_1 = \oplus_Z V_1(Z)$.

18.19 Lemma. *Let $\mathcal{C} \subseteq V_1$ be a cyclic code and $\mathcal{B} \subseteq \mathcal{P}$ the set of all polynomials $p(X)$ of degree $< n$ such that $tr(ev(p(X))) \in \mathcal{C}$. Let $p(X) = a_1 X^{z_1} + a_2 X^{z_2} + \dots \in \mathcal{B}$ where $z_1, z_2, \dots$ are representatives of different cyclotomic cosets. Then $p^{(l)}(X) = a_1 \alpha^{lz_1} X^{z_1} + a_2 \alpha^{lz_2} X^{z_2} + \dots \in \mathcal{B}$ for all l. The smallest cyclic code containing $tr(ev(p(X)))$ is the code spanned by the $tr(ev(p^{(l)}(X)))$ for all l.*

PROOF The entry of $tr(ev(p(X)))$ in coordinate α^i is $tr(a_1 \alpha^{iz_1} + a_2 \alpha^{iz_2} + \dots)$ After cyclic shift we obtain a codeword whose entry in coordinate α^i is $tr(a_1 \alpha^{(i+1)z_1} + a_2 \alpha^{(i+1)z_2} + \dots) = tr(a_1 \alpha^{z_1} \alpha^{iz_1} + a_2 \alpha^{z_2} \alpha^{iz_2} + \dots)$, which is the trace of the evaluation of $a_1 \alpha^{z_1} X^{z_1} + a_2 \alpha^{z_2} X^{z_2} + \dots$ The first claim follows by repeated application, and the second is obvious. $\square$

Let us complete the proof of Theorem 18.17. Let $\mathcal{C}$ be an irreducible cyclic code and $tr(ev(p(X))) \in \mathcal{C}$ where $p(X)$ is in standard form. Assume $p(X)$ is not a monomial. Let $\mathcal{B}$ be defined as in Lemma 18.19 and $p(X) = a_1 X^{z_1} +$

$a_2 X^{z_2} + \cdots \in \mathcal{B}$. Lemma 18.19 implies that $a_1 \alpha^{lz_1} X^{z_1} + a_1 \alpha^{lz_2} X^{z_2} \in \mathcal{B}$ for all l. Assume at first $|Z_1| \neq |Z_2|$. Choose l such that $\alpha^{lz_1} = 1$ and $\alpha^{lz_2} \neq 1$. By subtraction we have $a_2(\alpha^{lz_2} - 1)X^{z_2} + \cdots \in \mathcal{B}$. The cyclic code generated by the trace-evaluation of this polynomial has trivial projection to $V_1(Z_1)$. As this is not true of $\mathcal{C}$, it follows from irreducibility that this codeword must be the 0 word, hence $a_2(\alpha^{lz_2} - 1) \in L_2^{\perp}$. This implies $a_2 \in L_2^{\perp}$, a contradiction. Now assume $|Z_1| = |Z_2| = s$, and let $L = \mathbb{F}_{q^s}$. Use Lemma 18.19. Let $c_l \in \mathbb{F}_q$ such that $\sum_l c_l \alpha^{lz_1} = 0$. The same argument as above shows that $\sum_l c_l \alpha^{lz_2} = 0$. We have $\beta = \alpha^{z_1} \in L$ and L is the smallest subfield of F containing β. Also $\alpha^{z_2} = \beta^j$ where j is coprime to the order of β. Let $\sum_{l=0}^{s} c_l \beta^l$ be the minimal polynomial of β. We just saw that $\sum_{l=0}^{s} c_l \beta^{jl} = 0$. This shows that the mapping $x \mapsto x^j$ is a field automorphism of L. It implies that z_1 and z_2 are in the same cyclotomic coset, a contradiction.

We have seen that a polynomial in standard form whose trace-evaluation generates an irreducible cyclic code is necessarily a monomial aX^z where $a \notin L^{\perp}$ using the by now standard terminology ($z \in Z, |Z| = s, L = \mathbb{F}_{q^s}, \perp$ with respect to the trace form). Lemma 18.19 shows that $a\alpha^{lz} X^z \in \mathcal{B}$ for all l. As the α^{lz} generate L (see Lemma 18.18), it follows that $auX^z \in \mathcal{B}$ for all $u \in L$. By Lemma 18.19 again we have in fact $\mathcal{C} = tr(ev(aLX^z))$. Consider the mapping $u \mapsto tr(ev(auX^z))$ from L onto $\mathcal{C}$. The kernel of this mapping is 0, so $\mathcal{C}$ has dimension s. As it is contained in $V_1(Z)$ of dimension s, we have equality.

This completes the proof of Theorem 18.17. As a result, we obtain an algorithmic description of all codewords of all cyclic linear codes in the permutation sense of length n, where $\gcd(n, p) = 1$. The codes are in bijection with sets of cyclotomic cosets. Let $Z_1 \cup \cdots \cup Z_t$ be such a set. Let $z_k \in Z_k, s_k = |Z_k|, L_k = \mathbb{F}_{q^{s_k}}$. The dimension of the code is $\sum_k s_k$. Its generic codeword $w(x_1, \ldots, x_t)$ where $x_k \in L_k$ has entry

$$\sum_k tr_{L_k | \mathbb{F}_q}(x_k \alpha^{iz_k})$$

in coordinate i where $i = 0, \ldots, n - 1$. The cyclic shift of codeword $w((x_k)_k)$ is $w((x_k \alpha^{z_k})_k)$.

Cyclic additive codes

Now let $m \geq 1$ arbitrary. This is the case of additive not necessarily linear cyclic codes in the permutation sense. Recall that we still assume $\gcd(n, p) = 1$. We want to determine the irreducible cyclic codes in this case. The fact that we have dealt with case $m = 1$ already will be helpful. Let $\mathcal{C}$ be an irreducible cyclic code and $\pi_j, j = 1, \ldots, m$ the projections to the inner coordinates. If $\pi_j(\mathcal{C})$ is not identically zero, then $\pi_j(\mathcal{C})$ is a linear irreducible cyclic code. It is therefore described by a cyclotomic coset (see Theorem 18.17). It can be assumed that this happens for all j (otherwise we are dealing with a smaller value of m). Let $Z_1, \ldots, Z_m$ be the corresponding cyclotomic cosets. At first we show that they are identical.

18.20 Lemma. *Let C be an irreducible cyclic code. The cyclotomic cosets determined by the irreducible linear cyclic codes $\pi_j(C)$ are all identical.*

PROOF We can assume $m = 2$ and $tr(ev(p_1(X), p_2(X))) \in C$ where $p_1(X) = a_1 X^{z_1}, p_2(X) = a_2 X^{z_2}$ and $a_i \notin L_i^{\perp}$ with the notation used in the previous subsection. Assume z_1, z_2 are in different cyclotomic cosets. The proof is similar to the main portion of the proof of Theorem 18.17. Lemma 18.19 shows $(a_1 \alpha^{lz_1} X^{z_1}, a_2 \alpha^{lz_2} X^{z_2}) \in \mathcal{B}$ for all l, where $\mathcal{B}$ is the $\mathbb{F}_q$-linear space of tuples of polynomials whose trace-evaluation is in C. Assume z_1, z_2 are in different cyclotomic cosets Z_1, Z_2, of lengths s_1, s_2. If $s_1 \neq s_2$, the same argument applies as in the proof of Theorem 18.17. If $s_1 = s_2$, the argument used in the proof of Theorem 18.17 shows $Z_1 = Z_2$, another contradiction. $\square$

18.21 Theorem. *Let Z be a cyclotomic coset. Each nonzero codeword in $V_m(Z) = tr(ev(\mathcal{P}(Z), \ldots, \mathcal{P}(Z)))$ generates an irreducible cyclic code of dimension $s = |Z|$.*

PROOF Such a codeword can be written as $tr(ev(a_1 X^z, a_2 X^z, \ldots, a_m X^z))$. The entry in outer coordinate i and inner coordinate j is $tr(a_j \alpha^{iz})$. Lemma 18.19 shows that the cyclic code C generated by this codeword is the span of the codewords with entry $tr(a_j \alpha^{lz} \alpha^{iz})$ in the same position. As the α^{lz} generate $L = \mathbb{F}_{q^s}$ (see Lemma 18.18), it follows that C consists of the codewords with entry $tr(a_j u \alpha^{iz})$ in the (i, j) coordinate, where $u \in L$. It is now clear that this code is the cyclic code generated by any of its nonzero codewords; in other words, C is irreducible. By the transitivity of the trace, we have $tr(a_j u \alpha^{iz}) = tr_{L|\mathbb{F}_q}(b_j u \alpha^{iz})$ where $b_j = tr_{F|L}(a_j)$. The fact that the code is nonzero means that not all the b_j vanish. The mapping $u \mapsto tr(ev(a_1 u X^z, \ldots, a_m u X^z)))$ is injective, so $dim(C) = s$. $\square$

18.22 Corollary. *The total number of irreducible permutation cyclic q-linear q^m-ary codes of length n coprime to the characteristic is $\sum_Z ((q^{ms} - 1)/(q^s - 1))$ where Z varies over the cyclotomic cosets and $s = |Z|$.*

PROOF In fact, $V_m(Z)$ has $\mathbb{F}_q$-dimension ms, and each of the irreducible subcodes has $q^s - 1$ nonzero codewords. $\square$

Observe that this is true also in the linear case $m = 1$: the number of irreducible cyclic codes is the number of cyclotomic cosets in this case (see Theorem 18.17). More importantly, we obtain a parametric description of the irreducible codes. Such a code is described by the following data:

1. A cyclotomic coset Z. Let $s = |Z|, L = \mathbb{F}_{q^s}$, choose $z \in Z$.

2. A point $P = (b_1 : \cdots : b_m) \in PG(m-1, L)$.

The codewords are then parametrized by $x \in L$. The entry in outer coordinate i and inner coordinate j is $tr_{L|\mathbb{F}_q}(b_j x \alpha^{iz})$. Denote this code by $\mathcal{C}(z, P)$. This leads to a parametric description of all cyclic additive codes, not just the irreducible ones.

18.23 Definition. *let Z be a cyclotomic coset of length s, $z \in Z$, $L = \mathbb{F}_{q^s}$ and $U \subset V(m, L)$ a k-dimensional vector subspace (equivalently, a $PG(k-1, L) \subset PG(m-1, L)$). Let $P_1, \ldots, P_k$ be the projective points determined by a basis of U. Define $\mathcal{C}(z, U) = \mathcal{C}(z, P_1) \oplus \cdots \oplus \mathcal{C}(z, P_k)$.*

Refer to the $\mathcal{C}(z, U)$ as **constituent codes**. Here $\mathcal{C}(z, U)$ has dimension ks if $U \subseteq L^m$ has dimension k. Use an **encoding matrix** $B = (b_{lj})$, a (k, m) matrix with entries from L. The rows z_l of B form a basis of U. The codewords $w(x)$ are parametrized by $x = (x_l) \in L^k$. The entry of codeword $w(x)$ in outer coordinate i and inner coordinate $j = 1, \ldots, m$ is

$$tr_{L|\mathbb{F}_q}((x \cdot s_j)\alpha^{iz}) \tag{18.1}$$

where s_j is column j of B. The image of $w(x)$ under one cyclic shift is $w(\alpha^z x)$. What happens if we use a different basis (z_l') for the same subspace U? Then $z_l' = \sum_{r=1}^{k} a_{lr} z_r$ and $A = (a_{lr})$ is invertible. The image of $w(x)$ under this operation (replacing z_l by z_l') has (i, j) entry $tr_{L|\mathbb{F}_q}(\sum_{l=1}^{k} x_l \sum_{r=1}^{k} a_{lr} b_{rj} \alpha^{iz})$. This describes $w(x')$ where $x_l' = \sum_{r=1}^{k} x_r a_{rl}$, in other words $x' = xA \in L^k$. We have seen that $\mathcal{C}(z, U)$ is indeed independent of the choice of an encoding matrix. Basic properties of the trace show that the dependence on the choice of representative $z \in Z$ is given by $\mathcal{C}(qz, U^q) = \mathcal{C}(z, U)$.

Here is a concrete expression for the weights of constituent codes: codeword $w(x)$ has entry $w(x)_i = 0$ in outer coordinate i if and only if $x \cdot s_j \in (\alpha^{iz})^{\perp}$ for all j, with respect to the $tr_{L|\mathbb{F}_q}$ form.

Finally, each cyclic code can be written in a unique way as the direct sum of its constituent codes: $\mathcal{C} = \oplus_Z \mathcal{C}(z, U_Z)$ where Z varies over the cyclotomic cosets and z is a fixed representative of Z.

18.24 Definition. *Let $N(m, q)$ be the total number of subspaces of the vector space $\mathbb{F}_q^m$.*

In particular, $N(1, q) = 2, N(2, q) = q + 3, N(3, q) = 2(q^2 + q + 2)$.

18.25 Corollary. *The total number of permutation cyclic q-linear q^m-ary codes of length n coprime to the characteristic is $\prod_Z N(m, q^{|Z|})$.*

18.26 Example ($[7, 3.5, 4]_4$). *Let $q = 2, m = 2, n = 7$. The cyclotomic cosets have lengths $1, 3, 3$ (representatives $0, 1, -1$). There are $3 + 9 + 9 = 21$ irreducible cyclic, $5 \times 11 \times 11 = 605$ cyclic codes altogether. A $[7, 3.5, 4]_4$ code is obtained as $\mathcal{C}(1, (1, 0)\mathbb{F}_8) \oplus \mathcal{C}(-1, (0, 1)\mathbb{F}_8) \oplus \mathcal{C}(0, (1, 1)\mathbb{F}_2)$, where $\mathcal{C}(0, (1, 1)\mathbb{F}_2)$ simply is the repetition code $\langle (11)^7 \rangle$. In the language of Definition 18.1, we have $km = 2k = 7$ and the quaternary dimension is therefore $k = 3.5$. It may be checked that the minimum distance is indeed 4.*

18.27 Example ($[15, 4.5, 9]_4$). *Let $q = 2, m = 2, n = 15$. Representatives of the cyclotomic cosets are 0 (length 1), 5 (length 2) and $1, 3, 14$ (length 4 each). There are $3 + 5 + 3 \times 17 = 59$ irreducible cyclic codes and a total of $5 \times 7 \times 19^3$ cyclic codes. A $[15, 4.5, 9]_4$-code is obtained as $\mathcal{C}(1, (\epsilon, 1)\mathbb{F}_{16}) \oplus \mathcal{C}(3, (1, \epsilon^2)\mathbb{F}_{16}) \oplus \mathcal{C}(0, (1, 1)\mathbb{F}_2)$; see [19]. Here $F = \mathbb{F}_{16} = \mathbb{F}_2(\epsilon)$ and $\epsilon^4 = \epsilon + 1$.*

Equivalence

When are two additive cyclic codes $\oplus_Z \mathcal{C}(z, U_Z)$ and $\oplus_z \mathcal{C}(z, U'_Z)$ equivalent? Two such situations are easy to see. Here is the first.

18.28 Proposition. *Let $C = (c_{uj}) \in GL(m, q)$. Then the additive cyclic code $\mathcal{C} = \oplus_Z \mathcal{C}(z, U_Z)$ is monomially equivalent to $\oplus_Z \mathcal{C}(z, U_Z C)$.*

PROOF   Use the GL part of monomial equivalence. An equivalent code is obtained if we apply the matrix C to each entry of $\mathcal{C}$. This means that, for each constituent Z, the entry $w(x)_i(j) = tr_{L|\mathbb{F}_q}((x \cdot s_j)\alpha^{iz})$ is replaced by $\sum_{u=1}^{m} c_{uj} w(x)_i(u)$. As $c_{uj} \in \mathbb{F}_q$, this amounts to replacing the encoding matrix B_Z by $B_Z C$, and therefore the subspace U_Z by $U_Z C$. $\square$

As a special case of Proposition 18.28, consider the case of irreducible codes for $m = 2$. We have a fixed cyclotomic coset Z of length $s = |Z|$ and $U = (b_1 : b_2)$ is a point of the projective line $PG(1, L)$. Multiplication by C from the right amounts to applying a Möbius transformation, an element of $PGL(2, q)$. The number of nonequivalent irreducible codes belonging to cyclotomic coset Z with respect to the equivalence described by Proposition 18.28 equals the number of orbits of $PGL(2, q)$ on the projective line $PG(1, q^s)$.

Here is the second such situation.

18.29 Proposition. *Let $\gcd(t, n) = 1$. The additive permutation cyclic code $\mathcal{C} = \oplus_Z \mathcal{C}(z, U_Z)$ is permutation equivalent to $\oplus_Z \mathcal{C}(tz, U_Z)$.*

PROOF   The generic codeword $w(x)$ of the second code above has entry $tr_{L|\mathbb{F}_q}((x \cdot s_j)\alpha^{itz})$ in outer coordinate i, inner coordinate j (see Equation 18.1). This is the entry of the first code in outer coordinate i/t mod n and inner coordinate j. $\square$

Proposition 18.29 describes an action of the group of units in $\mathbb{Z}/n\mathbb{Z}$. Here is an example.

18.30 Example. *There are precisely three nonequivalent length 7 irreducible additive quaternary codes which are cyclic in the permutation sense.*

PROOF This is case $q = 2, m = 2, n = 7$. The total number of irreducible cyclic codes is $3 + 9 + 9 = 21$ (once $PG(1,2)$ and twice $PG(1,8)$). The number of inequivalent such codes is at most $1 + 2 = 3$. In fact, the three codes $\mathcal{C}(0, U)$ where $U \in PG(1,2)$ are all equivalent by Proposition 18.28. By Proposition 18.29 it suffices to consider $Z(1)$ for the remaining irreducible codes. The number of possibly inequivalent such irreducible codes is the number of orbits of $GL(2,2)$ on $PG(1,8)$. There are two such orbits. The corresponding codes are $\mathcal{C}(1, (0,1))$ and $\mathcal{C}(1, (1, \epsilon))$ where $\epsilon \notin \mathbb{F}_2$. It is in fact obvious that those irreducible codes are pairwise nonequivalent. Clearly the first one is inequivalent to the others, as it has binary dimension 1 (the repetition code $\langle (11)^7 \rangle$). The second code has $w(x)_i = (0, tr(x\epsilon^i))$, of constant weight 4. The third of those irreducible codes has $w(x)_i = (tr(x\epsilon^i), tr(x\epsilon^{i+1}))$, clearly not of constant weights. $\qquad\Box$

Cyclic additive codes in the monomial sense

Let $\mathcal{C}$ be a q^m-ary q-linear code of length n, which is cyclic in the monomial sense. Let $\gcd(n, p) = 1, m \geq 1$ and $A \in GL(m, q)$ of order u, where $\gcd(u, p) = 1$. Let $r = ord_n(q), r' = ord_{un}(q)$ and $F = \mathbb{F}_{q^r} \subseteq F' = \mathbb{F}_{q^{r'}}$ be the corresponding fields. Consider cylotomic cosets $Z \subset \mathbb{Z}/un\mathbb{Z}$. For fixed Z of length s, let $z \in Z, L = \mathbb{F}_{q^s}$. Also, let $\kappa = ord_u(q)$ and $K = \mathbb{F}_{q^\kappa}$. It follows from Proposition 18.13 that by monomial equivalence it suffices to consider cyclic codes in the monomial sense fixed under the action of $g = g(I, \ldots I, A, (0, 1, \ldots, n-1))$ of order un. Such codes are also known as **constacyclic** codes. Here the matrix A plays the role of a constant factor. Let $G = \langle g \rangle$. Also, let β be a generator of the group of order un in F' and $\alpha = \beta^u \in F$. Let $\gamma = \beta^n$ of order u. Then $\gamma \in K = \mathbb{F}_{q^\kappa}$.

18.31 Definition. *Let* $A \in GL(m, q), A^u = 1$. *Define the* **inflation** $I_A :$ $\mathbb{F}_q^{mn} \longrightarrow \mathbb{F}_q^{umn}$ *by*

$$I_A(\underbrace{x_0, \ldots, x_{n-1}}_{x}) = (x|Ax \ldots |A^{u-1}x).$$

The mapping $contr : (x|Ax \ldots |A^{u-1}x) \mapsto x \in \mathbb{F}_q^{mn}$ *is the* **contraction** : $I_A(\mathbb{F}_q^{mn}) \longrightarrow \mathbb{F}_q^{mn}$.

18.32 Lemma. *In the situation of Definition 18.31, the q-linear q^m-ary code $\mathcal{C}$ of length n is invariant under the (monomial) action of the cyclic group G of order un generated by $g(I, \ldots, I, A, (0, 1, \ldots, n-1))$ if and only if $I_A(\mathcal{C})$ is* *cyclic in the permutation sense under the permutation* $(0, 1, \ldots, un - 1)$. *The contraction* $contr(\mathcal{C})$ *is irreducible under the action of* G *if and only if the cyclic length* un *code* $\mathcal{C}$ *is irreducible.*

PROOF   Observe that $\mathcal{C}$ and $I_A(\mathcal{C})$ have the same dimension and $\mathcal{C}$ is recovered from $I_A(\mathcal{C})$ as the projection to the first n coordinates. The claims are by now obvious. $\quad\square$

18.2.2   The linear case $m = 1$

In this special case we have $g = g(1, \ldots, 1, \gamma, (0, 1, \ldots, n - 1))$ where the matrix A now is a constant $\gamma \in \mathbb{F}_q^*$ of order u where $u | q - 1$. Let $u = ord(\gamma)$. Write the elements of the ambient space $\mathbb{F}_q^{un}$ as (x_i) where $i \in \mathbb{Z}/un\mathbb{Z}$. Then $x \in \mathbb{F}_q^{un}$ is in $I_\gamma(\mathbb{F}_q^n)$ if and only if $x_{i+n} = \gamma x_i$ always holds.

18.33 Theorem. *Let* $\gcd(n, p) = 1$ *and* $\gamma \in \mathbb{F}_q^*, ord(\gamma) = u$. *The codes which are invariant under the group generated by* $g = g(1, \ldots, 1, \gamma, (0, 1, \ldots, n - 1))$ *are the contractions of the cyclic codes of length* un *in the permutation sense defined by cyclotomic cosets consisting of numbers which are 1 mod* u. *If there are* a *such cyclotomic cosets in* $\mathbb{Z}/un\mathbb{Z}$, *then the number of codes invariant under* g *is* 2^a.

PROOF   Observe $\gcd(un, p) = 1$. By Lemma 18.32 the codes in question are the contractions of the length un cyclic codes all of whose codewords (x_i) satisfy $x_{i+n} = \gamma x_i$ for all i. Let Z be a cyclotomic coset, $z \in Z, |Z| = s, L = \mathbb{F}_{q^s}$. A typical codeword of the irreducible cyclic code defined by Z has entry $x_i = tr_{L|\mathbb{F}_q}(a\beta^{iz})$, where $a \in L$. It follows $x_{i+n} = tr_{L|\mathbb{F}_q}(a\beta^{(i+n)z}) = \gamma^z x_i$. We must have $z = 1 \mod u$. $\quad\square$

Here is an algorithmic view concerning linear length n q-ary codes which are invariant under $g = g(1, \ldots, 1, \gamma, (0, 1, \ldots, n - 1))$ where $ord(\gamma) = u$. Let $r' = ord_{un}(q)$ and $F' = \mathbb{F}_{q^{r'}}$. Let $ord(\beta) = un$ such that $\gamma = \beta^n$. Let $\alpha = \beta^u$. Consider the cyclotomic cosets $Z_1, \ldots Z_a$ in $\mathbb{Z}/un\mathbb{Z}$ whose elements are 1 mod u. Observe that $|Z_1 \cup \cdots \cup Z_a| = n$. Let $s_k = |Z_k|, L_k = \mathbb{F}_{q^{s_k}} \subseteq F'$. Then $\mathcal{C} \subseteq \oplus \mathcal{C}_k$ where the $\mathcal{C}_k$ are the irreducible γ-constacyclic codes of length n. We have $dim(\mathcal{C}_k) = s_k$. The codewords of $\mathcal{C}_k$ are $w(x)$ where $x \in L_k$. The entry of $w(x)$ in coordinate i is $tr_{L_k|\mathbb{F}_q}(x\beta^{iz_k})$. The image of $w(x)$ under the monomial operation $g(1, \ldots, 1, \gamma, (0, 1, \ldots, n - 1))$ is $w(\beta^{z_k}x)$. After n applications, this leads to $w(\gamma^{z_k}x) = w(\gamma x) = \gamma w(x)$, as expected.

18.34 Example. *Consider* $q = 4, u = 3, n = 15$. *Then* $F = \mathbb{F}_{16}$, *whereas* $F' = \mathbb{F}_{4^6}$. *The cyclotomic cosets in* $\mathbb{Z}/45\mathbb{Z}$ *consisting of numbers divisible by 3 are in bijection with the nine cyclotomic cosets mod 15. The contractions of the cyclic length 45 codes defined by* Z *cosets all of whose elements are* *divisible by* 3 *reproduce precisely the length* 15 *cyclic codes in the permutation sense. There are only three cyclotomic cosets all of whose elements are* 1 *mod* 3. *They are*

$$\{1, 4, 16, 19, 31, 34\}, \{7, 13, 22, 28, 37, 43\}, \{10, 25, 40\}.$$

It follows that there are $2^3 = 8$ *constacyclic quaternary linear codes with constant* $\gamma = \beta^{15}$ *of order* 3.

18.35 Example. *Consider* $q = 8, n = 21$. *Then* $F' = \mathbb{F}_{2^{42}} = \mathbb{F}_{8^{14}}$. *The constant is* $\gamma = \beta^7$ *of order* 7. *We have* $un = 7 \times 21 = 147$ *and we have to consider the action of* $q = 8$ *(the Galois group of order* 14*) on the* 21 *elements which are* 1 *mod* 7. *The corresponding cyclotomic cosets are*

$$Z(1) = \{1, 8, 64, 71, -20, -13, 43, 50, -41, -34, 22, 29, 85, -55\}$$

of length 14 *and*

$$Z(15) = \{15, -27, 78, 36, -6, -48, 57\}$$

of length 7. *It follows that there are precisely two irreducible* γ-*constacyclic* $\mathbb{F}_8$-*linear codes of length* 21, *of dimensions* 14 *and* 7.

Here are some more interesting and well-known examples.

18.36 Example. *Let* $q = 4, u = 3, Q = 2^f$ *where* f *is odd, and* $n = (Q^f + 1)/3, F = \mathbb{F}_{4^f}$. *The irreducible constacyclic quaternary code defined by the cyclotomic coset generated by* 1 *has dimension* f *and dual distance* 5. *Its dual is therefore a linear* $[(2^f + 1)/3, (2^f + 1)/3 - f, 5]_4$ *code. This is the first family from [90]. We constructed them in Section 13.4.*

18.37 Example. *The second family of constacyclic quaternary codes from [90] occurs in the case* $u = 3, n = (4^f - 1)/3$ *for arbitrary* f *where the cyclotomic cosets are those generated by* 1 *and* -2. *This yields a* $2f$-*dimensional quaternary code of dual distance* 5 *and therefore* $[(4^f - 1)/3, (4^f - 1)/3 - 2f, 5]_4$ *codes for arbitrary* f.

The general case $m \geq 1$

18.38 Theorem. *With notation as introduced above, choose a representative* z *for each cyclotomic coset* Z *and* U_Z *the Eigenspace of the Eigenvalue* γ^z *of* A *in its action on* L^m. *Then each code stabilized by* G *is contained in* $\oplus_Z contr(\mathcal{C}(z, U_Z))$.

PROOF The additive cyclic length un codes are direct sums over cyclotomic cosets Z of codes parametrized by subspaces $U_Z \subset L^m$, using the customary terminology. Such a code is in $I_A(\mathbb{F}_q^{mn})$ if and only if each codeword $x = (x_i)$ (where $i = 0, \ldots, un - 1$ and $x_i \in \mathbb{F}_q^m$) satisfies $x_{i+n} = Ax_i$. Fix

Z and ask for which subspace U this is satisfied. Let $A = (a_{jj'}) \in GL(m, q)$ and $U = P = (b_1 : \cdots : b_m) \in PG(m-1, L)$ a point. Let $tr = tr_{L|\mathbb{F}_q}$. The condition is

$$tr(xb_j\beta^{(i+n)z}) = \sum_{j'} a_{jj'}tr(xb_{j'}\beta^{iz})$$

for all i, all j and $x \in L$. As $\beta^n = \alpha$, an equivalent condition is

$$tr(x\beta^{iz}(\gamma^z b_j - \sum_{j'} a_{jj'}b_{j'})) = 0,$$

which is equivalent to $\gamma^z b_j = \sum_{j'} a_{jj'}b_{j'}$; in other words, b (written as a column vector) is an Eigenvector for the Eigenvalue γ^z of A. $\quad\Box$

Here is an algorithmic view again. The constacyclic length n additive codes are direct sums of their constituents, where the constituents correspond to cyclotomic cosets in $\mathbb{Z}/un\mathbb{Z}$. Let Z be such a cyclotomic coset, $z \in Z, |Z| = s, L = \mathbb{F}_{q^s}$ and $\beta, \alpha, \gamma = \beta^n$ as usual. The irreducible subcodes are the contractions of $\mathcal{C}(z, P)$ where $P = (b_1, \ldots, b_m) \in L^m$ has to satisfy $AP = \gamma^z P$ (and we write P as a column vector). This irreducible length n constacyclic code $contr(\mathcal{C}(z, P))$ has dimension s. Its codewords are $w(x), x \in L$. The entry of $w(x)$ in outer coordinate $i = 0, 1, \ldots, n-1$ and inner coordinate j is $tr_{L|\mathbb{F}_q}(xb_j\beta^{iz})$. The effect of the generator $g = g(I, \ldots, I, A, (0, \ldots, n-1))$ of the cyclic group G is $g(w(x)) = w(\beta^z x)$.

18.39 Proposition. *Let $\mathcal{C}$ be a q-linear q^m-ary length n code, which is invariant under $g(I, \ldots, I, A, (0, 1, \ldots, n-1))$, where $\gcd(n, p) = 1$ and A of order $u = q^m - 1$ represents multiplication by a primitive element γ in $\mathbb{F}_{q^m}$. Then $\mathcal{C}$ is $\mathbb{F}_{q^m}$ linear.*

PROOF   It suffices to prove this for irreducible constacyclic codes $contr(\mathcal{C}(z, P))$. The generic codeword $w(x)$ has been described above. Applying matrix A in each outer coordinate $i = 0, \ldots, n-1$ yields the codeword $w(\gamma^z x)$, which is still in the code. $\quad\Box$

Proposition 18.39 applies in particular in the cases $q = 2, m = 2, u = 3$ and $q = 2, m = 3, u = 7$.

18.40 Corollary. *Each quaternary additive code which is cyclic in the monomial sense either is quaternary linear or is equivalent to cyclic in the permutation sense.*

PROOF   Use Proposition 18.13 to obtain a code which is invariant under $g = (I, \ldots, I, A, (0, \ldots, n-1))$ where $A \in GL(2, 2)$ and $u = det(A)$. If $u = 1$ or $u = 2$ or when $u = 3$ and n not divisible by 3, then the additive code is cyclic

in the permutation sense (see Proposition 18.13). In the case when $u = 3$, Proposition 18.39 shows that the code is a quaternary linear constacyclic code.
□

This shows that, in the quaternary case, the full theory of additive constacyclic codes does not produce anything useful, as in each case we are reduced to a more elementary theory, either quaternary linear constacyclic or additive and cyclic in the permutation sense.

18.3 Additive quaternary codes: The geometric approach

The geometric approach to linear codes was used in Chapter 17, where we saw that a linear q-ary k-dimensional code of length n may be equivalently described by a multiset of n points in the projective geometry $PG(k-1, q)$. The minimum distance and in fact the weights of codewords are determined by the distribution of the points on hyperplanes.

Consider now a q-linear q^m-ary code $[n, k]_{q^m}$ as in Definition 18.1. It is obvious that this may be equivalently described by a multiset of n subspaces $PG(m-1, q)$ in $PG(k-1, q)$. Here it is to be observed that those subspaces may also degenerate to subspaces of smaller dimension. In the present section we specialize to the quaternary case.

18.41 Theorem. *The following are equivalent:*

- *a quaternary additive $[n, k]_4$ code.*

- *a multiset of n lines (the **codelines**, which may degenerate to points) in $PG(2k-1, 2)$.*

Here the codewords are in bijection with the hyperplanes, and a codeword has weight w if and only if the corresponding hyperplane contains all but w of the codelines.

PROOF This is harder to state than to prove. Start from the code and describe it by a generator matrix. Observe that the code has binary dimension $2k$. Each of the n coordinate pairs yields a pair of points in $PG(2k-1, 2)$. Consider the multiset of lines thus obtained. These are the codelines. A hyperplane H yields a codeword with zero entry in the section described by codeline L_i if and only if $L_i \subset H$.
□

The strength of this code (equal to the minimum distance of the dual minus 1) is the largest number t such that each t codelines is in general position, equivalently, the projection of a generator matrix to any set of t codelines has binary rank $2t$.

The additive quaternary codes of small length and their minimum distances were first considered in Blokhuis and Brouwer [34]. Currently the maximum values of the minimum distances d are known for quaternary additive $[n, k, d]_4$ codes, $n \leq 13$, as follows:

$k \backslash n$	1	2	3	4	5	6	7	8	9	10	11	12	13
1	1	2	3	4	5	6	7	8	9	10	11	12	13
1.5		1	2	3	4	5	6	6	7	8	9	10	11
2		1	2	3	4	4	5	6	7	8	8	9	10
2.5			1	2	3	4	5	6	6	7	8	8	9
3			1	2	3	4	4	5	6	6	7	8	9
3.5				1	2	3	4	4	5	6	7	8	8
4				1	2	2	3	4	5	6	6	7	8
4.5					1	2	3	3	4	5	6	6	7
5					1	2	2	3	4	5	6	6	7
5.5						1	2	3	3	4	5	6	6
6						1	2	2	3	4	5	6	6
6.5							1	2	3	3	4	5	6
7							1	2	2	3	4	4	5
7.5								1	2	2	3	4	5
8								1	2	2	3	4	4
8.5, 9									1	2	2	3	4
9.5, 10										1	2	2	3
10.5, 11											1	2	2

Shortening and puncturing works just as in the linear case: an additive $[n, k, d]_4$ code implies additive codes with parameters $[n - 1, k - 1, d]_4$ and $[n - 1, k, d - 1]_4$. Observe the link to binary linear codes via concatenation:

18.42 Theorem (concatenation). *An additive quaternary $[n, k, d]_4$ code implies a binary linear $[3n, 2k, 2d]_2$ code.*

PROOF Simply apply concatenation, just as in Section 5.2, using the $[3, 2, 2]_2$ code. The additivity of the quaternary code implies that the concatenated code is linear. ⬜

In many situations it is profitable to think geometrically. Consider, for example, the case of dimension $k = 1.5$. The binary dimension is 3, which means that our codes are defined as multisets of lines in the Fano plane. As the hyperplanes of the Fano plane are the lines, it follows that the highest value of d is obtained if the multiplicities of lines are chosen as uniformly as possible. In the case of $n = 14$, for example, each of the seven lines of the Fano plane should have multiplicity 2. This yields an $[14, 1.5, 12]_4$ code. This elementary argument explains the entries in row $k = 1.5$.

In Chapter 17 we described the linear Simplex codes as the codes defined by the set of all points in the corresponding projective space. An analogous construction in the present setting is to choose all lines:

18.43 Theorem (Simplex codes). *For each integer $l \geq 3$ there is an additive $[(2^l - 1)(2^{l-1} - 1)/3, l/2, 2^{l-2}(2^{l-1} - 1)]_4$ code.*

PROOF The binary dimension is l, and the number of lines in $PG(l-1, 2)$ is $(2^l - 1)(2^{l-1} - 1)/3$. As each hyperplane contains $(2^{l-1} - 1)(2^{l-2} - 1)/3$ lines, we obtain a constant weight of $(2^l - 1)(2^{l-1} - 1)/3 - (2^{l-1} - 1)(2^{l-2} - 1)/3 = 2^{2l-3} - 2^{l-2}$. □

The codes of Theorem 18.43 are optimal, as concatenation yields a binary linear $[(2^l - 1)(2^{l-1} - 1), l, 2^{l-1}(2^{l-1} - 1)]_2$ code which meets the Griesmer bound with equality. The smallest instances of Theorem 18.43 are $[7, 1.5, 4]_4$ (the Fano plane) and $[35, 2, 28]_4$ (there exists a linear code with those parameters).

We will see that geometric objects like **partial spreads** (of lines) yield code constructions. Here a partial spread in a projective geometry is a set of lines which are pairwise skew. A **secundum** of a projective geometry is a hyperplane of a hyperplane (a subspace of codimension 2). It is sometimes convenient to use vector space dimension instead of projective dimension. Therefore denote $\mathbb{F}_2^i$ by V_i. A secundum in V_i is a V_{i-2}. The following geometric argument is from [36] .

18.44 Proposition. *Let $f \geq 2, q = 2^f$ and fix a secundum S in $PG(f+1, 2)$. Then $PG(f + 1, 2) \setminus S$ can be partitioned into q lines (the **spread lines**).*

Let H be a hyperplane. If $S \subset H$, then H contains none of the spread lines. If $S \not\subset H$, then H contains precisely $2^{f-2} = q/4$ spread lines.

PROOF Write the ambient space as $V_{f+2} = \{(a, x, y) \mid a \in F = \mathbb{F}_q, x, y \in \mathbb{F}_2\}$. Fix nonzero elements $a, b, c \in F$ such that $a + b + c = 0$. This is possible, as $f \geq 2$. As spread lines, we can choose the

$$L_x = \{(ax, 1, 0), (bx, 0, 1), (cx, 1, 1) \mid x \in F\}.$$

The union of those q disjoint lines consists precisely of the points in ambient space not contained in the secundum $S = \{(x, 0, 0) \mid 0 \neq x \in F\}$.

Each spread line together with S generates the whole ambient space. This shows that hyperplanes containing S do not contain any spread line. Let H be a hyperplane not containing S. Each spread line either is contained in H or meets H in one point. As there are $2^{f+1} - 2^{f-1} = 3 \times 2^{f-1} = q + 2^{f-1}$ points in H outside S, the number N of spread lines contained in H satisfies $3N + (q - N) = q + 2^{f-1}$, hence $N = q/4$. □

Using this in Theorem 18.41 yields an additive quaternary code of dimension $(f + 2)/2$ and length $q = 2^f$ all of whose codewords have weight either q or $q - q/4 = 3q/4$.

18.45 Corollary. *Proposition 18.44 yields an additive* $[2^f, 1 + f/2, 3 \times 2^{f-2}]_4$ *code.*

For $f = 3$ this yields an $[8, 2.5, 6]_4$ code. For $f = 4$ the parameters $[16, 3, 12]_4$ are not of independent interest. These are parameters of a quaternary linear code (geometrically, the affine plane $AG(2, 4)$). Case $f = 5$ yields $[32, 3.5, 24]_4$, by concatenation $[96, 7, 48]_2$. In fact, the parameters of the resulting binary codes (after concatenation) are identical to those of a family obtained via the $(u, u + v)$ construction in Section 5.2.

Proposition 18.44 also can be used for recursive constructions.

18.46 Corollary. *Let* $f \geq 2, q = 2^f$. *If there is an additive* $[m, f/2, d]_4$ *code, then there is an additive* $[q + m, 1 + f/2, Min(q, d + 3q/4)]_4$ *code.*

PROOF   Assume an additive $[m, f/2, d]_4$ code exists. Represent it geometrically on the secundum S used in Proposition 18.44. Together with the spread lines this yields a multiset of $q + m$ codelines. Outside each hyperplane H containing S there are precisely q of the codelines, the spread lines. Outside each of the remaining hyperplanes there are $3q/4$ spread lines and at least d of the lines in S.   ▯

As an example, using $[3, 1.5, 2]_4$ as an ingredient (geometrically, three different lines of the Fano plane), yields $[11, 2.5, 8]_4$.

Theorem 18.42 is useful to prove nonexistence as well. In fact, an additive $[n, k, d]_4$ cannot exist if no linear $[3n, 2k, 2d]_2$ code exists. As an example, an additive $[12, 2.5, 9]_4$ code cannot exist as concatenation would yield a binary $[36, 5, 18]_2$ code, which would contradict the Griesmer bound.

On the construction side we can use linear quaternary codes. In order to understand some of the entries in the table, the following linear codes are needed:

1. $[17, 4, 12]_4$ (geometrically, the elliptic quadric in $PG(3, 4)$); see Section 17.3.

2. $[21, 15, 5]_4$, from Section 13.4, which implies $[13, 7, 5]_4$.

3. $[18, 9, 8]_4$; see [143].

4. $[12, 6, 6]_4$ from Section 17.1.

5. $[12, 8, 4]_4$ (again from the elliptic quadric).

The following genuinely additive (nonlinear) codes are used in the table:

1. $[7, 3.5, 4]_4$ (see the cyclic Example 18.26).

2. A $[12, 3.5, 8]_4$ code is given in [34].

3. $[13, 6.5, 6]_4$, see Danielsen and Parker [64]. Those are cyclic codes.

4. $[13, 7.5, 5]_4$; see [29].

5. $[15, 4.5, 9]_4$ (see the cyclic Example 18.27).

The proofs of upper bounds require geometrical work. Intersection with hyperplanes shows that it is unavoidable to study mixed binary/quaternary codes:

18.47 Definition. *An* $[(l, r), k]_{(4,2)}$ *code is a binary 2k-dimensional vector space of binary* $(2l + r)$-*tuples, where the coordinates are divided into l pairs and r single coordinates. We view each codeword as an* $(l + r)$-*tuple, where the left coordinates are quaternary, and the right ones are binary.*

These mixed quaternary-binary codes also have some interest in their own right. The following nonexistence results are needed to complete our table of length ≤ 13 optimal code parameters:

- There is no $[8, 4.5, 4]_4$ code. This is proved by a computer search in [34]. A relatively short geometrical proof is available.

- There is no $[12, 4.5, 7]_4$ code (see [31]).

- There is no $[12, 7, 5]_4$ code (see [29]).

The smallest open problems concern the existence of codes with parameters $[14, 7.5, 6]_4$ and $[14, 8.5, 5]_4$.

Additive quaternary codes: Characterization of the quaternary linear case

It is not forbidden for a quaternary additive code to be in fact quaternary **linear**. The following natural characterization of the quaternary linear codes was given in Blokhuis and Brouwer [34].

18.48 Theorem. *Let C be an additive quaternary code described geometrically by a (multi-)set of lines, the codelines. Then the following are equivalent:*

- *C is an* $\mathbb{F}_4$-*linear code.*

- *Each set of codelines generates a binary vector space of even dimension.*

PROOF Let us make sure we understand the situation. First of all, it is clear that we can assume the code to be described by a set of lines, in other words, all nonzero multiplicities of lines can be assumed to be $= 1$. Also, one direction is obvious: if C is linear over $\mathbb{F}_4$, then all $\mathbb{F}_2$ dimensions in question will be even. We therefore assume this is the case and have to prove that C is $\mathbb{F}_4$ linear. In particular, the codelines are nondegenerate (they are really lines), any two codelines are skew (so the codelines form a partial spread) and so forth.

Let C be such an additive code, of binary dimension $2m$, and $\mathcal{L}$ the corresponding set of codelines. We work with binary dimensions. Recall the basic assumption that each subset of $\mathcal{L}$ generates a vector space of even dimension. Let $\mathbb{F}_4 = \{0, 1, \omega, \overline{\omega}\}$. It can be assumed that $L_i = \langle x_i, y_i \rangle, i = 1, 2, \ldots, m$ are codelines. Here the x_i, y_i form a basis of the ambient space $V = \mathbb{F}_2^{2m}$. There are precisely 2^m $\mathbb{F}_4$ structures of V such that the code described by the L_i is $\mathbb{F}_4$ linear. In fact, for each $i = 1, 2, \ldots, m$ there are precisely two ways to give L_i the structure of an $\mathbb{F}_4$ vector space: either multiplication by ω is the 3-cycle $x_i \mapsto y_i \mapsto x_i + y_i$ or it is the inverse of this 3-cycle. We have to prove that there is at least one such $\mathbb{F}_4$ structure on V such that each line in $\mathcal{L}$ is an $\mathbb{F}_4$ vector space. Let π_i be the projection : $V \longrightarrow L_i$. For $v \in V$, let $supp(v) = \{i \mid \pi_i(v) \neq 0\} \subseteq \{1, \ldots, m\}$ be the support of v. Let $M_1 \in \mathcal{L}$ be a codeline different from the L_i. It follows from the assumption that each $0 \neq v \in M_1$ has the same support, which we then call the support $supp(M_1)$ of M_1. This implies that $\mathcal{L} \cup \{M_1\}$ is $\mathbb{F}_4$ linear for some of the 2^m structures above: the $\mathbb{F}_4$ structure of L_i for some $i \in supp(M_1)$ determines the $\mathbb{F}_4$ structure for all $j \in supp(M_1)$. If, for example, $1, 2 \in supp(M_1)$ and $M_1 = \langle u, v \rangle$ where $u = x_1 + x_2 + \ldots, v = y_1 + x_2 + y_2 + \ldots$ then $\omega x_1 = y_1$ implies $\omega x_2 = x_2 + y_2$ (and, in the other case, $\omega x_1 = x_1 + y_1$ forces $\omega x_2 = y_2$). Now assume we have codelines $M_1, M_2 \in \mathcal{L}$ different from the L_i. A conflict can occur only if $|supp(M_1) \cap supp(M_2)| \geq 2$. Assume $1, 2 \in supp(M_1) \cap supp(M_2)$ and let $M_i = \langle u_i, v_i \rangle$ where $\pi_1(u_i) = x_1, \pi_1(v_i) = y_1, i = 1, 2$. Assume, for example, $\pi_2(u_1) = x_2, \pi_2(v_1) = y_2$. A conflict would occur, for example, if $\pi_2(u_2) = x_2, \pi_2(v_2) = x_2 + y_2$. However, this is impossible, as $M_1 + M_2 + L_3 + \cdots + L_m$ would have dimension $2m - 1$, contrary to our basic assumption. All cases are analogous and lead to the same contradiction. ⬜

Exercises 18.3

18.3.1. *Use geometric language to confirm the (obvious) fact that the largest minimum distance of an additive $[n, 1, d]_4$ code is $d = n$.*

18.3.2. *Determine the maximum minimum distance d of additive $[n, 1.5, d]_4$ codes, for all n.*

18.3.3. Let $PG(2,q) \subset PG(2,q^2)$: ($PG(2,q)$ *is known as a* **Baer subplane** *of* $PG(2,q^2)$). *Clearly* $PG(2,q^2)$ *contains many copies of* $PG(2,q)$. *We fix one such Baer subplane.*
Show that each line of $PG(2,q^2)$ *intersects the Baer subplane either in one point or in* $q+1$ *points. In particular, each line of* $PG(2,q^2)$ *meets the Baer subplane in at least one point (the Baer subplane is a* **blocking set**).

18.3.4. *Describe a construction of a* $[14,3,10]_4$ *code.*

18.3.5. *Describe a construction of a* $[13,4,8]_4$ *code.*

18.3.6. *Show that the binary linear codes obtained by concatenation from Corollary 18.45 meet the Griesmer bound with equality.*

18.3.7. *Give a geometric proof for the nonexistence of a linear* $[10,3,7]_4$ *code.*

18.3.8. *Give a geometric proof for the nonexistence of an additive* $[10,3,7]_4$ *code.*

18.3.9. *Use Proposition 18.44 to show the following: if* E *is a plane in* $PG(2m,2)$, *then* $PG(2m,2) \setminus E$ *can be partitioned into lines.*

18.3.10. *Show that* $PG(2m,2)$ *possesses a partial spread of* $1+8(2^{2m-2}-1)/3$ *lines.*

18.3.11. *Show that* 9 *is the maximal number of lines in* $PG(4,2)$ *which form a partial spread.*

18.4 Quantum codes

Quantum coding happens in Hilbert spaces. It is not obvious that finite structures like error-correcting codes come into the picture at all. Fortunately such a link has been found. In Calderbank et al. [39] it has been described how certain error-correcting codes in symplectic space can be used for the purposes of quantum error correction. This construction is known as the **stabilizer construction.** As the stabilizer construction seems to be the most important source for codes correcting quantum errors (whatever that means) it is not completely unjustified to call these symplectic codes **quantum codes.** More precisely, the definition is as follows:

18.49 Definition. *Let $E = \mathbb{F}_q^2$ be a symplectic two-dimensional space (a hyperbolic plane) with respect to the symplectic bilinear form $\langle , \rangle$, and E^n the 2n-dimensional q-ary symplectic space.*

*A quantum code $[[n, k, d]]_q$ is a q-linear q^2-ary code $\mathcal{Q} \subset E^n$ of dimension $(n + k)/2$ which contains its dual and such that every word in $\mathcal{Q} \setminus \mathcal{Q}^\perp$ has weight $\geq d$. A quantum code is **pure** if all of its nonzero codewords have weight $\geq d$.*

In the original setting of [39] only case $q = 2$ of Definition 18.49 was considered (these are quaternary codes). Ashikhmin and Knill [2] gave a physical justification for the general case of Definition 18.49. Still, it seems that the quaternary case is the most important.

If we compare with our general notion of additive codes, we see that quantum codes form a parametric subclass, the case when $m = 2$ and the underlying bilinear form $\langle , \rangle$ is symplectic (for the notion of a symplectic bilinear form, compare Section 17.2). The additional assumption is that the quantum code has to contain its dual. In Definition 18.49 we used the general terminology for additive codes. Recall that the dimension $(n + k)/2$ is the q^2-ary dimension. Code $\mathcal{Q}$ has q^{n+k} codewords. A pure $[[n, k, d]]_q$ is in particular an $(n, q^{n+k}, d)_{q^2}$ code.

An extremal case is $k = n$. We have that $\mathcal{Q} = E^n$ equals the ambient space, and clearly $d = 1$ in this case. This is an $[[n, n, 1]]_q$ code. In the other extremal case $k = 0$ we must have $\mathcal{Q} = \mathcal{Q}^\perp$. Only pure codes are interesting, so d simply is the minimum distance of the self-dual code $\mathcal{Q}$.

What is the contribution of linear q^2-ary codes? One source is codes, which are self-orthogonal with respect to Hermitian inner products. These Hermitian forms were not considered in Section 17.2, as they are not bilinear, but they are closely related to bilinear forms.

18.50 Definition. *Let $F = \mathbb{F}_{q^2}$ and T, N be the trace and norm down to $\mathbb{F}_q$. In analogy to complex conjugation, write $\overline{u} = u^q$ for $u \in F$. The Hermitian* *inner product (in its standard form) is defined on $V = F^n$ by*

$$(x, y) = \sum_{i=1}^{n} x_i \overline{y_i}.$$

Here $\overline{u}$ is the image of u under the nontrivial automorphism of $F \mid \mathbb{F}_q$, and $T(u) = u + u^q$, $N(u) = u^{q+1} \in \mathbb{F}_q$. Clearly the Hermitian inner product is biadditive and satisfies $(\lambda x, y) = \lambda(x, y)$, $(x, \lambda y) = \overline{\lambda}(x, y)$ for $\lambda \in F$. The second equality is what makes the Hermitian inner product not bilinear. It is a **sesquilinear** form.

Duality can be based not only on any nondegenerate bilinear form but also on nondegenerate sesquilinear forms. It should be clear by now what we see as the basic rules of duality. In particular, we can consider dual codes with respect to the Hermitian inner product. Here is a construction of quantum codes based on q^2-ary linear codes, which are self-orthogonal with respect to the Hermitian form.

18.51 Theorem. *Let C be a linear $[n, k, d]_{q^2}$ code containing its dual with respect to the Hermitian form. There is an $\mathbb{F}_q$ isomorphism from C to a pure $[[n, 2k - n, d]]_q$ quantum code Q.*

PROOF   Choose $0 \neq \gamma \in F = \mathbb{F}_{q^2}$ such that $T(\gamma) = 0$ and define $\Phi : F \longrightarrow \mathbb{F}_q$ by $\Phi(u) = T(\gamma u)$. The kernel of Φ is $\mathbb{F}_q$. Fix α such that $\Phi(\alpha) = 1$. As $\alpha + \alpha^q \in \mathbb{F}_q$, it follows that $\Phi(\overline{\alpha}) = -1$.

Let $0 \neq \omega_1 \in F$ arbitrary and define ω_2 by $\alpha = \omega_1 \overline{\omega_2}$. Then ω_1, ω_2 form a basis of $F \mid \mathbb{F}_q$ (if $\omega_2 = \lambda \omega_1$ with $\lambda \in \mathbb{F}_q$, then $\alpha \in \mathbb{F}_q$, hence $\Phi(\alpha) = 0$, a contradiction). Define projections $\pi_1, \pi_2 : F \to \mathbb{F}_q$ by $u = \pi_1(u)\omega_1 + \pi_2(u)\omega_2$. This defines an $\mathbb{F}_q$ isomorphism $\tau : \mathbb{F}_{q^2}^n \longrightarrow \mathbb{F}_q^{2n}$, where

$$\tau(x_1, \ldots, x_n) = (\pi_1(x_1), \pi_2(x_1), \ldots, \pi_1(x_n), \pi_2(x_n)).$$

Denote by $(,)$ the Hermitian form on $\mathbb{F}_{q^2}^n$ and define a bilinear form $\langle, \rangle$ on $\mathbb{F}_q^{2n}$ by $\langle \tau(x), \tau(y) \rangle = \Phi(x, y)$. It is clear from the definition that $\langle, \rangle$ is biadditive and bilinear. It is nondegenerate, as the Hermitian form is nondegenerate. If $y = x$, then $(x, x) \in \mathbb{F}_q$, hence $\Phi(x, x) = 0$. This shows that $\langle, \rangle$ is symplectic (compare Section 17.2). We see that $Q = \tau(C)$ contains its dual with respect to the symplectic form, as claimed. As τ acts coordinatewise, the weight of $x \in F^n$ equals the weight of $\tau(x) \in E^n$, where $E = \mathbb{F}_q^2$ is the hyperbolic plane. $\Box$

Once again consider the hexacode, one of our favorite codes. Generator matrices were first given in the problems of Section 3.4. We see that this $[6, 3, 4]_4$ code is self-dual with respect to the Hermitian form while it is not self-dual with respect to the dot product. In fact, there is no $[6, 3, 4]_4$-code which is self-dual with respect to the dot product. Application of Theorem 18.51 yields a $[[6, 0, 4]]_2$ quantum code.

Quantum codes: The cyclic theory

A major source for quantum codes is cyclic additive codes. We start in a more general setting, using the cyclic additive codes of Section 18.2.

Fix a nondegenerate bilinear form $\langle,\rangle$ on the vector space $E = \mathbb{F}_q^m$, and extend it to $V_m = E^n$, to F^m and to $V_{m,F} = (F^m)^n$ in the natural way. In coordinates this means that $\langle\rangle$ is described by coefficients $a_{kl} \in \mathbb{F}_q$ where $k, l = 1 \ldots, m$ such that $\langle (x_k), (y_l) \rangle = \sum_{k,l} a_{kl} x_k y_l$. Duality will be with respect to this bilinear form. Basic information is derived from the following fact:

18.52 Lemma. *Let* $W = \langle \alpha \rangle \subset F = \mathbb{F}_{q^r}$ *the cyclic subgroup of order* n, *and the integer* l *not a multiple of* n. *Then* $\sum_{i=0}^{n-1} \alpha^{il} = 0$.

PROOF Let $S = \sum_{i=0}^{n-1} \alpha^{il}$. Then $\alpha^l S = S$ and $\alpha^l \neq 1$. It follows that $(1 - \alpha^l)S = 0$, which implies $S = 0$. $\quad\square$

18.53 Proposition. $V_{m,F}(Z)^\perp = \oplus_{Z' \neq -Z} V_{m,F}(Z')$. $V_m(Z)^\perp = \oplus_{Z' \neq -Z} V_m(Z')$.

PROOF As the dimensions are right, it suffices to show that $V_{m,F}(Z)$ is orthogonal, with respect to $\langle\rangle$, to $V_{m,F}(Z')$ when $Z' \neq -Z$ (an analogous statement holds for the second equality). Concretely this means $\sum_{k,l=1}^{m} a_{kl} \sum_{i=0}^{n-1} p_k(\alpha^i) q_l(\alpha^i) = 0$ where $p_k(X) \in \mathcal{P}(Z), q_l(X) \in \mathcal{P}(Z')$. By bilinearity it suffices to prove this for fixed k, l and $p_k(X) = X^z, q_l(X) = X^{z'}$ where $z \in Z, z' \in Z'$. Let $j = z + z'$. Then j is not a multiple of n. We need to show $\sum_{i=0}^{n-1} \alpha^{iz + iz'} = 0$. This follows from Lemma 18.52. This proves the first equality. The proof of the second equality is analogous, using the definition of the trace. $\quad\square$

Observe that the result of Proposition 18.53 is independent of the bilinear form $\langle,\rangle$. The information contained in the bilinear form will determine the duality relations between subspaces of $V_{m,F}(Z)$ and $V_{m,F}(-Z)$ as well as between subspaces of $V_m(Z)$ and $V_m(-Z)$. In order to decide self-orthogonality, the following formula is helpful:

18.54 Lemma. $\sum_{u \in W} tr(au^z) tr(bu^{-z}) = n \times tr(a tr_{F|L}(b))$.

PROOF Let S be the sum on the left. Writing out the definition of the trace, $S = \sum_{j,k=0}^{r-1} a^{q^j} b^{q^j} \sum_{u \in W} u^{(q^i - q^j)z}$. The inner sum vanishes if $(q^i - q^j)z$ is not divisible by n. In the contrary case the inner sum is n. The latter case occurs if and only if $i = j + \rho s$ where $s = |L|$ and $\rho = 0, \ldots, s - 1$. It follows that $S = n \sum_i a^{q^i} \sum_\rho b^{q^i q^{\rho s}} = n tr(a tr_{F|L}(b))$. $\quad\square$

The cyclic quantum stabilizer codes correspond to the special case when $m = 2$ and the bilinear form is the symplectic form $\langle (x_1, x_2), (y_1, y_2) \rangle = x_1 y_2 - x_2 y_1$ (in particular, $\langle x, x \rangle = 0$ always). By definition, such a code is a (cyclic) q-ary quantum stabilizer code if it is self-orthogonal with respect to the symplectic form. The basic lemma to decide self-orthogonality is the following:

18.55 Proposition. *In the situation considered here ($m = 2$ and the symplectic form, $\gcd(n, p) = 1$) irreducible codes $\mathcal{C}(z, P)$ and $\mathcal{C}(-z, P')$ are orthogonal if and only if $P = P'$.*

PROOF   Observe that Z and $-Z$ have the same length $s, L = \mathbb{F}_{q^s}$ and $P, P' \in PG(1, L)$. It suffices to show that $\mathcal{C}(z, P)$ and $\mathcal{C}(-z, P)$ are orthogonal. Let $P = (v_1 : v_2)$. Writing out the symplectic product of typical vectors, we need to show

$$\sum_{u \in W} tr(av_1 u^z) tr(bv_2 u^{-z}) = \sum_{u \in W} tr(av_2 u^z) tr(bv_1 u^{-z})$$

for $a, b \in F$. Lemma 18.54 shows that the sum on the left equals
$n \times tr(av_1 tr_{F|L}(bv_2)) = n \times tr(av_1 v_2 tr_{F|L}(b)) = n \times tr_{L|\mathbb{F}_q}(v_1 v_2 tr_{F|L}(a) tr_{F|L}(b))$,
which, by symmetry, coincides with the sum on the right. □

18.56 Theorem. *The number of additive cyclic q^2-ary quantum stabilizer codes is*

$$\prod_{Z = -Z, s = 1} (q + 2) \prod_{Z = -Z, s > 1} (q^{s/2} + 2) \prod_{Z \neq -Z} (3q^s + 6).$$

The number of self-dual such codes is

$$\prod_{Z = -Z, s = 1} (q + 1) \prod_{Z = -Z, s > 1} (q^{s/2} + 1) \prod_{Z \neq -Z} (q^s + 3).$$

Here $s = |Z|$ and the last product is over all pairs $\{Z, -Z\}$ of cyclotomic cosets such that $Z \neq -Z$.

PROOF   Let $\mathcal{C} = \sum_Z S_Z$ where $S_Z = \mathcal{C}(z, U_Z)$. This is self-orthogonal if and only if S_Z and S_{-Z} are orthogonal for each Z. Consider at first the generic case $Z \neq -Z$. If S_Z or S_{-Z} is 0, then there is no restriction on the other. If $S_Z = \mathcal{C}(z, L^2)$, then $S_{-Z} = 0$.

Consider case $Z = -Z, s > 1$. Then $s = 2i$. Either $S_Z = 0$ or $S_Z = \mathcal{C}(z, P)$ is a self-orthogonal irreducible code where $P \in PG(1, L)$. The dual of S_Z in its constituent is $\mathcal{C}(-z, P)$ by Proposition 18.55. As $-z = z^{q^i}$, we have $\mathcal{C}(-z, P) = \mathcal{C}(z, P^{q^i})$. This equals S_Z if and only if $P \in PG(1, L')$ where $L' = \mathbb{F}_{q^i}$. There are $q^i + 1$ choices for P. Case $Z = -Z, s = 1$ contributes $q + 2$ self-orthogonal and $q + 1$ self-dual codes. □

18.57 Example. *In the case $n = 7$ we obtain $4 \times 30 = 120$ quantum codes altogether and $3 \times 11 = 33$ self-dual ones.*

18.58 Example. *For $n = 15$ the number of quantum codes is $4 \times 4 \times 6 \times 54$ and there are $3 \times 3 \times 5 \times 19$ self-dual codes.*

Binary quantum codes: The geometric theory

Recall that a quantum stabilizer code is defined by a dual pair $Q^\perp \subset Q$ of q^2-ary q-linear codes, where duality is with respect to the symplectic form. Now we restrict to the most interesting case when $q = 2$ and the additive code is therefore a quaternary code. Recall in particular that **binary** quantum codes are described by **quaternary** additive codes. Many examples are obtained by the cyclic theory as described in the previous subsection. In the present subsection we describe the geometric approach. It will turn out to be even more fruitful than the cyclic approach. At first let us get used to the relation between coding parameters and quantum parameters as in Definition 18.49. Here we concentrate on the pure case.

Start from a pure $[[n, k, d]]_2$ quantum code. By Definition 18.49, this is a dual pair of codes where the smaller code is a quaternary $[n, (n - k)/2]_4$ code which is contained in its symplectic dual, an $[n, (n + k)/2]_4$ code.

Let us do this the other way around: start from a quaternary additive $[n, m]_4$ code which is contained in its symplectic dual (an $[n, n - m]_4$ code). Observe that m need not be an integer but $2m$ is an integer. The smaller code has $4^m = 2^{2m}$ codewords. The quantum parameters are $[[n, n - 2m, d]]_2$. In the pure case the minimum distance d of the binary quantum code simply is the minimum distance of the larger of the two quaternary codes, equivalently, 1 more than the strength of the smaller code.

In the beginning of this section we saw the truly trivial example when $m = 0$: the smaller code is the 0 code and its dual therefore the ambient space. The corresponding quantum parameters are $[[n, n, 1]]_2$. Consider the next case $m = 1$. The smaller of the dual pair of corresponding quaternary codes has two codewords. Its binary generator matrix consists of a single row which we can choose as $(10, 10, \ldots, 10)$. Clearly this is contained in its symplectic dual, and the strength is 0. It follows that the quantum parameters are $[[n, n - 1, 1]]_2$.

Even the next case $m = 2$ is trivial. The smaller of the quaternary codes has binary dimension 2. Choose the two rows as $(10)^n$ and $(01)^n$. The strength is 1. The symplectic product of those two rows is $1 + 1 + \cdots = n \in \mathbb{F}_2$, which vanishes if and only if n is even. This yields pure quantum codes $[[n, n - 2, 2]]_2$ for even n.

Next we want to translate from the coding-theoretic (linear algebraic) setting into geometric language. This is from [32].

18.59 Theorem. *The following are equivalent:*

- *A pure* $[[n, n - r, t + 1]]_2$ *quantum code, and*

- *a set of* n *lines (the codelines) in* $PG(r - 1, 2)$ *any* t *of which are in general position and such that, for every secundum* S, *the number of codelines skew to* S *is even.*

Recall that a secundum in $PG(r - 1, 2)$ is a subspace of codimension 2 (a hyperplane of a hyperplane).

PROOF   Start from the coding theoretic description. The smaller code $Q^\perp$ has quaternary dimension $r/2$ and binary dimension r, and can therefore be described by a binary generator matrix H with r rows and n pairs of columns. See the columns as elements in $PG(r - 1, 2)$ and let the codelines be the lines determined by coordinate pairs. The quantum distance is $> t$ if and only if $Q^\perp$ has strength t, equivalently, the projection to any set of t column pairs has binary rank $2t$, equivalently, the corresponding t codelines are in general position.

It remains to express the quantum condition $(Q^\perp \subset Q)$ in geometric terms. An equivalent condition is: any two rows of the generator matrix H have symplectic product 0. Consider two codewords v_1, v_2 and the hyperplanes H_1, H_2 corresponding to them. Let $S = H_1 \cap H_2$ be the corresponding secundum. The coordinate sections corresponding to codelines L that intersect S nontrivially give no contribution to the symplectic product of v_1 and v_2; the others do. $\square$

As a first application, use Proposition 18.44 again. A repeated application shows that the points in $V_{f+2} \setminus V_{f-2i}$ can be partitioned into lines as long as $2i < f$. The number of lines in this partial spread is $(2^{f+2} - 2^{f-2i})/3$. Proposition 18.44 shows that the quantum condition of Theorem 18.59 is satisfied. The strength is 2. Letting $l = f+2$, we arrive at quantum parameters as follows:

18.60 Corollary. *Let* $g_{l,i} = (2^l - 2^{l-2i})/3$. *If* $l - 2i = 0$ *or* $l - 2i \geq 3$, *there is a pure* $[[g_{l,i}, g_{l,i} - l, 3]]_2$-*quantum code.*

This family contains the so-called quantum Hamming codes and Gottesman codes; see [39], Theorems 10 and 11. Some small examples are

$$[[5, 1, 3]]_2, [[8, 3, 3]]_2, [[21, 15, 3]]_2, [[40, 33, 3]]_2.$$

A recursive construction using Corollary 18.46 is also possible; see [32].

In order to simplify further, let us consider the situation when the quaternary code is $\mathbb{F}_4$ linear. It is, after all, not forbidden for the additive quaternary code to be in fact linear over $\mathbb{F}_4$. The following characterization is obtained:

18.61 Theorem. *The following are equivalent:*

- *A pure* $[[n, n - 2m, t + 1]]_2$ *quantum code which is* $\mathbb{F}_4$ *linear.*

- *A set of n points in* $PG(m - 1, 4)$ *any t of which are in general position and such that the intersection size with any hyperplane of* $PG(m - 1, 4)$ *has the same parity as n.*

- *A linear* $[n, m]_4$ *code of strength t all of whose weights are even.*

- *A linear* $[n, m]_4$ *code of strength t which is self-orthogonal with respect to the Hermitian form.*

PROOF   Most of those statements are by now obvious. Key is the equivalence of the first two items. Observe that a hyperplane of $PG(m - 1, 4)$ is a secundum of $PG(2m - 1, 2)$. The preceding theorem shows that the first implies the second. This leaves us with the task of showing that the second item implies the first.

Let S be a secundum in $PG(2m - 1, 2)$ which is not an $\mathbb{F}_4$ hyperplane. Let $\mathbb{F}_4 = \{0, 1, \omega, 1 + \omega\}$. Then the largest $\mathbb{F}_4$ subspace contained in S is $K = S \cap \omega S$, of binary codimension 4 in the ambient space $\mathbb{F}_2^{2m}$. Consider the factor space $PG(3, 2)$. The quaternary hyperplanes form a spread $\{S_1, S_2, S_3, S_4, S_5\}$ (five lines in $PG(3, 2)$ which partition the points). Also S/K is a line in $PG(3, 2)$ (none of the S_i). Let l be the number of codepoints (they are in fact lines in the binary space) which are contained in K, and a_i the number of codepoints in S_i but outside K. Then the number of codepoints is $n = l + \sum a_i$, and the condition of the second statement shows that $\sum_{j \neq i} a_j$ is even, for all i. It follows that all a_i have the same parity. The number of codepoints disjoint from S is the sum of two of those numbers and therefore even.   ☐

Let us apply Theorem 18.61 in the case of distance 4 (strength 3). The strength condition in Theorem 18.61 says that the point set in $PG(m - 1, 4)$ is a cap. If also the quantum condition is satisfied, we may speak of a **quantum cap**; see [33]. A small example is the hyperoval in $PG(2, 4)$. As it meets each line (hyperplane) in even cardinality, it satisfies the quantum condition of Theorem 18.61 and therefore defines a $[[6, 0, 4]]_2$ quantum code.

Binary quantum codes of distance 3

The classification of the parameters of binary distance 3 quantum codes is in [225]. The result is as follows:

18.62 Theorem. *Let* $r \geq 4$ *and* $n_{max}(r) = \begin{cases} (2^r - 8)/3 & \text{if r is odd,} \\ (2^r - 1)/3 & \text{if r is even.} \end{cases}$
Then quantum codes $[[n, n - r, 3]]_2$ *exist for all n such that* $r \leq n \leq n_{max}(r)$ *with the exceptions of* $n_{max}(r) - 1$ *for odd r and* $n_{max}(r) - \{1, 2, 3\}$ *for even r.*

The lower bound $n \geq r$ is obvious. In general, the constructive part of Theorem 18.62 is not hard. In fact, the preceding subsections contain many such constructions. In this subsection we want to concentrate on the non-existence part of the proof of Theorem 18.62. We have to prove that $[[n, n - r, 3]]_2$ do not exist in each of the following situations:

1. $r \geq 5$ odd, $n \in \{(2^r - 2)/3, (2^r - 5)/3, (2^r - 11)/3\}$

2. $r \geq 4$ even, $n \in \{(2^r - 4)/3, (2^r - 7)/3, (2^r - 10)/3\}$

Consider an $[[n, n - r, 3]]_2$ quantum code, $r \geq 4$. We concentrate on the generic case that the code is pure. Use Theorem 18.59 to obtain an explicit geometric representation. We work in $PG(r-1, 2)$ (the set of nonzero binary t-tuples). The strength is $t = 3 - 1 = 2$. Strength $t \geq 1$ means that our codelines are really lines (not degenerate to points), and the condition $t \geq 2$ implies that those lines are mutually skew (they do not intersect). It follows that those n lines form a partial spread in $PG(r - 1, 2)$. Only one additional condition remains to be satisfied for this spread to define a quantum $[[n, n - r, 3]]_2$ code, the **quantum condition** from Theorem 18.59: for every secundum S of $PG(r - 1, 2)$, the number of codelines disjoint from S is even. This is the geometric expression for the self-orthogonality condition of Definition 18.49.

Fix notation: let X be the set of points covered by the lines of the partial spread. Then $|X| = 3n$. Let Y be the complement of X. Then $|Y| = y = 2^r - 1 - 3n$. Observe that n and y have different parities (one is even, the other is odd). In the case r even, the upper bound $n \leq n_{max}(r) = (2^r - 1)/3$ is obvious and it is also easy to see that it can be reached. In fact it suffices to show that $PG(r - 1, 2)$ possesses a spread of lines (partitioning the point set into lines). The quantum condition is then automatically satisfied (see the Exercises). For odd r, Corollary 18.60 shows that $n = (2^r - 8)/3$ can be reached (corresponding to a partial spread covering all points outside a Fano plane, case $y = 7$). When n is close to the upper bound, the number y will be small. It will therefore be advantageous to work with the small point set Y. In terms of y, our claim above is the following:

1. For odd $r \geq 5$, values $y = 1, y = 4, y = 10$ are impossible.

2. For even $r \geq 4$, the values $y = 3, y = 6, y = 9$ are impossible.

Let G be a binary (r, y) matrix whose columns are the points from Y and C the binary linear code generated by G and denote its dimension by $l \leq k$. Let us see how the quantum condition restricts the structure of C.

18.63 Lemma. *C is a projective binary $[y, l]_2$ code which is self-orthogonal ($C \subseteq C^{\perp}$) with respect to the dot product.*

PROOF Let S be a secundum. The quantum condition says that an even number of codelines avoids S. Those which meet S intersect in one or in three points. It follows that $|S \cap X|$ has the same parity as n. As $|S| = 2^{r-2} - 1$ is odd, $|S \cap Y|$ has the parity of y. Let $H \supset S$ be a hyperplane and D the code whose generator matrix has as columns the elements of $H \cap Y$. We just saw that all nonzero codewords of D have weights of the same parity. Those weights must therefore all be even. It follows that $|H \cap Y| - |S \cap Y|$ is even whenever S is a secundum, H a hyperplane and $S \subset H$. In particular, all weights of codewords of C are even.

Let $u, v \in C$ be different nonzero codewords and a, b, c, d the numbers of co-ordinates where the entries in (u, v) are $(0, 0), (0, 1), (1, 0), (1, 1)$, respectively. We saw that a has the parity of y and b, c, d are even. The fact that d is even shows that $u \cdot v = 0$. $\qquad \square$

We can now prove our nonexistence claim. Observe that C is a binary $[y, l]_2$ code which is contained in its dual. In particular, the dimension of C is $l \leq y/2$. As C is projective, we have $y \leq 2^l - 1$. It follows that $l \geq 3$ and therefore $y \geq 6$, so values $1, 3, 4$ are excluded. As $d(C^\perp) \geq 3$ and $C \subseteq C^\perp$, it follows that $d(C) \geq 4$. Assume $y = 6$. Then $l = 3$ and C is a self-dual $[6, 3, 4]_2$ code, equivalently, a set of six points in the Fano plane which form a cap. This is nonsense.

Assume $y = 9$. Then $l = 4$. We have a $[9, 4, 4]_2$ code and a generator matrix uses as columns 9 of the 15 points of $PG(3, 2)$. As the Simplex code is self-orthogonal (see Exercise 2.5.5), the remaining six points of $PG(3, 2)$ also define a self-dual code. Its dimension is therefore 3. It must be that all points but one of the Fano plane $PG(2, 2)$ define a self-dual $[6, 3]_2$ code. This is not the case.

Finally, let $y = 10$. The argument of the preceding case shows that $l = 4$ is impossible. It follows that C is a self-dual projective $[10, 5, 4]_2$ code. Write a generator matrix in the form $(I|P)$. Then P is a $(5, 5)$ matrix with rows of odd weight with pairwise even intersection. The latter shows that no row of P has weight 5. The minimum weight shows that all rows of P have weight 3. This is impossible.

Exercises 18.4

18.4.1. *Show that there is no linear $[6, 3, 4]_4$ code which is self-dual with respect to the dot product.*

18.4.2. *Give a geometric description of a pure $[[5, 1, 3]]_2$ quantum code.*

18.4.3. *Show that the notion of a pure $[[n, n - r, 3]]_2$ quantum code does not make sense if $r < 4$.*

18.4.4. *Show that $PG(r-1,2)$ for even r possesses a spread of lines and determine the parameters of the corresponding binary quantum code.*

18.5   Network codes and subspace codes

18.64 Definition. *Let A, B be subspaces $PG(m-1,q)$ of a $PG(l-1,q)$. Define the **subspace distance** as $d_s(A,B) = l - dim(A \cap B)$, where dim is the vector space dimension over $\mathbb{F}_q$.*

Observe that $d_s(A,B) = dim(A+B) - m$ and $d_s(A,B)$ is an integer between 0 and m. It is left as an exercise to show that the subspace distance defines a metric on the set of subspaces $PG(m-1,q)$ contained in a given $PG(l-1,q)$.

18.65 Definition. *A constant-weight **subspace code** $S_q(l,n,d_s,m)$ is a set of n subspaces $PG(m-1,q)$ of $PG(l-1,q)$ where any two are at subspace distance $\geq d_s$.*

There is a slightly more general notion than Definition 18.65 where codewords are arbitrary subspaces of $PG(l-1,q)$ (dropping the restriction to vector space dimension m), but we will use only Definition 18.65, the most interesting case.

Motivation

One possible motivation comes from an analogy with our basic coding problem. There we had an underlying alphabet A and worked in A^n, equipped with the Hamming metric. A code is an arbitrary subset of A^n. In the analogous situation we replace A^n by the new ambient space $\mathbb{F}_q^l$. Codewords are vector subspaces of the ambient space and the subspace distance plays the role of Hamming distance. This also explains why we speak of constant-weight codes in the special case of Definition 18.65: the vector space dimension of the subspace corresponds in the analogy to the weight of the codeword, and we use fixed dimension m in Definition 18.65. This subject was popularized by a paper by Koetter and Kschischang [126].

Relation to additive codes

We encountered families of vector subspaces earlier in this chapter, in the geometric description of additive codes; see Section 18.3. In fact, we restricted to the quaternary $\mathbb{F}_2$-linear case at that time. Obviously the general case is as follows:

18.66 Theorem. *The following are equivalent:*

- *a q-linear q^m-ary additive $[n,k]$ code.*

- *a multiset of n subspaces $PG(m-1,q)$ (the **codespaces**, which may degenerate to smaller dimension) in $PG(l-1,q)$, where $l = km$.*

Here the codewords of the additive code are in bijection with the hyperplanes, and a codeword has weight w if and only if the corresponding hyperplane contains all but w of the codespaces.

Given a subspace code $S_q(l, n, d_s, m)$, we can associate with it the additive code (q-linear q^m-ary) of length n and dimension $k = l/m$ as in Theorem 18.66. However, the notions of distance are completely different. The weights of the codewords of the additive codes are determined by the distribution of the codespaces on hyperplanes. The distances in the subspace code are determined by the dimensions of the pairwise sums of subspaces. For example, in the case of additive codes, we restricted almost entirely to the quaternary cases, as this case is already general and difficult enough. Where subspace codes are concerned, the case $m = 2$ is almost trivial.

Linearized polynomials

18.67 Definition. *Let $\mathbb{F}_q \subset F = \mathbb{F}_{q^n}$. A **linearized polynomial** is a polynomial of the form $\sum_{i=0}^{n-1} a_i X^{q^i}$, where $a_i \in F$.*

The most important property of linearized polynomials is the obvious fact that they describe $\mathbb{F}_q$-linear mappings : $F \longrightarrow F$; see the discussion of the Frobenius automorphism and the Galois group in Chapter 12. Once a basis of $F|\mathbb{F}_q$ has been fixed, each linearized polynomial corresponds to an (n,n) matrix with entries in $\mathbb{F}_q$, and different linearized polynomials of the form in Definition 18.67 yield different matrices. As the total number of linearized polynomials in Definition 18.67 is $|F|^n = q^{n^2}$, it follows that this $\mathbb{F}_q$ vector space can be identified with the space of (n,n) matrices. This is the main application of those polynomials: they yield a different approach to the space of quadratic matrices over finite fields, an approach using the structure of the extension field $\mathbb{F}_{q^n}$. We could (and probably should) have introduced this concept much earlier. It is very useful in the theory of algebraic field extension.

Gabidulin codes

18.68 Definition. *Let $V = (\mathbb{F}_q)_{n,n}$ be the $\mathbb{F}_q$ vector space of (n,n) matrices. An $MRD(k,n,q)$ code is a subset $\mathcal{C} \subset V$ of size $|\mathcal{C}| = q^{kn}$ such that $rk(A - B) \geq n - k + 1$ whenever $A, B \in \mathcal{C}, A \neq B$.*

Here MRD stands for **maximal rank distance.** Another way of expressing Definition 18.68 is to work with the rank metric for matrices:

$$d(A, B) = rk(A - B).$$

In words: the distance between two matrices is defined as the rank of the difference. Projection onto the first k rows shows that the size of the code as given in Definition 18.68 certainly is optimal. The classical construction uses linearized polynomials:

18.69 Definition. *Let* $\mathbb{F}_q \subset F = \mathbb{F}_{q^n}$. *The kn-dimensional* **Gabidulin code** *(see [87]) is the space of linearized polynomials as in Definition 18.67 of degree* $\leq q^{k-1}$.

18.70 Proposition. *The kn-dimensional Gabidulin code over* $\mathbb{F}_q$ *is an* $MRD(k, n, q)$.

PROOF We know that the Gabidulin code is an $\mathbb{F}_q$ vector space of linearized polynomials. The dimension of this space is kn. We think of each of its elements as an $\mathbb{F}_q$ linear mapping : $F \longrightarrow F$ or, after introduction of a basis, as an (n, n) matrix. As a nonzero polynomial of degree q^{k-1} has at most q^{k-1} roots, it follows that the nonzero elements of the Gabidulin code have kernel of dimension at most $k - 1$. The rank is therefore $\geq n - k + 1$. ▢

Let us take a look at $MRD(2, 3, 2)$ in some more detail.

Let $B = \begin{pmatrix} 1 & 1 & 0 \\ 0 & 1 & 1 \\ 1 & 1 & 1 \end{pmatrix}$, an element of order 7 in the group $GL(3, 2)$. The

binary vector space generated by $I, B, B^{-1} = \begin{pmatrix} 0 & 1 & 1 \\ 1 & 1 & 1 \\ 1 & 0 & 1 \end{pmatrix}$ consists, aside from

the zero matrix, precisely of the powers of B. In fact, $B + I = B^5, B^{-1} + I = B^4, B + B^{-1} = B^2, B + B^{-1} + I = B^3$. This shows that $\{(a, b, Ba + B^{-1}b)\}$ is an $MRD(2, 3, 2)$ where a, b are arbitrary column vectors. As this is a vector space, it suffices to show that each such matrix for $(a, b) \neq (0, 0)$ has rank ≥ 2. This is in doubt only if either $a = 0$ or $b = 0$ or $a = b \neq 0$. In those cases the property of the matrix B provides a proof.

Subspace codes: The case of planes

We return to the constant-weight subspace codes $S_q(l, n, d_s, m)$ of Definition 18.65. Here $d_s \leq m$. The basic coding problem is to determine the maximum value of n when the remaining parameters are given.

We saw that the case $m = 1$ (points) is trivial and case $m = 2$ (lines) is almost trivial. The smallest interesting value of m is therefore $m = 3$. The code is a family of planes in the ambient space $PG(l - 1, q)$. Let us restrict to the binary case $q = 2$. Again $d_s = 1$ is trivial and $d_s = 3$ is close to being trivial. Concentrate on $d_s = 2$. Then the coding question in geometric terms is the following: what is the maximum number n of Fano planes in $PG(l-1, 2)$ such that any two of those code planes intersect at most in a point. For each value of l, we have one such coding problem to solve. Cases $l \leq 4$ are not interesting, as we cannot even find two such planes. Case $l = 5$ is expected to be somehow degenerate. In fact:

18.71 Proposition. *The largest n such that $S_2(5, n, 2, 3)$ exists is $n = 9$.*

PROOF   We are in $PG(4, 2)$ and search planes which pairwise intersect precisely in a point (it is impossible that they are disjoint). Simply use duality: the dual of a plane is a line. We are looking for a set of lines which are pairwise skew, a partial spread. We saw in an exercise that the largest size of a partial spread of lines in $PG(4, 2)$ is 9.   ⬚

Turn to the coding problem for $S_2(6, n, 2, 3)$. In an exercise it is shown that $n \leq 81$.

18.72 Proposition. *The largest n such that $S_2(6, n, 2, 3)$ exists satisfies $65 \leq n \leq 81$.*

PROOF   The construction of an $S_2(6, 64, 2, 3)$ uses a Gabidulin code. In fact, consider an $MRD(2, 3, 2)$. This is a space of 64 binary $(3, 3)$ matrices having the property that the difference of any two of them still has rank ≥ 2. Now use a construction which is often useful: consider the matrices $(I|A)$ where $A \in MRD(2, 3, 2)$ and the subspace $E(A) \in \mathbb{F}_2^6$ generated by its rows. The presence of the unit matrix shows that $E(A)$ has rank 3 and therefore is a Fano plane in $PG(5, 2)$. The defining property of an $MRD(2, 3, 2)$ shows that any two of those planes intersect in at most one point. We have an $S_2(6, 64, 2, 3)$. Together with the row space of matrix $(0|I)$ an $S_2(6, 65, 2, 3)$ is obtained.   ⬚

Using the concrete description of $MRD(2, 3, 2)$ given above, the $S_2(6, 64, 2, 3)$ looks as follows: it consists of the row spaces $E(a, b)$ of the matrices $(I|a, b, Ba + B^{-1}b)$ where a, b are arbitrary column vectors in $\mathbb{F}_2^3$ and B our fixed matrix of order 7 in $GL(3, 2)$. Call this space of 64 planes in $PG(5, 2)$ the **lifted Gabidulin code** $\{E(a, b)|a, b, \in \mathbb{F}_2^3\}$. Let S be the 65th plane, the row space of $(0|I)$. We use this concrete description to obtain an improvement of the lower bound. In an exercise it is shown that the lines contained in the planes forming the lifted Gabidulin code are precisely the lines which are skew to S. The construction idea is the following: find a space of 8 planes in the lifted Gabidulin code such that the set of $8 \times 7 = 56 = 4 \times 14$ lines it contains is also contained in a set of 14 planes each of which meets S in a point. Such a space is $U = \langle 111000, 000111, 010001 \rangle$. Our new subspace code will consist of the 56 planes $E(a, b)$ where $(a, b) \notin U$

and of 14 further planes. As an example, consider the lines $l(100, 010, a, b) = \langle (1, 0, 0, a_1, b_1, a_1 + a_2 + b_2 + b_3), (0, 1, 0, a_2, b_2, a_2 + a_3 + b_1 + b_2 + b_3) \rangle \subset E(a, b)$ where $a = (a_1, a_2, a_3), b = (b_1, b_2, b_3)$ and $(a, b) \in U$. It is immediately checked that each plane through $000100 \in S$ and one of those eight lines contains four more of those lines, so all those eight lines are in the union of two planes containing 000100. The same relation holds for the lines $l(100, 001, a, b)$ and point $000010 \in S$, lines $l(100, 011, a, b)$ and $000110 \in S$ and so forth. This defines a subspace code with $64 - 8 + 14 = 70$ planes.

A final extension is obtained by choosing a set of seven planes each of which contains a line of S (one new plane for each line of S which has no points outside S in common. Clearly this can be found and it results in a subspace code with 77 planes. This is in fact the maximum, as has been shown by a computer search [117].

Exercises 18.5

18.5.1. *Let Ω be the set of m-dimensional vector subspaces of a fixed l-dimensional space $\mathbb{F}_q^l$. Show that the subspace distance defines a metric on Ω.*

18.5.2. *Find the maximum length n of a subspace code $S_q(l, n, 1, m)$.*

18.5.3. *Show that the coding problem for subspace codes in the case $m = 2$ is essentially trivial.*

18.5.4. *Show that different linearized polynomials as in Definition 18.67 yield different linear mappings : $F \longrightarrow F$.*

18.5.5. *Use double counting to show that for $S_2(6, n, 2, 3)$ we have $n \leq 81$.*

18.5.6. *Show that the lines contained in the planes of the lifted Gabidulin code are precisely the lines which are skew to the special plane S.*

Part III

Codes and algebraic curves

Chapter 19

Introduction

19.1 Polynomial equations and function fields

Basic concepts

The objective of algebraic geometry is polynomial equations. We specialize to polynomials in two variables $h(X,Y)$ whose coefficients are in a ground field K. The corresponding equation is then $h(X,Y) = 0$. It is advantageous to work with homogeneous polynomials.

19.1 Definition. *Let $h(X,Y) \in K[X,Y]$ be a polynomial of degree n. Its* **homogenization** *is the polynomial $H(X,Y,Z)$ obtained by replacing each monomial cX^iY^j occurring in $h(X,Y)$ by $cX^iY^jZ^{n-i-j}$.*

By construction, $h(X,Y) = H(X,Y,1)$. Consider triples $(x,y,z) \in K^3$ satisfying $H(x,y,z) = 0$. The 0 triple always does the job and, in general, because of homogeneity, we have that, whenever $H(x,y,z) = 0$, then also $H(\lambda x, \lambda y, \lambda z) = 0$ for all $\lambda \in K$. It is therefore natural to consider as zeroes of $H(X,Y,Z)$ one-dimensional subspaces of the d-dimensional vector space K^3. These zeroes are therefore points of the projective plane $PG(2,K)$ (throw away the boring solution $(0,0,0)$ and identify each nonzero solution with all its scalar multiples).

Recall that projective planes were first introduced in Chapter 7. General projective geometries and homogeneous coordinates made their appearance in Chapter 17. The homogeneous coordinates $(a : b : c)$ describe the one-dimensional subspace generated by (a,b,c), which we call a point of $PG(2,K)$.

If $K = \mathbb{F}_q$ is a finite field, this leads to a natural counting problem: how many points of $PG(2,q)$ are zeroes of the homogeneous equation $H(x,y,z) = 0$? In fact, this leads to an infinity of counting problems: as the coefficients of $H(X,Y,Z)$ are in $\mathbb{F}_q$, they are also in each extension field $\mathbb{F}_{q^r}$, so we can ask how many points of $PG(2,q^r)$ are roots of our equation.

19.2 Definition. *Let $H(X,Y,Z) \in \mathbb{F}_q[X,Y,Z]$ be a homogeneous polynomial and $h(X,Y) = H(X,Y,1)$. Then $N_r(H) = N_r(h)$ is the number of points $(x : y : z) \in PG(2,q^r)$ such that $H(x,y,z) = 0$.*

The **affine solutions** *are those of the form* $(x : y : 1)$, *and* $h(x, y) = 0$ *is the* **affine equation.** *The nonaffine solutions* $(x : y : 0)$ *are the solutions* **at infinity.** *A* **rational point** *is a solution in* $PG(2, q)$ *(defined over the ground field* $\mathbb{F}_q$*).*

In general, we will use capital letters $X, Y, Z, \ldots$ for variables of polynomials and small letters $x, y, z \ldots$ if we think of field elements.

Here is a famous example:

19.3 Definition. *Let* $H(X, Y, Z) = X^3Y + Y^3Z + Z^3X$, *the* **Klein quartic.** *As all nonzero coefficients are 1, we can consider it over any ground field* K.

There are two solutions at infinity, $(1 : 0 : 0)$ and $(0 : 1 : 0)$. The affine equation given by the Klein quartic is

$$X^3Y + Y^3 + X = 0.$$

Over $\mathbb{F}_2$ the only affine solution is $(0 : 0 : 1)$. We count $N_1(h) = 3$ solutions of the Klein quadric over $\mathbb{F}_2$. Over $\mathbb{F}_4$ (first introduced in Section 3.2), only two additional solutions arise: $(\omega : \overline{\omega} : 1)$ and $(\overline{\omega} : \omega : 1)$. Over the extension field $\mathbb{F}_8$ (constructed in Chapter 4), things are more interesting: we obtain 21 solutions which were not defined over $\mathbb{F}_2$, and therefore $N_3(h) = 24$. Observe that $\mathbb{F}_4$ is not a subfield of $\mathbb{F}_8$ (see Proposition 12.1 and Exercise 3.2.3), and correspondingly the two solutions defined over $\mathbb{F}_4$ but not over $\mathbb{F}_2$ do not show up over $\mathbb{F}_8$.

Hyperelliptic equations

Let $char(K) \neq 2$. Equations

$$y^2 = f(x)$$

where $f(x) \in K(x)$ are called **hyperelliptic.** Observe that $f(x)$ need not be a polynomial; see Definition 19.7 below. However, it can be assumed that $f(x)$ is a square-free polynomial of degree ≥ 3 (see Exercise 19.2.3). Let $K = \mathbb{F}_q$. Observe that $f(x)$ is separable: over the algebraic closure it decomposes into different linear factors. Also, not much harm is done by assuming that $f(x)$ is monic. In fact, let $f(x) = cg(x)$ where $c \in K$. If $c = a^2$ is a square, then $f(\alpha)$ is square if and only if $g(\alpha)$ is. If c is a nonsquare, then $f(\alpha)$ is square if and only if $g(\alpha)$ is nonsquare. Let $N = N_1 = N(y^2 - f(x))$ be the number of rational points. How many points at infinity do we have? Let $f(x) = x^n + a_{n-1}x^{n-1} + \ldots$. The homogenization of $y^2 - f(x)$ is $y^2 z^{n-2} - x^n - a_{n-1}x^{n-1}z - \ldots$. The points at infinity are those where $z = 0$. The equation becomes $x^n = 0$. This implies $x = 0$. The only point at infinity is $(0 : 1 : 0)$. Clearly

$$N = 1 + 1 \times |\{\alpha \in K \mid f(\alpha) = 0\}| + 2 \times |\{\alpha \in K \mid 0 \neq f(\alpha) \text{ a square in } K\}|.$$

Knowledge of the number of solutions of hyperelliptic equations is equivalent to knowledge of the number of field elements $\alpha \in K$ such that $f(\alpha)$ is a square.

Artin-Schreier equations

In earlier chapters we used the additive version of Hilbert's theorem 90 only in characteristic 2. We need it for general finite fields:

19.4 Theorem. *Let* $tr : F = \mathbb{F}_{q^r} \longrightarrow \mathbb{F}_q$ *be the trace. For an element* $b \in F$ *we have that* $tr(b) = 0$ *if and only if there is* $a \in F$ *such that* $b = a - a^q$.

The reader is asked in Exercise 19.2.2 to provide the easy proof; compare Exercise 12.1.6. It helps us to understand the following class of examples:

19.5 Definition. *Consider the polynomial*

$$h(X, Y) = Y^q - Y - f(X)$$

defined over $F = \mathbb{F}_{q^r}$, *where* $f(X)$ *is a polynomial.*

This type of equation is interesting because of the link to the trace. Consider the solutions of this affine equation over $F = \mathbb{F}_{q^r}$. Whenever $f(x)$ does not have trace 0, there is no solution (x, y). Let $tr(f(x)) = 0$. Then, by Theorem 19.4, there is a solution $y_0 \in F$, and of course the whole coset $y_0 + \mathbb{F}_q$ yields solutions. As a nonzero polynomial of degree q cannot have more than q roots, this accounts for all affine solutions.

19.6 Proposition. *The number of solutions of the homogenization of the affine equation* $h(X, Y) = Y^q - Y - f(X)$ *from Definition 19.5 is* $1 + qt$, *where* t *is the number of* $x \in F$ *such that* $tr(f(x)) = 0$.

PROOF The number of solutions of the affine equation is qt. There is precisely one solution at infinity. Assume, for example, $n = deg(f(X)) > q$. After filling up with powers of Z and putting $Z = 0$, it follows that $X = 0$. The solution at infinity is then $(0 : 1 : 0)$. The other two cases are similar. ◻

Knowledge of the number of solutions of Artin-Schreier equations is equivalent to knowledge of the number of field elements $\alpha \in K$ such that $f(\alpha)$ has trace $= 0$.

Function fields

With a little luck, we can associate a field to a homogeneous polynomial. The starting point is the rational function field:

19.7 Definition. *Let* K *be a field. The* **rational function field** $K(x)$ *is the field whose elements are the quotients (rational functions)* $f(x)/g(x)$ *where* f, g *are polynomials with coefficients in* K *and* $g(x)$ *is not the 0 polynomial.*

Clearly $K(x)$ contains the polynomial ring $K[x]$ and therefore is an infinite-dimensional vector space over K.

Now consider the affine (not necessarily homogeneous) polynomial equation

$$h(x, y)(= H(x, y, 1)) = 0.$$

We can see $h(x, y)$ as a polynomial in one variable y with coefficients in the rational function field $K(x)$. Recall from Section 3.2 that an **irreducible polynomial** in one variable of degree r with coefficients in a field L defines a field extension of L : a field containing L, which is an r-dimensional vector space over L. If $h(x, y)$ is irreducible as a polynomial in the variable y (degree n, with coefficients in $K(x)$), we therefore obtain a degree n field extension of $K(x)$.

19.8 Definition. *Let $H(X, Y, Z)$ be a homogeneous polynomial over ground field K. Assume $h(x, y) = H(x, y, 1)$ is irreducible of degree n as a polynomial in the indeterminate y, with coefficients in $K(x)$. The* **function field** *of h is the degree n field extension of $K(x)$ defined by this irreducible polynomial.*

Because of this link to an algebraic object, the function field, we prefer polynomials $h(X, Y)$ which are irreducible in Y. Recall that we can consider h over any field extension of K, in particular over the algebraic closure $\overline{K}$. We would like this link to be available not only over K itself but over all those algebraic extension fields of K :

19.9 Definition. *The polynomial $h(X, Y) \in K[X, Y]$ is* **absolutely irreducible** *in Y if it is irreducible in the variable Y with coefficients in $\overline{K}(X)$.*

This link leads to a general definition of the algebraic objects we are interested in:

19.10 Definition. *Let K be a field. An* **algebraic function field in one variable (function field)** *over K is a finite extension of the rational function field $K(x)$.*

We saw that polynomials in two variables satisfying the irreducibility criterion of Definition 19.8 lead to a function field. How are we getting back? Given a function field F, by definition we can write it as a finite extension (degree n) of some rational function field $K(x)$. Hopefully, this extension is generated by one element, a **primitive element** y, meaning that $1, y, \ldots, y^{n-1}$ is a basis of F as a $K(x)$ vector space. The minimal polynomial of y will then have degree n, and so we end up with a polynomial $h(X, Y)$ which does satisfy the irreducibility criterion.

A little later we are going to see that the Artin-Schreier polynomials of Definition 19.5 are indeed irreducible provided either the degree of $f(X)$ is not a multiple of the characteristic p or there is an irreducible polynomial $g(X)$ such that the maximal exponent i with $g(X)^i$ dividing $f(X)$ is not a multiple of p. This will be quite easy to prove once we have developed the theory a little further.

19.2 Places of the rational function field

As function fields are by definition finite-dimensional extensions of the rational function field, we should try to understand the rational function field first.

Basic concepts

Recall that the elements of the rational function field $K(x)$ are the fractions $q(x) = f(x)/g(x)$, where f, g are polynomials and $g \neq 0$. Use also that each polynomial can be written as a product of irreducible polynomials in an essentially unique way. The representation of nonzero rational functions in the form $f(x)/g(x)$ becomes uniquely determined if we write it in reduced form $(\gcd(f(x), g(x)) = 1)$ and such that $g(x)$ is **monic** (highest coefficient $= 1$). Fix a monic irreducible polynomial $p(x)$. We record how often (to which exponent) $p(x)$ appears in the numerator or denominator of $q(x)$. Write $v_p(q(x)) = i \geq 0$ if $p(x)^i$ is the precise power of $p(x)$ dividing $f(x)$ and $v_p(q(x)) = -j < 0$ if $p(x)^j$ is the precise power of $p(x)$ dividing $g(x)$. Those mappings v_p have a number of interesting properties. We artificially define $v_p(0) = \infty$.

19.11 Definition. *For each irreducible polynomial $p(x) \in K[x]$, let $\mathcal{O}_p$ be the set of rational functions $q(x)$ such that $v_p(q(x)) \geq 0$ and $\mathcal{P}_p$ the set of rational functions $q(x)$ such that $v_p(q(x)) > 0$.*

Here are some obvious properties which hold for all irreducibles $p = p(x)$.

- $\mathcal{O}_p$ is a subring of $K(x)$ and $\mathcal{P}_p$ its uniquely determined maximal ideal.

- If $q \notin \mathcal{O}_p$, then $1/q \in \mathcal{O}_p$.

- If $v_p(q_1(x)) < v_p(q_2(x))$, then $v_p(q_1(x) + q_2(x)) = v_p(q_1(x))$.

- $v_p(q_1(x)q_2(x)) = v_p(q_1(x)) + v_p(q_2(x))$.

19.12 Proposition. *Let $p(x) \in K[x]$ be an irreducible polynomial of degree f. Then $\mathcal{O}_p/\mathcal{P}_p$ is an extension field of degree f of K.*

PROOF   At first restrict to the polynomial ring $K[x]$. Clearly $K[x] \subset \mathcal{O}_p$ and $K[x] \cap \mathcal{P}_p$ consists of the polynomials which are multiples of $p(x)$. This is the ideal $(p(x))$. The factor ring $K[x]/(p(x))$ is an extension field of degree f of K. This is a mechanism we saw early on, in Theorem 3.7. We claim that not only all polynomials but all rational functions in $\mathcal{O}_p$ are mapped to this field: let $q(x) = f(x)/g(x) \in \mathcal{O}_p$. If it is in $\mathcal{P}_p$, it is mapped to 0. If it is not in $\mathcal{P}_p$, then by definition $g(x)$ is coprime with $p(X)$. An important theoretical consequence of the Euclidean algorithm (we recalled and used it in Section 13.3) implies that there are polynomials $a(x), b(x)$ such that

$$1 = a(x)g(x) + b(x)p(x).$$

It follows that $q(x) = q(x)\cdot 1 = f(x)a(x) + q(x)b(x)p(x)$. The second summand maps to 0 and the image of $q(x)$ coincides with the image of the polynomial $f(x)a(x)$. □

The v_p are examples of **discrete valuations**, the $\mathcal{O}_p$ are called **valuation rings**, the $\mathcal{P}_p$ are **places** and $F_p = \mathcal{O}_p/\mathcal{P}_p$ is known as the **residue class field.** The degree $f = [F_p : K]$ is the degree $deg(P)$ of the place P, and in this case we know that the degree of the place defined by the polynomial $p(X)$ coincides with the degree of the polynomial $p(X)$. Aside from the valuations corresponding to irreducible polynomials, there is another valuation v_∞ **at infinity.**

19.13 Definition. $v_\infty(f(x)/g(x)) = deg(g(x)) - deg(f(x))$ *if $f(x) \neq 0$ and as usual $v_\infty(0) = \infty$.*

The places corresponding to v_p and v_∞ are all the places of the rational function field; see [198], Theorem 1.2.2. Clearly $q(x) = f(x)/g(x) \in \mathcal{O}_\infty \setminus \mathcal{P}_\infty$ if $f(x), g(x)$ have the same degree: $f(x) = a_n x^n + \ldots, g(X) = b_n x^n + \ldots$. The residue class field $\mathcal{O}_\infty/\mathcal{P}_\infty = K$ and the image of the $q(x)$ above is a_n/b_n.

Another important observation is that, for each $q(x)$, the sum $\sum_P v_P(q(X))deg(P)$ extended over the v_p and also v_∞ equals 0. In fact, the sum over all v_p equals of course $deg(f(X)) - deg(g(X))$, and the valuation at infinity balances this out.

In the next section we start a study of general function fields and their valuations. By definition they can be written as finite-dimensional extensions of rational function fields. The fact that we know precisely what happens in the case of the rational function field will be very helpful for the general theory.

Exercises 19.2

19.2.1. *List the 24 solutions of the Klein quartic over $\mathbb{F}_8$.*

19.2.2. *Prove Theorem 19.4, Hilbert's theorem 90 for finite fields.*

19.2.3. *Show that for a hyperelliptic equation* $y^2 = f(x)$, *it can be assumed without loss of generality that* $f(x)$ *is a polynomial without quadratic factors.*

19.2.4. *Show that the hyperelliptic equation* $y^2 = f(x)$ *where* $f(x)$ *is a polynomial of degree* ≥ 3 *has* $(0 : 1 : 0)$ *as the only point at infinity.*

19.2.5. *Explain why the equation* $y^2 = f(x)$ *is not interesting in characteristic* 2.

Chapter 20

Function fields, their places and valuations

20.1 General facts

Basic concepts

We encountered function fields (in one variable) in the context of polynomials $h(X, Y)$ with coefficients in a ground field K, which are irreducible with respect to the variable Y. This led to Definition 19.10.

20.1 Definition. *A* **function field** *is a pair of fields $F|K$ where F is a finite-dimensional extension of $K(x)$ for some $x \in F$ which is transcendental over the ground field K. Here x is* **transcendental** *over K if the powers $1, x, x^2 \ldots$ are linearly independent over K (equivalently, if x is not algebraic over K). The* **algebraic closure** *$\tilde{K}$ of K in F consists of the elements of F which are* **algebraic** *(not transcendental) over K. We call K* **algebraically closed** *in F if $\tilde{K} = K$.*

We start from this definition. For a while we are going to move in this ring theoretic context. Motivated by the rational function field, we introduce the notions of valuations, places and valuation rings. The definition of a valuation ring is rather general:

20.2 Definition. *A* **valuation ring** *of the function field $F|K$ is a proper subring $\mathcal{O} \subset F$ which properly contains K and satisfies the condition:*

- *If $z \in F, z \notin \mathcal{O}$, then $1/z \in \mathcal{O}$.*

Here are some basic properties:

20.3 Proposition. *Let $\mathcal{O}$ be a valuation ring of the function field $F|K$, $\mathcal{O}^*$ its group of units and $\mathcal{P}$ the set of nonunits of $\mathcal{O}$. Then the following hold:*

- *If $x \notin \mathcal{O}$, then $1/x \in \mathcal{P}$.*

- *$\mathcal{P}$ is an ideal of $\mathcal{O}$, its uniquely determined maximal ideal.*

- *The field $\tilde{K}$ of constants is contained in each valuation ring and $\tilde{K} \cap P = \{0\}$.*

PROOF   Let $x \notin \mathcal{O}$. Assume $1/x$ is a unit. Then $x \in \mathcal{O}$, a contradiction.

Let $x \in P, z \in \mathcal{O}$. As $\mathcal{O}$ is a ring, we have $xz \in \mathcal{O}$. If xz were a unit, also x would be a unit, a contradiction. This shows $xz \in P$. Let $x, y \in P$. We have to show $x + y \in P$. It can be assumed $xy \neq 0$ and $x/y \in \mathcal{O}$. We have $x + y = y(1 + x/y)$. As $y \in P$ and $1 + x/y \in \mathcal{O}$, the first property above shows $x + y \in P$. This shows that P is indeed an ideal. As no unit can be contained in a proper ideal, it follows that P is the uniquely determined maximal ideal of $\mathcal{O}$.

Let $x \in \tilde{K}$. By definition it satisfies a polynomial equation $x^n + a_{n-1}x^{n-1} + \cdots + a_1 x + a_0 = 0$ with $a_i \in K$. Assume $x \notin \mathcal{O}$. Then $z = 1/x \in P$ by our first part. It satisfies $1 + a_{n-1}z + \ldots a_0 z^n = 0$. As $a_{n-i}z^i \in P$ for all $i > 0$, it follows $1 \in P$, a contradiction. $\qquad \Box$

If x is transcendental over K, then the subfield $K(x) \subseteq F$ is a copy of the rational function field. Rings with a uniquely determined maximal ideal are called **local rings**. At first we clarify the structure of the valuation rings $\mathcal{O}$. As the basic theory is very well described in Stichtenoth's book [198], we will not always give proofs but rather quote the corresponding result from [198]. The starting point is the following lemma (I.1.7 of [198]), which is a special case of a result from [53].

20.4 Lemma. *Let $\mathcal{O}$ be a valuation ring and $(x_1 = x, x_2, \ldots, x_n)$ such that $x_i \in P$ for all i and $x_i/x_{i+1} \in P$ for all $i < n$. Then $x_1, \ldots, x_n$ are linearly independent over $K(x)$ and $n \leq [F : K(x)] < \infty$.*

The fact that F has finite degree over $K(x)$ follows from the fact that x is transcendental. A little calculation shows the linear independence. Lemma 20.4 is the main ingredient in the proof of the structure theorem, I.1.6 of [198]:

20.5 Theorem. *Let $\mathcal{O}$ be a valuation ring with maximal ideal P. Then the following hold:*

- *P is a principal ideal.*

- *Let $P = t\mathcal{O}$. Then every nonzero element $z \in F$ can be uniquely represented as $z = t^n u$, where n is an integer and u a unit of $\mathcal{O}$.*

- *The ideals of $\mathcal{O}$ are $t^n \mathcal{O}$, for $n = 1, 2, \ldots$.*

Rings with these properties are called **discrete valuation rings**. The exponent n appearing there allows us to express the situation in terms of valuations.

20.6 Definition. *A* **place** *of a function field $F|K$ is the maximal ideal of some valuation ring $\mathcal{O}$. An element t such that $\mathcal{P} = t\mathcal{O}$ is a* **local parameter.**

Clearly $\mathcal{O}$ and $\mathcal{P}$ contain the same information: if $\mathcal{O}$ is known, then $\mathcal{P}$ is its maximal ideal; if $\mathcal{P}$ is known, then $\mathcal{O} = \{z \in F | 1/z \notin \mathcal{P}\}$. A third equivalent description is in terms of valuations:

20.7 Definition. *A* **discrete valuation** *is a mapping $v : F \longrightarrow \mathbb{Z} \cup \{\infty\}$ such that the following hold:*

- $v(x) = \infty$ *if and only if $x = 0$.*

- $v(xy) = v(x) + v(y)$ *for all $x, y \in F$.*

- $v(x + y) \geq min\{v(x), v(y)\}$ *for all $x, y \in F$.*

- *There exists $t \in F$ such that $v(t) = 1$.*

- $v(K) = 0$.

Here the symbol ∞ is to be interpreted in the obvious way. It is an immediate observation that the third property can be strengthened:

20.8 Lemma. *If $v(x) < v(y)$, then $v(x + y) = v(x)$.*

This is the **strict triangle inequality**, Lemma I.1.11 of [198]. Given a valuation ring $\mathcal{O}$, write $\mathcal{P} = t\mathcal{O}$ and use the unique representation $z = t^n u$ for $0 \neq z \in F$ (Theorem 20.5). A discrete valuation is defined by $v(z) = n$. If a discrete valuation v is given, then $\mathcal{O} = \{z \in F | v(z) \geq 0\}$ defines a valuation ring with maximal ideal $\mathcal{P} = \{z \in F | v(z) > 0\}$. The units are the elements of valuation $= 0$, of course. This is the promised third equivalent way of expressing the same concept (valuation rings, places or discrete valuations).

20.9 Corollary. *A valuation ring $\mathcal{O}$ of $F|K$ is a* **maximal** *proper subring of F.*

PROOF Let $y \notin \mathcal{O}$. Recall $v(y) < 0$. We have to show that the subring $\mathcal{O}[y]$ is all of F. Let $z \in F$. Choose n large enough such that $v(zy^{-n}) \geq 0$. This means $zy^{-n} \in \mathcal{O}$ and $z \in \mathcal{O}y^n \subset \mathcal{O}[y]$. $\square$

The term **triangle inequality** suggests a link between discrete valuations and metrics. This is an important fact. Choose a real number $0 < c < 1$ and define

$$\|z\| = c^{v(z)}$$

for $0 \neq z \in F$, $\|0\| = 0$. Then $d_v(x, y) = \|x - y\|$ defines a metric on F and our triangle inequality for the discrete valuation implies the familiar triangle inequality for metrics (see Definition 1.5).

As $c < 1$, we see that x, y are close if the integer $v(x - y)$ is large. It will be important later on to develop this kind of intuition: the elements which are close to 0 with respect to the metric given by v are those in $t^n \mathcal{O}$ for high values of n.

20.10 Definition. *The **residue class field** $F_{\mathcal{P}}$ of the place $\mathcal{P}$ is the field $\mathcal{O}/\mathcal{P}$. Its dimension over K is the **degree** of the place: $\deg(\mathcal{P}) = [F_{\mathcal{P}} : K]$.*

Recall that the factor ring $R/I = \overline{R}$ of a commutative ring R with respect to a maximal ideal I is indeed a field. Let $x \in R, x \notin I$. Then the ideal $\overline{x}\overline{R} = \overline{R}$ because of the maximality of I, so $\overline{x} = x + I$ is invertible in $\overline{R}$.

20.11 Proposition. *Let $\mathcal{P}$ be a place and $0 \neq x \in \mathcal{P}$. Elements in $\mathcal{O}$ whose images are independent in $F_{\mathcal{P}}$ are independent over $F(x)$. Consequently, $\deg(\mathcal{P}) \leq [F : K(x)] < \infty$.*

This is easy to see (Proposition I.1.15 of [198]). Recall from Proposition 20.3 that $\tilde{K}$ is a subfield of $F_{\mathcal{P}}$, for all $\mathcal{P}$.

If you wonder about the terminology which at first may sound bizarre, it is borrowed from the theory of Riemann surfaces and their fields of meromorphic functions, with the function field in the role of the field of meromorphic functions. The places (equivalently, discrete valuations, valuation rings) are to be seen as the points of the Riemann surface. So we started from the functions (elements of the function field) and constructed the set they are acting on (the places) only as a second step. When writing $f(P)$ and saying *the value of function f at place $\mathcal{P}$* what is meant concretely is, $\mathcal{P}$ is a place , $f \in \mathcal{O}$ (the corresponding valuation ring) and the value $f(\mathcal{P})$ is the element $f + \mathcal{P}$ of the residue class field. The ring $\mathcal{O}$ consists of the functions f which are **defined** at $\mathcal{P}$ (equivalently $v_{\mathcal{P}}(f) \geq 0$). If $v_{\mathcal{P}}(f) = n > 0$, then f has a zero of degree n; if $n < 0$, then $\mathcal{P}$ is a **pole** of degree n of f.

Next it has to be shown that places always exist in abundance. The basis of this is of course that F is a finite extension of a rational function field and we do know many places of $K(x)$. In fact, we know all of them; see Theorem 20.14.

20.12 Theorem. *Let R be a subring of the function field $F|K$ (possibly $R = F$) containing K, and let I be a nonzero proper ideal of R. Then R is contained in some valuation ring whose maximal ideal (place) contains I.*

This is Theorem I.1.18 of [198]. The proof is a typical ring theoretic little argument using Zorn's lemma. It is, however, a very important tool and will be used repeatedly later on to prove the existence of places.

20.13 Corollary. *Each transcendental element has a zero and a pole. The field $\tilde{K}$ of constants is a finite dimensional extension of K.*

PROOF　　Let $y \in F$ be transcendental. Apply Theorem 20.12 to $K(y)$ and its ideal generated by y. This shows the existence of a zero. Considering $1/y$

yields the existence of a pole. The fact that $\tilde{K}$ is embedded in each residue class field shows $[\tilde{K} : K] < \infty$. ☐

20.14 Theorem. *The places of the rational function field $K(x)$ are those corresponding to irreducible polynomials and the place at infinity. They are in bijection with the projective line $PG(1, K)$.*

This is now easy to prove (Theorem I.2.2 of [198]).

20.15 Theorem (weak approximation theorem). *Given different places $\mathcal{P}_i$, elements $x_i \in F$ and integers r_i (i = 1, ..., n), there exists an $x \in F$ such that*

$$v_{\mathcal{P}_i} = r_i \text{ for all } i.$$

The proof of the weak approximation theorem (Theorem I.3.1 of [198]) is a combinatorial case by case analysis using the strict triangle inequality.

20.16 Corollary. *Each function field has infinitely many places.*

PROOF If only n places existed, by the weak approximation theorem we could find $x \in F$ which has all those places as zeroes. Then x would not have any poles, contradicting Corollary 20.13. ☐

The epithet *weak* in Theorem 20.15 is due to the fact that only a finite number of places is involved. As we think of a function field F as a finite-dimensional extension of a rational function field $K(x)$, we are interested in the degree $[F : K(x)]$ of this extension. A typical application of the weak approximation theorem gives a good lower bound (Proposition I.3.3 of [198]):

20.17 Proposition. *If $\mathcal{P}_1, ..., \mathcal{P}_r$ are zeroes of x, then*

$$[F : K(x)] \geq \sum_{i=1}^{r} v_{\mathcal{P}_i}(x) \cdot deg(\mathcal{P}_i).$$

20.18 Corollary. *Each element of a function field has only finitely many zeroes and poles.*

20.2 Divisors and the genus

We have our meromorphic functions (the elements of the function field $F|K$) and our *Riemann surface*, so to speak, consisting of the places, equivalently, the valuation rings. The basic question to ask now is the following: if, for each place P, we prescribe an integer n_P, can we find a function z such that

$v_P(z) = n_P$ for all P? In other words, we want to prescribe the behavior of the function at all places, what pole and zero orders are concerned. As there are an infinity of places (Corollary 20.18) and $z \in F$ has only finitely many poles and zeroes, the answer to the question above can be affirmative only if $n_P = 0$ for almost all P. It is therefore natural to work with such tuples:

20.19 Definition. *The divisor group $\mathcal{D}_F$ is the free abelian group generated by the places. We write its elements, the divisors, as $A = \sum_P n_P P$ and also write $n_P = v_P(A)$. The corresponding module is defined as*

$$L(A) = \{x \in F | v_P(x) + v_P(A) \geq 0 \text{ for all } P\}.$$

*Further, $dim(A) = dim(L(A))$ is the **dimension** of divisor A.*

Each divisor prescribes the behavior at each place, and at almost all places it is prescribed that the function be defined at P. The use of the ≥ 0 sign in the definition of the space $L(A)$ weakens the original question. This is motivated by the obvious fact, following from the triangle inequality, that $L(A)$ is a vector space over K.

A divisor D is **effective** $(D \geq 0)$ if all its entries are ≥ 0. We write $D_1 \leq D_2$ if $D_2 - D_1 \geq 0$. Divisors $D = 1 \cdot P$ are also called **prime divisors.** As an example, when is $z \in L(P)$? This means that z is defined at all places $\neq P$ and z is either defined at P or has a pole there of order at most 1 $(v_P(z) \geq -1)$. The **support** of D consists of the places P where $v_P(D) \neq 0$. Each divisor D can be written in a unique way as $D = D_0 - D_\infty$, where D_0 (the zero divisor) and D_∞ (the pole divisor) are effective.

20.20 Definition. *Let $x \in F$. The **principal divisor** is defined as*

$$(x) = \sum_P v_P(x)P.$$

Consider also the zero divisor $(x)_0$ and the pole divisor $(x)_\infty$.

The principal divisors form a subgroup of the divisor group.

20.21 Definition. *The **divisor class group** $\mathcal{C}_F$ is the factor group of $\mathcal{D}_F$ mod the group of principal divisors. This defines an equivalence relation on divisors: $A \sim B$ if these divisors yield the same element of $\mathcal{C}_F$, equivalently if $A - B = (z)$ is a principal divisor.*

We know that the elements of $\tilde{K}$, the constants, are precisely those that do not have zeroes or poles. It follows that $L(0) = \tilde{K}$. As we know that $\tilde{K}$ is a finite-dimensional extension of K, it will not do much harm to assume $\tilde{K} = K$.

20.22 Proposition. *$L(0) = K$, $dim(0) = 1$, where it is assumed for the remainder of the text that $\tilde{K} = K$ is the full constant field.*

What does this mean in practice? Consider a function field defined over the field K of constants. Then one of the first problems to solve is to determine the exact field $\tilde{K}$ of constants. In many cases it will turn out $\tilde{K} = K$. Once $\tilde{K}$ is known, we view this as the field of constants and will call it K. In the general theory it is assumed that this step has been taken already.

20.23 Definition. *The* **degree** *of a divisor is defined as*

$$deg(D) = \sum_P v_P(D)deg(P).$$

.

Clearly $A \leq B$ implies $L(A) \subseteq L(B)$. If $A < 0$, then $L(A) \subseteq K$. However, the constants are not contained in $L(A)$, as they have no zeroes, so $dim(A) = 0$. If we want to control the dimensions of effective divisors we can start from 0 and go step by step, adding prime divisors. We need to control the dimensional increase when a prime divisor is added:

20.24 Lemma. $dim(A + P) - dim(A) \leq deg(P)$.
In words, the increase in dimension is bounded by the increase in degree.

PROOF   Let $x \in L(A + P)$ and t be a local parameter at P (meaning $v_P(t) = 1$). Then $xt^{v_P(A)+1}$ is defined at P and we can evaluate. Map $x \mapsto xt^{v_P(A)+1}(P)$. This is a K-linear mapping : $L(A + P) \longrightarrow F_P$ whose kernel is precisely $L(A)$. ∎

This also shows the reason behind the definition of the degree of a divisor. Important relations follow directly: if $A \leq B$, then $dim(B) - dim(A) \leq deg(B - A)$ and of course $dim(A) \leq deg(A) + 1$ for effective divisors A. In particular, $dim(A)$ is finite for each divisor A. The introduction of the divisor class group is justified by the following:

20.25 Proposition. *Equivalent divisors have the same dimension.*

PROOF   This is shamefully trivial: let $A' = A + (z)$. Then $x \in L(A)$ is equivalent to $(x) + A \geq 0$ equivalently $(x/z) + A' \geq 0$, or $x/z \in L(A')$. The isomorphism is given by $x \mapsto x/z$. ∎

Very often we can restrict to effective divisors:

20.26 Proposition. *The following are equivalent:*

- $dim(A) > 0$.

- A *is equivalent to some effective divisor.*

PROOF   If $0 \neq x \in L(A)$, then by definition $(x) + A \geq 0$, so we have an effective divisor $A' = (x) + A$, which is equivalent to A.   ⬜

It follows that divisors of negative degree have dimension 0. Also, iterative use of Lemma 20.24 shows:

20.27 Proposition. *If A_+ is the zero divisor of A, then*

$$dim(A) \leq deg(A_+) + 1 \leq deg(A) + 1.$$

Reed-Solomon codes

We are not very far into the theory, but already it is possible to understand the Reed-Solomon codes (see Chapter 4) from our new point of view. Consider the rational function field $\mathbb{F}_q(x)$ over the finite field. The places of degree 1 are Q_∞ (the place at infinity) and the places Q_α corresponding to irreducible polynomials $X - \alpha$ of degree 1. These are indexed by the elements $\alpha \in \mathbb{F}_q$. Consider $L((k-1)Q_\infty)$. In general, the dimension is $\leq k$, and in fact we know $dim((k-1)Q_\infty) = k$: this space consists of the polynomials of degree $\leq k-1$. In particular, we can evaluate at all the q places Q_α, $\alpha \in \mathbb{F}_q$. These are simply the evaluations of those polynomials. We obtain a code of dimension k and length n, the codewords being indexed by our polynomials, the coordinates by the elements of the field. The entries are the evaluations. This is precisely the description of $\mathcal{RS}(k,q)$ given in Chapter 4. How do we know from the general theory these are MDS codes? Assume a nonzero codeword has k zeroes $\alpha_1, \ldots, \alpha_k$. Then the corresponding polynomial z satisfies $(z) + (k-1)Q_\infty - \alpha_1 Q_{\alpha_1} - \cdots - \alpha_k Q_{\alpha_k} \geq 0$, so $z \in L((k-1)Q_\infty - \alpha_1 Q_{\alpha_1} - \cdots - \alpha_k Q_{\alpha_k})$, which is impossible, as this divisor has degree -1.

We continue with the general theory. In the case of the rational function field, we saw that all principal divisors have degree 0. This is true in general, and in fact the degree of the zero and the pole divisor of (x) carry an important piece of information:

20.28 Theorem. *The degree $[F : K(x)]$ equals the degree of the zero divisor and also the degree of the pole divisor of x. In particular, principal divisors have degree 0.*

This is Theorem I.4.11 of [198]. One direction follows from Proposition 20.17; the other direction is a typical case of a proof using the basic properties of divisor dimensions. We reproduce it here: let $n = [F : K(x)], u_1, \ldots, u_n$ be a basis of F over $K(x)$ and B the pole divisor of x. We want to prove $n = deg(B)$. Proposition 20.17 shows $deg(B) \leq n$.

Choose a divisor C such that $(u_i) + C \geq 0$ for all i. By definition of the u_i, the $x^j u_i$ are all independent over K. Fix some natural number l. Then $x^j u_i \in L(lB + C)$ when $j = 0, \ldots, l; i = 1, \ldots, n$. This shows $dim(lB + C) \geq n(l + 1)$. The only general fact we currently know about those dimensions is Proposition 20.27: $n(l + 1) \leq 1 + deg(C) + l \cdot deg(B)$, so $l(deg(B) - n) \geq n - 1 - deg(C)$ for all l. This is only possible if $deg(B) \geq n$.

20.29 Corollary. *For a divisor A of degree 0, the following are equivalent:*

- *A is principal.*

- *$dim(A) \geq 1$.*

- *$dim(A) = 1$.*

PROOF This is trivial by now. A principal divisor is equivalent to the 0 divisor, so it has dimension 1. A divisor of positive dimension is equivalent to an effective divisor (Proposition 20.26). As A has degree 0, it must be equivalent to the 0 divisor then. $\square$

The most important invariant of a function field is its genus. The following proposition (Proposition I.4.14 of [198]) paves the way to an elementary definition.

20.30 Proposition. *There is an integer γ depending only on the function field $F|K$ with the property*

$$deg(A) - dim(A) \leq \gamma$$

for all divisors A.

PROOF Lemma 20.24 shows that the expression $deg(A) - dim(A)$ only increases if A is replaced by $B \geq A$. Use the setup of the proof of Theorem 20.28. By now we know $deg(B) = n = [F : K(x)]$. As $n(l + 1) \leq dim(lB + C) \leq dim(lB) + deg(C)$, it follows that $dim(lB) \geq n(l + 1) - deg(C)$ and then

$$deg(lB) - dim(lB) \leq deg(C) - n.$$

The right side is a constant not depending on l. Let A be an arbitrary divisor. Choose an effective divisor $A_1 \geq A$. Because of the monotonicity property above, it suffices to show that A_1 is equivalent to some $D \leq lB$ for some l.

The standard inequalities and the fact that the statement has been proved for lB already shows $dim(lB - A_1) > 0$ when l is large enough. There is therefore some $x \in F$ such that $(x) + lB - A_1 \geq 0$, and $A_1 - (x)$ is the divisor equivalent to A_1, which is $\leq lB$. $\qquad\qquad\qquad\qquad\qquad\qquad\qquad\qquad\qquad\qquad\qquad\square$

Proposition 20.30 allows us to define the genus:

20.31 Definition. *The* **genus** *$g = g(F|K)$ is the maximum of $deg(A) + 1 - dim(A)$, over all divisors.*

The choice $A = 0$ shows that g is a nonnegative integer. By definition we have

$$dim(A) \geq deg(A) + 1 - g$$

for all divisors A. This is the **Riemann inequality.** Equality is reached for all divisors A with sufficiently high degree. This implies that the rational function field has genus $g = 0$. In fact, let $A = lP_\infty$. With high enough l, we have then $dim(A) = l + 1 - g$. On the other hand $dim(A) = l + 1$, as $L(A)$ consists of the polynomials of degree $\leq l$.

20.3 The Riemann-Roch theorem

The Riemann-Roch theorem turns the Riemann inequality

$$dim(A) \geq deg(A) + 1 - g$$

into an equality, so to speak. It explains the missing term.

20.32 Definition. *The* **index of speciality** *of the divisor A is*

$$i(A) = dim(A) - deg(A) + g - 1.$$

In other words,
$$dim(A) = deg(A) + 1 - g + i(A)$$

By Riemann's inequality, $i(A) \geq 0$ and $i(A) = 0$ if $deg(A)$ is large enough. Observe $i(0) = g$. The Riemann-Roch theorem provides an interpretation of $i(A)$ as the dimension of a certain vector space.

We know that $x \in F$ has only finitely many poles. The following definition generalizes the notion of a principal divisor:

20.33 Definition. *An* **adele** *is a tuple (α_P) of elements $\alpha_P \in F$, where P varies over the places and $v_P(\alpha_P) \geq 0$ for almost all P. The set $\mathcal{A}_F$, the* **adele space,** *is the set of all adeles. For each divisor A, let $\mathcal{A}_F(A)$ be the space of adeles satisfying $v_P(\alpha_P) + v_P(A) \geq 0$ for all P.*

The principal adele (x), with $x_P = x \in F$ for all P, is in fact an adele (it satisfies the additional condition). This defines the **diagonal embedding** $F \subset \mathcal{A}_F$. The spaces $\mathcal{A}_F(A)$ generalize the spaces $L(A)$ that we are mostly interested in. The first interpretation of $i(A)$ is given by the following (see Theorem I.5.4 of [198] and its rather elementary two page proof):

20.34 Theorem. *For each divisor* A,

$$i(A) = dim(\mathcal{A}_F/(\mathcal{A}_F(A) + F)).$$

In particular, those factor spaces have finite dimension (over K), although $\mathcal{A}_F/\mathcal{A}_F(A)$ and $\mathcal{A}_F/F$ are of course infinite dimensional. As a consequence,

$$g = dim(\mathcal{A}_F/(\mathcal{A}_F(0) + F)).$$

Recall the first fundamental question: given a divisor $A = (n_P)$, is there a function z such that $v_P(z) \geq -n_P$ for all P? Those functions form the space $L(A)$. The adeles and the corresponding modules are motivated by a second fundamental question: given $\alpha_P \in F$ for all P (an adele) and a divisor A, is there a function z such that $v_P(z - \alpha_P) \geq -n_P$? The $\alpha = (\alpha_P)$ which can be represented in this way form the space $\mathcal{A}_F(A) + F$. If $i(A) = 0$, then all adeles can be represented in that way. Consider the case $F = \mathbb{C}(x)$. The places not at infinity correspond to elements $a \in \mathbb{C}$ and we write $z \in F$ as a Laurent series

$$z = \sum_{k=r}^{\infty} c_k(x - a)^k.$$

The second fundamental question is then: given A, can we find $u \in F$ when we also prescribe the beginning of its Laurent series about each a, the first terms up to $c_{m-1}(x - a)^{m-1}$, where $m = -v_a(\alpha_a)$?

Clearly we can also take the dual point of view:

20.35 Definition. *A* **Weil differential** *of* $F|K$ *is a* K-*linear map* $\omega :$ $\mathcal{A}_F \longrightarrow K$ *vanishing on some* $\mathcal{A}_F(A) + F$. *The set of all Weil differentials is denoted* Ω_F; *those vanishing on* $\mathcal{A}_F(A) + F$ *form* $\Omega_F(A)$.

Obviously then $i(A) = dim(\Omega_F(A))$. Clearly Ω_F is an infinite-dimensional K vector space, and in fact just as obviously is an F vector space. Here the module structure is defined by $(x\omega)(\alpha) = \omega(x\alpha)$. A basic property is

20.36 Proposition. Ω_F *is a one-dimensional* F *vector space.*

This is Proposition I.5.9 of [198]. Let ω be a Weil differential and consider divisors A, B such that ω annihilates $\mathcal{A}_F(A) + F$ and $\mathcal{A}_F(B) + F$. It is easy to see that ω also annihilates $\mathcal{A}_F(max(A, B)) + F$. This shows that there is a uniquely determined maximal divisor A with this property. As this divisor is canonically associated with ω, it is denoted by $A = (\omega)$, the divisor of the Weil differential ω.

20.37 Definition. *The divisor (ω) of the Weil differential ω is the uniquely determined largest divisor A such that ω sends $\mathcal{A}_F(A)$ to 0.*

It follows directly from the definition that, for $x \in F$, we have $(x\omega) = (x) + (\omega)$. Because of Proposition 20.36, it follows that the divisors of nonzero Weil differentials form a uniquely determined equivalence class of divisors:

20.38 Definition. *The **canonical divisors** are the equivalence class of divisors of nonzero Weil differentials.*

The space $\Omega_F(A)$ consists of those differentials ω satisfying $A \leq (\omega)$. In the case $A = 0$, this leads to the g-dimensional space $\Omega_F(0)$. This is another characterization of the genus g, as the dimension of the space of holomorphic differentials.

The standard form of the Riemann-Roch theorem is obtained by the formal observation that, for a canonical divisor W, the spaces $\Omega_F(A)$ and $L(W - A)$ have the same dimension (which is $i(A)$). This is a duality theorem. An isomorphism simply sends $x \in L(W - A)$ to $x\omega$ where of course ω is chosen such that $(\omega) = W$. This follows directly from the definitions. Here is the central theorem:

20.39 Theorem (Riemann-Roch theorem). *Let W be a canonical divisor. Then for every divisor A, we have*

$$dim(A) = deg(A) + 1 - g + dim(W - A).$$

Some consequences are immediate: the special case $A = 0$ shows $1 = dim(0) = 0 + 1 - g + dim(W)$. Consequently, $dim(W) = g$. This is not really new. In fact we just saw the duality theorem which implies that W and $\Omega_F(0)$ have the same dimension. Use Riemann-Roch again, this time applying it to W. This shows $deg(W) = 2g - 2$. It is easy to see that the canonical divisor class is characterized by these properties:

20.40 Proposition. *A divisor W is canonical if and only if it has $deg(W) = 2g - 2$, $dim(W) = g$.*

Another important application of the Riemann-Roch theorem and its proof is the following strengthening of Theorem 20.15, Theorem 1.6.5 of [198]:

20.41 Theorem (strong approximation theorem). *Let S be a proper set of places (so there must be at least one place outside S) and $P_1, \ldots, P_r \in S$. Let $x_1, \ldots, x_r \in F$ and integers $n_1, \ldots, n_r$ be given. Then there exists an element $x \in F$ satisfying the following:*

- $v_{P_i}(x - x_i) = n_i, \ i = 1, 2, \ldots, n.$

- $v_P(x) \geq 0$ *for all remaining* $P \in S.$

20.4   Some hyperelliptic equations

Let $char(K) \neq 2$ and the function field F be defined as a quadratic extension of $K(x)$ by an equation

$$y^2 = f(x) \in K[x]$$

where $f(x)$ is a polynomial of degree $n = 2m + 1 \geq 3$.

First of all: as $deg(f)$ is odd, $f(x)$ is not a square in $K(x)$, so $y^2 - f(x)$ is an irreducible equation (over $K(x)$) and we really have a function field $F = K(x, y)$ of degree 2 over $K(x)$.

Consider F as a degree 2 extension of $K(x) = K(1/x)$ and use the basic formula

$$deg((x)_\infty) = deg((x)_0) = \sum v_{P_i}(x)deg(P_i) = 2.$$

Let Q_∞ be the place at infinity for $K(x)$. In principle there are three possibilities where the places P of F over Q_∞ are concerned:

- There are two such places P_1, P_2 and $v_{P_i}(x) = -1, [F_{P_i} : K] = 1$.

- There is only one such extension place P_∞ and $v_P(x) = -1, [F_P : K] = 2$.

- There is only one such extension place P_∞ and $v_P(x) = -2, F_P = K$.

In our original ring-theoretic language, a place P of F over the place Q_∞ of $K(x)$ corresponds to a valuation ring $\mathcal{O} \subset F$ with maximal ideal P such that $P \cap K(x) = Q_\infty$. We will not use that terminology any more and rather speak of places and functions.

Let P over Q_∞ be one of those places. The equation shows $v_P(y^2) = 2v_P(y) = nv_P(x)$. As n is odd, we must have $v_P(x) = -2$. This decides the case: the last of the scenarios above is what happens. We also have $v_P(y) = -n$. In particular, P is a rational place ($F_P = K$). It follows that $\tilde{K} = K$ is the full field of constants. Also, whenever x is defined at a place, then so is y. It follows that P is the only place where y is not defined, and $(y)_\infty = nP$.

Consider $L(2r)$. For large r we have, by Riemann-Roch, $dim(2rP) = 2r + 1 - g$. On the other hand, $L(2rP)$ contains $1, x, \ldots, x^r, y, xy, \ldots, x^{r-m-1}y$. These are $2r + 1 - m$ linear independent elements:

20.42 Lemma. *Let $x_1, \ldots, x_n$ be elements of a function field F and P a place such that the $v_P(x_i)$ are all different. Then $x_1, \ldots, x_n$ are linearly independent over the constant field K.*

The proof of Lemma 20.42 is obvious. In the present situation we conclude $2r + 1 - g \geq 2r + 1 - m$, so $g \leq m$. We have seen the following:

20.43 Theorem. *Let K be a field of characteristic $\neq 2$ and $F = K(x, y)$ the quadratic extension of $K(x)$ defined by $y^2 = f(x)$, where $f(x)$ is a polynomial of degree $2m + 1$. Then the following hold:*

1. *K is the full constant field of F.*

2. *There is exactly one place P which is a pole of x and this place is also the only pole of y.*

3. *The genus of F satisfies $g \leq m$.*

This describes a special case of **hyperelliptic** function fields. Assume the right side $f(x)$ is not a polynomial. Let $g(x)$ be an irreducible factor of the denominator. With $y' = yg(x)$ we have $F = K(x, y')$ and $y'^2 = f(x)g(x)^2$. Repeating this process we can assume without restriction that $f(x)$ is a polynomial. Now assume $g(x)$ is an irreducible polynomial such that $g(x)^2$ divides $f(x)$. Then we can replace y by $y/g(x)$ and obtain an equation with a smaller value of m. In particular, equality $g = m$ is possible only if such quadratic factors do not occur.

To sum up: given a hyperelliptic equation $y^2 = f(x)$ where $f(x) \in K(x)$ (and $char(K) \neq 2$), we can transform the equation, keeping the function field unchanged, such that $f(x)$ is a **square-free polynomial.** What we have seen in the argument above is that, in the special case when $deg(f(x)) = 2m + 1$ is odd, we have $g \leq m$. We will show later that indeed $g = m$. Also the case of an even degree polynomial $f(x)$ will be considered a little later. The special case when $f(x)$ has degree 3 and is square-free describes elliptic function fields.

Elliptic function fields in odd characteristic

A function field is **elliptic** if $g = 1$. Let $char(K) \neq 2$ and assume F is elliptic with a rational point P. By Proposition 20.40, the 0 divisor is canonical. Riemann-Roch implies $dim(A) = deg(A)$ whenever $deg(A) > 0$. Use this for the divisors $A = nP$ where $n \leq 6$. In particular, $dim(P) = 1$ and $L(P) = K$. Let $L(2P) = \langle 1, x \rangle$. Then x is an element which has P as its only pole. In particular, $(x)_\infty = 2P$ and $[F : K(x)] = 2$. Next $L(3P) = \langle 1, x, y \rangle$. As $v_P(y) = -3$, we have that $y \notin K(x)$. It follows that we can choose $1, y$ as a basis for $F \mid K(x)$. This is the first structural consequence: F is a quadratic extension of a rational function field. The next Riemann-Roch spaces are obvious:

$$L(4P) = \langle 1, x, x^2, y \rangle, \quad L(5P) = \langle 1, x, x^2, y, xy \rangle, \quad L(6P) = \langle 1, x, x^2, x^3, y, xy \rangle.$$

However, the last of those spaces also contains y^2. It follows that y^2 is a linear combination over K of $1, x, x^2, x^3, y, xy$. Observe $v_P(y^2) = -6 = v_P(x^3)$. Lemma 20.42 shows that the coefficient at x^3 must be nonzero. Also, we can replace by an arbitrary nonconstant $\alpha x + \beta \in L(2P)$ and y by an arbitrary element in $L(3P) \setminus L(2P)$. By completing the square (observe that we are in odd characteritic), it can be assumed $y^2 = a_0 + a_1 x + a_2 x^2 + a_3 x^3$ where $a_3 \neq 0$. As $y \mapsto \alpha y$ and $x \mapsto \beta x$ leads to a coefficient $a_3 \alpha^2 \beta^3$ at x^3, it can be assumed that $a_3 = 1$. The equation takes on the form $y^2 = x^3 + a_2 x^2 + a_1 x + a_0$. The substitution $x \mapsto x + c$ leads to a quadratic term $(3cx^2 + a_2)x^2$. This can be made to vanish if $char(K) \neq 2, 3$. We have seen the following:

20.44 Theorem. *Let $F|K$ be an elliptic function field with K as the full constant field and possessing a rational place. Assume also $char(K) \neq 2, 3$. Then $F = K(x, y)$ is quadratic over $K(x)$ with minimal polynomial $y^2 = x^3 + ax + b$ of y over $K(x)$.*

In the beginning of this section we saw that F has indeed genus $g \leq 1$. When do we have $g = 1$? Clearly, if the right side has a quadratic factor, then F is the rational function field. Assume therefore that $f(x)$ is square-free. We want to show $g = 1$.

Let $q(x)$ be an irreducible polynomial dividing $f(x)$ and P a place of F over $Q_{q(x)}$. We have $2v_P(y) = v_P(q(x)) = e$. It follows $e = 2$ and $v_P(y) = 1$.

Let P_i be the places of F over $Q_{q_i(x)}$, where $q_i(x)$ varies over the irreducible divisors of $f(x)$. We have seen that there is precisely one such place P_i over $Q_{q_i(x)}$ and $deg(P_i) = deg(Q_{q_i(x)}) = deg(q_i(x))$. It follows $(y) = \sum_i P_i - 3P_\infty$ and $(x)_\infty = 2P_\infty$. Write v_∞ for v_{P_∞}.

Assume now $g = 0$. By Riemann Roch, $dim(kP_\infty) = k + 1$ for all $k \geq 0$. Let $L(P_\infty) = \langle 1, z \rangle$. Then $L(2P_\infty) = \langle 1, x, z \rangle$ and it contains z^2. By completing the square, we can choose z such that $z^2 = c_0 + c_1 x$ where $c_0, c_1 \in K$ and $c_1 \neq 0$. Let $z = a(x)y + b(x)$. Then $z^2 = a(x)^2 f(x) + b(x)^2 + 2a(x)b(x)y = c_0 + c_1 x$. It follows $a(x)b(x) = 0$, hence $b(x) = 0$ and $z = a(x)y$, $z^2 = a(x)^2 f(x) = c_0 + c_1 x$. This is impossible, as $f(x)$ is square-free and has degree 3. We have seen the following:

20.45 Theorem. *Let $char(K) \neq 2$ and $F = K(x, y)$ be defined by $y^2 = f(x) = x^3 + ax + b$ where $f(x)$ is square-free. Then F is elliptic.*

A general bound

Here is a very useful general bound. Its proof relies entirely on Riemann-Roch and is similar to the proof of Theorem 20.43. Its hypothesis is satisfied when we start from a homogeneous polynomial $H(X, Y, Z)$ of degree n and derive from it the irreducible polynomial $h(x, y) = H(x, y, 1)$. It is Proposition 3.10.5 of [198].

20.46 Theorem (Plücker formula). *Let $F = K(x, y)$ of degree n over $K(x)$ be defined by the irreducible equation*

$$y^n + f_1(x)y^{n-1} + \cdots + f_{n-1}(x)y + f_n(x) = 0$$

where $f_j(x)$ is a polynomial of degree $\leq j$. Then $g(F) \leq \binom{n-1}{2}$.

PROOF   Let $A = (x)_\infty$, an effective divisor of degree n. By definition, $x \in L(A)$. It is easy to see that $y \in L(A)$. In fact, let P be such that $v_P(x) = -m < 0$. We want to show that $v_P(y) \geq -m$. Divide the irreducible equation by x^n and work with $z = y/x$. We want to show that z is defined at P. In fact,

$$z^n + \frac{f_1(x)}{x} z^{n-1} + \cdots + \frac{f_{n-1}(x)}{x^{n-1}} z + \frac{f_n(x)}{x^n} = 0.$$

Then $f_j(x)/x^j$ is a polynomial in $1/x$ and therefore satisfies $v_P(f_j(x)/x^j) \geq 0$. This shows the claim.

Let $l > n$ be large enough such that $dim(lA) = ln + 1 - g$. Then lA contains the linearly independent elements $x^i y^j$ where $j = 0, \ldots, n-1$ and $0 \leq i \leq l-j$. Comparison yields the claim. $\qquad \square$

Chapter 21

Determining the genus

The determination of the genus proceeds in steps: write your function field F as an algebraic extension of a function field $E \subset F$ whose genus is known. The Riemann-Hurwitz formula relates the canonical divisor of F to that of E and therefore $2g(F) - 2$ to $2g(E) - 2$. A typical application of this method uses $E = K(x)$, a rational function field. It follows that we need to study the relationship between the basic notions (places, divisors, residue class fields, genus, etc.) of E and of $F \supset E$.

21.1 Algebraic extensions of function fields

21.1 Definition. *An algebraic function field $F' \mid K'$ is a **finite extension** of a function field $F \mid K$ if $F' \mid F$ is algebraic and $K' \supseteq K$.*

There are two extremal cases: it may be that $K' = K$. This is the case we had in mind in the motivation above. The other extreme is when $F' = FK'$. In this case one speaks of a **constant field extension.** We can think of F' as described by the same equations as F, only that the constants are interpreted not as elements of K but as elements of K'. Take, for example, the case when $F = \mathbb{F}_2(x, y)$ and $K = \mathbb{F}_2 = \tilde{K}$. The coefficients of a minimal polynomial of y are elements in $\mathbb{F}_2(x)$. These are also in $\mathbb{F}_4(x)$. This defines a function field $F' \mid \mathbb{F}_4$, obtained from $F \mid \mathbb{F}_2$ by what is called **extension of constants.** Clearly each finite extension can be thought of as arising in (at most) two steps: at first a constant field extension, followed by an extension where the constant field does not change.

Consider the general situation, $F' \mid K'$ a finite extension of $F \mid K$. Some relations are obvious:

21.2 Definition. *A place P' of F' is **over** a place P of F if $P \subset P'$. Write $P' \mid P$ in this case.*

- Each place P' of F' is over precisely one place P, namely, $P = P' \cap F$.

- For each place P of F there is a finite nonzero number of places $P' \mid P$.

421

An important piece of data is the following: Let $P' \mid P$. Then there is (obviously) a constant $e > 0$ such that $v_{P'}(x) = e \cdot v_P(x)$ for all $x \in F$.

21.3 Definition. *The number $e = e(P' \mid P)$ such that $v_{P'}(x) = e \cdot v_P(x)$ for all $x \in F$ is the **ramification index** of $P' \mid P$. The number $f = f(P' \mid P) = [F'_{P'} : F_P]$, the degree of the residue class field of P' as a vector space over the residue class field of P, is the **relative degree**.*

Also, both e and f are multiplicative in towers.

21.4 Theorem (fundamental equality). *Let P be a place of F and $P_1, \dots, P_m$ the places of F' over P. Let e_i be the ramification indices and f_i the relative degrees. Then*

$$\sum_{i=1}^{m} e_i f_i = n = [F' : F].$$

This is a basic tool. It generalizes Theorem 20.28, which is the special case when the smaller of the two function fields is $K(x)$. The proof of Theorem 21.4 is an application of Theorem 20.28. A direct application is the following criterion of irreducibility. It is one of the occasions where we get the feeling that the theory developed is not only self-referential but generates useful output. Clearly this criterion, Proposition III.1.14 of [198], generalizes the famous Eisenstein criterion (where the function field F is chosen as the rational function field):

21.5 Proposition. *Let $F \mid K$ be a function field and*

$$\phi(T) = a_n T^n + \cdots + a_1 T + a_0 \in F[T].$$

Assume there is a place P such that one of the following holds:

- $v_P(a_n) = 0, v_P(a_i) \geq v_P(a_0) > 0$ *for* $i = 1, \dots, n - 1$
  and $\gcd(n, v_P(a_0)) = 1$.

- $v_P(a_n) = 0, v_P(a_i) \geq 0$ *for* $i = 1, \dots, n - 1, v_P(a_0) < 0$
  and $\gcd(n, v_P(a_0)) = 1$.

Then $\phi(T)$ is irreducible.

The proof starts from considering an irreducible factor of $\phi(T)$, the corresponding field extension $F' = F(y)$ and a place $P' \mid P$. The conditions make sure that the ramification index e is a multiple of n. By the fundamental inequality, this is only possible if $e = n$ and $\phi(T)$ is irreducible.

The following result is important in deciding which and how many places are above a given place of the smaller field:

21.6 Theorem (Kummer's theorem). *Let $F' = F(y)$, with minimal polynomial*

$$\phi(T) \in O_P[T]$$

where P is a place of F. Consider the polynomial $\overline{\phi}(T) \in F_P[T]$ obtained from $\phi(T)$ by reducing the coefficients mod P. Let

$$\overline{\phi}(T) = \prod_{i=1}^{r} \gamma_i(T)^{\epsilon_i}$$

be the decomposition into monic irreducibles of degree n_i. Then, for each i, there is a place $P_i \mid P$ where $f(P_i \mid P) \geq n_i$ and those places P_i are pairwise different.

If, moreover, $\epsilon_i = 1$ for all i, then those are all the places over P, $\epsilon(P_i \mid P) = 1$ and $f(P_i \mid P) = n_i$.

The proof is essentially a careful bookkeeping exercise based on Theorem 20.12.

Exercises 21.1

21.1.1. *Use the generalized Eisenstein criterion (Proposition 21.5) to prove that the Artin-Schreier polynomial $Y^q - Y - f(X)$ of Definition 19.5 is irreducible as a polynomial in Y provided the degree of the polynomial $f(X)$ is not a multiple of the characteristic p.*

21.1.2. *Find instances when the Artin-Schreier polynomial is not irreducible.*

21.2 The hyperelliptic case

We return to the hyperelliptic equations in odd characteristic

$$y^2 = f(x) = x^n + a_{n-1}x^{n-1} + \cdots + a_1 x + a_0$$

where $f(x)$ is a square-free polynomial. Consider the case when $n = 2m + 2$ is even. Comparing with the arguments in the case when n is odd a difficulty arises when determining the places above Q_∞. Using Kummer's theorem, we can solve this case now. Use the element $x' = 1/x$. Then $K(x') = K(x)$ and the equation is $y^2 = f(x)x'^n/x'^n = (a_n + a_{n-1}x' + \cdots + a_0x'^n)/x'^n$. Now use $y' = yx'^{n/2}$. Then we are still talking about the same function field $F = K(x,y) = K(x',y')$ and the equation is

$$y'^2 = 1 + a_{n-1}x' + \cdots + a_0 x'^n.$$

In the special case when $a_0 = 0$, this gives us a result already: we are in the case of a polynomial of odd degree $n - 1 = 2m + 1$ and know that $g \leq m$. The place $Q = Q_\infty$ is the place where $x'(Q) = 0$. The polynomial to consider in Kummer's theorem is $T^2 - 1$. Kummer implies that there are two rational places $P \mid Q_\infty$.

21.7 Proposition. *Let* $y^2 = f(x)$ *in odd characteristic, where* $f(x)$ *is a square-free monic polynomial of degree* n. *Then the following hold:*

1. *If* n *is odd, then there is one place* $P_\infty \mid Q_\infty$, *and* P_∞ *is rational (case* $e = 2, f = 1$*).*

2. *If* n *is even, then there are two rational places* $P \mid Q_\infty$.

The rest of the argument works as in Section 20.4. Let P_1, P_2 be the places above Q_∞. Then $(x)_\infty = P_1 + P_2$ and $(y)_\infty = (n/2)(P_1 + P_2)$. It follows that $L(r(P_1 + P_2))$ contains $1, x, \ldots, x^r, y, xy, \ldots, x^{r-n/2}y$. As these elements are linear independent, Riemann-Roch implies $dim(r(P_1 + P_2)) = 2r + 1 - g \geq 2r + 2 - n/2$ for large enough r. This shows $g \leq n/2 - 1$.

21.8 Theorem. *Let* K *be a field of characteristic* $\neq 2$ *and* $F = K(x, y)$ *the quadratic extension of* $K(x)$ *defined by* $y^2 = f(x)$, *where* $f(x)$ *is a square-free monic polynomial of degree* $2m + 2$. *Then the following hold:*

1. K *is the full constant field of* F.

2. *The genus of* F *satisfies* $g \leq m$.

21.3 The Kloosterman codes and curves

This is our first example where the theory of function fields is applied not to the construction of codes but to the analysis of the structure of codes.

21.9 Definition. *Let* $F = \mathbb{F}_q$, *where* $q = 2^r$. *Let* $tr : F \longrightarrow \mathbb{F}_2$ *be the absolute trace. The* **Kloosterman code** *of length* $q - 1$ *is the* $2r$-*dimensional binary linear code whose codewords* $c(a, b)$ *are parametrized by a pair* $a, b \in F$, *whose coordinates are parametrized by* $0 \neq x \in F$ *and with entry* $c(a, b)_x = tr(ax + b/x) \in \mathbb{F}_2$.

We know these codes. They are cyclic codes. Their duals are the Mélas codes with defining exponents $1, -1$. The reader has been asked in Section 15.2 to prove that the Mélas codes have minimum distance 5 provided r is odd. It was observed that this leads to almost perfectly linear functions on F. Here we want to study the individual codeword weights of the Kloosterman codes.

Obviously $wt(c(a,0)) = wt(c(0,b)) = q/2 - 1$ when $ab \neq 0$. Consider $wt(c(a,b))$ when $ab \neq 0$. Using $x' = ax$, it is clear that $wt(c(a,b)) = wt(c(1,v))$ where $v = ab$. In order to determine $wt(c(1,v))$, we need to know for how many $0 \neq x \in F$ the expression $x + v/x$ has trace 0. This is related to the Artin-Schreier curve

$$y^2 + y = x + v/x.$$

21.10 Definition. *Let* $F = \mathbb{F}_q$, *where* $q = 2^r$ *and* $0 \neq v \in \mathbb{F}_q$. *The Kloosterman function field is defined as a quadratic extension of* $\mathbb{F}_q(x)$ *by the equation* $y^2 + y = x + v/x$.

Consider this curve (for a fixed v). Let $v = w^2$. Then $x + v/x = (x+w)^2/x$. The places Q_∞, Q_0, Q_w play a special role. Let $P \mid Q_\infty$ or $P \mid Q_0$. The usual argument shows that, in each case, P is uniquely determined and $e = 2$, so P_∞ and P_0 are rational places. In particular, $\mathbb{F}_q$ is the full field of constants. Let $P \mid Q_w$. By Kummer there are two places P_w, P'_w over Q_w and y vanishes at one of them, say P_w. Then we have the following divisors:

$$(x) = -2P_\infty + 2P_0.$$

$$(y) = -P_\infty - P_0 + 2P_w.$$

$$(y+1) = -P_\infty - P_0 + 2P'_w.$$

Then $L(2rP_\infty)$ contains $1, x, \ldots, x^r, yx, \ldots, yx^{r-1}$. Riemann-Roch implies $g \leq 1$.

21.11 Theorem. *The Kloosterman curves have* $\mathbb{F}_q$ *as the full field of constants and* $g = 1$.

PROOF We know $g \leq 1$. Assume $g = 0$. The presence of rational places shows that F must be a rational function field. Its number of rational places is therefore $q + 1$. On the other hand, we know four rational places by name. By Kummer's theorem, the remaining rational places come in pairs and their total number is therefore even, a contradiction. □

The relation between the number of rational places of the Kloosterman curves and the corresponding codeword weights of the Kloosterman codes is best expressed by the following numbers:

21.12 Definition. *Let* $0 \neq v \in \mathbb{F}_q$, *where* $q = 2^r$. *Let* p_v *be the number of elements* $0 \neq x \in \mathbb{F}_q$ *such that* $tr(x) = tr(v/x) = 1$.

Always counting $0 \neq x \in \mathbb{F}_q$, we have p_v values where $tr(x) = tr(v/x) = 1$. It follows that there are $q/2 - p_v$ values where $tr(x) = 1, tr(v/x) = 0$ and equally many where $tr(x) = 0, tr(v/x) = 1$. There remain $p_v - 1$ values where $tr(x) = tr(v/x) = 0$. We count $2p_v - 1$ values of x where $tr(x + v/x) = 0$, which leads to $4p_v - 2$ rational points. Together with the rational points P_∞ and P_0, we obtain $N = 4p_v$ rational points for the Kloosterman curve.

21.13 Proposition. *The weights of the generic codewords of the Kloosterman code are* $wt(c(1, v)) = q - 2p_v = q - N/2$, *where N is the number of rational points of the corresponding Kloosterman curve.*

Looking ahead, we can use the famous Hasse bound $N \leq q + 1 + 2\sqrt{q}$ on the number of rational points N of an elliptic function field defined over $\mathbb{F}_q$. This implies the following:

21.14 Theorem. *Let $q = 2^r$. The Kloosterman code is a binary linear code $[2^r - 1, 2r, d]_2$, where $d \geq 2^{r-1} - 2^{r/2} - 1/2$.*

In the case $r = 2m$ even this yields the family of code parameters

$$[4^m - 1, 4m, 2^m(2^{m-1} - 1)]_2 \text{ for } m \geq 2.$$

In particular, this yields an effective construction for codes $[63, 12, 24]_2$, $[255, 16, 112]_2$ and $[1023, 20, 480]_2$.

21.4 Subrings and integrality

The definition of a subring of a function field is surprisingly simple:

21.15 Definition. *A **subring** of $F \mid K$ is a ring R where $K \subset R \subset F$ and R is not a field.*

Use is made of a standard notion of commutative algebra, integrality:

21.16 Definition. *An element $z \in F$ is **integral** over a subring R if it satisfies a monic polynomial equation (an **integral equation**)*

$$z^n + a_{n-1}z^{n-1} + \cdots + a_1 z + a_0 = 0$$

where $a_i \in R$.
*The **integral closure** $ic_F(R)$ is the set of all elements of F which are integral over R. A ring is called **integrally closed** if it coincides with its integral closure in its quotient field.*

The paradigmatic example of an integrally closed ring is the ring $\mathbb{Z}$ of integers.

21.17 Definition. *Let S be a proper subset of places. The corresponding holomorphy ring is*

$$\mathcal{O}_S = \cap_{P \in S} \mathcal{O}_P.$$

These are subrings, and $\mathcal{O}_S = \mathcal{O}_T$ if and only if $S = T$. They are integrally closed. The holomorphy rings are characterized among the subrings by the property of being integrally closed (Corollary 3.2.8 of [198]). Another important property is that $\mathcal{O}_S$ is a principal ideal domain if S is a finite set of places.

21.5   The Riemann-Hurwitz formula

The general strategy in deriving a formula that relates the genus of F and of its finite extension F' is to start from a canonical divisor for F (of degree $2g_F - 2$) and embed it in the divisor group of F' such that a canonical divisor of F' (of degree $2g_{F'} - 2$) results. One important instrument in the construction is the trace $tr : F' \longrightarrow F$. The trace is a nonzero F linear mapping if and only if the extension $F' \mid F$ is separable. It is therefore assumed that $F' \mid F$ is separable. The trace defines a nondegenerate symmetric bilinear form, the trace form $\langle x, y \rangle$ by $\langle x, y \rangle = tr(xy)$. Here F' is seen as a vector space over F, but in the definition of the trace form multiplication in F is used. Recall that bilinear forms were introduced in Section 16.2. Whenever a basis $\{z_1, \ldots, z_n\}$ of $F' \mid F$ is given, there is also the **dual basis** $\{z*_1, \ldots, z*_n\}$ defined by $tr(z_i z*_j) = \delta_{ij}$. This is used to prove the following (Corollary 3.3.5 of [198]):

21.18 Theorem. *Let $F' \mid F$ be a finite separable extension of the function field $F \mid K$ and P a place of F. Then the integral closure $\mathcal{O}'_P$ in F' is the intersection*

$$\mathcal{O}'_P = \cap_{P' \mid P} \mathcal{O}_{P'}.$$

*There exists a basis $\{u_1, u_2, \ldots, u_n\}$ of $F' \mid F$ (an **integral basis**) such that*

$$\mathcal{O}'_P = \sum_{i=1}^{n} \mathcal{O}_P u_i.$$

Moreover, each basis of $F' \mid P$ is an integral basis for almost all places P.

There is an obvious way to embed the divisor group of F into the divisor group of F' :

21.19 Definition. *Let P be a place of F. The **conorm** of P is*

$$Con(P) = \sum_{P' \mid P} e(P' \mid P)P'.$$

This mapping is extended by linearity to a mapping Con from divisors of F to divisors of F'.

One justification for this definition is that *con* maps principal divisors to principal divisors. Also, it is transitive in towers: $Con_{F'' \mid F}(A) = Con_{F'' \mid F'}(Con_{F' \mid F}(A))$. The degrees are easily controlled (Corollary 3.1.14 of [198]):

21.20 Theorem.

$$deg(Con_{F' \mid F}(A)) = \frac{[F' : F]}{[K' : K]} deg(A).$$

It would be a naive hope that the conorm of a canonical divisor is a canonical divisor. This is not true. The term that separates it from the truth is the **different**. In order to define it, one proceeds as follows:

Let P be a place of F and $\mathcal{O}'_P = \cap_{P'|P} \mathcal{O}_{P'}$ its integral closure. The **complementary module** $\mathcal{C}_P$ is defined by duality:

$$\mathcal{C}_P = \{z \in F' \mid tr(z\mathcal{O}'_P) \subseteq \mathcal{O}_P.$$

Then there is an element $t \in F'$ such that $\mathcal{C}_P = t\mathcal{O}'_P$. With this notation, the different can be defined:

21.21 Definition. *The **local different exponent** is $d(P' \mid P) = d(P') = -v_{P'}(t)$ where t is defined by $\mathcal{C}_P = t\mathcal{O}'_P$ as above. The **different** is $Diff(F' \mid F) = \sum_{P'} d(P')P'$.*

Why this definition and what are the main properties? Let $\{u_i\}$ be an integral basis: $\mathcal{O}'_P = \sum_{i=1}^n \mathcal{O}_P u_i$. Then the complementary module is simply $\mathcal{D}_P = \sum_{i=1}^n \mathcal{O}_P u *_i$. We write $\mathcal{D}_P = t\mathcal{O}'_P$. How is it recognized if $z \in F$ satisfies $z \in \mathcal{D}_P$? This is equivalent to $z/t \in \mathcal{O}'_P$, or, $v_{P'}(z) + d(P') \geq 0$ for all $P' \mid P$. As $t\mathcal{O}'_P = \mathcal{D}_P$, we have $\frac{1}{t}\mathcal{D}_P = \mathcal{O}'_P$, in particular, $1/t \in \mathcal{O}'_P$. This means $-v_{P'} = d(P') \geq 0$ for all P' : the different is an effective divisor.

21.22 Theorem (Riemann-Hurwitz formula). *Let F' be a finite separable extension of the function field $F \mid K$. If W is a canonical divisor of F, then $Con(W) + Diff(F' \mid F)$ is a canonical divisor of $F' \mid K'$.*

Consider the generic case when $K' = K$, and let $n = [F' : F]$. We assume we know g_F and wish to determine $g_{F'}$. Theorem 21.22 shows that $2g_{F'} - 2 = n(2g_F - 2) + deg(Diff(F' \mid F))$. We see that knowledge of $g_{F'}$ is equivalent to knowledge of the degree of the different. This is the famous Riemann-Hurwitz formula.

We want to understand at least the key step in the proof. Let ω be the given Weil differential of F and $W = (\omega)$ its divisor. Recall what this means: the maximal divisor A such that ω annihilates $\mathcal{A}_F(A)$ is $A = W$. Here an adele $\alpha = (\alpha_P)$ satisfies $\alpha \in \mathcal{A}_F(W)$ if $v_P(\alpha_P) + v_P(W) \geq 0$ for all P. Consider not the adele space $\mathcal{A}_{F'}$ of F' but rather the subspace $\mathcal{A}_{F'|F}$ of adapted adeles that are constant on the set of all places over a given place P of F. The reason for considering this space is that we can take the trace $tr : \mathcal{A}_{F'|F} \longrightarrow \mathcal{A}_F$ in the obvious way. Consider the concatenation $\omega_1 = \omega \circ tr : \mathcal{A}_{F'|F} \longrightarrow K$. Then ω_1 comes close to being a Weil differential of F'. In fact, checking that it can be extended to a mapping defined on the full adele space $\mathcal{A}_{F'}$ is routine, just as is the fact that it can be used to define a mapping with image K' (instead of K). Also, it is obvious that principal ideals are annihilated as required. What we want to understand is that ω_1 annihilates $\mathcal{A}_{F'|F}(W')$, where $W' = Con(W) + Diff(F' \mid F)$. Let $\alpha \in \mathcal{A}_{F'|F}(W')$. We have to show that ω annihilates $tr(\alpha)$. By definition of $W = (\omega)$, this means $tr(\alpha) + W \geq 0$.

Why is this true and why is W' the maximal divisor satisfying it for all $\alpha \in \mathcal{A}_{F'|F}(W')$?

Fix some $x \in F$ such that $v_P(x) = v_P(W)$. Let $P' \mid P$ and $e = e(P' \mid P)$. Then

$$v_{P'}(x\alpha_{P'}) = e \cdot v_P(W) + v_{P'}(\alpha_{P'}) \geq e \cdot v_P(W) - v_{P'}(W').$$

By definition of W', this shows $v_{P'}(x\alpha_{P'}) \geq -d(P')$, which, as we know, is equivalent to $x\alpha_{P'} \in \mathcal{D}_P$. Now use the definition of the complementary module: $tr(x\alpha_{P'}) = x\,tr(\alpha_{P'}) \in \mathcal{O}_P$. As this is true for all P', the claim $(tr(\alpha)) + W \geq 0$ follows. The details of the proof are in Section 3.4 of [198].

There is a close relationship between the ramification index $e(P' \mid P)$ and the different exponent $d(P' \mid P) = d(P')$. We use only the most fundamental statement:

21.23 Theorem (Dedekind's different theorem). *For each finite separable extension $F' \mid F$ of the function field F and each $P' \mid P$, the following hold:*

- $d(P') \geq e(P' \mid P)$.

- *Equality holds if and only if $e(P' \mid P)$ is not divisible by the characteristic.*

Chapter 22

AG codes, Weierstraß points and universal hashing

22.1 The basic construction

Here is the basic construction of codes from function fields:

22.1 Definition. *Let F be a function field with constant field $\mathbb{F}_q$, let $P_1, P_2, \ldots, P_n$ be rational points, $D = P_1 + P_2 + \cdots + P_n$ and G a divisor such that $v_{P_i}(G) = 0$ for $i = 1, 2, \ldots, n$. Then $\mathcal{C}(D, G)$ is the $\mathbb{F}_q$ linear code of length n whose coordinates are indexed by the P_i, whose codewords are indexed by $f \in L(G)$ and with entry $f(P_i)$ in coordinate i.*

The mapping $ev : L(G) \longrightarrow \mathbb{F}_q^n$ is the **evaluation map,** clearly an $\mathbb{F}_q$ linear map from the $dim(G)$-dimensional space $L(G)$ into the n-dimensional space $\mathbb{F}_q^n$. Our code, the algebraic-geometric code $\mathcal{C}(D, G)$, is the image of the evaluation map. Here $dim(G)$ is determined by Riemann-Roch: $dim(G) = deg(G) + 1 - g + dim(W - G)$, where g is the genus of F and W is a canonical divisor (of degree $2g - 2$).

22.2 Lemma. *The evaluation map is injective if and only if $L(G - D) = 0$. This is the case if $deg(G) < n$.*

PROOF Let $f \in L(G)$. By definition, $ev(f) = 0$ if and only if $f \in L(G - D)$. ☐

22.3 Theorem. *The $\mathbb{F}_q$ linear code $\mathcal{C}(D, G)$ has length n. If $deg(G) < n$, then the code dimension is $k = dim(G) = deg(G) + 1 - g + dim(W - G)$ and the minimum weight is $d \geq n - deg(G)$.*

PROOF It follows from Lemma 22.2 that $\mathcal{C}(D, G)$ has dimension $k = dim(G)$ if $deg(G) < n$. Because of Riemann-Roch, this simplifies to $k = deg(G) + 1 - g$ if $deg(G) > 2g - 2$. How about the minimum weight d? Assume $0 \neq f \in L(G)$ such that $f(P_1) = \cdots = f(P_a) = 0$. Then $0 \neq f \in L(G - (P_1 + \cdots + P_a))$. This implies $deg(G) \geq a$. As f has at most $deg(G)$ zeroes, it has at least $n - deg(G)$ nonzeroes, in other words, $wt(ev(f)) \geq n - deg(G)$. ☐

431

The most frequently used special case is when $G = mP_0$ is a multiple of a rational point P_0. This also has the advantage that the corresponding codes form a chain

$$\mathcal{C}(D, P_0) \subseteq \mathcal{C}(D, 2P_0) \subseteq \mathcal{C}(D, 3P_0) \subseteq \cdots$$

These are the **one-point codes.** An important feature of Theorem 22.3 is that $\mathcal{C}(D, G)$ satisfies $n + 1 - (k + d) \leq g$: the Singleton defect is bounded by the genus. As a special case we obtain the familiar fact that the Reed-Solomon codes are MDS codes (Singleton defect 0). For low-dimensional codes the Singleton defect (also called the **genus** of the code) will in fact be less than g. This leads to the topic of the next section.

22.2   Pole orders

Let P be a rational point and consider the spaces

$$L(0) \subseteq L(P) \subseteq L(2P) \subseteq \cdots \subseteq L(mP) \subseteq L((m+1)P) \ldots$$

At each step the codimension is ≤ 1 and we have $dim(0) = 1$, $dim(mP) = m + 1 - g$ for $m > 2g - 2$. It follows that the sequence of dimensions "stutters" precisely g times. The values of m where this happens are the gap numbers.

22.4 Definition. *An integer $m \geq 0$ is a **pole number** of P if there is $x \in F$ such that $(x)_\infty = mP$. Otherwise m is a **gap number** of P.*

What we saw above is that there are precisely g gap numbers. Also, m is a pole number if and only if $dim(mP) > dim((m - 1)P)$. Let the pole numbers of $P = P_0$ be $u_1 = 0, u_2, \ldots, u_l \ldots$ Then $dim(u_j P) = j$ and $u_g = 2g - 1, u_{g+1} = 2g, \ldots$ This yields a chain of codes

$$[n, 1, n]_q \subset [n, 2, n - u_2]_q \subset [n, 3, n - u_3]_q \ldots [n, g, n + 1 - 2g]_q \ldots$$

where the Singleton defect starts with 0 (the repetition code) and increases by 1 at each dimension until it reaches g.

Exercises 22.2

22.2.1. *Show that the pole orders form a **semigroup**: a subset of the natural numbers which is closed under addition.*

22.2.2. *Let S be the semigroup generated by q and $q + 1$. Show that there are precisely $\binom{q}{2}$ natural numbers which are not in S.*

22.2.3. *Let S be the semigroup of the preceding exercise. Order its elements:*
$u_1 = 0, u_2 = q, u_3 = q+1, u_4 = 2q, u_5 = 2q+1, u_6 = 2q+2, u_7 = 3q, \ldots$ *Show*
that $u_{\binom{i+1}{2}+1} = iq$ provided that $i \leq q$.

22.3 Examples of function fields and projective equations

Back to the hyperelliptic equation $y^2 = f(x)$ in characteristic $\neq 2$. We saw
in Section 20.4 that the following can be assumed without restriction: $f(x)$ *is*
a polynomial, product of different irreducibles. Let $deg(f(x)) = n = 2m + 1$
or $= 2m + 2$. We showed $g \leq m$ and promised to prove $g = m$.

We want to determine the genus, using the Riemann-Hurwitz formula and
Dedekind's different theorem. Let $f(x) = \prod_{i=1}^{r} f_i(x)$, where $f_i(x)$ is irreducible of degree n_i. Denote by Q_i the place of $K(x)$ defined by $f_i(x), P_i$
an extension place and e_i, f_i, v_i as usual. Then $2v_i(y) = e_i$. It follows $e_i =
2, f_i = 1$ and P_i is the unique place above Q_i. As there is precisely one ra-
tional place $P_\infty \mid Q_\infty$ ($e = 2, f = 1$) when n is odd and there are two
rational places over Q_∞ when n is even, it follows that the different degree is
n when $n + 1$ is odd and n when n is even. The Riemann-Hurwitz formula
is $2g - 2 = 2(0 - 2) + 2m + 2$ in both cases, hence $g = m$. In fact, it would
not have been necessary to know the situation at the place at infinity. It
would have sufficed to observe that the corresponding contribution is ≤ 1. It
is uniquely determined by this information, as the different degree is even.

22.5 Theorem. *Let $char(K) \neq 2$ and $F = K(x, y)$ defined by*

$$y^2 = f(x)$$

*where $f(x) \in K[x]$ a square-free monic polynomial of degree n. Write $n =
2m + 1$ if n is odd, $n = 2m + 2$ if n is even. Then F has genus m. The
constant field of F is K. The place at infinity of $K(x)$ is ramified if and only
if n is odd. The places of $K(x)$ corresponding to divisors of $f(x)$ are ramified.*

The relation between solutions and rational points

As an illustration, we consider the case of the hyperelliptic equation $y^2 =
f(x)$ in characteristic $\neq 2$. What is the relation between the number $N(y^2 -
f(x))$ of solutions of the corresponding homogeneous **equation** and the num-
ber $N(F)$ of **rational places** of the corresponding hyperelliptic function field
$F \mid K(x)$? Where the equation is concerned, we have

$$N(y^2 - f(x)) = 1 + 1 \times |\{\alpha \in K \mid f(\alpha) = 0\}|$$

$$+2 \times |\{\alpha \in K \mid 0 \neq f(\alpha) \text{ a square in } K\}|$$

(see Section 19.1). Where the rational places are concerned, the answer is provided by [198], Corollary 3.3.8. Let at first the degree of $f(x)$ be odd. Then

$$N(F) = 1 + 1 \times |\{\alpha \in K \mid f(\alpha) = 0\}| + 2 \times |\{\alpha \in K \mid 0 \neq f(\alpha) \text{ a square in } K\}|.$$

Why this? We know that there is precisely one place $P \mid Q_\infty$ and this place is rational (we have $e = 2, f = 1$). This explains the first term. If $f(\alpha) = 0$, then again there is precisely one place $P \mid Q_\alpha$ and P is rational. If $f(\alpha)$ is a nonsquare, then, by Corollary 3.3.8, there is again precisely one place $P \mid Q_\alpha$, but we have $e = 1, f = 2$ in this case, so P is not a rational place. Finally, if $f(\alpha)$ is a nonzero square in K, then there are two different rational places $P \mid Q_\alpha$. By comparison, we see $N(F) = N(y^2 - f(x))$ in this case. This is the ideal situation: rational places of the function field are in bijection with the solutions of the underlying homogeneous equation.

The case when $deg(f(x))$ is even shows that this is not always the case. The reader is asked to provide the details:

22.6 Theorem. *Consider the equation* $y^2 = f(x)$ *over* $\mathbb{F}_q$, *where* q *is odd and* $f(x)$ *a square-free polynomial of degree* $n \geq 3$. *Let* $F \mid \mathbb{F}_q(x)$ *be the corresponding function field. If* n *is odd, then* $N(F) = N(y^2 - f(x))$. *If* n *is even, then* $N(F) = N(y^2 - f(x)) + 1$.

The Klein quartic

It is easiest to remember in its homogeneous form

$$X^3 Y + Y^3 Z + Z^3 X.$$

The corresponding affine equation is

$$x^3 y + y^3 + x = 0.$$

Multiply by x^6, use $-x^2 y$ as a variable instead of x and x instead of y. The equation becomes

$$y^7 = x^3/(1 - x).$$

This is an irreducible equation in y. In particular, it defines a function field F, for any field K of constants. Assume $char(K) \neq 7$ (in characteristic 7 the extension $F \mid K(x)$ is purely inseparable). In order to determine the genus, we need to study ramification. Theorem 3.3.7 of [198] (Kummer's theorem) provides the necessary information. Clearly, for each of Q_∞, Q_0, Q_1, there is precisely one extension place of F and it is totally ramified ($e = 7, f = 1$). In particular, those places P_∞, P_0, P_1 are rational places and K is the full constant field. The Riemann-Hurwitz formula now yields $g \geq 3$. It suffices to invoke Theorem 20.46 to conclude $g = 3$. As an exercise, we continue with the details of the Riemann-Hurwitz formula. Let $Q = Q_{p(x)}$ be another place of

$K(x)$. Then $x^3/(1 - x)$ is defined and nonzero at Q. Use the following facts from elementary Galois theory: the polynomial $X^p - c \in K[X]$ (for a prime p)

- is irreducible if it has no zero in K,

- is a product of two irreducibles of degrees 1 and $p - 1$ if it has a zero in K and K does not contain primitive p-th roots of unity,

- splits completely if it has a zero in K and K contains primitive p-th roots of unity.

By [198],Theorem 3.3.7, this implies that $P \mid Q$ is not ramified. By Dedekind's theorem, we conclude that the different has degree $3 \times (7 - 1) = 18$ and so the genus is 3 :

22.7 Proposition. *The Klein quartic over a field K of constants whose characteristic is different from 7 has genus 3. Its full field of constants is K.*

Now let $K = \mathbb{F}_q$ (in characteristic $\neq 7$) and $Q = Q_\alpha$ where $\alpha \in K \setminus \{0, 1\}$. By [198],Corollary 3.3.8, the number of rational places $P \mid Q_\alpha$ is

- 0 if $\alpha^3/(1 - \alpha)$ has no seventh root in K,

- 1 if $\alpha^3/(1 - \alpha)$ has seventh roots in K and $p \nmid q - 1$,

- 7 if $\alpha^3/(1 - \alpha)$ has seventh roots in K and $p \mid q - 1$.

In particular, we have the following expression for the number $N(F)$ of rational places:

$$N(F) = 3 + |\{\alpha \in K \setminus \{0, 1\} \mid \alpha^3/(1 - \alpha) \in K^7\}|$$

if $7 \nmid q - 1$ and

$$N(F) = 3 + 7 \times |\{\alpha \in K \setminus \{0, 1\} \mid \alpha^3/(1 - \alpha) \in K^7\}|$$

if $7 \mid q - 1$.

Finally, compare this to the number $N(y^7 - x^3/(1 - x))$ of solutions to the homogeneous polynomial equation $Y^7 Z - XY^7 - X^3 Z^5 - 0$. We leave it as an exercise to show that $N(y^7 - x^3/(1 - x)) = N(F)$.

22.8 Theorem. *The number $N(F)$ of rational places of the function field F defined by the Klein quartic with constant field $K = \mathbb{F}_q$ in characteristic $\neq 7$ agrees with the number of projective solutions to the equation $Y^7 Z - XY^7 - X^3 Z^5 - 0$.*

Now consider $K = \mathbb{F}_8$, where we construct $\mathbb{F}_8$ as $\mathbb{F}_2(\epsilon)$, $\epsilon^3 + \epsilon^2 + 1 = 0$. The seventh power in $\mathbb{F}_8$ is 1. The equation $\alpha^3 = 1 + \alpha$ has three solutions: $\alpha \in \{\epsilon^3, \epsilon^5, \epsilon^6\}$. This describes the 24 rational places:

$$P_\infty, P_0, P_1, P(\epsilon^3, \beta), P(\epsilon^5, \beta), P(\epsilon^6, \beta)$$

where $0 \neq \beta \in \mathbb{F}_8$. Some principal divisors are clearly visible:

$$(x) = 7P_0 - 7P_\infty, \quad (x+1) = 7P_1 - 7P_\infty, \quad (y) = 3P_0 - 2P_\infty - P_1.$$

Here is a first result concerning good concrete codes: let the rational points over $\mathbb{F}_8$ be $P_0, \ldots, P_{23}$, use $G = 10P_0$ and $D = P_1 + \cdots + P_{23}$. This one-point code has length $n = 23$, dimension $dim(10P_0) = 10 + 1 - g = 8$ and minimum distance ≥ 13 (check: $8 + 13 = 24 - g$). We have a code $[23, 8, 13]_8$. This is the best known distance for a linear 8-ary code. Apply concatenation with $[4, 3, 2]_2$ as an inner code. This yields a binary linear code $[92, 24, 26]_2$, again representing the best known value of d. This result is from van Lint and van der Geer [213].

Back to the original description of the Klein quartic, by the equation

$$y^3 + x^3 y + x = 0.$$

Consider the places Q_0, Q_∞ of $K(x)$ and their extensions. Let $P \mid Q_0$. The values at P of the three terms are $v(y) + 3e$, $3v(y)$, e. It follows $3v(y) = e$, hence $e = 3, f = 1$: there is a unique place $P_0 \mid Q_0$ and $v_0(x) = 3, v_0(y) = 1$. Let $P \mid Q_\infty$. Assume $v(y) \leq 0$. Then $e = 2, f = 1$. In particular, there must be a place where $v(y) > 0$. For that place $v(y) = 2e$. As $[F : K(y)] = 3$, we have $e = 1$ in this case. To sum up: there are two places $P_{\infty,1}$ and $P_{\infty,2}$ above Q_∞, and

$$e_{\infty,1} = 2, \quad v_{\infty,1}(y) = -3, \quad e_{\infty,2} = 1, \quad v_{\infty,2}(y) = 2.$$

We know the divisors of x and y:

$$(x) = 3P_0 - 2P_{\infty,1} - P_{\infty,2}$$

$$(y) = P_0 - 3P_{\infty,1} + 2P_{\infty,2}$$

Here are some more examples of codes constructed from the Klein curve. These are from Moreno [153]. Consider $B = P_0 + P_{\infty,1} + P_{\infty,2}$ of degree 3 and $G = mB$, $D = P_1 + \cdots + P_{21}$, the sum of the remaining rational places over $\mathbb{F}_8$. By Riemann-Roch, $dim(mB) = 3m - 2$ for $m \geq 2$. Consider $m = 2$. It follows that $L(2B) = \langle 1, x, 1/y, y/x \rangle$. The corresponding code is a $[21, 4, 15]_8$ code. As $dim(3B) = 7$, we see that

$$L(3B) = \langle 1, x, 1/y, y/x, y, 1/x, x/y \rangle.$$

This yields a $[21, 10, 9]_8$ code.

The Hermitian codes

As an illustration, consider the **Hermitian curves** and the Hermitian codes derived from them. We describe how to obtain generator matrices for the Hermitian codes. Consider the field extension $\mathbb{F}_{q^2} \mid \mathbb{F}_q$ and the corresponding trace tr and norm N, where $tr(x) = x + x^q$, $N(x) = x^{q+1}$. Our codes are defined over $\mathbb{F}_{q^2}$ and have length q^3.

The coordinates are parametrized by the pairs (α, β), where $N(\alpha) = tr(\beta)$. So we need to calculate traces and norms of all elements in the field and to list all these pairs in some order. There are q^3 such pairs.

The general build-up: We construct a $(q^3 - g, q^3)$ matrix G with entries from $\mathbb{F}_{q^2}$. Here $g = \binom{q}{2}$. The first k rows of G generate the k dimensional Hermitian code. It has parameters

$$[q^3, k, \geq q^3 - k + 1 - g]_{q^2}.$$

The pole order test: For $n = 0, 1, 2, \ldots$ we have to decide if n is a **pole order** or not. If n is a pole order, we determine its **coordinate vector** (i, j). This is done as follows: let r be the remainder of $n \bmod q$, where $0 \leq r \leq q - 1$ and $-s$ the (negative) remainder of $n \bmod q + 1$, where $0 \leq s \leq q$. Then n is a pole order if and only if

$$x = \frac{n - r}{q} \geq \frac{n + s}{q + 1} = y.$$

If $n \geq 2g$, then the pole order test does not need to be performed. Every such number is a pole order. If n is a pole order, then $n = (q+1)i + qj$, where $i = (x - y)q + r, j = s$. The coordinate vector of n is (i, j).

Constructing the rows of G: Let $u_1 = 0, u_2 = q, u_3 = q + 1 \ldots$ be the first pole orders. If u_k has coordinate vector (i, j), then the entry of row k of G in coordinate (α, β) is $\beta^i \alpha^j$.

We conclude that the use of Hermitian codes requires the usual field arithmetic, just as Reed-Solomon codes.

Suzuki codes

The **Suzuki groups** are an infinite family of finite simple groups. The **Suzuki curves** were discovered much later. They bear this name, as they admit the Suzuki groups as their groups of automorphisms. In particular, the Suzuki curves are highly symmetric objects. Here we illustrate the general mechanism of AG codes by describing the codes that result from the Suzuki curve $\mathcal{S}_8$ defined over $\mathbb{F}_8$. It is defined by the equation

$$X^2(Z^8 + ZX^7) = Y^2(Y^8 + YX^7). \tag{22.1}$$

It can be checked that there are precisely 65 rational points. One is $P_\infty = (0 : 0 : 1)$. The remaining 64 rational points can be parametrized by pairs of points from $\mathbb{F}_8$. They are

$$P(y, z) = (1 : y : z), \text{ where } y, z \in \mathbb{F}_8. \tag{22.2}$$

The reader is invited to check that these points do indeed satisfy Equation 22.1. We could have used any of the 65 rational points in the role of P_∞. The Suzuki codes defined over $\mathbb{F}_8$ have length 64. Their coordinates are parametrized by the rational points $P(y, z)$, hence by pairs (y, z) of elements from $\mathbb{F}_8$. We can choose the following "functions" (elements of the function field):

$$f_1 = 1, \ f_2 = y, \ f_3 = z, \qquad f_4 = y^5 + z^4, \qquad f_5 = yz^4 + y^6 + z^2,$$
$$f_6 = y^2, \ f_7 = yz, \qquad f_8 = z^2, \qquad f_9 = y^2 z^4 + y^7 + yz^2,$$
$$f_{10} = zy^5 + z^5, \ f_{11} = yz^5 + y^6 z + z^3, \ f_{12} = y^3,$$
$$f_{13} = y^4 + y^5 z^2 + z^6, \ f_{14} = y^2 z.$$

Let v_i be the codeword defined by f_i. For example, v_1 is the all-1 word, v_2 has entry y in coordinate (y, z) and v_5 has entry $yz^4 + y^6 + z^2$ in that same coordinate. Define $\mathcal{C}_k$ to be the k-dimensional code generated by $v_1, v_2, \ldots, v_k$. These are the Suzuki codes. They form an ascending chain

$$\mathcal{C}_1 \subset \mathcal{C}_2 \subset \cdots \subset \mathcal{C}_k \subset .$$

The genus of the Suzuki curve $\mathcal{S}_8$ is $g = 14$. It follows that $\mathcal{C}_k$ has parameters $[64, k, \geq 50 - k]_8$. In many cases this describes the precise minimum distance. For low dimensions the theory gives better bounds, of course. The true parameters for the first of those codes are

$$[64, 1, 61]_8 \subset [64, 2, 56]_8 \subset [64, 3, 56]_8 \subset [64, 4, 52]_8 \subset [64, 5, 51]_8 \subset$$
$$[64, 6, 48]_8 \subset [64, 7, 46]_8 \subset [64, 8, 44]_8 \subset [64, 9, 43]_8 \subset [64, 10, 42]_8.$$

With the exception of $\mathcal{C}_3$, those minimum distances meet the lower bound, which follows from the theory. Code $\mathcal{C}_3$ meets the Griesmer bound with equality. The theory yields only $d \geq 54$. These parameters are not interesting, however. In fact, by Chapter 17, such a code is equivalent to a multiset of 64 points in $PG(2, 8)$ at most 8 of which are collinear. The affine plane $AG(2, 8)$ is an example of such a point set.

An interesting case is $\mathcal{C}_{11}$. It is a $[64, 11, 42]_8$ code. The theory states $d \geq 41$. Even more interesting codes can be constructed using one of our standard methods, construction X. Recall that chains of codes are exactly the ingredients needed to apply construction X. The following code parameters are examples:

$$[65, 5, 52]_8, \ [70, 5, 56]_8, \ [67, 6, 51]_8, \ [66, 11, 43]_8, \ [68, 11, 44]_8,$$
$$[66, 12, 41]_8, \ [69, 12, 43]_8, \ [65, 13, 40]_8, \ [68, 13, 42]_8, \ [70, 14, 42]_8,$$
$$[70, 17, 38]_8, \ [72, 17, 39]_8.$$

The curves mentioned in this section all belong to a highly interesting family, the **Deligne-Lusztig curves.** They were discovered in the wake of a revolutionary approach to the representation theory of classical finite simple groups, the groups of Lie type. For an introduction, see *Finite Groups of Lie Type* by R. W. Carter [45]. The starting point of the Deligne-Lusztig theory is the study of the action of the group (in fact, of a somewhat larger group) on the cohomology groups of a certain variety. This variety is canonically associated with the group in question. It was then observed that, in the case of groups of Lie rank 1, the corresponding variety is an algebraic curve and, as was to be expected, the codes constructed from curves with such a rich structure turned out to be exceptionally good as well. The family consists of our good friends the extended Reed-Solomon codes (with $PSL_2(q)$ as underlying groups), the Hermitian codes, the Suzuki codes (only in characteristic 2) and the Ree codes (only in characteristic 3).

In the 1980s, coding theory was revolutionized by Goppa and Manin. An early triumph of the AG code method was the construction [210], by Tsfsasman, Vladut and Zink, of families of codes which asymptotically beat the Gilbert-Varshamov bound, thus shattering an old belief in the coding community that the Gilbert-Varshamov bound would never be beaten. This improvement is proved for q-ary linear codes whenever $q \geq 49$ is a square (recall Definition 9.1). In the binary case, no improvement of the asymptotic GV bound is known. The construction of [210] uses a certain family of modular curves, thus using deeper theory than the theory of function fields alone. In this form, it is very hard to derive explicit equations. This difficulty was overcome by Garcia-Stichtenoth [89], who derive the same result using only completely explicit simple equations. The parameters of these asymptotically optimal AG codes are derived by applying the basic mechanism to the following family of curves:

22.9 Theorem. *Let q be a prime-power and $e \geq 1$ a rational number. Then there is an infinite set $\mathcal{M}$ of natural numbers such that em is a natural number for all $m \in \mathcal{M}$, and for every $m \in \mathcal{M}$ there is an algebraic curve $\mathcal{K}_m$ defined over $\mathbb{F}_{q^m}$ with more than q^{em} rational points and genus g_m, such that $\lim_{m \in \mathcal{M}} g_m / q^{em} = 0$.*

It follows that the corresponding codes $\mathcal{C}_m$ have parameters

$$[q^{em}, Rq^{em}, (1-R)q^{em} - g_m]_{q^m}.$$

The Zyablov bound

[228] states that the line $R = (1 - H_q(\mu))(1 - \frac{\delta}{\mu})$ is asymptotically reachable for all $0 \leq \mu \leq (q-1)/q$. In itself this is not terribly interesting as those lines are under the Gilbert-Varshamov bound. The point is that these codes can be efficiently constructed. In [228] this result is derived by concatenation

with a Reed-Solomon code as an outer code and a code reaching the Gilbert-Varshamov bound as the inner code. This procedure may be called semi-constructive. The Justesen bound from Theorem 9.14 is a special case of the Zyablov bound. In fact, Justesen's basic observation, when applied to the AG codes of Theorem 22.9, yields another proof of the Zyablov bound, which is free of conducting a search for codes meeting the GV bound; see Shen [186].

Exercises 22.3

22.3.1. *Show that the number of rational places of $F \mid \mathbb{F}_q(x)$ defined by the Klein quartic in characteristic $\neq 7$ equals the number of points in $PG(2, q)$ satisfying $Y^7 Z - XY^7 - X^3 Z^5 = 0$.*

22.3.2. *Let $B = P_0 + P_{\infty,1} + P_{\infty,2}$ (the Klein curve). Find a basis of $L(4B)$.*

22.3.3. *Find a generator matrix of a $[21, 4, 15]_8$ code.*

22.3.4. *Show that $dim(B) = 1$.*

22.3.5. *Prove the existence of a $[65, 5, 52]_8$ code.*

22.3.6. *Show that a $[70, 5, 56]_8$ code exists.*

22.4   The automorphism group

22.10 Definition. *The automorphism group $Aut(F \mid K)$ is the group of automorphisms of the function field F which leave K elementwise invariant.*

Automorphisms are often best described as polynomial invariants:

22.11 Definition. *Let $h(X, Y, Z) \in K[X, Y, Z]$ be a homogeneous polynomial and $\sigma \in GL(3, K)$. Define the image of h under σ as $(\sigma h)(X, Y, Z) = h(\sigma(X), \sigma(Y), \sigma(Z))$. Call $h(X, Y, Z)$ a **relative invariant** of σ if there is a constant such that $(\sigma h)(X, Y, Z) = c \cdot h(X, Y, Z)$.*

It is clear that $(\sigma h)(X, Y, Z)$ always is a homogeneous polynomial of the same degree as $h(X, Y, Z)$.

22.12 Theorem. *Let $F = K(x, y)$ be a function field and $h(X, Y, Z)$ the homogenization of the minimal polynomial of y over $K(x)$. Then each relative invariant σ of $h(X, Y, Z)$ defines an automorphism of $F \mid K$.*

PROOF The step from $h(X, Y, Z)$ to the minimal polynomial $\phi(x, y)$ simply is the following: $\phi(x, y) = h(x, y, 1)$, where $x = X/Z, y = Y/Z$. Define $\sigma(x) = \sigma(X)/\sigma(Z), \sigma(y) = \sigma(Y)/\sigma(Z)$. The fact that σ is a relative invariant of $h(X, Y, Z)$ shows that the images $\sigma(x), \sigma(y) \in F$ satisfy $\phi(\sigma(x), \sigma(y)) = 0$.
▯

In order to obtain curves which admit the action of a prescribed matrix group, it is often useful to reverse this process: start from a subgroup $G \subset GL(3, K)$ and study its relative invariants. The corresponding affine equation, provided it is irreducible, will then define a curve which admits G as an automorphism group. The theory of (relative) invariants of matrix groups has a rich history.

The Klein quartic

The description of the Klein quartic started from the homogeneous polynomial

$$h(X, Y, Z) = X^3 Y + Y^3 Z + Z^3 X.$$

The corresponding affine equation is $x^3 y + y^3 + x = 0$. One invariant of $h(X, Y, Z)$ is obvious, the cyclic permutation $\rho = (X, Y, Z)$. The corresponding

matrix is $\rho = \begin{pmatrix} 0\,1\,0 \\ 0\,0\,1 \\ 1\,1\,0 \end{pmatrix}$ and the corresponding action on F is

$$\sigma(x) = \sigma(X)/\sigma(Z) = Y/X = y/x, \ \ \sigma(y) = \sigma(Y)/\sigma(Z) = Z/X = 1/x.$$

It is now easy to see that this really describes an automorphism of F.

Assume now that K contains a seventh root of unity $\epsilon \neq 1$. Then $\sigma = diag(\epsilon, \epsilon^4, \epsilon^2)$ is an invariant of $h(X, Y, Z)$. The corresponding automorphism of F is $\sigma(x) = \epsilon^6 x, \ \sigma(y) = \epsilon^2 y$.

Assume now that K has characteristic 2 and contains $\mathbb{F}_8$. As usual, describe $\mathbb{F}_8$ as $\mathbb{F}_2(\epsilon)$ where $e^3 + \epsilon^2 + 1 = 0$. There is an automorphism of order 2 of $h(X, Y, Z)$. It is described by the matrix

$$\alpha = \begin{pmatrix} \epsilon^6 & \epsilon^4 & 1 \\ \epsilon^4 & 1 & \epsilon^6 \\ 1 & \epsilon^6 & \epsilon^4 \end{pmatrix}.$$

It normalizes $\langle \rho \rangle$ ($\rho^\alpha = \rho^2$). If we want to construct α using this condition, the information obtained is that α has the form $\begin{pmatrix} a\,b\,c \\ b\,c\,a \\ c\,a\,b \end{pmatrix}$. The fact that α is an involution shows $ab + bc + ac = 0, a^2 + b^2 + c^2 \neq 0$. It is then easy to find α as given above. The group $\langle \sigma, \rho, \alpha \rangle$ is the simple group $GL(3, 2)$ of order 168.

AG codes and concatenation

The mechanism of algebraic-geometry codes has been slightly generalized in [224, 71], using not only rational points but also points of degree > 1. We illustrate with some examples based on the Klein quartic. As we consider $\mathbb{F}_8$ as ground field and want to make use of points of degree 2, the field with 64 elements will be needed.

The field $\mathbb{F}_{64}$

We construct it as a quadratic extension of $\mathbb{F}_8$ using the primitive polynomial $X^2 + X + \epsilon$, where $\epsilon^3 + \epsilon^2 + 1 = 0$. Therefore, let $\mathbb{F}_{64} = \mathbb{F}_8(\rho)$ where $\rho^2 = \rho + \epsilon$. Then $\rho^9 = \epsilon$, which confirms that ρ is a primitive element. Let $\omega = \rho^{21} = \rho + \epsilon^3 = \omega \in \mathbb{F}_4$.

Back to the Klein quartic over $\mathbb{F}_8$,

its 24 rational points and its large group of automorphisms. We know a place $(\omega : \overline{\omega} : 1) = Q_0$ of degree 2. Application of our automorphism of order 7 yields an orbit of seven points of degree 2. These are $Q_i = (\omega/\epsilon^i, \epsilon^{2i}\overline{\omega} : 1)$ where the subscript i is taken mod 7. We wish to evaluate at all 24 rational points and also at 2 of the points Q_i of degree 2. We therefore choose $G = (k+2)(A - B)$ where A is a point of degree $m + 1$, and B is a point of degree m where $m \geq 2$. Then $deg(G) = k + 2$. Consider $L(G)$. Riemann-Roch shows that $dim(G) = k$ if $k > 4$. We can evaluate $f \in L(G)$ at all rational points. Let us also evaluate at Q_0 and Q_1 (we choose B different from Q_0, Q_1). Those values are in $\mathbb{F}_{64}$, which we concatenate with an inner code $[3, 2, 2]_8$. The values of the modified evaluation are in $\mathbb{F}_8^{24+2\times 3} = \mathbb{F}_8^{30}$. Let us do this for $7 \leq k \leq 9$. Our basic mechanism shows that the resulting codes have parameters $[30, 7, 19]_8, [30, 8, 18]_8$ and $[30, 9, 17]_8$, respectively. If we choose $G = mQ_2$ instead (of degree $2m$), we obtain, for example, $[30, 8, 18]_8$ and $[30, 10, 16]_8$. Those codes form a chain, as G is an effective divisor. We can apply construction X and obtain $[33, 10, 18]_8$.

Exercises 22.4

22.4.1. *Let K be a field containing an element ϵ of order 7. Show that the automorphisms σ of order 7 and ρ of order 3 of the Klein quartic together generate a group of order 21. Determine the conjugate $\rho^{-1}\sigma\rho$.*

22.4.2. *Show that the Klein quartic over $\mathbb{F}_8$ contains $GL(3,2)$ as a group of automorphisms.*

22.5   AG codes and universal hashing

Recall Chapter 6, where we saw that error-correcting codes and ϵ-universal hash families are identical objects: an $(n, M, d)_q$-code yields an ϵ-universal $(n, M)_q$ hash family where $\epsilon = 1 - d/n$ (Theorem 6.4). We saw in Section 16.7 how ϵ-universal $(n, M)_q$ hash families can be used in the construction of authentication codes.

In the present section we want to interpret AG codes as universal hash families. In fact, we consider only the Hermitian one-point codes. The ground field is $\mathbb{F}_{q^2}$, and the number of rational points is $q^3 + 1$. As we use one of those rational points P for the one-point code construction, our codes have length $n = q^3$. Instead of the generic bound (Singleton defect at most g), we now use more precise information. Let $D = P_1 + \cdots + P_{q^3}$, where $P_1, \ldots, P_{q^3}$ are the rational points different from P. The pole orders form the semigroup S generated by q and $q+1$. In fact, we saw in an exercise that there are precisely $\binom{q}{2}$ natural numbers outside S. This is the genus of the Hermitian curve, as it has to be.

Now let

$$u_1 = 0, u_2 = q, u_3 = q + 1, u_4 = 2q, u_5 = 2q + 1, u_6 = 2q + 2, u_7 = 3q, \ldots$$

be the pole orders in ascending order. We have $dim(u_j P) = j$. The corresponding $\mathbb{F}_{q^2}$ linear code $\mathcal{C}(D, u_j P)$ has length q^3, dimension j and minimum distance $\geq q^3 - u_j$. Specialize now to $j = \binom{i+1}{2} + 1$ and use $u_{\binom{i+1}{2}+1} = iq$, which was proved in an easy exercise. This yields a code $[q^3, \binom{i+1}{2} + 1, q^3 - iq]_{q^2}$. The relative minimum distance is $1 - i/q^2$ where $i \leq q$. In terms of universal hashing, this is an i/q^2-universal $(q^3, q^{i^2+i+2})_{q^2}$ hash family.

Exercises 22.5

22.5.1. *Construct linear codes* $[27, 4, 21]_9$ *and* $[27, 7, 18]_9$.

22.5.2. *Which* $\mathbb{F}_{16}$ *linear codes are obtained?*

22.6 The Hasse-Weil bound

The celebrated Hasse-Weil bound states the following:

22.13 Theorem. *Let* F *be a function field with field of constants* $\mathbb{F}_q$ *and genus g. Then its number* N *of rational places satisfies*

$$|N - (q + 1)| \leq 2g\sqrt{q}.$$

The original proof is due to Weil. A greatly simplified proof based on Bombieri's method constitutes the major part of a long chapter in Stichtenoth [198]. A slight improvement due to Serre states

$$|N - (q + 1)| \leq g\lfloor 2\sqrt{q} \rfloor.$$

We will sketch just one of many applications of the Hasse-Weil bound. It concerns the minimum distance of cyclic codes.

Let p be a prime and $tr(\mathcal{B}(A))$ a p-ary cyclic code of length $n = p^r - 1$ (this is the primitive case), in the terminology of Section 13.2. Observe that the coordinates of the code are indexed by elements $0 \neq x \in F = \mathbb{F}_{p^r}$. What can we say about the minimum distance $d = d(tr(\mathcal{B}(A)))$? Observe that $tr(\mathcal{B}(A))$ is the dual of $\mathcal{C}(A)$. The BCH bound from Section 13.2 states the following: let B be the complement of the Galois closure of A and t the largest length of an interval in B. Then $d \geq t + 1$. In the present situation we want to use Artin-Schreier curves and the Hasse-Weil bound to obtain a completely different lower bound on d. In fact, assume $0 \notin A$. We may replace A by its Galois closure, so there is no harm in assuming that A is Galois closed. Then A is a union of cyclotomic cosets. Represent each such cyclotomic coset by the smallest integer between 1 and $n - 1$ belonging to it, and let I be the set of those representatives. Each $i \in I$ is coprime to p. From the description of cyclic codes, we know that each codeword is described by a polynomial $f(X) = \sum_{i \in I} a_i X^i$ where $a_i \in F$ and the codeword in question has entry $tr(f(x))$ in coordinate x. Observe that $f(0) = 0$.

Consider the $a_i \in F$ as fixed, not all $= 0$. Let ν be the number of entries 0 in this codeword. Whenever $tr(f(x)) = 0$, we know from theorem 90 that this yields p solutions Y to the equation $Y^p - Y - f(x) = 0$. The total number of solutions $(x, y) \in F \times F$ of the equation $y^p - y - f(x) = 0$ is therefore $p(1 + \nu)$. Consider the Artin-Scheier polynomial $Y^p - Y - f(X) \in F[X, Y]$. As $deg(f(X))$ is not divisible by the characteristic p we know from Exercise 21.1.1 that $Y^p - Y - f(X)$ is absolutely irreducible and therefore defines a function field $K(X, Y)$, an Artin-Schreier extension of $K(X)$. We also know that its number of rational places is $N = 1 + pu$ where u is the number of affine solutions; in other words, $u = 1 + \nu$ (see Proposition 19.6). Now use the Hasse-Weil bound. It shows $|N - (p^r + 1)| = |pu - p^r| \leq 2gp^{r/2}$, where g is the genus. This implies $pu = p + p\nu \leq p^r + 2gp^{r/2}$, or $\nu \leq p^{r-1} + 2gp^{r/2-1} - 1$. Recall that ν is the number of zeroes in our codeword. The weight is therefore $n - \nu$, hence

$$d \geq p^r - p^{r-1} - 2gp^{r/2-1}.$$

All we need to do in order to make this explicit is the determination of the genus of this Artin-Schreier curve.

For more details and more general results see Wolfmann [223] and Güneri and Özbudak [101].

Chapter 23

The last chapter

23.1 List decoding

Block codes of minimum distance d can correct e errors per block, where e is the largest number such that $2e < d$. This is the classical philosophy which we observed in Section 1.3. In Section 13.3 we finally encountered a usable decoder. It is based on the Euclidean algorithm.

It is natural to ask what happens if more than e errors occur. Shall we abandon all hope then? Anyway, decoding will be done automatically. We should make sure the decoder does its best to generate good quality, no matter what happens in the channel. The best we can hope for is that, upon reception of an n-tuple y, the decoder produces the list of all codewords in a ball of some radius e' around y. Here e' should be chosen such that this list of codewords is not empty and in most of the cases consists of exactly one codeword. This is then the decoder's best guess. Important progress has been made fairly recently in the list decoding of Reed-Solomon codes. These methods can be extended to cover other classes of codes, including AG codes. In the present section we concentrate on Reed-Solomon codes. The exposition follows the 1999 paper [102] by Guruswami and Sudan. As we shall see, this is a nice application of some classical algebra.

The definition of the Reed-Solomon codes as evaluations of polynomials (see Chapter 4) shows that list decoding of these codes and their projections (truncations) can be formulated as follows:

The input consists of a field $F = \mathbb{F}_q$, a length $n \leq q$, pairs (x_i, y_i) for $i = 1, \ldots, n$ (where $x_i, y_i \in F$ and the x_i are pairwise different), a maximum degree k and a radius $n - t$. The output we wish to produce is the list of all polynomials $f(X) \in F[X]$ of degree $\leq k$ satisfying $f(x_i) = y_i$ for at least t of the indices i. In other words, we use a truncation of the Reed-Solomon code $\mathcal{RS}(k+1, q)$, the (x_i, y_i) describe a received word and we want to produce the **list** of all codewords at distance $\leq n - t$ from the received word. The main point is to find a fast algorithm.

The first idea is to work with polynomials in two variables. Here we view $Q(X, Y)$ as a polynomial with just one variable Y whose coefficients are in the field of rational functions (polynomials divided by polynomials) over F. We encountered this field $F(X)$ in the preceding chapter and called it the rational function field there. The strategy is to construct a suitable polynomial $Q(X, Y)$ with the property that every polynomial $f(X)$ on the **list** contributes a linear factor $Y - f(X)$ of $Q(X, Y)$. This will be the case if $f(X)$ is a root of $Q(X, Y)$, equivalently, if $Q(X, f(X))$ is the 0 polynomial.

Here is the trick: fix a **multiplicity** r and choose $Q(X, Y)$ such that the following conditions are satisfied. For each pair (x_i, y_i) we demand that the shifted polynomial defined by $Q^{(i)}(X, Y) = Q(X + x_i, Y + y_i)$ does not have terms of degree less than r. In particular, we demand $Q(x_i, y_i) = 0$. As the number of such terms is $\binom{r+1}{2}$, this gives $n\binom{r+1}{2}$ linear equations for the coefficients of $Q(X, Y)$ to satisfy. Let us see what the consequences are when these conditions are satisfied.

For fixed i, let $f(X)$ be a polynomial satisfying $f(x_i) = y_i$ and consider the polynomial $g(X) = Q(X, f(X))$ (which we wish to be the 0 polynomial if $f(X)$ is on the **list**). We have

$$g(X + x_i) = Q(X + x_i, f(X + x_i)) = Q^{(i)}(X, f(X + x_i) - y_i).$$

As $f(X + x_i) - y_i$ is a multiple of X, it follows that X^r divides $g(X + x_i)$ and therefore that $(X - x_i)^r$ divides $g(X)$. Now let $f(X)$ be on the **list**. Then $g(X)$ is a multiple of $(X - x_i)^r$ for all i such that $f(x_i) = y_i$. There are at least t such indices i. If we can manage that the degree of $g(X)$ is $< rt$, it follows that $g(X)$ is the 0 polynomial (as the 0 polynomial is the only polynomial of degree $\leq n$ with more than n roots, for any n).

Recall that $f(X)$ has degree $\leq k$. It follows that the degree of $g(X)$ is guaranteed to be $< rt$ if all monomials $X^i Y^j$ occurring in $Q(X, Y)$ satisfy $i + kj < rt$. This is it. We just need to write down the conditions explicitly and to work out when they can be satisfied.

The number of monomials in question equals the number of nonnegative solutions (i, j) of the inequality $i + kj \leq l$, where $l = rt - 1$. This number is

$$(\lfloor l/k \rfloor + 1)(2l + 2 - k\lfloor l/k \rfloor)/2$$

(see Exercise 23.1.1). The first factor $\lfloor l/k \rfloor + 1$ is $(l + k)/k$ if k divides l; it is $\geq (l + 1)/k$ otherwise. This gives us a lower bound of $(l + 1)(l + 2)/(2k)$ on the number of monomials. The condition $l < rt$ will be satisfied if we choose $l = rt - 1$. Only one condition remains to be satisfied. The polynomial $Q(X, Y)$ must be nonzero such that its coefficients satisfy some $n\binom{r+1}{2}$ linear equations. This can be done provided the number of monomials is larger than the number of constraints. Using $l = rt - 1$, the condition becomes $rt(rt + 1)/k > nr(r + 1)$, equivalently, $r(t^2 - kn) > kn - t$. This can be done as long as $t^2 > kn$.

We sum up: assume $t^2 > kn$, equivalently $t > \sqrt{kn}$. Dividing by n we obtain the equivalent condition $t/n = 1 - \delta > \sqrt{R}$, where R is the information rate and δ the relative minimum distance. Use multiplicity r such that $r > (kn - t)/(t^2 - kn)$. Find a nonzero polynomial $Q(X, Y) = \sum_{i,j} q_{ij} X^i Y^j$ with coefficients $q_{ij} \in F$ where the sum is over all pairs i, j such that $i + kj \leq rt - 1$

and such that the $Q(X + x_i, Y + y_j)$ have no monomials of degree less than r. This can be done by Gauß elimination, finding a nontrivial solution (q_{ij}) to a system of $n\binom{r+1}{2}$ linear equations. View $Q(X, Y)$ as a polynomial in the variable Y. Determine all the zeroes of $Q(X, Y)$, equivalently, all linear factors $Y - f(X)$ of $Q(X, Y)$. Each polynomial $f(X)$ on the **list** does contribute such a linear factor. In this way the determination of the list is reduced to the solution of a system of linear equations and to the determination of all roots of a polynomial of bounded degree. These operations are efficient enough to allow list decoding in practice.

Exercises 23.1

23.1.1. *Show that the number of solutions (i, j) in nonnegative integers to $i + kj \leq l$ is*

$$(\lfloor l/k \rfloor + 1)(2l + 2 - k\lfloor l/k \rfloor)/2.$$

23.2   Expander codes

The idea is to translate the problem of constructing good linear codes into the language of graph theory. Let Γ be a **bipartite graph.** A graph consists of a set of points (also called vertices) and a family of edges: the graph structure is determined by deciding which pairs of points $\{x, y\}$ form an **edge.** In this case we also say that y is a neighbor of x (and x is a neighbor of y). A graph is bipartite if its points can be partitioned into two subsets, the left points and the right points, such that each edge contains a left point and a right point (equivalently, there are no edges consisting only of left points or only of right points).

Let Γ have n left points and m right points, where $n > m$. Assume that each left point has r neighbors and each right point has l neighbors. Let $\mathcal{D}$ be an $[l, k_0, d_0]_q$ code, the auxiliary code. We construct a code $\mathcal{C} = \mathcal{C}(\Gamma, \mathcal{D})$ as follows: for each right point $R_j, j = 1, \ldots, m$, fix an order of the neighbors of R_j. A tuple $(x_1, \ldots, x_n) \in \mathbb{F}_q^n$ is in $\mathcal{C}$ if the projection to the neighbors of R_j is in $\mathcal{D}$, for all j. As each right point imposes $l - k_0$ linear conditions, it follows that $\mathcal{C}$ has dimension $k \geq n - m(l - k_0)$. Which graph-theoretic condition must be satisfied to guarantee a high minimum distance? Assume $\mathcal{C}$ has a codeword of weight w. Let W be the corresponding set of left points. There are wr edges emanating from points of W. Let $N(W)$ be the set of neighbors of elements from W. There must be some $R \in N(W)$ which is on at most $wr/|N(W)|$ of those edges. This implies that $\mathcal{D}$ contains a nonzero codeword of weight $\leq wr/|N(W)|$. It follows that $|N(W)| \leq wr/d_0$. This shows which graph theoretic property must be satisfied in order to guarantee that the minimum distance of $\mathcal{C}$ is at least d. We must have that any set of $w \leq d$ left points has more than wr/d_0 neighbors. In other words, if W is a set of left points and $|W| = w \leq d$, then $|N(W)|/|W| > c$, where $c = r/d_0$. This is an expansion property: every set of left points (provided it is not too large) has many neighbors. Loosely speaking, expander graphs are graphs with good expansion properties. There are many variants of this concept and there is a broad range of applications. What is more, strong algebraic and number-theoretic mechanisms have been developed to construct graphs with good expansion properties, making this area one of the best developed fields of discrete mathematics. For an introduction see P. Sarnak's *Some Applications of Modular Forms* [179]. The resulting graphs are known as **magnifiers, expanders, dispersers** but there are even more variations of this concept. The starting point is a link to Eigenvalues.

Let Γ be a connected graph on n vertices. The **Laplacian** Δ of Γ is described as a real symmetric (n, n) matrix whose rows and columns are indexed by the vertices. It has the degrees (numbers of neighbors) of the vertices on the main diagonal and entries 0 or -1 off the main diagonal. The entry is -1 if the corresponding pair of vertices forms an edge, 0 otherwise. As we know from Chapter 17, a symmetric matrix describes a quadratic form. The quadratic form described by the Laplacian is positive semidefinite. The Eigenvalue 0 has multiplicity 1, as Γ is connected. The main observation concerns then the second-smallest Eigenvalue: if it is far away from 0, then the graph has all sorts of extremal properties, for example, high connectivity. Most importantly in our context, Γ has good expansion properties if and only if the second-smallest Eigenvalue is much larger than 0. **Ramanujan graphs** are defined as graphs whose second Eigenvalue meets an asymptotic bound. In a celebrated paper [136], Lubotzky, Phillips and Sarnak construct Ramanujan graphs from the simple groups $PSL(2, q)$. These Ramanujan graphs need not be bipartite. There are mechanisms that allow the construction of unbalanced bipartite expanding graphs, as they are needed for the standard construction out of those Ramanujan graphs. One method is to identify the vertices of the underlying graph with the right vertices of the bipartite graph to be constructed, and to let the left vertices correspond to paths of a certain fixed length. It is explained in Sipser and Spielman [192] how an application of the basic construction to this situation, with auxiliary codes meeting the Gilbert-Varshamov bound, leads to asymptotically nontrivial families of codes.

LDPC codes

Another fruitful variant of this idea is much older. It goes back to Gallager [88]. The starting point is the observation that codes constructed by the bipartite graph method possess effective simple decoding algorithms. The best known variant of these algorithms is known nowadays as **belief propagation.** It is important that the number of edges is small. In the simplest and most effective variant, the inner code is the sum zero code, so each right vertex simply describes a parity check. In order to have a small number of edges, the degree of the vertices is seen as a small constant. Such codes are known as **low density parity check codes** or short LDPC codes. It is clear that each linear code can be described by a bipartite graph in various ways. The LDPC condition says that the dual code must be generated by words of small weight. The second idea behind LDPC codes is the use of randomness. If an LDPC code is constructed at random, subject to the degree constraints, there is a high probability that it has good properties, leading to good error correction, even though the minimum distance is not necessarily high. These ideas led to what are known as **turbo codes,** construction procedures for codes that almost achieve channel capacity. Recent constructions for such capacity achieving families use an additional degree of liberty. The bipartite graphs need not be regular. Rather, the degrees should be chosen in some very special way; see A. Shokrollahi [187].

23.3 tms-nets

tms-nets were defined by Niederreiter [157], based on earlier work [195] by Sobol, in the context of quasi-Monte Carlo methods of numerical integration. Close links to combinatorial structures had been noticed right from the start and were further explored by Lawrence [131], Mullen and Schmid [155, 180], among others. Our description is inspired by Rosenbloom and Tsfasman [177]. It is best to start from a technical description as the close link to orthogonal arrays will become obvious.

Let $\Omega = \Omega^{(T,s)}$ be a set of Ts elements, partitioned into s **blocks** $B_i, i = 1, 2 \ldots, s$, where $B_i = \{\omega_1^{(i)}, \ldots, \omega_T^{(i)}\}$. Each block carries a total ordering:

$$\omega_1^{(i)} < \omega_2^{(i)} \cdots < \omega_T^{(i)}.$$

This gives Ω the structure of a partially ordered set, the union of s totally ordered sets (chains) of T points each. We consider Ω as a basis of a Ts dimensional vector space $\mathbb{F}_q^{(T,s)}$. An **ideal** in Ω is a set of elements closed under predecessors. An **antiideal** is a subset closed under followers, equivalently, the complement of an ideal.

Visualize elements $x = (x_j^{(i)}) \in \mathbb{F}_q^{(T,s)}, i = 1, \ldots, s; j = 1, \ldots, T$ as matrices with T rows and s columns. The **Rosenbloom-Tsfasman metric** is defined as follows: if, in each block, the leading zeroes evaporate, the number of remaining cells is the **weight** $\rho(x)$. The **distance** between x and y is then $\rho(x-y)$. The **minimum weight** (minimum distance) of a subspace $\mathcal{C} \subseteq \mathbb{F}_q^{(T,s)}$ is the minimum among the weights of its nonzero members.

A linear subspace (code) $\mathcal{C} \subseteq \mathbb{F}_q^{(T,s)}$ has **strength** $k = k(\mathcal{C})$ if k is maximal such that the projection from $\mathcal{C}$ to any ideal of size k is onto. We also call such a subspace an **ordered orthogonal array** (OOA), which is q linear, has **length** s, **depth** T, **dimension** $m = dim(\mathcal{C})$ and strength k. Duality is defined with respect to the symmetric bilinear form

$$\langle x, y \rangle = \sum_{i=1}^{s} x_1^{(i)} y_T^{(i)} + x_2^{(i)} y_{T-1}^{(i)} + \cdots + x_T^{(i)} y_1^{(i)}.$$

These notions are justified by the following generalization of the duality principle for linear codes.

23.1 Theorem. *Let $\mathcal{C} \subseteq \mathbb{F}_q^{(T,s)}$ be a linear subspace (code). Then*

$$\rho(\mathcal{C}^\perp) = k(\mathcal{C}) + 1.$$

Call $\mathbb{F}_q^{(T,s)}$ with these notions of duality, strength and distance the **NRT space**, which is short for Niederreiter-Rosenbloom-Tsfasman space. Observe that, in the case $T = 1$, we obtain the familiar Hamming space of coding theory.

In order to understand the application to quasi-Monte Carlo methods of numerical integration, interpret $x \in \mathbb{F}_q^{(T,s)}$ as a point in the s-dimensional unit cube by reading the $x_j^{(i)}$ for fixed i as the T first digits of the $q-$ary expansion of a real number between 0 and 1. As an example, the point $\begin{array}{|c|c|c|c|} \hline 0 & 0 & 1 & 1 \\ \hline 1 & 1 & 1 & 0 \\ \hline 1 & 0 & 0 & 1 \\ \hline \end{array}$ in $\mathbb{F}_2^{(3,4)}$ is mapped to the point $(\frac{3}{8}, \frac{1}{4}, \frac{3}{4}, \frac{5}{8}) \in [0,1)^4$. This also motivates the hierarchical ordering inside the blocks.

The aim is to construct point sets in the unit cube $[0,1)^s$ of s-dimensional Euclidean space, which in a sense are almost uniformly distributed. Roughly speaking, the idea of the Monte Carlo method is to choose this set of points at random, whereas quasi-Monte Carlo tries to reach or even beat the performance of a random choice by a deterministic choice.

A code of strength k in NRT space corresponds to a set of points in the s-dimensional unit cube, which is almost uniformly distributed: we control the first k digits in q-adic expansion. This is why **tms-nets** are the most important objects in this theory. By definition, a linear $(t, m, s)_q$-net is equivalent to an m-dimensional code $\mathcal{C} \subseteq \mathbb{F}_q^{(k,s)}$ of strength k, where $t = m - k$.

As NRT space is a generalization of Hamming space, and NRT space of depth 1 is identical to Hamming space, it is a natural strategy to generalize methods from coding theory. This is the point of view taken in [177] and in various other papers. See [27] for generalizations of the Gilbert-Varshamov bound and the $(u, u+v)$ construction as well as for a useful bibliography. The central parameter is the strength. This is why tms-nets should be considered as generalizations of linear orthogonal arrays. Sometimes it is profitable to take the dual point of view and make use of the distance in NRT space. As an example, this happens in the proof of the $(u, u + v)$ construction.

A natural problem is **net embeddability:** given an m-dimensional linear orthogonal array of strength k and length s (the dual of an $[s, s - m, k + 1]_q$ code if $s > m$), is it possible to construct an m-dimensional OOA of strength k in NRT space of depth k (equivalently, a tms-net) which projects to the given OA? Results of the Gilbert-Varshamov type give affirmative answers to this question based only on the numerical values of the parameters. It is desirable to obtain explicit constructions instead. Some explicit embeddings of BCH codes are in [21, 79].

Where bounds are concerned, the Rao bound on orthogonal arrays was generalized to tms-nets by Martin and Stinson [146]. Martin [145] finally succeeded in generalizing the linear programming bound as well. It was an important insight that the algebraic-geometric construction of linear codes generalizes to tms-nets; see Niederreiter and Xing [158, 159].

A net embedding

We give an example of a net embedding of binary BCH codes. For an arbitrary natural number r, let $q = 2^r$ and start from the binary BCH code of length $s = q + 1$ with $A = \{0, 1\}$ as the defining set of exponents. Here we use the theory developed in Chapter 13. As the Galois closure of A contains the interval $\{-2, -1, 0, 1, 2\}$, the corresponding BCH codes have parameters $[2^r + 1, 2^r - 2r, 6]_2$, by the BCH bound. In other words, the set of vectors $(w, 1) \in \mathbb{F}_2^{2r+1}$, where w varies through the s-th roots of unity in $F = \mathbb{F}_{q^2}$, is 5-independent. Here we view $w \in F$ as an element of $\mathbb{F}_2^{2r}$. The idea of net embedding is to use the $(w, 1)$ as first elements of block B_w and to search for a second, ... , fifth element such that a family of strength 5 in $\mathbb{F}_2^{(5,s)}$ is obtained. We have not been able to construct this net embedding. However, we do obtain a net embedding after introducing an additional coordinate.

23.2 Definition. *Let W be the subgroup of order $s = q + 1$ of $F = \mathbb{F}_{q^2}$. Choose $\alpha \in \mathbb{F}_q \setminus \mathbb{F}_2$ such that $\alpha^2 + 1$ is not of the form $w' + 1/w'$ for $w' \in W$. For each $w \in W$, define the block $X(w)$ as follows:*

$$X_1(w) = (w, 1, 0), \ X_2(w) = (\alpha w, 0, 1), \ X_3(w) = (\mathbf{0}, 1, 1),$$

$$X_4(w) = (\mathbf{0}, 0, 1) \text{ and } X_5(w) = X_1(w')$$

for some $w' \neq w$.

The reason for the choice of α will become apparent in the proof. We have to prove that any family F of five of the vectors $X_i(w) \in \mathbb{F}_2^{2r+2}$ is linearly independent provided F consists of the first n_1 vectors from some block, the first n_2 vectors from some other block and so forth, where $n_1 + n_2 + \cdots = 5$. Order the n_i such that $n_1 \geq n_2 \geq \ldots$ and call $(n_1, n_2, \ldots)$ the **type** of family F. The possible types are

$$(1,1,1,1,1), \ (2,1,1,1), \ (2,2,1), \ (3,1,1), \ (3,2), \ (4,1), \ (5).$$

As we chose the BCH code as the point of departure, type $(1,1,1,1,1)$ is independent. The last coordinate shows that type $(2,1,1,1)$ is independent as well. Also, by the choice of $X_5(w)$, type (5) reduces to type $(4,1)$. It suffices to prove that families of types $(2,2,1)$ through $(4,1)$ are independent.

- Type $(2,2,1)$

Assume there is a linear combination of

$$(\alpha w_1, 0, 1), \ (\alpha w_2, 0, 1), \ (w_1, 1, 0), \ (w_2, 1, 0), \ (w_3, 1, 0)$$

with coefficients $\lambda_1, \ldots, \lambda_5$. As type $(2,1,1,1)$ has been considered already, we can assume $\lambda_1 = \lambda_2 = 1$. The middle coordinate shows $\lambda_3 + \lambda_4 + \lambda_5 = 0$. Assume at first $\lambda_5 = 0$. Clearly $\lambda_3 = \lambda_4 = 1$. The first coordinate section yields the contradiction $w_1 = w_2$. We can therefore assume $\lambda_3 = \lambda_5 = 1, \lambda_4 = 0$. The equation is

$$(\alpha + 1)w_1 + \alpha w_2 + w_3 = 0.$$

Multiplication by w_1^{-1} shows that we can assume $w_1 = 1$. We have $\alpha + 1 = \alpha w_2 + w_3$. Raising to power q yields $\alpha + 1 = \alpha/w_2 + 1/w_3$, after multiplication $\alpha^2 + 1 = \alpha^2 + 1 + \alpha(w_2/w_3 + w_3/w_2)$. Let $x = w_2/w_3$. Then $1 \neq x \in W$ and $x + 1/x = 0$. This yields the contradiction $x^2 = 1$, hence $x = 1$.

- Type $(3, 2)$

The first coordinate section shows that there is no nontrivial linear combination of

$$(\mathbf{0}, 1, 1), \ (\alpha w_1, 0, 1), \ (\alpha w_2, 0, 1), \ (w_1, 1, 0), \ (w_2, 1, 0).$$

- Type $(3, 1, 1)$

Consider a linear combination of

$$(\mathbf{0}, 1, 1), \ (\alpha w_1, 0, 1), \ (w_1, 1, 0), \ (w_2, 1, 0), \ (w_3, 1, 0).$$

Clearly $\lambda_1 = 1$. The last coordinate shows $\lambda_2 = 1$. The first coordinate section shows $\lambda_4 = \lambda_5 = 1$. The middle coordinate yields $\lambda_3 = 1$. We have $(\alpha + 1)w_1 = w_2 + w_3$. As before, we can assume $w_1 = 1$. Raising to power $q + 1$, we obtain $(\alpha + 1)^2 = w_2/w_3 + w_3/w_2$. Let $x = w_2/w_3$. By choice of α a contradiction is reached.

Type $(4, 1)$ is easy to check.

In order to complete the proof, it remains to show that α can be chosen as required in Definition 23.2. This is left to the reader; see Exercise 23.3.1.

We have shown that there is a $(2r + 2)$-dimensional subspace of $\mathbb{F}_2^{(5,s)}$ of strength 5. In the tms terminology this is a linear $(2r - 3, 2r + 2, 2^r + 1)_2$ net. It gives us a set of 2^{2r+2} points in the unit cube $[0, 1)^s$, which in some technical sense are uniformly distributed. The quality of the distribution is measured by the strength $m - t$, which in our case is 5.

Exercises 23.3

23.3.1. *Prove that α can be chosen as required in Definition 23.2.*

23.4   Sphere packings

The theory of sphere packings in Euclidean space can be seen in analogy to coding theory, in a different context. Both settings of the problem of reliable information transmission are discussed right from the start, in Shannon's treatise [183]. In the case of the **Gaussian channel,** everything happens in n-dimensional Euclidean space $E = \mathbb{R}^n$ (n-tupels of real numbers with Euclidean distance). Codes in this space are discrete subsets $\Gamma \subset \mathbb{R}^n$. In the analogy we use Euclidean distance in the place of Hamming distance. The basic parameter corresponding to the minimum distance for a code is the minimum Euclidean distance between any two different points from Γ. In order to avoid square roots, it is customary to use the **norm** $\mu(\Gamma)$, the square of the minimum distance, instead. We imagine each point from $\Gamma \subset E$ as the center of a sphere (or ball) of radius $\rho = \sqrt{\mu}/2$, the maximum radius such that these balls may touch one another but do not penetrate each other, much like billiard balls in dimension n. This is why such a set Γ is called a **sphere packing.** The parameter ρ is the **packing radius.**

The second basic parameter, corresponding to the number of codewords in Hamming space, is a little harder to describe. In order to do that, we restrict at first to a special class of sphere packings. These **lattice packings** may be seen as analogs of linear codes. They are defined as follows: pick a basis $B = \{v_1, v_2, \ldots, v_n\}$ of E. The lattice $\Gamma(B)$ generated by B consists of all linear combinations with integer coefficients:

$$\Gamma(B) = \{\sum_{i=1}^{n} \lambda_i v_i | \lambda_i \in \mathbb{Z}\}.$$

Observe that the origin is contained in each lattice. In number theory, $\Gamma(B)$ would be called an integer lattice. Just as in the case of linear codes, the minimum distance between any two lattice points equals the minimum distance between a nonzero lattice point and the origin. Also, a lattice is itself a group, where the group operation is addition.

The first case of interest is $n = 2$. The best of all two-dimensional lattices is the **hexagonal lattice** $\mathcal{H}$. The idea is: use the points of a regular hexagon to generate the lattice. As $cos(60°) = 1/2$, these points are, in counterclockwise order,

$$(1,0), \ (1/2, \frac{\sqrt{3}}{2}), \ (-1/2, \frac{\sqrt{3}}{2}),$$

$$(-1,0), \ (-1/2, -\frac{\sqrt{3}}{2}), \ (1/2, -\frac{\sqrt{3}}{2}).$$

Let $v_1 = (1,0), v_2 = (-1/2, \sqrt{3}/2))$. The points of the hexagon are then

$$v_1, \ v_1 + v_2, \ v_2, \ -v_1, \ -v_1 - v_2, \ -v_2.$$

In particular, they are all contained in the lattice $\mathcal{H}$ generated by v_1 and v_2. The minimum norm of $\mathcal{H}$ is of course $\mu = 1$. Consider the **fundamental parallelotope** $\{c_1 v_1 + c_2 v_2 | 0 \leq c_i < 1\}$ spanned by v_1 and v_2. The **volume** $vol(\Gamma(B))$ is defined as the volume of the fundamental parallelotope. In the case of the hexagonal lattice, the volume is $\sqrt{3}/2$, the same as the determinant of the $(2,2)$ matrix $\begin{pmatrix} 1 & 0 \\ -1/2 & \sqrt{3}/2 \end{pmatrix}$ with v_1, v_2 as rows. This is not a coincidence. It is part of the idea behind the definition of the determinant. The absolute value of the determinant is a volume factor: an area of volume V is mapped by M to an area of volume $V |det(M)|$. It follows that $vol(\Gamma)$ is a measure for the density of Γ. Small volume means high density. We can think of $1/vol(\Gamma)$ as kind of an analog for the number of codewords. The objective is to construct lattices (or, more general, sphere packings) of small volume and large minimum distance.

The situation is, however, different from the basic problem of coding theory in Hamming space as we have the possibility of scaling sphere packings. In fact, multiplication by a positive real number a produces a lattice $a\Gamma$ which we want to consider as equivalent to Γ. We can therefore decide to work with lattices of volume 1. The problem is then to maximize the minimum distance. Scaling indicates how to define the figure of merit we are interested in. Under multiplication by a, the volume picks up a factor of a^n. As we consider Γ and $a\Gamma$ as equivalent, the basic parameter should be defined as $\rho(\Gamma)^n/vol(\Gamma)$. This is the **center density.**

23.3 Definition. *A lattice $\Gamma \subset \mathbb{R}^n$ consists of all linear combinations with integer coefficients of a basis $\{v_1, v_2, \ldots, v_n\}$ of $E = \mathbb{R}^n$. The **volume** $vol(\Gamma)$ is the volume of the fundamental parallelotope spanned by the v_i, equivalently, the absolute value of the determinant of the (n,n) matrix having the v_i as rows. The **packing radius** $\rho(\Gamma)$ is half the minimum Euclidean distance between two different lattice points.*

*The **center density** is*

$$\delta(\Gamma) = \rho(\Gamma)^n/vol(\Gamma). \tag{23.1}$$

The basic problem is the construction of an n-dimensional lattice with maximal center density.

Consider the hexagonal lattice. We constructed it such that $\rho = 1/2$. The volume is $\sqrt{3}/2$. It follows that its center density is $(1/4)/(\sqrt{3}/2) = 1/(2\sqrt{3}) \approx 0.28868$. The hexagonal lattice is the uniquely determined densest packing in dimension 2. Even in dimensions 2 and 3 these problems are not easy to solve. The maximum density of a packing in three-dimensional space was determined only in 1998. Some of the history and an infinity of anecdotes are in George G. Szpiro's book *Kepler's Conjecture* [204]. In all higher dimensions the maximum packing density is unknown.

Lattices and quadratic forms

In Chapter 17 we studied bilinear and quadratic forms over finite fields. Lattices in Euclidean space can be represented by quadratic forms as well. This link to classical number theory is a main reason why the area is rather well developed.

Let Γ be the lattice described by the basis $v_1, \ldots, v_n$. The elements of Γ are $\sum_{i=1}^{n} x_i v_i$, where $x_i \in \mathbb{Z}$. They are parametrized by the n-tuples $x = (x_1, \ldots, x_n) \in \mathbb{Z}^n$. The norm (square of the distance from the origin) of this lattice element is

$$\left(\sum_i x_i v_i\right) \cdot \left(\sum_i x_i v_i\right) = \sum_{i,j} x_i x_j v_i \cdot v_j.$$

Let M be the matrix with the v_i as rows and $A = MM^t$. The entries of A are $a_{i,j} = v_i \cdot v_j = a_{j,i}$. In particular, A is a symmetric matrix and we know that the theory of symmetric matrices is the theory of quadratic forms. In fact, the norm of our lattice point can be written as $Q(x) = xAx^t$, a quadratic form. The norms of lattice points are precisely the values of the quadratic form Q applied to integer vectors x.

In the case of the hexagonal lattice we took $v_1 = (1,0)$, $v_2 = (-1/2, \sqrt{3}/2))$ as basis, hence $M = \begin{pmatrix} 1 & 0 \\ -1/2 & \sqrt{3}/2 \end{pmatrix}$ and $A = MM^t = \begin{pmatrix} 1 & -1/2 \\ -1/2 & 1 \end{pmatrix}$. The corresponding quadratic form is

$$Q(x_1, x_2) = (x_1, x_2)A(x_1, x_2)^t = x_1^2 + x_2^2 - x_1 x_2.$$

The equivalent bilinear form is then given by

$$\langle x, y \rangle = xAy^t = x_1 y_1 + x_2 y_2 - \frac{1}{2} x_1 y_2 - \frac{1}{2} x_2 y_1.$$

As we are free in the choice of a basis of the lattice, it gives rise to many different quadratic forms. This leads to a notion of equivalence among integral quadratic forms.

If the quadratic form Q arises from an integral lattice, we have that $Q(x)$, being a norm, is always nonnegative, and it is 0 if and only if $x = 0$. In other words, these quadratic forms are positive definite.

The E_8 lattice

There is a close connection between codes and sphere packings. We illustrate with a famous eight-dimensional lattice, which is known as the E_8 lattice. The name derives from a link with group theory which we do not want to explore here.

Remember one of our favorite codes, the extended Hamming code $[8, 4, 4]_2$. Denote it by $\mathcal{C}$ for the moment. The lattice Γ will be a sublattice of $\mathbb{Z}^8$. If, in each coordinate, we calculate mod 2, we obtain $\mathbb{F}_2^8$ ($\mathbb{Z}^8/(2\mathbb{Z})^8 \cong \mathbb{F}_2^8$). In this way each 8-tuple $x = (x_1, x_2, \ldots, x_8)$ gives us a word $\overline{x} = (\overline{x_1}, \ldots, \overline{x_8}) \in \mathbb{F}_2^8$.

The E_8 lattice is defined as follows:

$$\Gamma = \{x \in \mathbb{Z}^8 | \overline{x} \in \mathcal{C}\}.$$

In particular all the information is contained in the code.

We simply represent the ambient vector space $\mathbb{F}_2^8$ as the top layer $\mathbb{Z}^8/(2\mathbb{Z})^8$ of the integer lattice $\mathbb{Z}^8$ and use as lattice points of Γ those points which project to a codeword of $\mathcal{C}$. This is the most direct link between codes and lattices one could imagine. It is a special case of what is known as **construction A.**

What are the basic parameters of the E_8 lattice Γ? Clearly, $\mathbb{Z}^8$ has volume 1, $(2\mathbb{Z})^8$ has volume 2^8 and Γ, being a union of 2^4 cosets of $(2\mathbb{Z})^8$, has volume $2^8/2^4 = 16$. As $\mathcal{C}$ is a linear code, Γ is an additive subgroup, hence a lattice. The smallest norm of a nonzero point in $(2\mathbb{Z})^8$ is 4. There are 16 such points. Each coset of a codeword of weight 4 contains vectors of the type $(\pm 1)^4 0^4$, of norm 4, and there are 16 such minimum norm points in each such coset. The points in the coset given by the all-1 word all have norm ≥ 8. We conclude that the minimum norm of Γ is 4, and there are exactly $16 + 14 \times 16 = 240$ points of norm 4 in Γ. The minimum distance is therefore 2, and the packing radius is $\rho = 1$. It follows that the center density is $\delta = 1/16 = 0.0625$. The corresponding integer quadratic form Q in eight variables looks rather complicated. To prove that Q is positive definite, has 4 as its minimum nonzero value and that there are precisely 240 integer vectors x such that $Q(x) = 4$ would require lengthy direct calculations. Using the coding approach, this was easy enough for us.

Coset codes

We describe a generalization of construction A. In general, it produces sphere packings which are not necessarily lattices. The construction is known by the relatively absurd name **coset codes.** The first ingredient is a chain

$$\mathcal{A}_0 \supset \mathcal{A}_1 \supset \cdots \supset \mathcal{A}_l$$

of m-dimensional lattices. The factor groups $A_i = \mathcal{A}_{i-1}/\mathcal{A}_i$ are therefore finite. Denote by a_i the order of this factor group.

The second ingredient consists of a_i-ary codes C_i, all of the same length n, with M_i codewords and minimum distance d_i. For each i choose representatives $\alpha_{ij}, j = 1, \ldots, a_i$ for the cosets of $\mathcal{A}_i$ in $\mathcal{A}_{i-1}$. Choose $\alpha_{i1} = 0$. Use $B_i = \{\alpha_{ij}, j = 1, \ldots, a_i\}$ as the alphabet over which code C_i is defined. It is convenient to assume that the all-0 word belongs to C_i.

The $N = nm$-dimensional packing

$$\Gamma = \Gamma(\mathcal{A}_0 \supset \mathcal{A}_1 \supset \cdots \supset \mathcal{A}_l; C_1, C_2, \ldots, C_l)$$

is defined as the union of $M_1 M_2 \ldots M_l$ cosets of the sublattice $(\mathcal{A}_l)^n$. The cosets are parametrized by l-tupels of codewords $(v_1, v_2, \ldots, v_l)$, with $v_i \in C_i$. Let $v_i = (v_{i1}, \ldots, v_{in})$, where $v_{ij} \in B_i$. Then the coset $N(v_1, v_2, \ldots, v_l)$ is defined as

$$N(v_1, v_2, \ldots, v_l) = (\sum_{i=1}^{l} v_{ij})_{j=1}^{n} + (\mathcal{A}_l)^n.$$

Observe that $N(0, 0, \ldots, 0) = (\mathcal{A}_l)^n$.

Let us see what we can say about the basic parameters of this sphere packing. First of all, we are in dimension $N = nm$. The sublattice $(\mathcal{A}_l)^n$ has volume $vol(\mathcal{A}_l)^n$. Our sphere packing is defined as the union of $M_1 M_2 \ldots M_l$ cosets of this sublattice. As all these cosets have the same volume, it is clear that the volume of Γ is

$$vol(\Gamma) = \frac{vol(\mathcal{A}_l)^n}{M_1 \ldots M_l}.$$

How about the minimum distance?

If $x, y \in \Gamma$ belong to the same coset, then their difference is in $(\mathcal{A}_l)^n$. It follows $\|x - y\| \geq \sqrt{\mu(\mathcal{A}_l)}$. So assume x, y are in different cosets, $x \in N(v_1, v_2, \ldots, v_l)$, $y \in N(v_1', v_2', \ldots, v_l')$ and i minimal such that $v_i \neq v_i'$. As C_i has minimum distance d_i, it follows that $x - y$ has in d_i of its n components an entry in $\mathcal{A}_{i-1} \setminus \mathcal{A}_i$. It follows $\|x - y\| \geq \sqrt{d_i \cdot \mu(\mathcal{A}_{i-1})}$. We can calculate the center density.

Construction A as used for the E_8 lattice is an extremal special case, where $m = 1$, the chain of lattices is $\mathcal{A}_0 = \mathbb{Z} \supset \mathcal{A}_1 = 2\mathbb{Z}$ and there is just one binary code. The coset code construction was first introduced by Forney [85]. We follow the description from [28].

The Kschischang-Pasupathy lattice

The method of coset codes can also be applied in case $m = 2$. It is natural then to base oneself on the hexagonal lattice $\mathcal{H}$. The sublattice $2\mathcal{H}$ has index 4 in $\mathcal{H}$. Its volume is therefore $4vol(\mathcal{H}) = 2\sqrt{3}$, and the minimum norm (squared distance from the origin) is 4. Consider the sublattice $\mathcal{L}$ of $\mathcal{H} = \langle v_1, v_2 \rangle$ generated by $3v_1 = (3, 0)$ and $v_2 + 2v_1 = (3/2, \sqrt{3}/2)$. As it is obtained by application of a linear mapping with determinant 3 to $\mathcal{H}$, its volume is $3vol(\mathcal{H}) = 3\sqrt{3}/2$. The minimum norm of this lattice is the norm of $v_2 + 2v_1$, which is 3. As our first ingredient, we choose the chain

$$\mathcal{A}_0 = \mathcal{H} = \langle v_1, v_2 \rangle \supset \mathcal{A}_1 = 2\mathcal{H} \supset \mathcal{A}_2 = \langle 6v_1, 2v_2 + 4v_1 \rangle.$$

Observe that $\mathcal{A}_2 = 2\mathcal{L}$. The volumes of our chain of lattices are $\sqrt{3}/2, \ 2\sqrt{3}, \ 6\sqrt{3}$, the packing radii are $1/2, \ 1, \ \sqrt{3}$.

We have $a_1 = 4$, $a_2 = 3$. In order to obtain a 36-dimensional sphere packing, we need a quaternary code C_1 and a ternary code C_2, both of length 18. It is easy to see that a lattice will be obtained if C_1 and C_2 are linear codes. In the case of C_1, it is sufficient for this purpose that C_1 is additive.

We choose C_2 to be the ternary sum-0 code $[18, 17, 2]_3$. For C_1 we use a linear $[18, 9, 8]_4$ code (see also MacWilliams et al. [143]). Consider the cyclic quaternary codes of length 17 (see our theory in Chapter 13). The cyclotomic cosets are

$$\{0\}, \{1, 4, 13, 16\}, \{2, 8, 9, 15\}, \{3, 5, 12, 14\}, \{6, 7, 10, 11\}.$$

The intervals $\{6, 7, 8, 9, 10, 11\}$ and $\{12, 14, 16, 1, 3, 5\}$ show that the cyclic codes $\mathcal{C}(C(2) \cup C(6))$ and $\mathcal{C}(C(1) \cup C(3))$ are $[17, 9, 7]_4$ codes. The Roos bound (see Exercise 13.2.10), can be used to show that $\mathcal{C}(\{0\} \cup C(1) \cup C(3))$ has $d = 8$. In fact, $I = \{13, 14\}$ is an interval, $J = \{0, 4, 8, 16, 3, 7\}$ is an interval minus a point (12) and $I + J = \{0\} \cup C(1) \cup C(3)$. We have a chain $[17, 8, 8]_4 \subset [17, 9, 7]_4$. Application of construction X produces an $[18, 9, 8]_4$ code.

As $\mathcal{A}_2$ has volume $6\sqrt{3}$ and the Kschischang-Pasupathy lattice Γ is the union of $M_1 M_2 = 4^9 3^{17}$ cosets of $\mathcal{A}_2^{18}$, it has volume

$$vol(\Gamma) = 6^{18} 3^9 / (4^9 3^{17}) = 3^{10}.$$

How about the packing radius ρ? Let us go over the argument from the preceding subsection again. Let $x, y \in \Gamma, x \neq y$.

If x, y are in the same coset of $\mathcal{A}_2^{18}$, their distance is at least the minimum distance of $\mathcal{A}_2$, which is $2\sqrt{3}$.

Let $x \in N(v_1, v_2)$, $y \in N(v_1', v_2')$. If $v_1 \neq v_1'$, it follows that $x - y$ has at least $d_1 = 8$ coordinates with nonzero entries from $\mathcal{A}_0$. The distance between x and y is at least $\sqrt{8 \times 1} = 2\sqrt{2}$.

If, finally, $v_1 = v_1'$, but $v_2 \neq v_2'$, then $x - y$ has at least $d_2 = 2$ coordinates with nonzero entries from $\mathcal{A}_1$. In this case, x, y are at distance $\geq \sqrt{2 \times 4} = 2\sqrt{2}$.

We conclude that Γ has minimum distance $2\sqrt{2}$ and packing radius $\sqrt{2}$. It follows from Equation 23.1 that the center density is

$$\delta(\Gamma) = \frac{\rho^{36}}{vol(\Gamma)} = 2^{18} / 3^{10}.$$

The original paper is Kschischang and Pasupathy [129].

The Leech lattice

The 24-dimensional Leech lattice Λ is an extraordinary object. It is closely related to the extended binary Golay code $[24, 12, 8]_2$. We will construct the Leech lattice by a variant of the coset code construction. Start from the one-dimensional chain

$$\mathbb{Z} \supset 2\mathbb{Z} \supset 4\mathbb{Z} \supset 8\mathbb{Z}$$

where we use $\{0, 1\}$ as representatives for $2\mathbb{Z}$ in $\mathbb{Z}$, $\{0, -2\}$ as representatives of $4\mathbb{Z}$ in $2\mathbb{Z}$ and $\{0, 4\}$ as representatives of $8\mathbb{Z}$ in $4\mathbb{Z}$.

Consider the repetition code $\{0, 1\}$ in the top layer, the extended binary Golay code $\mathcal{C}$ with parameters $[24, 12, 8]_2$ in the middle layer (over the alphabet $\{0, -2\}$) and the sum-0 code $[24, 23, 2]_2$ in the third layer, with alphabet $\{0, 4\}$. With the same terminology as in the coset code construction, we define Λ as the union of the following cosets of $(8\mathbb{Z})^{24}$ in $\mathbb{Z}^{24}$:

- $N(0, v, w)$, where $v \in \mathcal{C}$ and w has even weight,

- $N(1, v, w)$, where $v \in \mathcal{C}$ and w has odd weight.

It is left as an exercise to show that this set of cosets is closed under addition. This shows that Λ is a lattice. As we have $2^{12}2^{24}$ cosets, it follows $vol(\Lambda) = 8^{24}/2^{36} = 2^{36}$.

Nonzero points in $(8\mathbb{Z}^{24})$ have length at least 8. The shortest length of a point in $N(0, 0, w)$ where $w \neq 0$ is $\sqrt{32}$. These are the vectors with entries ± 4 in two coordinates, 0 in all other coordinates. The shortest length of a point in $N(0, v, w)$ where $v \neq 0$ is $\sqrt{32}$. These are obtained when $wt(v) = 8$ and the support of w is contained in the support of v. Finally, the shortest length of vectors in some $N(1, v, w)$ is $\sqrt{32}$ as well. In order to obtain such vectors, we can use $v \in \mathcal{C}$ arbitrary and w of weight 1.

In particular, the minimum distance of Λ is $4\sqrt{2}$, and the packing radius is therefore $\rho = 2\sqrt{2}$. Equation 23.1 shows that the Leech lattice has center density $\delta = 1$.

The Leech lattice was constructed by J. Leech [132]. Our construction follows Vardy [214, 215] and [28]. The twisted version of the coset code construction can be generalized to produce dense sphere packings in certain dimensions. The densest known packings in all even dimensions between 18 and 30 are of this type.

The kissing number

The Newton number or kissing number problem is a local variant of the sphere packing problem. Fix a sphere of radius a in Euclidean n-space. The Newton number τ_n is the maximal number of spheres of radius a which touch the given sphere and do not penetrate each other. Because of scaling, this does not depend on the radius a. Consider an n-dimensional integer lattice. Each point has the same number of closest neighbors. This number is then a lower bound on τ_n. The E_8 lattice and the Leech lattice are exceptional even in this respect. In fact, we saw that each point in the E_8 lattice has 240 neighbors. It can be proved that higher kissing numbers are impossible: $\tau_8 = 240$.

Even more impressive is the behavior of the Leech lattice. It can be proved that $\tau_{24} \leq 196,560$. Count the closest neighbors of the origin in the Leech lattice. We saw three types of points of lowest norm 32 in the preceding subsection. We count $4 \times \binom{24}{2} = 1104$ such points of type $((\pm 4)^2 0^{22})$. As the number of minimum weight vectors of the Golay code is $A_8 = 759$, there are $759 \times 2^7 = 97,152$ minimum weight points in cosets $N(0, v, w)$, where $v \neq 0$. Finally, there are $24 \times 2^{12} = 98,304$ minimum weight points in cosets $N(1, v, w)$ ($v \in \mathcal{C}$ is arbitrary and w arbitrary of weight 1; each such coset contains precisely one minimum weight point). The grand total is $196,560$, miraculously. We conclude $\tau_{24} = 196,560$.

What is the analog of the kissing number problem in Hamming space, when we work in the space A^n of n-tuples with entries from a q-letter alphabet A instead of Euclidean n-space, using Hamming distance instead of Euclidean distance? Well, we fix a tuple which we have no reason not to choose as the zero tuple. All words of the code we are looking for must have the same weight w and their pairwise distance is lower bounded by d. In coding theory these are known as **constant weight codes.** Not surprisingly constant weight codes are essential ingredients in constructions of "kissing" configurations in Euclidean space. We did not cover constant weight codes in this text. One of the reasons is that not much is known on this topic, almost nothing in the nonbinary case. In Edel, Rains and Sloane [82] an elementary construction involving binary constant weight codes is used to construct kissing configurations in certain dimensions which are much better than what the known lattices yield.

Further links

We noted that lattices are number-theoretic objects with a rich history. There are even deeper links for certain subfamilies. Just as in the case of codes, duality is an important concept. Let Γ be an n-dimensional lattice. Its dual $\Gamma^\perp$ consists of all tupels $(y_1, \ldots, y_n)$ satisfying $x \cdot y \in \mathbb{Z}$ for all $x \in \Gamma$. We have $vol(\Gamma^\perp) = 1/vol(\Gamma)$. A lattice is **unimodular** if $\Gamma^\perp = \Gamma$. This notion is analogous to the notion of a self-dual linear code. Γ is **even** if $x \cdot x$ (the norm of x) is an even integer for all $x \in \Gamma$. If Γ is an even lattice, we are interested in the numbers w_m of lattice vectors having norm $2m$. Knowing the w_m for all m is analogous to the determination of the weight enumerator of a self-dual code. In the case of linear codes, we found it convenient to use the weight numbers for the construction of a polynomial, the weight polynomial, as it was possible then to describe the weight polynomial of the dual code by an algebraic operation, the MacWilliams transform. Something analogous happens in the case of lattices. This time the object defined by the w_m will be infinite, a power series. It is known as the **theta series** of Γ:

$$\Theta_\Gamma = \sum_{m=0}^{\infty} w_m q^m,$$

where q is a variable. Now write $q = e^{2\pi i \tau}$, where τ is a complex parameter. The theta series is now a complex function:

$$\Theta_\Gamma(\tau) = \sum_{m=0}^{\infty} w_m e^{2\pi i m \tau}.$$

Here we let τ vary in the complex upper half plane, the complex numbers with positive imaginary part. By construction we have $\Theta_\Gamma(\tau+1) = \Theta_\Gamma(\tau)$. Recall that unimodular lattices are in analogy with self-dual codes. In the case of self-dual codes, the MacWilliams formula is an involutory operation, which leaves the weight enumerator invariant. The same principle applies in the setting of lattices. The classical Poisson summation formula plays the role of the MacWilliams formula and we obtain an identity relating $\Theta_\Gamma(\tau)$ and $\Theta_\Gamma(-1/\tau)$. All in all one obtains an action of a group, which in a sense leaves the theta series invariant. This group is the **modular group** $SL(2, \mathbb{Z})$, the group of all 2×2 matrices with integer entries and determinant ± 1. Invariants under this kind of action of the modular group (or of subgroups of finite index) are known as **elliptic modular forms.** This is a classical and highly developed topic of number theory and complex analysis. The upshot is that modular forms form vector spaces of finite dimension. These dimensions are known and bases are known.

Return to the Leech lattice. After scaling such that the volume is 1 (factor $1/(2\sqrt{2})$ with respect to the representation given earlier), it is an even unimodular lattice of dimension 24, with minimum distance 2, hence minimum norm 4. In the present terminology we have $w_0 = 1, w_1 = 0, w_2 = 196,560$. The space of modular forms containing its theta series has dimension 2. One needs only two pieces of information to completely determine the theta series. It follows that the theta series is determined by $w_0 = 1, w_1 = 0$. The modular forms give the information $w_2 = 196,560$ for free, so to speak. The next coefficients of the theta series are $w_3 = 16,773,120$ and $w_4 = 398,034,000$. The combinatorial information contained in these numbers is so strong that it is easy to construct the Leech lattice on the basis of this information alone. It follows that the Leech lattice is the uniquely determined even unimodular 24-dimensional lattice without vectors of norm 2. What we just described is the approach used by John H. Conway in his celebrated paper [60].

This leads to yet another link, group theory.

Simple groups

A finite group with a nontrivial normal subgroup can be thought of as composed from the subgroup and the factor group. The "atoms" of this decomposition, the groups which cannot be further decomposed, are the groups without a nontrivial normal subgroup. These are known as **simple groups,** although their theory is far from being simple. We use the term "simple group" short for "nonabelian finite simple group." Many infinite series of simple groups are known. The alternating groups form one such family; the isometry groups of the various types of bilinear, sesquilinear and quadratic forms are essentially simple groups. A more general structure theory of classical simple groups is based on the classification of simple complex Lie algebras.

The classical simple groups can be described as algebraic groups related to those Lie algebras, defined over finite fields (these are the Chevalley groups) as well as fixed groups of those under certain symmetries (the Steinberg groups). The classical simple groups are also known as **groups of Lie type.** There is a general theoretical framework for the classical simple groups; see Roger W. Carter, *Simple Groups of Lie Type* [44].

What makes things more complicated and interesting is the presence of certain finite simple groups which are not classical. In 1982 it was announced that the classification of finite simple groups was complete. If we are inclined to believe this, the picture is as follows: aside from the classical simple groups there are exactly 26 "sporadic" simple groups. Most of those exotic creatures are highly interesting. The earliest sporadic groups are the Mathieu groups $M_{24}, M_{23}, M_{22}, M_{12}, M_{11}$; see E. Mathieu [150, 151]. The subscript indicates the degree of the smallest, the natural permutation representation. M_{24} is a subgroup of the symmetric group S_{24}. This simple group of order

$$|M_{24}| = 24 \times 23 \times 22 \times 21 \times 20 \times 48$$

is the automorphism group of the Golay code $[24, 12, 8]_2$ and of the large Witt design $S(5, 8, 24)$. The best way to understand M_{24} is in terms of $S(5, 8, 24)$, which itself is best understood in terms of the projective plane of order 4. In fact, start from a putative $S(5, 8, 24)$. Fix a set of three points (the points at infinity). The blocks containing all three points at infinity form an $S(2, 5, 21)$. Now $PG(2, 4)$ is an $S(2, 5, 21)$ and it is easy to see that it is the only $S(2, 5, 21)$. It is now a matter of combinatorial counting to check that the $S(5, 8, 24)$ can be constructed in exactly one way using the plane of order 4. The ingredients used are hyperovals in $PG(2, 4)$ (they yield blocks containing two points at infinity), Fano subplanes (those yield blocks with precisely one point at infinity) and lines of the $PG(2, 4)$. This shows that $S(2, 5, 24)$ exists and is uniquely determined. The same reasoning also shows that the group of automorphisms is 5-transitive and that the stabilizer of three points is the simple group $PSL(3, 4)$. This approach is used in Hughes and Piper [118].

The automorphism group of the ternary Golay code $[12, 6, 6]_3$ (the stabilizer under the action of the monomial group) is $Z_2 \times M_{12}$, as we saw in Chapter 17. Conway's construction of the Leech lattice as described above yields immediately the order of the automorphism group. What is more, after factoring out -1, a sporadic simple group is obtained, the large Conway group Co_1 of order

$$|Co_1| = 2^{21} \cdot 3^9 \cdot 5^4 \cdot 7^2 \cdot 11 \cdot 13 \cdot 23$$

(see Conway [59]). Some of the story leading from the origins of coding theory to the Leech lattice and Conway's groups is told in T. M. Thompson's book *From Error-Correcting Codes Through Sphere Packings to Simple Groups* [208].

23.5   Permutation codes

Here is yet another variant of the coding problem.

23.4 Definition. *A* **permutation code** *is a subset C of the symmetric group S_n on n letters, where S_n is equipped with the Hamming distance.*

Observe that we have two ways of writing permutations. One is the cycle notation, for example, $(1, 2, 4) \in S_4$, which maps $1 \mapsto 2 \mapsto 4 \mapsto 1, 3 \mapsto 3$. From a coding perspective it is more natural to write this permutation as a quadruple 2431. The (Hamming) distance $d(\pi, \pi')$ is the number of letters i such that $\pi(i) \neq \pi'(i)$.

Clearly, permutation groups are examples of permutation codes. We concentrate on large Hamming distances here. The largest distance is n. What is the largest cardinality of a permutation code $C \subset S_n$ of distance n?

23.5 Proposition. *The largest size of a permutation code $C \subset S_n$ of distance n is n. Such codes are equivalent to Latin squares of order n.*

PROOF   Imagine the elements of the code written as the rows of an array. Pairwise distance n means that in each column each entry appears at most once. This shows $|C| \leq n$. Clearly, the extremal case corresponds to Latin squares. We know that Latin squares exist for each order n, an example being the cyclic group. ⬚

This leads to another view on the Latin squares which we encountered in Section 3.4. We see them as sharply 1-transitive sets of permutations here.

What happens if we consider the coding problem in distance $n - 1$? This is in fact a much more interesting case. It leads to the existence problem for projective planes.

23.6 Definition. *A* **projective plane** *of order n is a geometry of points and lines satisfying the following:*

1. *There are $n^2 + n + 1$ points.*

2. *There are $n^2 + n + 1$ lines.*

3. *Each line has $n + 1$ points.*

4. *Each point is on $n + 1$ lines.*

5. *Any two different lines meet in precisely one point.*

6. *Any two different points are on precisely one common line.*

Observe that the axioms are chosen in a self-dual way: if we use the points of a projective plane of order n as the lines of a new structure and the lines of our projective plane as points of the new structure, then the new structure is a projective plane again: the dual of a projective plane is a projective plane.

The axioms given in Definition 23.6 are far from being minimal: a small subset suffices and the remaining properties can then be proved to hold. Case $n = 1$ is degenerate: this projective plane is a triangle. In fact, projective planes are also known as **generalized triangles.**

The terminology has been chosen such that the classical projective planes $PG(2, q)$ (points: the 1-dim subspaces of $\mathbb{F}_q^3$, lines: the 2-dim subspaces of $\mathbb{F}_q^3$) which we encountered repeatedly are indeed projective planes of order q. This solves the existence question of projective planes of order q (a prime-power). It is a famous open problem to decide if there are any projective planes whose order n is a composite number. It is generally conjectured that this is not the case. Definitively there are no projective planes of order $6, 10, 14, 22$. The smallest open problem concerns $n = 12$. As we consider only the coding problem relative to distance $n - 1$, we use the following.

23.7 Definition. *A permutation code $C \subset S_n$ is* **sharply 2-transitive** *if for any two pairs of letters $(i, j), (k, l)$ where $i \neq j, k \neq l$ there is precisely one element of C mapping $i \mapsto k, j \mapsto l$.*

23.8 Proposition. *The size of a distance $n - 1$ permutation code $C \subset S_n$ is at most $n(n - 1)$, with equality if and only if C is sharply 2-transitive. A sharply 2-transitive permutation code exists if and only if a projective plane of order n exists.*

PROOF Write our permutation code as an array again. The pairs of entries in the first two columns must be different in each row. This yields the upper bound $n(n - 1)$ and shows that the code is sharply 2-transitive in the case of equality. Assume now a projective plane of order n exists. Pick any two of its lines. They meet in a point ∞. Label the remaining points of the first line $P_1, \ldots, P_n$, and those on the second line $Q_1, \ldots, Q_n$. The number of points not on the union of those two lines is $n^2 + n + 1 - (1 + 2n) = n(n - 1)$. To each such point R, we associate a permutation as follows: if the line RP_i meets the second line in Q_j, then the permutation maps $i \mapsto j$. The geometric axioms of a projective plane show that the resulting permutation code is sharply 2-transitive. It is easy to see that this process can be reversed: a sharply 2-transitive permutation code on n letters allows the construction of a projective plane of order n. $\qquad\square$

Exercises 23.5

23.5.1. *The alternating group A_4 is sharply 2-transitive on 4 letters. Use it to construct a projective plane of order 4.*

23.6 Designs

Designs $t - (v, k, \lambda)$ and Steiner systems $S(t, k, v)$ were defined in Chapter 7, where we also saw the single most famous design, the large Witt design $S(5, 8, 24)$. It is closely related to the binary Golay code (see Witt's classical papers [221] and [222]) and to the Mathieu group M_{24}. We encountered many more examples of t designs in connection with codes, starting from the projective and affine geometries which are 2-designs. As t is the central parameter, it is a natural question to ask for which t it is possible to construct t-designs. The set of all k subsets of a v set clearly is a t design for all $t \leq k$. These designs are called **complete designs** and are not terribly interesting. Noncomplete designs are also called nontrivial.

In the definition of designs it is customary to admit also repeated blocks. In this terminology the designs of our Definition 7.3 would be called **simple designs.** Allowing block repetition really simplifies the construction problem.

It was shown by R. M. Wilson [220], using methods from linear algebra, that nonsimple designs exist in abundance, for all t. The case of simple non-trivial designs is much harder. For a long time not a single example of a simple nontrivial t design for $t > 5$ was known and it was natural to conjecture they did not exist. The reason is a link with permutation groups. It is obvious that each t transitive permutation group on v points produces simple t designs for all block sizes $k < v$. Just choose an arbitrary k subset and choose as blocks the orbit of this base block under the group. In fact, it suffices for the group to be t homogeneous (transitive on unordered subsets of size t). In this way the symmetric and alternating groups produce the complete designs and the Mathieu groups produce, among others, the Witt designs.

As an important consequence of the classification of the finite simple groups, all t homogeneous permutation groups (simple or not) are known (if we really believe the classification of simple groups). It turns out that, aside from the symmetric and alternating groups, there is not a single t homogeneous permutation group for $t > 5$. This and the fact that no nontrivial simple t designs for $t > 5$ were known raised the suspicion that possibly such designs could not exist because for some unknown reason t designs could only be constructed from t homogeneous permutation groups.

An indication that this was unjustified came from coding theory. The

Assmus-Mattson theorem creates a link from codes to t designs. It allowed the construction of a large number of simple 5-designs which are not derived from 5 homogeneous permutation groups.

Consider a q-ary linear code $\mathcal{C}$ of length n, minimum distance d and weight distribution (A_i). The weight distribution of the dual code is (A_i'). By Section 15.2 (and earlier: it is the MacWilliams transform from Section 3.5):

$$A_k' = \frac{1}{|\mathcal{C}|} \sum_{i=0}^{n} A_i K_k(i).$$

All we need to know about the Kravchouk polynomials is that $K_k(X)$ has degree k. It follows that A_k' is a linear combination of the $\sum_{j=0}^{n} j^i A_j$ for $i = 0 \ldots, k$. The explicit formulas of $\sum_{j=0}^{n} j^k A_j$ in terms of the $A_0', \ldots, A_k'$ and the Krawchouk polynomials are known as the *Pless power moments* (see V. Pless [167]).

23.9 Proposition. *The weight distribution of a linear code is completely determined if the following are known:*

- *the A_i', $i \leq s - 1$, for some s, and*

- *all but s of the A_i.*

PROOF Consider the equations expressing A_k' in terms of the $\sum_j j^i A_j$ for $k < s$. These are s linear equations for s unknowns. The coefficient matrix is a Vandermonde matrix. As we know from Chapter 4, Vandermonde matrices are invertible over any field. It follows that the solution is uniquely determined. □

Proposition 23.9 is the main ingredient in the proof of the Assmus-Mattson theorem (see Pless [167], Assmus and Mattson [3]).

23.10 Theorem. *Let $\mathcal{C}$ be a linear q-ary code of length n and minimum distance d. Denote by s the number of indices i such that $0 < i \leq n - t$ and $A_i' \neq 0$. Assume $s \leq d - t$.*

Then, for every $i \leq n - t$ such that $A_i' \neq 0$, the supports of the words of weight i in $\mathcal{C}^\perp$ form a t design.

The supports of vectors of weight d in $\mathcal{C}$ form a t-design.

PROOF Let T be a set of t coordinates, I its complement in the set of all coordinates. Denote by $\mathcal{C}(I)$ the projection of $\mathcal{C}$ to I. As $t < d$, we have $dim(\mathcal{C}(I)) = dim(\mathcal{C}) = k$ (this argument was used in the proof of the Singleton bound, Chapter 4). Let $\mathcal{D}(I)$ be the projection to I of the subcode of $\mathcal{C}^\perp$ vanishing at the coordinates in T. As $\mathcal{D}(I)$ is orthogonal to $\mathcal{C}(I)$ and $dim(\mathcal{D}(I)) \geq n-k-t$, we have equality and $\mathcal{D}(I) = \mathcal{C}(I)^\perp$. By our assumption, at most s weight numbers of $\mathcal{D}(I)$ are in doubt (the others are $= 0$), and the first s weight numbers of $\mathcal{C}(I)$ are known (they are $1, 0, \ldots, 0$). It follows from Proposition 23.9 that the weight distribution of $\mathcal{C}(I)$ and of $\mathcal{D}(I)$ is uniquely determined, independent of the choice of T.

This shows that the coordinates with entry $= 0$ of the weight i vectors of $C^{\perp}$ form a t design. It follows that the complements of those sets (block size $n - i$) form a t-design as well. In the binary case, these designs are simple by construction. In cases where $q > 2$ one has to add an additional assumption in order to obtain simple designs. It has to be guaranteed that different codewords having the same support are necessarily scalar multiples of one another.

A similar proof works for the weight d words in $\mathcal{C}$. This design is obviously simple. ⬚

The Assmus-Mattson theorem is from Assmus and Mattson [3], where it was used to construct new 5-designs. More 5-designs were derived in Pless [166] from a class of ternary codes. As an example, the fact that the octads of the large Witt design, the words of weight 8 in the Golay code $[24, 12, 8]_2$, form a 5-design follows also from the Assmus-Mattson theorem.

As there were no 5-homogeneous permutation groups underlying these 5-designs derived from codes, it became clear that the link from permutation groups to designs was not as tight as one could have suspected.

The next step was taken with the help of computer results. The first simple 6-designs were constructed by Magliveras-Leavitt [144]. The method was to start from groups $PGL_2(q)$ in their 3-transitive action on the $q + 1$ points of the projective line $PG(1, q)$ and construct t designs with $t > 3$ as unions of orbits. This led to $6 - (33, 8, 36)$ designs in [144], to $6 - (14, 7, 4)$ designs in Kreher-Radziszowski [127] and to some more explicit examples of simple t designs for high t.

The last step was taken by Luc Teirlinck, who, in [206, 207], used a purely combinatorial machinery to construct simple t designs for every t. An important open question is the existence of Steiner systems $S(t, k, v)$ for $t > 5$. No example is known.

23.7 Nonlinear codes

We encountered many nonlinear codes along the way. The additive codes of Chapter 18 are nonlinear but it could be objected that they are not very serious about their nonlinearity. A similar phenomenon is represented by the so-called Z_4 linear binary codes. More precisely, one should speak of images under the Gray map of Z_4 linear codes. The only nonlinear element in their construction is the Gray map. This gives rise to several families of very good binary codes which in many respects behave like linear codes. The single most famous of the Z_4 linear binary codes is the Nordstrom-Robinson code $\mathcal{NR}$. In Section 16.5 we described it precisely from this Z_4-linear point of view.

It was first constructed when Robinson gave an introductory talk on coding theory for high school students, where he used as an illustration the open problem of the existence of a $(15, 256, 5)_2$ (equivalent to $(16, 256, 6)_2$). The corresponding linear codes cannot exist; see Exercise 16.5.4. Nordstrom was one of the students. He picked up the challenge and succeeded in constructing the code.

The Nordstrom-Robinson code can be seen from various different angles. It is a member of two families of Z_4 linear binary codes, the **Kerdock codes** and the **Preparata codes.** It is also closely related to the Reed-Muller codes and to Galois geometries, in particular quadratic forms. We start from a description of the Reed-Muller codes and the Kerdock codes from a geometric point of view.

Reed-Muller codes and Kerdock codes

We constructed the Reed-Muller codes $\mathcal{RM}(a, r)$ using the $(u, u + v)$ construction in Exercise 5.2.9 of Section 5.2, and as lengthened primitive cyclic codes in Section 14.1. Recall that $\mathcal{RM}(a, r)$ is a $[2^r, \sum_{i=0}^{a} \binom{r}{i}, 2^{r-a}]_2$ code. In this section we want to get a better understanding from a different point of view.

23.11 Definition. *The set $\mathcal{F}_r$ of all mappings $f : \mathbb{F}_2^r \longrightarrow \mathbb{F}_2$ is a binary vector space of dimension 2^r. Here addition is defined by $(f + g)(x) = f(x) + g(x)$. A basis consists of the characteristic functions χ_a for $a \in \mathbb{F}_2^r$ defined by $\chi_a(a) = 1, \chi_a(b) = 0$ for $b \neq a$.*

This is one way of representing $\mathbb{F}_2^q$, where $q = 2^r$. Each polynomial in r variables $x_1, x_2, \ldots, x_r$ with coefficients in $\mathbb{F}_2$ describes an element of $\mathcal{F}_r$. Exponents higher than 1 of x_i are not needed, as x_i and x_i^2 describe the same mapping on $\mathbb{F}_2$. So we can restrict to monomials, which are products of different x_i. Such a monomial is described by a subset of the index set $\{1, 2, \ldots, r\}$. For example, the subset $\{1, 2, 4\}$ describes the monomial $x_1 x_2 x_4$. As simple as that. How many different such polynomials are there? The set $\{1, 2, \ldots, r\}$ has $q = 2^r$ subsets, so this is the number of different monomials of that type. In order to show that these monomials form a basis of $\mathcal{F}_r$, it suffices to represent the characteristic functions χ_a in terms of polynomials.

This is easy to see. In fact,

$$\chi_a = \prod_{i=1}^{r}(1 + x_i + a_i).$$

We have proved the following:

23.12 Theorem. *Let $V = \mathbb{F}_2^r$. Each mapping $f : V \longrightarrow \mathbb{F}_2$ can be written in a unique way as a polynomial in r variables $x_1, x_2, \ldots, x_r$ with coefficients in $\mathbb{F}_2$ such that in each monomial each variable occurs with exponent ≤ 1. This representation of f is known as the* **algebraic normal form** *(ANF).*

For example, the ANF of the mapping $00 \mapsto 0$, $10 \mapsto 0$, $01 \mapsto 1$, $11 \mapsto 1$ simply is x_2, and the ANF of $00 \mapsto 1$, $10 \mapsto 0$, $01 \mapsto 1$, $11 \mapsto 0$ is $x_1 + 1$. The algebraic normal form plays an important role in information transmission and cryptology.

If we want to study the behavior of functions $f \in \mathcal{F}_r$, it is natural to construct a list of all function values. That is, we order the elements of V in some way and consider the 2^r-tuple $(f(a))_{a \in V}$. This is how functions defined on V are identified with codewords of length $q = 2^r$.

23.13 Definition. *Let $V = \mathbb{F}_2^r$ and $f : V \longrightarrow \mathbb{F}_2$. Order the elements of V in some way: $V = \{a_1, \ldots, a_q\}$, where $q = 2^r$. Then*

$$L_f = (f(a_1), f(a_2), \ldots, f(a_q)).$$

Recall that each $f \in \mathcal{F}_r$ has an algebraic normal form. We consider the degree of this ANF. For low degrees this leads us to familiar objects. If f is homogeneous of degree 2, then f is a binary quadratic form.

23.14 Definition. *The* **Reed-Muller code** $\mathcal{RM}(a, r)$ *has as codewords the L_f, where $f \in \mathcal{F}_a$ varies over the mappings whose algebraic normal form has degree $\leq a$.*

As addition of polynomials does not increase the degree, $\mathcal{RM}(a, r)$ is a linear code. Its length is by definition $q = 2^r$. The number of monomials of degree i is $\binom{r}{i}$. This shows that $\mathcal{RM}(a, r)$ has dimension $\sum_{i=0}^{a}\binom{r}{i}$. Clearly, $\mathcal{RM}(0, r)$ is the repetition code. Concentrate on small a. We have $dim(\mathcal{RM}(1, r)) = r + 1$, and this code consists of L_f, where f is a linear mapping $: V \longrightarrow \mathbb{F}_2$ (these are the 2^r linear combinations of $x_1, x_2, \ldots, x_m$) and the complementary tuples. Each nontrivial linear function f has 2^{r-1} zeroes. This shows $wt(L_f) = 2^r - 2^{r-1} = 2^{r-1}$. As $wt(L_f + 1) = q - wt(L_f)$, we see that each element of $\mathcal{RM}(1, r) \setminus \mathcal{RM}(0, r)$ has weight 2^{r-1}. This is the beginning of an inductive argument which can be used to determine the minimum distance.

23.15 Theorem. $\mathcal{RM}(a, r)$ *is a binary linear code of length 2^r, dimension $\sum_{i=0}^{a}\binom{r}{i}$ and minimum distance 2^{r-a}.*

PROOF Only the minimum distance is still in doubt. Let $f \in \mathcal{F}_a$. We can assume by induction that the degree is precisely a. Choose a variable which occurs in the ANF, without restriction x_1, and write

$$f(x_1, \ldots, x_m) = x_1 g(x_2, \ldots, x_m) + h(x_2, \ldots, x_m),$$

where g is not identically zero. View g, h as mappings : $W = \mathbb{F}_2^{r-1} \longrightarrow \mathbb{F}_2$. The elements of V are $(0, b)$ and $(1, b)$, where $b \in W$. We have $f(0, b) = h(b)$ and $f(1, b) = g(b) + h(b)$. Observe that g has degree $\leq a - 1$. By induction there are at least $2^{(r-1)-(a-1)} = 2^{r-a}$ points $b \in W$ such that $g(b) = 1$. For each such b we have that either $f(0, b) = 1$ or $f(1, b) = 1$. □

Consider $\mathcal{RM}(2, r)$. Let $f \in \mathcal{F}_2$ of degree 2. If the constant term is 0, then f is a sum of terms of the form $x_i x_j$ for $i \neq j$ and of linear terms x_i. We can write x_i^2 instead of x_i and obtain the same mapping f. This shows that f simply is a binary quadratic form in dimension r. We studied quadratic forms in Chapter 17. The weight $wt(L_f)$ equals the number of vectors v where $f(v) = 1$. These representation numbers have been determined.

As we know, the underlying symplectic form is determined by the terms $x_i x_j$ for $i \neq j$. Different choices of quadratic terms x_i^2 lead to different quadratic forms belonging to the same symplectic bilinear form. Consider the case when the symplectic form is nondegenerate. Then the quadratic form will automatically be nondegenerate, for every choice of the diagonal terms. This can happen only when $r = 2m$ is even. We know the representation numbers of these quadratic forms; see Section 17.2, where we found out the formulas given in Theorems 17.41 and 17.42 for the odd characteristic case are valid in characteristic 2 as well. The numbers are

$$e_m(0) = 2^{2m-1} - 2^{m-1}, \ e_m(1) = 2^{2m-1} + 2^{m-1}$$

$$h_m(0) = 2^{2m-1} + 2^{m-1}, \ h_m(1) = 2^{2m-1} - 2^{m-1}.$$

Whenever f is elliptic, we have $wt(L_f) = e_m(0)$, $wt(L_f + \mathbf{1}) = e_m(1)$. If f is hyperbolic, then $wt(L_f) = h_m(0)$, $wt(L_f + \mathbf{1}) = h_m(1)$. We see that only two different weights occur. Fixing the symplectic form means fixing a coset of $\mathcal{RM}(1, 2m)$ in $\mathcal{RM}(2, 2m)$. Whenever that coset is described by a nondegenerate symplectic form, all codewords for f in that coset will have weights $2^{2m-1} \pm 2^{m-1}$.

This leads to the following strategy to construct good codes: find a family of symplectic bilinear forms on $V(2m, 2)$ such that, for any two different forms, their difference is nondegenerate. The union of the corresponding cosets of $\mathcal{R}(1, 2m)$ in $\mathcal{R}(2, 2m)$ will then be a code of minimum distance $2^{2m-1} - 2^{m-1}$. How many cosets can we expect in such a set? As we know, each symplectic form on $V(2m, 2)$ is determined by a symmetric $(2m, 2m)$ matrix with zeroes on the diagonal. How many of those matrices can we find such that the difference between any two is nondegenerate? Certainly the first rows of these matrices must be different (if they are equal, the difference starts with the 0 row and therefore is degenerate). As the first row starts with a 0, there are only 2^{2m-1} possible first rows. So this is a bound on the size of such a family of matrices.

23.16 Definition. *A **Kerdock set** of $(2m, 2m)$ matrices is a family of 2^{2m-1} binary symmetric matrices with zeroes on the main diagonal (symplectic matrices), such that the difference (sum) of any two different matrices is nondegenerate (equivalently, has nonzero determinant).*

23.17 Theorem. *If a Kerdock set of binary symplectic forms exists, we can construct a binary code (the corresponding **Kerdock code**) of length 2^{2m} with 2^{4m} codewords and minimum distance $2^{2m-1} - 2^{m-1}$.*

Kerdock sets can always be constructed. We cannot give details here, as this would lead us too deep into Galois geometries. In the case $m = 2$ the Kerdock code is a $(16, 256, 6)_2$-code. These are the parameters of $\mathcal{NR}$. This code is a union of 8 cosets of the $[16, 5, 8]_2$ code $\mathcal{RM}(1, 4)$. The original construction is in Kerdock [124]. Our presentation follows Cameron and van Lint [40].

Galois rings and Z_4 linear codes

In Section 3.2 we constructed finite fields as extensions of prime fields $\mathbb{F}_p$, using irreducible polynomials. There is a similar mechanism producing extension rings of the rings $Z_n = \mathbb{Z}/n\mathbb{Z}$. These finite rings are known as Galois rings and were first studied by W. Krull [128]. In coding theory the case when the underlying ring is Z_4 has proved particularly fruitful. We concentrate on this case. The corresponding Galois rings are $GR(m, Z_4)$.

A readable introduction is to be found in Wan [217]. In order to define $GR(m, Z_4)$, one should start from the field $F = \mathbb{F}_q$, where $q = 2^m$. As we know, F can be described by a primitive polynomial $h_2(X)$ of degree m with coefficients in $\mathbb{F}_2$. Consider liftings $h(X)$ of $h_2(X)$, meaning that $h(X) \in Z_4[X]$ is monic of degree m and $\overline{h}(X) = h_2(X)$. Here the mapping $a \mapsto \overline{a}$ from Z_4 to $Z_2 = \mathbb{F}_2$ maps $0, 1 \mapsto 0$, $1, 3 \mapsto 1$ (see Section 16.5). If we choose such a lifting $h(X)$, then $R = Z_4[X]/(h(X))$ is the Galois ring $GR(m, Z_4)$. This ring is uniquely determined, independent of the choice of $h(X)$ (and it would have sufficed to lift an arbitrary irreducible polynomial $h_2(X)$). Among all liftings of the primitive polynomial $h_2(X)$, there is a uniquely determined special one with the property that $h(X) \mid X^{q-1} - 1$. This is the **Hensel lift.**

The recipe is the following: write $X^{q-1} - 1 \in \mathbb{F}_2[X]$ as a product of irreducibles. Recall that we gathered a lot of information on this factorization in the course of our study of trace codes, subfield codes and cyclic codes. For example, the number of irreducible polynomials in this product equals the number of cyclotomic cosets in length $n = q - 1$. Lift this to a factorization of $X^{q-1} - 1$ in $Z_4[X]$. This can be done in a unique way, and the corresponding liftings of the irreducibles in $\mathbb{F}_2[X]$ are the Hensel liftings. There is a computationally efficient way, Graeffe's method, to determine these Hensel liftings.

As an example, consider the case $n = 7$. We described $\mathbb{F}_8$ by way of the primitive polynomial $X^3 + X^2 + 1$ but observed that we could as well have chosen $X^3 + X + 1$. The factorization of $X^7 - 1$ over $\mathbb{F}_2$ is

$$X^7 - 1 = (X - 1)(X^3 + X^2 + 1)(X^3 + X + 1).$$

This reflects the cyclotomic cosets

$$\{0\}, \ \{1, 2, 4\}, \ \{3, 5, 6\}.$$

The corresponding factorization over Z_4 is

$$X^7 - 1 = (X - 1)(X^3 - X^2 + 2X - 1)(X^3 + 2X^2 + X - 1).$$

The Hensel lifting of $X^3 + X^2 + 1$ is therefore $X^3 - X^2 + 2X - 1$, and the Hensel lifting of $X^3 + X + 1$ is $X^3 + 2X^2 + X - 1$.

Each element of $GR(m, Z_4) = Z_4[X]/(h(X))$ is uniquely represented by a polynomial of degree $< m$. It follows that the Galois ring $R = GR(m, Z_4)$ has $4^m = q^2$ elements. If $h(X)$ is the Hensel lifting of a primitive polynomial, a zero ϵ of $h(X)$ is an element of order $q - 1$. Denote by $\mathcal{T}$ the set of all powers of ϵ together with 0. Observe that $\mathcal{T}$ is multiplicatively closed. Then $|\mathcal{T}| = q$. The zero divisors of R are precisely the multiples of 2. They form the ideal $2R$, which is a maximal ideal, and we have that the factor ring, which of course is a field, is the finite field with q elements: $R/2R \cong \mathbb{F}_q$. The $q^2 - q = q(q - 1)$ elements not in $2R$ are units. Each element of the form $1 + 2r$ has order 2. It follows that $1 + 2R$ is an elementary abelian group of order q. This clarifies the structure of the group of units: it is the direct product of the cyclic group of order $q - 1$ generated by ϵ and the elementary abelian group $1 + 2R$.

We can think of $R = GR(m, Z_4)$ as built up from two copies of $\mathbb{F}_q$. On one hand, we have $R/2R \cong \mathbb{F}_q$. On the other hand, multiplication by 2 maps R to $2R$, with $2R$ as kernel. It follows that $2R \cong R/2R$ can itself be considered as a copy of $\mathbb{F}_q$.

Each element $r \in R$ can be written uniquely in the form $r = a + 2b$ where $a, b \in \mathcal{T}$. This is the 2-adic representation of elements of R. The mapping $\phi : a + 2b \mapsto a^2 + 2b^2$ defines a ring automorphism of R, with only the elements of Z_4 as fixed elements. Clearly, ϕ has order m. It can be proved that this describes all ring automorphisms of R. The **Galois group $G = \langle \phi \rangle$** is therefore cyclic of order m. We used the Galois group of a field extension to define the trace, which in turn gave us methods to derive linear codes over the ground field from linear codes over extension fields. The same phenomenon happens over Galois rings. The trace $T : GR(m, Z_4) \longrightarrow Z_4$ is defined as $T(r) = \sum_{g \in G} g(r)$, just as in the case of field extensions. The theory of Z_4 linear cyclic codes can be developed along the lines we know from the theory of linear (or additive) cyclic codes. The additional feature is that we are really interested in binary codes. The last step, after constructing a Z_4 linear code by way of the cyclic mechanism, is the application of the Gray map. This leads to a binary code which is twice as long as the quaternary parent code and typically will be nonlinear.

In the case $m = 3$, describe $GR(3, Z_4)$ by the Hensel lifting $X^3 - X^2 + 2X - 1$. Let ϵ be a root of this polynomial. The fundamental equation is therefore

$$\epsilon^3 = \epsilon^2 + 2\epsilon + 1.$$

The remaining powers of ϵ are

$$\epsilon^4 = -\epsilon^2 - \epsilon + 1, \quad \epsilon^5 = 2\epsilon^2 - \epsilon - 1, \quad \epsilon^6 = \epsilon^2 - \epsilon + 2$$

and finally $\epsilon^7 = 1$, checking the correctness of the calculations. Consider the Z_4 linear code whose check matrix has as columns the $(1, \epsilon^i)^t$ with an additional first column such that the row sums are 0. Expressing the ϵ^i in terms of the basis $1, \epsilon, \epsilon^2$ this leads to the following generator matrix:

$$G = \begin{pmatrix} 1\,1\,1\,1\,1\,1\,1\,1 \\ 0\,1\,0\,0\,1\,1\,3\,2 \\ 0\,0\,1\,0\,2\,3\,3\,3 \\ 0\,0\,0\,1\,1\,3\,2\,1 \end{pmatrix}.$$

After a little Gauß elimination (subtract the sum of the remaining rows from the first row and reorder the columns in the right half such that the entries 2 end on the main diagonal) the matrix is

$$\begin{pmatrix} 1\,0\,0\,0 & 2\,3\,1\,1 \\ 0\,1\,0\,0 & 1\,2\,1\,3 \\ 0\,0\,1\,0 & 3\,3\,2\,3 \\ 0\,0\,0\,1 & 3\,1\,1\,2 \end{pmatrix}.$$

Some more row and column operations lead to the generator matrix of the octacode, as given in Section 16.5. We have rediscovered the octacode, a lifting of the $[8, 4, 4]_2$ binary Hamming code, which, after application of the Gray map, produces the Nordstrom-Robinson code.

This procedure generalizes in an obvious way. Let G_m be the quaternary $(m + 1, 2^m)$ matrix whose columns are $(1, 0)^t$ and the $(1, \epsilon^i)^t$ where $\epsilon \in GR(m, Z_4)$ is an element of order $q - 1$. Here the ϵ^i are expanded with respect to a basis, probably $1, \epsilon, \dots, \epsilon^{m-1}$. The matrix G above, which generates a version of the octacode, is G_3. There are two obvious ways to continue, by considering G_m either as a generator matrix or as a check matrix of a quaternary code. The quaternary code generated by G_m is a lifting of the Reed-Muller code $\mathcal{RM}(1, m)$. The generator matrix variant produces a $(q, 4^{m+1})_4$ code. The image under the Gray map is therefore a $(2q, 4^{m+1})_2$ code. In the case of odd m, it can be shown that the minimum distance is $q - 2^{(m-1)/2}$.

With a slight shift in notation, this leads exactly to the parameters of the Kerdock codes from Theorem 23.17.

The quaternary code with G_m as check matrix has length q and 4^{q-m-1} codewords. In the case of odd $m \geq 3$, its Gray image has minimum distance 6. This produces $(4^l, 4^{2^{2l-1}-2l}, 6)_2$ codes for all $l \geq 2$ (where $m = 2l-1$). These are the parameters of the **Preparata codes.** Codes with these parameters were first constructed by Preparata [169]. It was observed that they look like linear codes in several respects. The number of codewords is a power of 2 and the weight distribution is distance invariant (the distribution of distances from a given codeword is independent of the choice of the codeword). More than that, the Preparata codes look like the duals of the Kerdock codes. In fact, the weight distribution of both families of codes shows that they are indeed formal duals in the sense that they are Kravchuk transforms of one another. On the other hand, these codes are not linear. The business of Z_4 linearity finally provided an explanation for this phenomenon. These codes are formal duals of each other because they are images under the Gray map of dual Z_4 linear codes. This interesting observation was made in Hammons et al. [104]. The same mechanism produces Z_4 linear constructions of further families of nonlinear binary codes which were originally constructed by Delsarte and Goethals [68].

Optimum binary nonlinear codes

Seriously nonlinear codes made their appearance repeatedly in this text. In Theorem 3.16 we note that orthogonal arrays of strength 2 and index 1 are equivalent to sets of mutually orthogonal Latin squares. Theorem 4.8 states the equivalence of MDS codes and OA of index 1. Proposition 9.5 gives a construction of nonlinear codes out of OA of strength 2 and arbitrary index.

The **Best code** is a $(10, 40, 4)_2$ code. The linear program (see Section 15.2) which shows that this is indeed best in the sense that a $(10, 41, 4)_2$ code cannot exist, is easily computed by hand. The original construction is in Best [15]. Conway and Sloane [61] have given a quaternary construction as follows: A quaternary code $\mathcal{B}_0$ with entries in Z_4 has as entries the quintuples $(a - b, c, a, b, a + c)$ and their cyclic shifts, where $a, b, c = \pm 1$. Code $\mathcal{B}_0$ is known as the **pentacode.** The Best code is the image of $\mathcal{B}_0$ under the Gray map γ; see Definition 16.21. It is easy to see that the Best code does have minimum distance 4. Observe that $a - b, a + b \in \{0, 2\}$. The Best code yields the densest known sphere packing in dimension 10, simply by construction A (see Section 23.4).

Another optimum nonlinear code of minimum distance 4 can be derived from the small Witt design $S(5, 6, 12)$ (see the construction of the ternary Golay code in Section 17.1). Identify each block with its support and hence with an element of weight 6 in $\mathbb{F}_2^{12}$. As there are 132 blocks which pairwise intersect in at most 4 points, this yields a $(12, 132, 4)_2$-code. There are several ways of obtaining a code with even more codewords. Recall that the complement of a block is a block. Fix a pair B_1, B_2 of complementary blocks. Partition the points of B_i into three pairs of points. Consider the six words of weight 2, whose support is one part of the partition. These six words and their complements (of weight 10) complement the small Witt design to a $(12, 144, 4)_2$ code.

Another object of classical combinatorial theory is a source of optimum nonlinear binary codes: a **Hadamard matrix** H of order n is an (n, n) matrix with (real) entries ± 1 such that any two different columns are orthogonal. We describe Hadamard matrices in terms of orthogonal arrays. When the Bose-Bush bound Theorem 9.6 is applied to an $OA_\lambda(2, n - 1, 2)$, it says that the number of rows is $b = 4\lambda \geq n$. Let $\mathcal{A}$ be such an OA meeting the Bose-Bush bound with equality. In particular, $\mathcal{A}$ is a binary $(n, n - 1)$ array all of whose columns have weight $n/2$. If we add the all-1 column and change the entries according to the rule $0 \mapsto 1, 1 \mapsto -1$ (seen as a mapping : $\mathbb{F}_2 \longrightarrow \mathbb{R}$) we obtain a Hadamard matrix of order n. As $\lambda = n/4$, we see that, for $n > 2$, we must have that n is a multiple of 4. It is easy to see that this process is reversible:

23.18 Theorem. *Let $n > 2$. A Hadamard matrix of order n exists if and only if an $OA_{n/4}(2, n - 1, 2)$ exists.*

Assume an $OA_{n/4}(2, n - 1, 2)$ exists. Add the all-1 column. Consider the n columns of the resulting matrix and their complements as a binary code. This **Hadamard code** has parameters $(n, 2n, n/2)_2$. It follows from the quadratic bound Theorem 15.37 that Hadamard codes are optimal in the sense that no code with more codewords can exist.

23.19 Definition. *Let $A_q(n, d)$ be the maximum number of words in an $(n, M, d)_q$ code.*

The complete determination of the numbers $A_q(n, d)$ is too ambitious an aim. In the binary case, it suffices to consider even minimum distances, for reasons which by now are obvious. The Hadamard codes show $A_2(n, n/2) = 2n$ whenever a Hadamard matrix of order n exists. It is a famous conjecture that these matrices exist for all $n > 2$ which are multiples of 4. Trivially, $A_2(n, 2) = 2^{n-1}$. The $A_2(n, 4)$ are known for $n \leq 16$. For $n = 10$ and $n = 9$ the Best code and its shortening are optimal. The values $A_2(12, 4) = 144$, $A_2(11, 4) = 72$ were established by Östergård, Baicheva and Kolev [161]. The Nordstrom-Robinson code is responsible for $A_2(16, 6) = 256$ and all the way down to $A_2(13, 6) = 32$. In order to establish the upper bound, it suffices to show that $(13, 33, 6)_2$ cannot exist, which is easily done by hand using the Simplex algorithm of linear programming.

23.8   Some highly symmetric codes

In the case of linear codes, we use Definition 12.19: the automorphism group of a linear q-ary length n code $\mathcal{C}$ is its stabilizer in the group of motions, the monomial group of order $(q-1)^n n!$ The $[16, 4, 12]_8$ code used for illustration purposes in Section 12.4 belongs to a family considered in [30]. Its automorphism group is isomorphic to $E_4 \times A_4$, of order 48. Some more very symmetric codes with interesting parameters are constructed in [30]. The parameters are $[39, 13, 17]_4, [28, 7, 18]_8, [32, 8, 20]_8$. One of the $[28, 7, 18]_8$ codes has an automorphism group of order $8 \times 9 \times 49$.

The extended binary Hamming code $[8, 4, 4]_2$ has an automorphism group of order 8×168 (see Exercise 12.4.5), the semidirect product of an elementary-abelian group of order 8 and the simple group $GL(3, 2)$. In Section 17.1 we studied the ternary Golay code (the uniquely determined $[12, 6, 6]_3$ code) and saw that its automorphism group has order $2 \times 12 \times 11 \times 10 \times 9 \times 8$. This group contains a center of order 2 and the factor group mod the center is the 5-transitive simple Mathieu group $M_{12} \subset S_{12}$ which acts on the Steiner system $S(5, 6, 12)$. Large caps seem to have the tendency to admit large groups of automorphisms. The Hill cap (the uniquely determined 56-cap in $PG(5, 3)$) has an automorphism group of order 8! containing the simple group $PSL(3, 4)$ as a subgroup of index 2. The Calderbank-Fishburn 236-cap in $AG(7, 3)$ has $E_{32} \cdot S_5$ as a group of automorphisms (see [26]). Glynn's 126-cap in $PG(5, 4)$ admits $PGL(3, 4)$ as a group of automorphisms (see [18]). The largest cap in $AG(4, 4)$ has 40 points and is uniquely determined. Its automorphism group has order 960 and contains the smallest simple group A_5; see [80].

The automorphism group of the (nonlinear) Nordstrom-Robinson code is the semidirect product of E_{16} and A_7.

The small Witt design and the Pace code

We close this section with a description of a recently discovered M_{12}-invariant ternary code. The paradigmatic example of such a code is the ternary Golay code $[12, 6, 6]$; see Section 17.1 where we also saw the small Witt design $S(5, 6, 12)$ and the group M_{12}. It came as a surprise when Nicola Pace constructed an M_{12}-invariant $[66, 10, 36]_3$ code in 2014. We will describe it by a generator matrix G. As we will see, this generator matrix is closely related to the small Witt design.

Recall the basic features of $S(5, 6, 12)$: denote the underlying set by $\Omega = \{1, 2, \ldots, 10, \infty, *\}$. The design is a family of 132 6-subsets of Ω (the **blocks**) satisfying the axiom that each 5-subset of Ω is in precisely one block. The complement of a block is a block. We saw in Section 17.1 that the 5-transitive group M_{12} is a group of automorphisms of the ternary Golay code and also of the small Witt design. Let $\mathcal{B}$ be the set of 66 blocks which do not contain the symbol $* \in \Omega$. Each block $B \in \mathcal{B}$ will contribute a column to the generator matrix G. Here we distinguish between blocks $B \in \mathcal{B}$ which do not contain the symbol ∞ (there are 30 such blocks) and the 36 blocks $B \in \mathcal{B}$ that do contain ∞.

Let $B \in \mathcal{B}$ such that $B \subset \{1, 2, \ldots, 10\}$. Identify the rows $z_1, \ldots, z_{10}$ of G with the corresponding elements of Ω. The column indexed by B is the characteristic function of B: the entry in row i and column B is 1 if $i \in B$ and 0 otherwise. This leads to the following $(10, 30)$ submatrix of G:

```
111111111111111111000000000000
111111111100000000111111110000
111110000011111000111110001110
110001110011100110111001101101
001101101011010101110101011011
001011011010101011101010111011
100100011101101101100111100111
100011101110011110011100110111
010111000100110111010011111110
011000110101011011001111011101
```

To give two examples, the first column corresponds to block $\{1, 2, 3, 4, 7, 8\}$ and the last of those 30 columns corresponds to block $\{4, 5, 6, 7, 8, 10\}$.

Now let $B \in \mathcal{B}$ such that $\infty \in B$. The column indexed by B is defined as follows: the entry in row i and column B is 1 if $i \notin B$ and 0 otherwise. The corresponding right half of the generator matrix G is

```
000000000000000001111111111111111111
000000001111111110000000000011111111
000111110000011111000001111100000111
011001110011100011001110001100011001
101010110101101100010110110001100001
011111001100100101100100101100011010
110011011101010010101011100000110010
101101101011010100111000101001001100
111100010110011001100101011011010000
110110101010101010011101000110100100
```

Here the first column corresponds to block $\{1, 2, 3, 4, 6, \infty\}$ and the last of those columns corresponds to $\{6, 7, 8, 9, 10, \infty\}$. We claim that the ternary code generated by G has parameters $[66, 10, 36]_3$. Observe that all entries of G are 0 or 1. The entry $2 = -1$ does not occur. Each row of G has weight 36. Also the codeword $\sum_{i=1}^{10} z_i = (0^{30}2^{36})$ of weight 36.

Let A, B be disjoint subsets of Ω such that $|A| = a$, $|B| = b$ and $a + b \leq 5$. As $S(5, 6, 12)$ is a 5-design, the number $i(a, b)$ of blocks which contain A and are disjoint from B is uniquely determined, independent of the choice of A, B. As the complements of blocks are blocks, we have $i(b, a) = i(a, b)$. Those numbers are determined by elementary counting:

a	b	$i(a,b)$
1	0	66
2	0	30
1	1	36
3	0	12
2	1	18
4	0	4
3	1	8
2	2	10
5	0	1
4	1	3
3	2	5

Let us collect information concerning the weights of codewords.

23.20 Definition. *Let* $U, V \subset \{1, \ldots, 10\}, U \cap V = \emptyset, |U| = u, |V| = v$. *Let* $\nu(U, V)$ *be the number of blocks* $B \in \mathcal{B}$ *which satisfy that* $|B \cap U|$ *and* $|B \cap V|$ *have the same congruence* c *mod* 3. *Here* $c \in \{0, 1, 2\}$.

The weights of arbitrary codewords are described as follows:

23.21 Lemma. *Let* $z_i, 1 \leq i \leq 10$ *be the rows of the generator matrix* G. *The codeword* $\sum_{i \in U} z_i - \sum_{j \in V} z_j$ *has weight* $66 - \nu(U, V)$.

PROOF The number of zeroes in codeword $\sum_{i \in U} z_i - \sum_{j \in V} z_j$ is by definition $\alpha + \beta$ where α is the number of blocks B such that $\infty, * \notin B$ and $|B \cap U|$ and $|B \cap V|$ have the same congruence mod 3 and β is the number of blocks B such that $* \notin B, \infty \in B$ and $|\overline{B} \cap U|$ and $|\overline{B} \cap V|$ have the same congruence mod 3. Working with the complement, this shows that $\alpha + \beta$ is the number of blocks such that $\infty \notin B$ and $|B \cap U|$ and $|B \cap V|$ have the same congruence. Replacing ∞ by $*$, we arrive at the statement of the lemma. ∎

In order to show that the minimum distance is $d = 36$, we need to show that $\nu(U, V) \leq 30$ for all $(U, V) \neq (\emptyset, \emptyset)$. We will consider here only some cases when the codeword weights can be determined directly. Case $u = 1, v = 0$ corresponds to the rows z_i. We have $\nu(U, V) = i(0, 3) + i(1, 2) = 12 + 18 = 30$. This confirms the observation that the rows of G have weight 36. The sums of two rows correspond to $u = 2, v = 0$. In this case, $\nu(U, V) = i(1, 3) + i(0, 4) = 8 + 4 = 12$, so the sums of two rows have weight 54. Let $u = 8, v = 2$. Then $\nu(U, V) = 2i(2, 2) + i(0, 4) = 20 + 4 = 24$, so those codewords have weight 42. In the case $u = 7, v = 3$, we obtain $\nu(U, V) = 3i(3, 2) + 1 + i(3, 2) = 1 + 4 \times 5 = 21$, leading to codewords of weight 45. Not in all cases are the codeword weights easily determined.

Why is this code M_{12} invariant? In order to see this, we start from an M_{12}-invariant code and identify it with the code just constructed. We start from the simple group M_{12} in its 5-transitive action on Ω. Consider the natural module of M_{12} on a 12-dimensional ternary vector space $V(12,3)$ by choosing a basis $b_i, i \in \Omega$ and letting M_{12} act by permuting this basis. This module contains two obvious submodules, the 11-dimensional module $I = \{\sum_{i \in \Omega} a_i b_i \mid \sum a_i = 0\}$ and the one-dimensional submodule generated by the **diagonal** $\Delta = \sum_{i \in \Omega} b_i$. A basis of I is the $u_i = b_i - b_*, i = 1, 2, \ldots, 10, \infty$. As we are in characteristic 3, we have $\Delta \in I$ and M_{12} acts on the 10-dimensional factor space $V = I/\langle \Delta \rangle$. A natural basis of V consists of the $v_i = \overline{u_i}, i = 1, \ldots, 10$, where $\overline{u_\infty} = -\sum_{i=1}^{10} v_i$. This defines an embedding of M_{12} in $GL(10, 3)$. For each orbit we consider the 10-dimensional ternary code whose generator matrix has representatives of the projective points of the orbit as columns. This is then an M_{12} invariant code.

For each 6-subset $B \subset \Omega$, let $u_B = \sum_{i \in B} b_i$. We have $u_B \in I$ as $|B| = 6$. Let $v_B = \overline{u_B} \in V$. The group M_{12} permutes the v_B where B varies over the blocks of the $S(5, 6, 12)$. If $\overline{B}$ is the complement of B in Ω, then $u_B + u_{\overline{B}} = \Delta$. It follows that the projective points generated by the v_B where B varies over the blocks form an M_{12} orbit of length 66. The Pace code as defined in [163] is equivalent to the $[66, 10]_3$ code whose generator matrix has as columns representatives of this orbit. In order to identify it with the code generated by G, it suffices to express the v_B where B varies over the blocks not containing $*$ in terms of the basis $v_i, i = 1, \ldots, 10$ of V. If $B \subset \{1, 2, \ldots 10\}$, we have $u_B = \sum_{i \in B} b_i = \sum_{i \in B} u_i$ and we obtain the corresponding column of G. Let B be a block such that $\infty \in B, * \notin B$. Then

$$u_B = \sum_{i \in B} b_i = \sum_{i \in B} u_i = \sum_{i \in B, i \neq \infty} u_i - \sum_{i=1}^{10} u_i = - \sum_{i \leq 10, i \notin B} u_i.$$

This is the negative of the corresponding column of our matrix G.

23.9   Small fields

Here is where we encountered concrete small fields in the text.

$\mathbb{F}_4$

The smallest of all nonprime fields appears first at the end of Section 3.2.

$\mathbb{F}_8$

is first constructed in Exercise 3.2.2. A detailed construction is in Chapter 4, using $X^3 + X^2 + 1$ as an irreducible polynomial.

$\mathbb{F}_9$

is constructed in Exercise 3.2.6, using $X^2 - X - 1$ as an irreducible polynomial over $\mathbb{F}_3$.

$\mathbb{F}_{27}$

The construction of this field is Exercise 12.1.15, with irreducible polynomial $X^3 - X^2 + 1$.

$\mathbb{F}_{16}$

A detailed construction is in Section 13.1, using $X^4 + X + 1$ as an irreducible polynomial over $\mathbb{F}_2$.

$\mathbb{F}_{64}$

has been constructed in Section 22.4 as a quadratic extension $\mathbb{F}_8(\rho)$ of $\mathbb{F}_8$, with equation $\rho^2 = \rho + \epsilon$ and $\epsilon \in \mathbb{F}_8$ such that $\epsilon^3 + \epsilon^2 + 1 = 0$.

$\mathbb{F}_{81}$

is constructed in Section 13.2, using $X^4 - X - 1$ as an irreducible polynomial.

23.10   Short codes

Binary linear codes

The (extended) Hamming code $[8, 4, 4]_2$ (Exercises 2.1.2 and 2.4.3, also a Reed-Muller code), code $[10, 3, 5]_2$ (Exercise 2.1.5), $[16, 11, 4], [32, 26, 4]_2 \ldots$ (Section 2.5, Theorem 5.5), the binary Simplex codes $[15, 4, 8]_2, [31, 5, 16]_2, \ldots$ (Section 2.5),
$[24, 7, 10]_2$ (Exercise 5.1.5), $[12, 4, 6]_2, [32, 9, 12]_2, [32, 12, 10]_2, [32, 18, 6]_2$ and $[80, 28, 20]_2, [80, 32, 18]_2, [80, 44, 12]_2, [80, 48, 10]_2$ (Exercise 5.2.10), the binary extended Golay code $[24, 12, 8]_2$ (Section 5.1, Chapter 7),
$[12, 4, 6]_2, [24, 5, 12]_2, [48, 6, 24], \ldots, [128, 12, 56]_2$ (Section 5.2),
$[56, 12, 20]_2, [18, 6, 8]_2$ (Exercises 5.2.1 and 5.2.2),

$[16, 5, 8]_2$ (Section 5.2 or: a Reed-Muller code, also Section 13.1, Section 13.2), $[32, 6, 16]_2, [32, 16, 8]_2, [64, 7, 32]_2, [64, 42, 8]_2, [128, 8, 64]_2$ (Reed-Muller codes, Exercise 5.2.11), $[28, 5, 14]_2, [56, 6, 28]_2, [112, 7, 56]_2, ...$ (Exercise 5.2.12), $[240, 8, 120]_2$ (Exercise 5.2.13), $[14, 3, 8]_2, [18, 3, 10]_2$ (Chapter 10) $[15, 10, 4]_2$ (Section 13.1), $[64, 51, 6]_2, [33, 22, 6]_2$ (Section 13.2), $[18, 9, 6]_2, [66, 53, 6]_2, [258, 241, 6]_2$ (Section 13.2), $[32, 16, 8]_2, [64, 36, 12]_2$ (Section 13.2, Exercise 13.2.3), $[21, 5, 10]_2, [28, 5, 14]_2, [32, 6, 16]_2$ (Section 14.1), $[448, 3, 256]_2, [448, 9, 224]_2, [448, 12, 220]_2, [228, 11, 110]_2, [224, 11, 108]_2,$ $[452, 12, 222], [455, 12, 224]_2, [159, 12, 72]_2, [126, 36, 34]_2, [133, 37, 34]_2$ (Section 14.1), $[240, 13, 112]_2$ (Chapter 14), $[129, 114, 6]_2$ (Exercise 16.2.8), $[70, 9, 32]_2, [196, 9, 96]_2$ (Exercises 17.1.24 and 18.1.9), $[63, 12, 24]_2, \ [255, 16, 112]_2, [1023, 20, 480]_2$ (Kloosterman codes of Section 21.3), $[92, 24, 26]_2$ (Section 22.3).

Ternary linear codes

The extended ternary Golay code $[12, 6, 6]_3$ (Exercise 3.6.3, Section 17.1), $[27, 10, 10]_3$ (Section 5.2), $[13, 7, 5]_3, [121, 111, 5]_3, [41, 33, 5]_3$ (Section 13.2), $[28, 8, 15]_3, [28, 14, 9]_3$ (Sections 13.4 and 14.1), $[22, 6, 12]_3, [58, 10, 30]_3, [31, 7, 17]_3, [33, 7, 18]_3, [33, 8, 17]_3, [35, 8, 18]_3$ (Section 14.1), the hyperbolic quadric $[16, 4, 9]_3$ (Exercise 17.2.14), the Hill cap $[56, 50, 4]_3$ and the dual code $[56, 6, 36]_3$, the Pellegrino cap $[20, 15, 4]_3$ (Section 17.3, Section 3.6), $[84, 6, 54]_3$ (Exercise 18.1.8), the Pace code $[66, 10, 36]_3$ (Section 23.8).

Quaternary linear codes

The hexacode $[6, 3, 4]_4$ (Exercise 3.7.4, Section 17.1), $[12, 6, 6]_4$ (Exercise 17.1.10), the ovoid code $[17, 4, 12]_4$ (Section 17.3), $[21, 18, 3]_4$ (the projective plane $PG(2, 4)$, Exercise 3.4.3), $[43, 36, 5]_4, [21, 15, 5]_4,$ $[85, 77, 5]_4$ (Section 13.4), $[53, 6, 36]_4, [197, 10, 132]_4, [70, 7, 48]_4, [65, 8, 44]_4, [70, 8, 46]_4, [73, 8, 48]_4$ (Section 14.1), $[87, 56, 13]_4$ (Chapter 9), the Glynn cap $[126, 121, 4]_4$ and its dual, an $[126, 6, 88]_4$-code (Section 17.3), $[260, 6, 192]_4$ (Exercise 18.1.8).

Linear codes over larger fields

$[21, 14, 6]_5, [43, 36, 6]_7$ (Section 13.2), the ovoid codes $[26, 4, 20]_5, [50, 4, 42]_7, [65, 4, 56]_8, ...$ (Section 17.3), $[106, 6, 80]_5$ (Section 14.1), $[10, 3, 8]_8$ (Section 17.1), Barlotti codes $[11, 3, 8]_5, [16, 3, 12]_5, [22, 3, 18]_7, [29, 3, 24]_7$ (exercises in Section

17.1),
$[28, 3, 24]_8$ (Exercise 10.9), $[16, 4, 12]_8$ (Section 12.4),
$[23, 8, 13]_8, [21, 4, 15]_8, [21, 10, 9]_8,$
the hyperbolic quadric $[36, 4, 25]_5$ (Exercise 17.2.14),
Suzuki codes $[64, 5, 51]_8, [64, 6, 48]_8, [64, 7, 46]_8, [64, 8, 44]_8, [64, 9, 43]_8, [64, 11, 42]_8,$
codes $[65, 5, 52]_8, [70, 5, 56]_8, [67, 6, 51]_8, [66, 11, 43]_8, [68, 11, 44]_8, [66, 12, 41]_8,$
$[69, 12, 43]_8, [65, 13, 40]_8, [68, 13, 42]_8, [70, 14, 42]_8, [70, 17, 38]_8, [72, 17, 39]_8,$ (Section 22.3),
$[30, 7, 19]_8, [30, 8, 18]_8, [30, 9, 17]_8,$ and $[33, 10, 18]_8$ (Section 22.4),
$[27, 4, 21]_9, [27, 7, 18]_9, [64, 4, 56]_{16}, [64, 7, 52]_{16}, [64, 11, 48]_{16}$ (Section 22.5),
$[200, 5, 168]_8$ (Section 17.3).

Additive quaternary codes

$[65, 59.5, 4]_4, [1025, 1016.5, 4]_4$ (Section 18.1), $[7, 3.5, 4]_4, [15, 4.5, 9]_4, [13, 6.5, 6]_4,$ (Section 18.2), $[8, 2.5, 6]_4, [32, 3.5, 24]_4, [11, 2.5, 8]_4$ (Section 18.3).

Additive codes over larger fields

$[18, 14\frac{1}{3}, 4]_8, [257, 252, 4]_8, [1025, 1019, 4]_8, [34, 30.5, 4]_{16}, [1025, 1020\frac{1}{4}, 4]_{16}$ (Section 18.1).

Nonlinear nonadditive codes

The Best code $(10, 40, 4)_2$. The Nordstrom-Robinson $(16, 256, 6)_2$ code (Section 16.5), $(31, 25, 24)_4$ (Chapter 9).

References

[1] W. O. Alltop: *A method for extending binary linear codes,*
IEEE Transactions on Information Theory **30** (1984), 871-872.

[2] A. Ashikhmin and E. Knill: *Nonbinary quantum stabilizer codes,*
IEEE Transactions on Information Theory **47** (2001), 3065-3072.

[3] E. F. Assmus, Jr and H. F. Mattson, Jr: *New 5-designs,*
Journal of Combinatorial Theory **6** (1969), 122-151.

[4] S. Ball, A. Blokhuis and F. Mazzocca:
Maximal arcs in desarguesian planes of odd order do not exist,
Combinatorica **17** (1997), 31-47.

[5] J. Barát, Y. Edel, R. Hill, L. Storme: *On complete caps in the projective
geometries over $\mathbb{F}_3$, part II: new improvements,*
Journal of Comb. Mathematics and Comb. Computing **49** (2004), 9-31.

[6] A. Barlotti: *Some topics in finite geometrical structures,*
Institute of Statistics, University of Carolina, Mimeo Series **439** (1965).

[7] D. Bartoli, J. Bierbrauer, G. Faina, S. Marcugini and F. Pambianco:
The nonexistence of an additive quaternary $[15, 5, 9]$-code, Finite Fields
and Their Applications **36** (2015), 29-40.

[8] B. I. Belov, V. N. Logachev and V. P. Sandimirov:
*Construction of a class of linear binary codes achieving the Varshamov-
Griesmer bound,*
Problems of Information Transmission **10** (1974), 211-217.

[9] C. H. Bennett, G. Brassard, C. Crépeau, U. M. Maurer:
Generalized privacy amplification,
IEEE Transactions on Information Theory **41** (1995), 1915-1923.

[10] C. H. Bennett, G. Brassard and J.-M. Roberts:
Privacy amplification by public discussion,
SIAM Journal of Computation **17** (1988), 210-229.

[11] E. R. Berlekamp: *Algebraic Coding Theory,*
McGraw-Hill, New York 1968.

[12] E. R. Berlekamp:
On decoding binary Bose-Chaudhuri-Hocquenghem codes,
IEEE Transaction on Information Theory **11** (1965), 577-579.

488   *References*

[13] E. R. Berlekamp (ed.): Key Papers in the Development of Coding The-
      ory, *IEEE Press*, New York, 1974.

[14] M. R. Best, A.E. Brouwer, F.J. Mac Williams, A.M. Odlyzko, N.J.A.
      Sloane: *Bounds for binary codes of length less than 25, IEEE Transac-
      tions on Information Theory* **24** (1978), 81-93.

[15] M. R. Best: *Binary codes with a minimum distance of four,*
      IEEE Transactions on Information Theory **26** (1980), 738-742.

[16] J. Bierbrauer: *The theory of cyclic codes and a generalization to additive
      codes,* Designs, Codes and Cryptography **25** (2002), 189-206.

[17] J. Bierbrauer: *Direct constructions of additive codes,*
      Journal of Combinatorial Designs **10** (2002), 207-216.

[18] J. Bierbrauer: *Large caps,* Journal of Geometry **76** (2003), 16-51.

[19] J. Bierbrauer: *Cyclic additive and quantum stabilizer codes,* Arithmetic
      of Finite Fields WAIFI, Madrid 2007 (C. Carlet and B. Sunar (eds),
      Lecture Notes in Computer Science **4547** (2007), 276-283.

[20] J. Bierbrauer: *Cyclic additive codes,*
      Journal of Algebra **372** (2012), 661-672.

[21] J. Bierbrauer and Y. Edel: *Construction of digital nets from BCH-codes,*
      Monte Carlo and Quasi-Monte Carlo Methods 1996,
      Lecture Notes in Statistics **127** (1997), 221-231.

[22] J. Bierbrauer and Y. Edel: *Some codes related to BCH-codes of low
      dimension,* Discrete Mathematics **205** (1999), 57-64, reprinted in
      Discrete Mathematics, Editor's Choice, 1999.

[23] J. Bierbrauer and Y. Edel: *A family of caps in projective 4-space in odd
      characteristic,* Finite Fields and Their Applications **6** (2000), 283-293.

[24] J. Bierbrauer and Y. Edel: *Bounds on affine caps,*
      Journal of Combinatorial Designs **10** (2002), 111-115.

[25] J. Bierbrauer and Y. Edel: *Quantum twisted codes,*
      Journal of Combinatorial Designs **8** (2000), 174-188.

[26] J. Bierbrauer and Y. Edel: *Large caps in projective Galois spaces,* in:
      Current research topics in Galois geometry, J. de Beule and L. Storme
      (eds), Nova Science Publishers 2011 (2nd quarter), pp. 81-94.

[27] J. Bierbrauer, Y. Edel and W. Ch. Schmid: *Coding-theoretic construc-
      tions for* (t, m, s)-*nets and ordered orthogonal arrays,*
      Journal of Combinatorial Designs **10** (2002), 403-418.

[28] J. Bierbrauer and Y. Edel: *Dense sphere packings from new codes,*
      Journal of Algebraic Combinatorics **11** (2000), 95-100.

[29] J. Bierbrauer, Y. Edel, G. Faina, S. Marcugini and F. Pambianco: *Short additive quaternary codes, IEEE IT Transactions* **55** (2009), 952-954.

[30] J. Bierbrauer, S. Marcugini and F. Pambianco: *A family of highly symmetric codes, IEEE Transactions on Information Theory* **51** (2005), 3665-3668.

[31] J. Bierbrauer, S. Marcugini and F. Pambianco: *A geometric non-existence proof of an extremal additive code,* Journal of Combinatorial Theory, Series A, **117** (2010), 128-137.

[32] J. Bierbrauer, G. Faina, M. Giulietti, S. Marcugini and F. Pambianco: *The geometry of quantum codes,* Innovations in Incidence Geometry **6-7** (2009), 53-71.

[33] J. Bierbrauer, D. Bartoli, G. Faina, S. Marcugini, F. Pambianco and Y. Edel: *The structure of quaternary quantum caps,* Designs, Codes and Cryptography **72** (2014), 733-747.

[34] A. Blokhuis and A. E. Brouwer, *Small additive quaternary codes,* European Journal of Combinatorics **25** (2004), 161-167.

[35] D. Boneh and J. Shaw: *Collusion-secure fingerprinting for digital data,* IEEE Transactions on Information Theory **44** (1998), 1897-1905.

[36] R. C. Bose and K. A. Bush: *Orthogonal arrays of strength two and three,* Annals of Mathematical Statistics **23** (1952), 508-524.

[37] R. C. Bose and D. K. Ray-Chaudhuri: *On a class of error-correcting binary group-codes,* Information and Control **3** (1960), 68-79.

[38] A. R. Calderbank and P. C. Fishburn: *Maximal three-independent subsets of* $\{0, 1, 2\}^n$, Designs, Codes and Cryptography **4** (1994), 203-211.

[39] A. R. Calderbank, E. M. Rains, P. W. Shor and N. J. A. Sloane: *Quantum error correction via codes over* $GF(4)$, IEEE Transactions on Information Theory **44** (1998), 1369-1387.

[40] P. J. Cameron and J. H. van Lint: *Designs, Graphs, Codes and their Links,* Cambridge University Press, 1991.

[41] P. Camion and A. Canteaut: *Correlation-immune and resilient functions over a finite alphabet and their applications in cryptography,* Designs, Codes and Cryptography **16** (1999), 121-149.

[42] C. Carlet, P. Charpin and V. Zinoviev: *Codes, bent functions and permutations suitable for DES-like cryptosystems,* Designs, Codes and Cryptography **15** (1998), 125-156.

[43] J. L. Carter and M. N. Wegman: *Universal classes of hash functions,* Journal of Computer and System Science **18** (1979), 143-154.

[44] R. W. Carter: *Simple Groups of Lie Type,* Wiley, 1972.

[45] R. W. Carter: *Finite Groups of Lie Type,* Wiley, 1993.

[46] S. C. Chang and J. K. Wolf: *A simple derivation of the MacWilliams' identity for linear codes,* IEEE Transactions on Information Theory **26** (1980), 476-477.

[47] M. Chateauneuf and D. L. Kreher: *On the state of strength-three covering arrays,* Journal of Combinatorial Designs **10** (2002), 217-238.

[48] C. L. Chen: *Byte-oriented error-correcting codes for semiconductor memory systems,* IEEE Transactions on Computers C **35** (1986), 646-648.

[49] C. L. Chen: *Symbol error-correcting codes for computer memory systems,* IEEE Transactions on Computers **41** (1992), 252-256.

[50] C. L. Chen and B. W. Curran: *Switching codes for delta-I noise reduction,* IEEE Transactions on Computers **45** (1996), 1017-1021.

[51] C. L. Chen and L. E. Grosbach: *Fault-tolerant memory design in the IBM Application System/400,* Digest of Papers, The 21-st International Symposium on Fault-Tolerant Computing (1991), 393-400.

[52] C. L. Chen and M. Y. Hsiao: *Error-correcting codes for semiconductor memory applications: a state-of-the-art review,* IBM Journal of Research and Development **2** (1984), 124-134.

[53] C. Chevalley: *Introduction to the Theory of Algebraic Functions of One Variable,* American Mathematical Society, 1951.

[54] B. Chor, O. Goldreich, J. Håstad, J. Friedman, S. Rudich, R. Smolensky: *The bit extraction problem or t-resilient functions,* IEEE Symposium on Foundations of Computer Science **26** (1985), 396-407.

[55] G. Cohen, I. Honkala, S. Litsyn and A. Lobstein: *Covering Codes,* North-Holland, 1997.

[56] G. Cohen, S. Litsyn and G. Zémor: *Binary B_2-sequences: A new upper bound,* Journal of Combinatorial Theory A **94** (2001), 152-155.

[57] Charles J. Colbourn: *Projective planes and congestion-free networks,* Discrete Applied Mathematics **122** (2002), 117-126.

[58] L. Collatz and W. Wetterling: *Optimierungsaufgaben,* Springer 1966.

[59] J. H. Conway: *A group of order 8,315,553,613,086,720,000,*
Bulletin of the London Mathematical Society **1** (1969), 79-88.

[60] J. H. Conway: *A characterization of Leech's lattice,*
Inventiones Mathematicae **7** (1969), 137-142.

[61] J. H. Conway and N. J. A. Sloane: *Quaternary constructions for the
binary single-error-correcting codes of Julin, Best and others,*
Designs, Codes and Cryptography **41** (1994), 31-42.

[62] T. M. Cover and J. A. Thomas: *Elements of Information Theory,*
Wiley, 1991.

[63] D. Danev and J. Olsson:
On a sequence of cyclic codes with minimum distance 6,
IEEE Transactions on Information Theory **46** (2000), 673-674.

[64] Lars Eirik Danielsen and Matthew G. Parker: *Directed graph representation of half-rate additive codes over GF*(4), manuscript.

[65] B. L. Davis and D. Maclagan: *The card game set,* The Mathematical
Intelligencer **25** (2003), 33-40.

[66] P. Delsarte: *Four fundamental parameters of a code and their combinatorial significance,* Information and Control **23**(1973),407-438.

[67] P. Delsarte and J. M. Goethals:
Tri-weight codes and generalized Hadamard matrices,
Information and Control **15** (1969), 196-206.

[68] P. Delsarte and J. M. Goethals: *Alternating bilinear forms over GF*(*q*),
Journal of Combinatorial Theory A **19** (1975), 26-50.

[69] R. H. F. Denniston: *Some maximal arcs in finite projective planes,*
Journal of Combinatorial Theory **6** (1969), 317-319.

[70] W. Diffie and M. E. Hellman: *New directions in cryptography,*
IEEE Transaction on Information Theory **22** (1976), 644-654.

[71] C. Ding, H. Niederreiter and C. Xing: *Some new codes from algebraic
curves,* IEEE Transactions on Information Theory **46** (2000), 2638-2642.

[72] S. Dodunekov: *Minimal block length of a linear q-ary code with specified
dimension and code distance,* Problems Inform. Transmission **20** (1984),
239-249.

[73] I. Dumer and V. A. Zinoviev: *Some new maximal codes over GF*(4),
Probl. Peredach. Inform **14** (1978), 24-34, translation in
Problems in Information Transmission (1979), 174-181.

[74] Y. Edel: *Extensions of generalized product caps,*
Designs, Codes and Cryptography **31** (2004), 5-14.

[75] Y. Edel and J. Bierbrauer: *Lengthening and the Gilbert-Varshamov bound,* IEEE Transactions on Information Theory **43** (1997), 991-992.

[76] Y. Edel and J. Bierbrauer, 41 *is the largest size of a cap in* $PG(4,4)$,
Designs, Codes and Cryptography **16** (1999), 151-160.

[77] Y. Edel and J. Bierbrauer: *Recursive constructions for large caps,*
Bulletin of the Belgian Mathematical Society - Simon Stevin **6** (1999), 249-258.

[78] Y. Edel and J. Bierbrauer: *Large caps in small spaces,*
Designs, Codes and Cryptography **23** (2001), 197-212.

[79] Y. Edel and J. Bierbrauer: *Families of ternary* (t, m, s)*-nets related to BCH-codes,* Monatshefte für Mathematik **132** (2001), 99-103.

[80] Y. Edel and J. Bierbrauer: *The largest cap in* $AG(4,4)$ *and its uniqueness,* Designs, Codes and Cryptography **29** (2003), 99-104.

[81] Y. Edel and J. Bierbrauer: *Caps of order* $3q^2$ *in affine 4-space in characteristic* 2, Finite Fields and Their Applications **10** (2004), 168-182.

[82] Y. Edel, E. M. Rains and N. J. A. Sloane: *On kissing numbers in dimensions* 32 *to* 128, Electronic Journal of Combinatorics **5** (1998), R22.

[83] P. G. Farrell: *Linear binary anticodes,*
Electronics Letters **6** (1970), 419-421.

[84] W. Feller: *An Introduction to Probability Theory and Its Applications I,*
Wiley, 1950.

[85] G. D. Forney: *Coset Codes, IEEE Transaction on Information Theory,*
part I: Introduction and geometrical classification,
part II: Binary lattices and related codes **34** (1988), 1123-1187.

[86] Q. Fu, R. Li, L. Guo and G. Xu: *Large caps in projective space* $PG(r, 4)$,
Finite Fields and Their Applications **35** (2015), 231-246.

[87] E. M. Gabidulin: *The theory with maximal rank metric distance,* Probl. Inform. Transm. **21** (1985), 1-12.

[88] R. G. Gallager: *Low Density Parity-Check Codes,* MIT Press, 1963.

[89] A. Garcia and H. Stichtenoth: *A tower of Artin-Schreier extensions of function fields attaining the Drinfeld-Vladut bound,*
Inventiones Mathematicae **121**(1995), 211-222.

[90] D. N. Gevorkyan, A. M. Avetisyan and G. A. Tigranyan: *On the structure of two-error-correcting in Hamming metric over Galois fields,* in: Computational Techniques (in Russian) 3, Kuibyshev, 1975, 19-21.

[91] E. N. Gilbert: *A comparison of signalling alphabets, Bell System Technical Journal* **31** (1952), 504-522.

[92] M. Giorgetti and A. Previtali: *Galois invariance, trace codes and subfield codes, Finite Fields and Their Applications* **16** (2010), 96-99.

[93] D. Glynn: *A 126-cap of $PG(5,4)$ and its corresponding $[126,6,88]$ code, Utilitas Mathematica* **55** (1999), 201-210.

[94] M. J. E. Golay: *Notes on Digital Coding,*
Proceedings of the IRE, June 1949.

[95] O. Goldreich: *Modern Cryptography, Probabilistic Proofs and Pseudorandomness,* Springer, 1999.

[96] S. W. Golomb: *Shift Register Sequences,*
Holden Day, San Francisco, 1967.

[97] V.D. Goppa:
Codes on algebraic curves,
Soviet Math. Dokl. **24** (1981), 170-172.

[98] K. Gopalakrishnan and D. R. Stinson:
A simple analysis of the error probability of two-point based sampling,
Information Processing Letters **60** (1996), 91-96.

[99] J. H. Griesmer: *A bound for error-correcting codes,*
IBM Journal Research Development **4** (1960), 532-542.

[100] B. Groneick and S. Grosse: *New binary codes,*
IEEE Transactions on Information Theory **40** (1994), 510-512.

[101] C. Güneri and F. Özbudak:
Weil-Serre type bounds for cyclic codes,
IEEE Transactions on Information Theory **54** (2008), 5381-5395.

[102] V. Guruswami and M. Sudan:
Improved decoding of Reed-Solomon and algebraic-geometry codes,
IEEE Transactions on Information Theory **45** (1999), 1757-1767.

[103] N. Hamada and F. Tamari: *Construction of optimal codes and optimal fractional factorial designs using linear programming,*
Annals of Discrete Mathematics **6** (1980), 175-188.

[104] A. R. Hammons, Jr, P. V. Kumar, A. R. Calderbank, N. J. A. Sloane and P. Solé: *The Z_4-linearity of Kerdock, Preparata, Goethals and related codes, IEEE Transactions on Information Theory* **40** (1994), 301-319.

[105] S. Han and J. L. Kim: *Formally self-dual additive codes over $\mathbb{F}_4$, Journal of Symbolic Computation* **45** (2010), 787-799.

[106] C. J. Colbourn and J. H. Dinitz (eds.):
The CRC Handbook of Combinatorial Designs, Second Edition, CRC Press, 2006.

[107] A. J. Menezes, P. C. van Oorschot and S. A. Vanstone:
Handbook of Applied Cryptography, CRC Press, 1997.

[108] M. Hattori, R. J. McEliece and G. Solomon:
Subspace subcodes of Reed-Solomon codes,
IEEE Transactions on Information Theory **44** (1998), 1861-1880.

[109] A. S. Hedayat, N. J. A. Sloane and J. Stufken:
Orthogonal Arrays: Theory and Applications, Springer, 1999.

[110] T. Helleseth and P. V. Kumar: *Sequences with low correlation*, in *Handbook of Coding Theory*
(V. S. Pless, W. C. Huffman, eds), Elsevier, 1998.

[111] R. Hill: *On the largest size of cap in $S_{5,3}$*,
Atti Accad. Naz. Lincei Rendiconti **54** (1973), 378-384.

[112] R. Hill: *A First Course in Coding Theory*, Clarendon 1986.

[113] J. W. P. Hirschfeld: *Projective Geometries over Finite Fields*,
Oxford University Press, ^{2}1998.

[114] J. W. P. Hirschfeld: *Finite Projective Spaces of Three Dimensions*, Clarendon, Oxford, 1985.

[115] A. Hocquenghem: *Codes correcteurs d'erreurs*,
Chiffres (Paris) **2** (1959), 147-158.

[116] D. G. Hoffman, D. A. Leonard, C. C. Lindner, K. T. Phelps, C. A. Rodger, J. R. Wall: *Coding Theory - The Essentials*,
Marcel Dekker, New York, 1991.

[117] T. Honold, M. Kiermaier and S. Kurz: *Optimal binary subspace codes of length* 6, *constant dimension* 3 *and minimum distance* 4,
Contemporary Mathematics **632** (2015), 157-176.

[118] D. R. Hughes and F. C. Piper: *Design Theory*, Cambridge University Press, 1985.

[119] D. B. Jaffe: *Looking inside codes*, Proceedings of the Second International Workshop on Optimal Codes and Related Topics, Sozopol (Bulgaria) (1998), 137-143.

[120] J. Justesen: *A class of asymptotically good algebraic codes*,
IEEE Transactions on Information Theory **18** (1972), 652-656.

[121] G. Kabatianskii, B. Smeets and T. Johansson: *On the cardinality of systematic authentication codes via Error-Correcting Codes*,
IEEE Transactions on Information Theory **42** (1991), 583-602.

[122] D. Kahn: *The Codebreakers*, Macmillan 1967, Scribner 1996.

[123] S. Kaneda and E. Fujiwara: *Single-byte error-correcting - double byte error detecting codes for memory systems*,
IEEE Transactions on Information Theory **31** (1982), 596-602.

[124] A. M. Kerdock: *A class of low-rate nonlinear codes*,
Information and Control **20** (1972), 182-187.

[125] T. Klove, *Classification of permutation codes of length 6 and minimum distance 5*, Internat. Symposium on Info. Theory and its Applications, Honolulu, 2000, 465-468.

[126] R. Koetter and F. Kschischang: *Coding for errors and erasures in random network coding*,
IEEE Transactions on Information Theory **54** (2008), 3579-3591.

[127] D. L. Kreher and S. P. Radziszowski:
The existence of simple $6 - (14, 7, 4)$ designs,
Journal of Combinatorial Theory A **43** (1986), 237-243.

[128] W. Krull: *Algebraische Theorie der Ringe*,
Mathematische Annalen **92** (1924), 183-213.

[129] F. R. Kschischang and S. Pasupathy: *Some ternary and quaternary codes and associated sphere packings*,
IEEE Transactions on Information Theory **38** (1992), 227-246.

[130] S. Lang: *Algebra*, Revised 3rd ed, New York, Springer 2002.

[131] K. M. Lawrence: *A combinatorial characterization of (t, m, s)-nets in base b*, Journal of Combinatorial Designs **4** (1996), 275-293.

[132] J. Leech: *Notes on sphere packings*,
Canadian Journal of Mathematics **19** (1967), 251-267.

[133] R. Li, Q. Fu, L. Guo and X. Li: *Construction of quantum caps in projective space $PG(r, 4)$ and quantum codes of distance 4*,
Quantum Information Processing **15** (2016), 689-720.

[134] R. Lidl and H. Niederreiter: *Introduction to Finite Fields and Their Applications*, Cambridge University Press [1] 1986, revised edition 1994.

[135] B. Lindström: *Determination of two vectors from the sum*,
Journal of Combinatorial Theory **6** (1969), 402-407.

[136] A. Lubotzky, R. Phillips and P. Sarnak: *Ramanujan graphs*,
Combinatorica **8** (1988), 261-277.

[137] M. Luby: *A simple parallel algorithm for the maximal independent set problem*, SIAM Journal of Computation **15** (1986), 1036-1053.

[138] Y.I. Manin: *What is the maximum number of points on a curve over* $\mathbb{F}_2$?, *J. Fac. Sci. Univ. Tokyo Sect IA Math.* **28** (1981), 715-720.

[139] R. J. MacEliece: *A public-key cryptosystem based on algebraic coding theory*, *DSN Progress Report* **42-44** (1978), 114-116.

[140] R. J. MacEliece, E. R. Rodemich, H. Rumsey and L. R. Welch: *New upper bounds on the rate of a code via the Delsarte-MacWilliams inequalities*, *IEEE Transactions on Information Theory* **23** (1977), 157-166.

[141] F. J. MacWilliams: *A theorem on the distribution of weights in a systematic code*, *Bell System Technical Journal* **42** (1963), 79-94.

[142] F. J. MacWilliams and N. J. A. Sloane:
The Theory of Error-Correcting Codes, North-Holland, 1977.

[143] F. J. MacWilliams, A. M. Odlyzko, N. J. A. Sloane and H. N. Ward:
Self-dual codes over $GF(4)$,
Journal of Combinatorial Theory, Series A, **25** (1978), 288-318.

[144] S. S. Magliveras and D. W. Leavitt: *Simple six designs exist*, Proc. 14th Southeastern Conference on Combinatorics, Graph Theory, Computing, *Congressus Numerantium* **40** (1983), 195-205.

[145] W. J. Martin: *Linear programming bounds for ordered orthogonal arrays and* (t, m, s)-*nets*, Monte Carlo and quasi-Monte Carlo methods 1998, Proceedings of a Conference held at Claremont, CA, 1998, Springer, 2000, 368-376.

[146] W. J. Martin and D. R. Stinson: *A generalized Rao bound for ordered orthogonal arrays and (t,m,s)-nets*,
Canadian Mathematical Bulletin **42** (1999), 359-370.

[147] W. J. Martin and D. R. Stinson: *Association schemes for ordered orthogonal arrays and (t,m,s)-nets*,
Canadian Journal of Mathematics **51** (1999), 326-346.

[148] U. M. Maurer and J. L. Massey: *Local randomness in pseudorandom sequences*, *Journal of Cryptology* **4** (1991), 135-149.

[149] J. L. Massey: *Shift register synthesis and BCH decoding*,
IEEE Transactions on Information Theory **15** (1969), 122-127.

[150] E. Mathieu: *Sur les fonctions cinq fois transitives de* 24 *quantités*, Journal Math. Pures et Appliquées **6** (1861), 241-243.

[151] E. Mathieu: *Mémoire sur l'étude des fonctions de plusieurs quantités*, *Journal Math. Pures et Appl.* **18** (1873), 25-46.

[152] R. Meshulam: *On subsets of finite abelian groups with no* 3-*term arithmetic progression*,
Journal of Combinatorial Theory A **71** (1995), 168-172.

[153] Carlos J. Moreno: *Algebraic Curves over Finite Fields,* Cambridge University Press, 1991.

[154] A. C. Mukhopadhyay: *Lower bounds on $m_t(r, s)$,* Journal of Combinatorial Theory A **25** (1978), 1-13.

[155] G. L. Mullen and W. Ch. Schmid: *An equivalence between (t, m, s)-nets and strongly orthogonal hypercubes,* Journal of Combinatorial Theory A **76** (1996), 164-174.

[156] D. E. Muller: *Application of boolean algebra to switching circuit design and to error detection,* IRE Transactions on Electronic Computers **3** (1954), 6-12.

[157] H. Niederreiter: *Point sets and sequences with small discrepancy,* Monatshefte Mathematik **104** (1987), 273-337.

[158] H. Niederreiter and C. P. Xing: *Low discrepancy sequences obtained from algebraic function fields over finite fields,* Acta Arithmetica **72** (1995), 281-298.

[159] H. Niederreiter and C. P. Xing: *Low discrepancy sequences and global function fields with many rational places,* Finite Fields and Their Applications **2** (1996), 241-273.

[160] A. W. Nordstrom and J. P. Robinson: *An optimum nonlinear code,* Information and Control **11** (1967), 613-616.

[161] P. R. J. Östergård, T. Baicheva and E. Kolev: *Optimal binary one-error-correcting codes of length 10 have 72 codewords,* IEEE Transactions on Information Theory **45** (1999), 1229-1231.

[162] L. H. Ozarow and A. D. Wyner: *Wire-Tap Channel II,* AT&T Bell Laboratories Technical Journal **63**(1984), 2135-2157.

[163] N. Pace: *New ternary linear codes from projectivity groups,* Discrete Mathematics **331** (2014), 22-26.

[164] G. Pellegrino: *Sul massimo ordine delle calotte in $S_{4,3}$,* Matematiche (Catania) **25** (1970), 1-9.

[165] Vera Pless: *The Theory of Error-Correcting Codes,* Wiley, 1989.

[166] Vera Pless: *Symmetry codes over $GF(3)$ and new five-designs,* Journal of Combinatorial Theory **12** (1972), 119-142.

[167] Vera Pless: *Power moment identities on weight distributions in error-correcting codes,* Information and Control **6** (1963), 147-152.

[168] A. Potechin, *Maximal caps in $AG(6,3)$,* Designs, Codes and Cryptography **46** (2008), 243-259.

[169] F. P. Preparata: *A class of optimum nonlinear double-error-correcting codes*, Information and Control **13** (1968), 378-400.

[170] Oliver Pretzel: *Error-Correcting Codes and Finite Fields*, Clarendon Press, Oxford, 1992.

[171] J. Quistorff: *A survey on packing and covering problems in the Hamming permutation space*, Electron. J. Comb. (2006), A1.

[172] I. S. Reed: *A class of multiple-error-correcting codes and the decoding scheme*, IRE Transactions on Information Theory **4** (1954), 38-49.

[173] I. S. Reed and G. Solomon: *Polynomial codes over certain finite fields*, J. Soc. Ind. Appl. Math **8** (1960), 300-304.

[174] A. Rényi: *Foundations of Probability*, Wiley, 1971.

[175] R. L. Rivest, A. Shamir and L. Adleman: *A method for obtaining digital signatures and public key cryptosystems*, Communications of the ACM **21** (1978), 120-126.

[176] C. Roos: *A new lower bound for the minimum distance of a cyclic code*, IEEE Transactions on Information Theory **29** (1983), 330-332.

[177] M. Yu. Rosenbloom and M. A. Tsfasman: *Codes for the m-metric*, Problems of Information Transmission **33** (1997), 45-52, translated from Problemy Peredachi Informatsii **33** (1996), 55-63.

[178] R. A. Rueppel: *Analysis and Design of Stream Ciphers*, Springer, 1986.

[179] P. Sarnak: *Some Applications of Modular Forms*, Cambridge University Press, 1991.

[180] W. Ch. Schmid: *(T, M, S)-nets: Digital construction and combinatorial aspects*, PhD dissertation, Salzburg (Austria), 1995.

[181] C. P. Schnorr and S. Vaudenay: *Black box cryptanalysis of hash networks based on multipermutations*, Advances in Cryptology - EUROCRYPT 94 (A. de Santis, ed.), Lecture Notes in Computer Science **950** (1995), 47-57.

[182] B. Segre: *Le geometrie di Galois*, Annali di Matematica Pura ed Applicata **48** (1959), 1-97.

[183] C. E. Shannon: *A mathematical theory of communication*, Bell System Technical Journal **27** (1948), 379-423, 623-656.

[184] C. E. Shannon: *Communication theory of secrecy systems*, Bell System Technical Journal **28** (1949), 656-715.

[185] V. Rijmen, J. Daemen, B. Preneel, A. Bosselaers and E. de Win: *The cipher SHARK*, Fast Software Encryption (D. Gollmann, ed.), Lecture Notes in Computer Science **1039** (1996).

[186] B. Z. Shen: *A Justesen construction of binary concatenated codes that asymptotically meet the Zyablov bound for low rate,*
IEEE Transactions on Information Theory **39** (1993), 239-242.

[187] A. Shokrollahi: *Codes and graphs, Proceedings of STACS 2000,*
Lecture Notes in Computer Science **1770** (2000), 1-12.

[188] J. Daemen, L. Knudsen and V. Rijmen: *The block cipher SQUARE,*
Fast Software Encryption (E. Biham, ed.),
Lecture Notes in Computer Science **1267** (1997).

[189] T. Siegenthaler: *Correlation-immunity of nonlinear combining functions for cryptographic applications,*
IEEE Transactions on Information Theory **30** (1984), 776-780.

[190] R. C. Singleton: *Maximum distance q-nary codes,*
IEEE Transactions on Information Theory **10** (1964), 116-118.

[191] G. J. Simmons: *Authentication theory/coding theory,*
Advances in Cryptology,
Lecture Notes in Computer Science **196** (1985), 411-431.

[192] M. Sipser and D. A. Spielman: *Expander codes,*
IEEE Transactions on Information Theory **42** (1996), 1710-1722.

[193] N. J. A. Sloane, S .M. Reddy and C. L. Chen: *New binary codes,*
IEEE Transactions on Information Theory **18** (1972), 503-510.

[194] N. J. A. Sloane and D. S. Whitehead:
A new family of single-error correcting codes,
IEEE Transactions on Information Theory **16** (1970), 717-719.

[195] I. M. Sobol: *Distribution of points in a cube and the approximate evaluation of integrals* (in Russian), Zh. Vychisl. Mat. i Mat. Fiz **7** (1967), 784-802.
English translation in USSR Comput. Math. Math. Phys **7** (1967), 86-112.

[196] G. Solomon and J. J. Stiffler: *Algebraically punctured cyclic codes,*
Information and Control **8** (1965), 170-179.

[197] T. H. Spenser: *Provably good pattern generators for a random pattern test,* Algorithmica **11** (1994), 429-442.

[198] H. Stichtenoth: *Algebraic Function Fields and Codes,* Springer, 1993.

[199] D. R. Stinson: *A short proof of the nonexistence of a pair of orthogonal Latin squares of order six,*
Journal of Combinatorial Theory A **36** (1984), 373-376.

[200] D. R. Stinson: *Cryptography - Theory and Practice,*
CRC Press, 1995.

[201] D. R. Stinson: *Cryptography - Theory and Practice*, third edition, Chapman and Hall, CRC Press, 2006.

[202] D. R. Stinson: *Resilient functions and large sets of orthogonal arrays*, *Congressus Numerantium* **92** (1993), 105-110.

[203] Y. Sugiyama, M. Kasahara, S. Hirasawa and T. Namekawa: *A method for solving key equation for decoding Goppa codes*, *Information and Control* **27** (1975), 87-99.

[204] G. G. Szpiro: *Kepler's Conjecture*, Wiley, 2002.

[205] G. Tarry: *Le problème des 36 officiers*, *Comptes Rendus de l'Académie des Sciences, Paris* **2** (1901), 170-203.

[206] L. Teirlinck: *Non-trivial t-designs without repeated blocks exist for all t*, *Discrete Mathematics* **65** (1987), 301-311.

[207] L. Teirlinck: *Locally trivial t-designs and t-designs without repeated blocks*, *Discrete Mathematics* **77** (1989), 345-356.

[208] T. M. Thompson: *From error-correcting codes through sphere packings to simple groups*, The Mathematical Association of America, Washington, DC, 1983.

[209] W. Trappe and L. C. Washington: *Introduction to Cryptography with Coding Theory*, Prentice Hall, 2002.

[210] M. A. Tsfasman, S. G. Vladut and T. Zink: *Modular curves, Shimura curves and Goppa codes better than Varshamov-Gilbert bound*, *Mathematische Nachrichten* **104** (1982), 13-28.

[211] J. H. van Lint: *Introduction to Coding Theory*, Springer Verlag, 1982.

[212] J. H. van Lint and R. M. Wilson: *On the minimum distance of cyclic codes*, *IEEE Transactions on Information Theory* **32** (1986), 23-40.

[213] J. H. van Lint and G. van der Geer: *Introduction to Coding Theory and Algebraic Geometry*, Birkhäuser, 1988.

[214] A. Vardy: *A new sphere packing in 20 dimensions*, *Inventiones Mathematicae* **121** (1995), 119-134.

[215] A. Vardy: *Density doubling, double-circulants, and new sphere packings*, *Transactions of the Americal Mathematical Society* **351** (1999), 271-283.

[216] R. R. Varshamov: *Estimate of the number of signals in error correcting codes*, *Dokl. Acad. Nauk SSSR* **117** (1957), 739-741.

[217] Z. X. Wan: *Quaternary codes*, World Scientific, 1997.

[218] M. N. Wegman and J. L. Carter: *New hash functions and their use in authentication and set equality*, *Journal of Computer and System Science* **22** (1981), 265-279.

[219] S. B. Wicker and V. K. Bhargava: *Reed-Solomon Codes and Their Applications,* IEEE Press, 1994.

[220] R. M. Wilson: *The necessary conditions for t-designs are sufficient for something,* Utilitas Mathematica **4** (1973), 207-215.

[221] E. Witt: *Die 5-fach transitiven Gruppen von Mathieu,* Abhandlungen aus dem Mathematischen Seminar der Universität Hamburg **12** (1938), 256-264.

[222] E. Witt: *Über Steinersche Systeme,* Abhandlungen aus dem Mathematischen Seminar der Universität Hamburg **12** (1938), 265-274.

[223] J. Wolfmann: *New bounds on cyclic codes from algebraic curves,* Lecture Notes in Computer Science **388** (1989), 47-62.

[224] C. P. Xing, H. Niederreiter and K .Y. Lam: *A generalization of algebraic-geometry codes,* IEEE Transactions on Information Theory **45** (1996), 2498-2501.

[225] S. Yu, J. Bierbrauer, Y. Dong, Q. Chen and C.H. Oh: *All the stabilizer codes of distance 3,* IEEE IT Transactions **59** (2013), 5179-5185.

[226] N. Zierler and W. H. Mills: *Products of linear recurring sequences,* Journal of Algebra **27** (1973), 147-157.

[227] V. A. Zinoviev: *Generalized concatenated codes,* Problemy Peredachi Informatsii **12** (1976), 5-15.

[228] V. V. Zyablov: *An estimate of the complexity of constructing binary linear cascade codes,* Problems in Information Transmission **7** (1971), 3-10.

Index